FOUNDATIONS FOR GUIDED-WAVE OPTICS

BICENTENNIAL
1807
☼WILEY
2007
BICENTENNIAL

THE WILEY BICENTENNIAL–KNOWLEDGE FOR GENERATIONS

$\mathcal{E}$ach generation has its unique needs and aspirations. When Charles Wiley first opened his small printing shop in lower Manhattan in 1807, it was a generation of boundless potential searching for an identity. And we were there, helping to define a new American literary tradition. Over half a century later, in the midst of the Second Industrial Revolution, it was a generation focused on building the future. Once again, we were there, supplying the critical scientific, technical, and engineering knowledge that helped frame the world. Throughout the 20th Century, and into the new millennium, nations began to reach out beyond their own borders and a new international community was born. Wiley was there, expanding its operations around the world to enable a global exchange of ideas, opinions, and know-how.

For 200 years, Wiley has been an integral part of each generation's journey, enabling the flow of information and understanding necessary to meet their needs and fulfill their aspirations. Today, bold new technologies are changing the way we live and learn. Wiley will be there, providing you the must-have knowledge you need to imagine new worlds, new possibilities, and new opportunities.

Generations come and go, but you can always count on Wiley to provide you the knowledge you need, when and where you need it!

WILLIAM J. PESCE
PRESIDENT AND CHIEF EXECUTIVE OFFICER

PETER BOOTH WILEY
CHAIRMAN OF THE BOARD

FOUNDATIONS FOR GUIDED-WAVE OPTICS

Chin-Lin Chen
Purdue University
West Lafayette, Indiana

BICENTENNIAL
BICENTENNIAL
1807
BICENTENNIAL
WILEY
2007
BICENTENNIAL
BICENTENNIAL

WILEY-INTERSCIENCE
A JOHN WILEY & SONS, INC., PUBLICATION

For general information on our other products and services or for technical support, please contact our
Customer Care Department within the United States at (800) 762-2974, outside the United States at
(317) 572-3993 or fax (317) 572-4002.

Wiley also publishes its books in a variety of electronic formats. Some content that appears in print
may not be available in electronic formats. For more information about Wiley products, visit our web
site at www.wiley.com.

Library of Congress Cataloging-in-Publication Data:

Chen, Chin-Lin.
    Foundations for guided-wave optics / by Chin-Lin Chen.
        p. cm.
    Includes index.
    ISBN-13 978-0-471-75687-3 (cloth)
    ISBN-10 0-471-75687-3 (cloth)
1. Optical wave guides.   I.   Title.
    TA1750.C473   2006

    621.36'92—dc22

                                                2006000881

Printed in the United States of America.

10 9 8 7 6 5 4 3 2 1

CONTENTS

PREFACE

Over the last 50 years, we have witnessed an extraordinary evolution and progress in optical science and engineering. When lasers were invented as light sources in 1960, cladded glass rods were proposed as transmission media in 1966; few, if any, would foresee their impact on the daily life of the modern society. Today, lasers and fibers are the key building blocks of optical communication systems that touch all walks of life. Photonic components are also used in consumer products, entertainment and medical equipment, not to mention the scientific and engineering instrumentation. Optical devices may be in the "bulk" or guided-wave optic forms. Guided-wave optic components are relatively new and much remains to be accomplished or realized. Thus engineering students and graduates should acquire a basic knowledge of principles, capabilities and limitations of guided-wave optic devices and systems in their education even if their specialization is not optics or photonics. The purpose of this book is to present an intermediate and in-depth treatment of integrated and fiber optics. In addition to the basic transmission properties of dielectric waveguides and optical fibers, the book also covers the basic principles of directional couplers, guided-wave gratings, arrayed-waveguide gratings and fiber optic polarization components. In short, the book examines most topics of interest to engineers and scientists.

The main objective of Chapter 1 is to introduce the nomenclature and notations. The rest of the book treats three major topics. They are the integrated optics (Chapters 2 to 8), fiber optics (Chapters 9 to 12) and the pulse evolution and broadening in optical waveguides (Chapters 13 and 14). Attempts are made to keep each chapter sufficiently independent and self-contained.

The book is written primarily as a textbook for advanced seniors, first-year graduate students, and recent graduates of engineering or physics. It is also useful for self-study. Like all textbooks, materials contained herein may be found in journal articles, research monographs and/or other textbooks. My aim is to assemble relevant materials in a single volume and to present them in a cohesive and unified fashion. It is not meant to be a comprehensive treatise that contains all topics of integrated optics and fiber optics. Nevertheless, most important elements of guided-wave optics are covered in this book.

Each subject selected is treated from the first principles. A rigorous analysis is given to establish its validity and limitation. Whenever possible, elementary mathematics is used to analyze the subject matter. Detailed steps and manipulations are provided so that readers can follow the development on their own. Extensions or generalizations are noted following the initial discussion. If possible, final results are cast in terms of normalized parameters. Results or conclusions based on numerical calculations or experimental observations are explicitly identified. Convoluted theories that can't be established in simple mathematics are clearly stated without proof. Pertinent references are given so that readers can pursue the subject on their own. In spite of the analysis and mathematic manipulations, the emphasis of this book is physical concepts.

The book is based on the lecture notes written for a graduate course on integrated and fiber optics taught several times over many years at Purdue University. I apologize to students, past and present, who endured typos, corrections, and inconsistencies in various versions of the class notes. Their questions and comments helped immensely in shaping the book to the final form. I also like to thank my colleagues at Purdue for their encouragement, free advice, and consultation. I wish to acknowledge 3 friends in particular. They are Professors Daniel S. Elliott, Eric C. Furgason and George C. S. Lee.

Finally and more importantly, I like to express my sincere appreciation to my wife, Ching-Fong, for her enormous patience, constant encouragement, and steady support. She is the true force bonding three generations of Chen's together and the main pillar sustaining our family. Because of her, I am healthier and happier. I am forever indebted to her.

CHIN-LIN CHEN

West Lafayette, Indiana
September 2006

1

BRIEF REVIEW OF ELECTROMAGNETICS AND GUIDED WAVES

1.1 INTRODUCTION

The telecommunication systems are a key infrastructure of all modern societies, and the optical fiber communications is the backbone of the telecommunication systems. An optical communication system is comprised of many optical, electrical, and electronic devices and components. The optical devices may be in the "bulk," integrated, or fiber-optic form. Therefore, an understanding of the operation principles of these optical devices is of crucial importance to electrical engineers and electrical engineering students. This book is on integrated and fiber optics for optical communication applications. The subject matter beyond the introductory chapter is grouped into three parts. They are the integrated optics (Chapters 2–8), fiber optics (Chapters 9–12), and the propagation and evolution of optical pulses in linear and nonlinear fibers (Chapters 13 and 14).

The main purpose of Chapter 1 is to introduce the nomenclature and notations. In the process, we also review the basics of electromagnetism and essential theories of guided waves.

The theory of thin-film waveguides with constant index regions is relatively simple and complete and it is presented in Chapter 2. Most quantities of interest are

Foundations for Guided-Wave Optics, by Chin-Lin Chen
Copyright © 2007 John Wiley & Sons, Inc.

expressed in elementary functions. Examples include field components, the dispersion relation, the confinement factor, and power transported in each region. We also express the quantities in terms of the generalized parameters to facilitate comparison. In short, we use the step-index thin-film waveguides to illustrate the notion of guided waves and the basic properties of optical waveguides.

Many dielectric waveguides have a graded-index profile. Examples include optical waveguides built on semiconductors and lithium niobates. While the basic properties of graded-index waveguides are similar to that of step-index waveguides, subtle differences exist. In Chapter 3, we first analyze modes guided by linearly and exponentially tapered dielectric waveguides. We obtain closed-form expressions for fields and the dispersion relations for these waveguides. Then we apply the WKB (Wentzel, Kramers, and Brillouin) method and a numerical method to study optical waveguides with an arbitrary index profile.

So far, we have considered ideal waveguides made of loss-free materials and having a perfect geometry and index profile. While loss in dielectric materials may be very small, it is not zero. Obviously, no real waveguide structure or index profile is perfect either. As a result, waves decay as they propagate in real waveguides. In Chapter 4, we examine the effects of dielectric loss on the propagation and attenuation of guided modes and the perturbation on the waveguide properties by the presence of metallic films near or over the waveguide regions. The use of metal-clad waveguides as waveguide polarizers or mode filters is also discussed.

In practical applications, we may wish to pack the components densely so as to make the most efficient use of the available "real estate." It is then necessary to reduce interaction between waveguides and to minimize cross talks. For this purpose, it is necessary to confine fields in the waveguide regions. To confine fields in the two transverse directions, geometric boundaries and/or index discontinuities are introduced in the transverse directions. This leads to three-dimensional waveguides such as channel waveguides and ridge waveguides. In Chapter 5, we examine the modes guided by three-dimensional waveguides with rectangular geometries. Two approximate methods are used to establish the dispersion of modes guided by rectangular dielectric waveguides. They are the Marcatili method and the effective index method. A detailed comparison of the two methods is presented in the last section.

Having discussed the propagation, attenuation, and fields of modes guided by isolated waveguides, we turn our attention to three classes of passive guided-wave components: the directional coupler devices, the waveguide grating devices and arrayed waveguide gratings. In Chapter 6, we discuss Marcatili's improved coupled-mode equations for co-propagating modes and use these equations to establish the essential characteristics of directional coupling. Then we consider the switched $\Delta\beta$ directional couplers and their applications as switches, optical filters, and modulators.

Waveguide gratings are periodic topological structures or index variations built permanently on the waveguides. Periodic index variations induced by electrooptic or acoustooptic effects onto the waveguides are also waveguide gratings. These gratings are the building blocks of guided-wave components. Coupled-mode equations are developed in Chapter 7 to describe the interaction of contrapropagating modes in the grating structures. Then we use these equations to study the operation of grating reflectors, grating filters, and distributed feedback lasers. Arrayed-waveguide gratings are briefly discussed in Chapter 8.

The transmission and input/output properties of single-mode fibers are discussed in Chapters 9 and 10. In Chapter 9, we study the transmission properties of linearly polarized (LP) modes in weakly guiding step-index fibers with a circular core. For these fibers, a rigorous analysis of fields is possible, and we obtain closed-form expressions for several quantities of interest. We discuss the phase velocity, group velocity, and the group velocity dispersion of LP modes. For obvious reasons, we are particularly interested in the intramodal dispersion of single-mode fibers. The generalized parameters of step-index fibers are also used in the discussion.

In most applications, it would be necessary to couple light into and out of fibers. Therefore, the input and output characteristics of fibers are of practical interest. In Chapter 10, we suppose that fibers are truncated, and we examine the fields radiated by LP modes from the truncated fibers. We also examine the excitation of LP modes in step-index fibers by uniform plane waves and Gaussian beams.

Ideal fibers would have a circular cross section and a rotationally symmetric index profile and are free from mechanical, electric, and magnetic disturbances. But no ideal fiber exists because of the fabrication imperfection and postfabrication disturbances. Real fibers are birefringent. In Chapter 11, we begin by tracing the physical origins of the fiber birefringence. Then we estimate the fiber birefringence due to noncircular cross section and that induced by mechanical, electrical, and magnetic disturbances. Lastly, we use Jones matrices to describe the birefringent effects in fibers under various conditions.

Most manufactured fibers have graded-index profiles. Naturally, we are interested in the propagation and dispersion of the modes guided by graded-index fibers. In Chapter 12, we concentrate on fibers having a radially inhomogeneous and angularly independent index profile. Of particular interest to telecommunications is the fundamental mode guided by the graded-index fibers. The notion of the mode field radius or spot size is discussed.

All fibers, ideal or real, are dispersive. As a result, optical pulses evolve as they propagate in linear fibers. In Chapter 13, we study the propagation and evolution of pulses in linear, dispersive waveguides and fibers. Three approaches are used to analyze the pulse broadening and distortion problems. The first approach is a straightforward application of the Fourier and inverse Fourier transforms. The concept of the impulse response of a transmission medium is then introduced. Finally, we recognize that fields are the product of the carrier sinusoids and the pulse envelope. The propagation of carrier sinusoids is simple and well understood. Our attention is mainly on the slow evolution of the pulse envelope. A linear envelope equation is developed to describe the evolution of the pulse envelope. A general discussion of the envelope distortion and frequency chirping is then presented.

In nonlinear dispersive fibers, both the pulse shape and spectrum evolve as the pulse propagates. However, if the nonlinear fibers have anomalous group velocity dispersion and if the input pulse shape, temporal width, and amplitude satisfy a well-defined relationship, the pulses either propagate indefinitely without distortion or they reproduce the original pulse shape, width, and peak amplitude periodically. These pulses are known as solitary waves or optical solitons. Naturally, the formation and propagation of the optical solitons are of interest to telecommunications and we study these subjects in Chapter 14. A nonlinear envelope equation, often referred to as the nonlinear Schrödinger equation, is developed to describe the evolution of pulses on nonlinear dispersive fibers. We rely on a simple and straightforward method to derive an expression for the fundamental solitons. From the expression for fundamental solitons, we extract the key properties and the basic parameters of the fundamental solitons. Higher-order solitons and interaction of fundamental solitons are briefly discussed.

1.2   MAXWELL'S EQUATIONS

To study the waves guided by optical waveguides and fibers, we begin with the time-dependent, source-free Maxwell equations:

$$\nabla \times \mathcal{E}(\mathbf{r};t) = -\frac{\partial \mathcal{B}(\mathbf{r};t)}{\partial t} \tag{1.1}$$

$$\nabla \times \mathcal{H}(\mathbf{r};t) = \frac{\partial \mathcal{D}(\mathbf{r};t)}{\partial t} \tag{1.2}$$

$$\nabla \cdot \mathcal{B}(\mathbf{r};t) = 0 \tag{1.3}$$

$$\nabla \cdot \mathcal{D}(\mathbf{r};t) = 0 \tag{1.4}$$

where $\mathcal{E}(\mathbf{r};t)$, $\mathcal{D}(\mathbf{r};t)$, $\mathcal{H}(\mathbf{r};t)$, and $\mathcal{B}(\mathbf{r};t)$ are the *electric field intensity* (V/m), *electric flux density* (C/m^2), *magnetic field intensity* (A/m), and *magnetic flux density* (T or W/m^2), respectively. They are real functions of position $\mathbf{r}$ and time t. Although some dielectric waveguides and fibers may contain anisotropic materials, most optical waveguides and fibers of interest are made of isotropic, nonmagnetic dielectric materials. We confine our discussion in this book to isotropic, nonmagnetic materials only. For nonmagnetic and isotropic materials, the *constitutive relations* are

$$\mathcal{B}(\mathbf{r};t) = \mu_0 \mathcal{H}(\mathbf{r};t) \tag{1.5}$$

and

$$\mathcal{D}(\mathbf{r};t) = \varepsilon_0 \mathcal{E}(\mathbf{r};t) + \mathcal{P}(\mathbf{r};t) \tag{1.6}$$

where ε_0 ($\approx 1/36\pi \times 10^{-9}$ F/m) and μ_0 ($= 4\pi \times 10^{-7}$ H/m) are the vacuum *permittivity* and *permeability*. $\mathcal{P}(\mathbf{r};t)$ is the *electric polarization* of the medium [1–3].

It is convenient to use phasors to describe fields that vary sinusoidally in time. In the frequency domain, the Maxwell equations are

$$\nabla \times \mathbf{E}(\mathbf{r}; \omega) = -j\omega \mathbf{B}(\mathbf{r}; \omega) \tag{1.7}$$

$$\nabla \times \mathbf{H}(\mathbf{r}; \omega) = j\omega \mathbf{D}(\mathbf{r}; \omega) \tag{1.8}$$

$$\nabla \cdot \mathbf{B}(\mathbf{r}; \omega) = 0 \tag{1.9}$$

$$\nabla \cdot \mathbf{D}(\mathbf{r}; \omega) = 0 \tag{1.10}$$

where $\mathbf{E}(\mathbf{r}; \omega)$, $\mathbf{H}(\mathbf{r}; \omega)$, and so forth are the phasor representation of $\mathcal{E}(\mathbf{r}; t)$, $\mathcal{H}(\mathbf{r}; t)$, and so forth and ω is the angular frequency. In general, $\mathbf{E}(\mathbf{r}; \omega)$, $\mathbf{D}(\mathbf{r}; \omega)$, $\mathbf{H}(\mathbf{r}; \omega)$, and $\mathbf{B}(\mathbf{r}; \omega)$ are complex functions of $\mathbf{r}$ and ω. The time-domain field vectors and the corresponding frequency-domain quantities are related. For example,

$$\mathcal{E}(\mathbf{r}; t) = \mathrm{Re}[\mathbf{E}(\mathbf{r}; \omega)e^{j\omega t}] \tag{1.11}$$

The constitutive relations in the frequency domain are, in lieu of (1.5) and (1.6),

$$\mathbf{B}(\mathbf{r}; \omega) = \mu_0 \mathbf{H}(\mathbf{r}; \omega) \tag{1.12}$$

$$\mathbf{D}(\mathbf{r}; \omega) = \varepsilon_0 \mathbf{E}(\mathbf{r}; \omega) + \mathbf{P}(\mathbf{r}; \omega) \tag{1.13}$$

We assume that the fields are weak enough that the nonlinear response of the medium is negligibly small. We will not be concerned with the second- and third-order polarizations until Chapter 14. In the first 13 chapters, we take the media as linear media. In simple, isotropic and linear media, $\mathbf{P}(\mathbf{r}; \omega)$ is proportional to and in parallel with $\mathbf{E}(\mathbf{r}; \omega)$. Then the electric flux density can be written as

$$\mathbf{D}(\mathbf{r}; \omega) = \varepsilon_0[1 + \chi^{(1)}(\mathbf{r}; \omega)]\mathbf{E}(\mathbf{r}; \omega) = \varepsilon_0 \varepsilon_r(\mathbf{r}; \omega)\mathbf{E}(\mathbf{r}; \omega) \tag{1.14}$$

where $\chi^{(1)}(\mathbf{r}; \omega)$ is the *electric susceptibility*, and $\varepsilon_r(\mathbf{r}; \omega)$ is the *relative dielectric constant*. In optics literature, we often rewrite the above equation in terms of a *refractive index* $n(\mathbf{r}; \omega)$:

$$\mathbf{D}(\mathbf{r}; \omega) = \varepsilon_0 n^2(\mathbf{r}; \omega)\mathbf{E}(\mathbf{r}; \omega) \tag{1.15}$$

The relative dielectric constant and the refractive index may be functions of position and frequency. For example, different waveguide regions may have different ε_r and n.

To find the waves guided by a waveguide amounts to solving the Maxwell equations subject to the usual boundary conditions. Consider the boundary between media 1 and 2 as shown in Figure 1.1. Let $\mathbf{E}_i(\mathbf{r}; \omega)$, $\mathbf{H}_i(\mathbf{r}; \omega)$, and so forth be the field vectors in region i with an index n_i where $i = 1$ or 2. $\hat{\mathbf{n}}_i$ is the *unit vector* normal to the boundary separating the two media and pointing in the outward direction

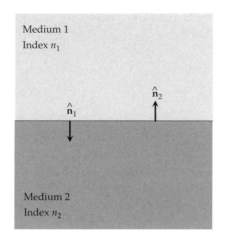

Figure 1.1   Unit vectors $\hat{\mathbf{n}}_1$ and $\hat{\mathbf{n}}_2$ normal to the boundary.

relative to region i. In the absence of the surface charge density and surface current density, the boundary conditions are as follows:

1. The tangential components of $\mathbf{E}(\mathbf{r};\omega)$ and $\mathbf{H}(\mathbf{r};\omega)$ are continuous at the boundary:

$$\hat{\mathbf{n}}_1 \times \mathbf{E}_1(\mathbf{r};\omega) + \hat{\mathbf{n}}_2 \times \mathbf{E}_2(\mathbf{r};\omega) = 0 \qquad (1.16)$$

$$\hat{\mathbf{n}}_1 \times \mathbf{H}_1(\mathbf{r};\omega) + \hat{\mathbf{n}}_2 \times \mathbf{H}_2(\mathbf{r};\omega) = 0 \qquad (1.17)$$

2. The normal components of $\mathbf{D}(\mathbf{r};\omega)$ and $\mathbf{B}(\mathbf{r};\omega)$ are also continuous at the boundary:

$$\hat{\mathbf{n}}_1 \cdot \mathbf{D}_1(\mathbf{r};\omega) + \hat{\mathbf{n}}_2 \cdot \mathbf{D}_2(\mathbf{r};\omega) = 0 \qquad (1.18)$$

$$\hat{\mathbf{n}}_1 \cdot \mathbf{B}_1(\mathbf{r};\omega) + \hat{\mathbf{n}}_2 \cdot \mathbf{B}_2(\mathbf{r};\omega) = 0 \qquad (1.19)$$

For brevity, we will drop the arguments $(\mathbf{r};t)$ and $(\mathbf{r};\omega)$ in the remaining discussion. In other words, we simply write $\mathcal{E}$, $\mathcal{H}$, $\mathbf{E}$, $\mathbf{H}$, and so forth in lieu of $\mathcal{E}(\mathbf{r};\omega)$, $\mathcal{H}(\mathbf{r};\omega)$, $\mathbf{E}(\mathbf{r};\omega)$, and $\mathbf{H}(\mathbf{r};\omega)$.

1.3   UNIFORM PLANE WAVES IN ISOTROPIC MEDIA

Waves are labeled as *plane waves* if the constant phase surfaces of the waves are planes. If the wave amplitude is the same everywhere on the constant phase plane, waves are identified as *uniform plane waves*. Consider uniform plane waves propagating in an arbitrary direction $\hat{\mathbf{k}}$ in free space. The electric and magnetic field intensities can be expressed as

$$\mathbf{E} = \mathbf{E}_0 e^{-j\mathbf{k}\cdot\mathbf{r}} \qquad (1.20)$$

$$\mathbf{H} = \mathbf{H}_0 e^{-j\mathbf{k}\cdot\mathbf{r}} \qquad (1.21)$$

where $\mathbf{k}$ is the *wave vector* in free space. $\mathbf{E}_0$ and $\mathbf{H}_0$ are the amplitudes of the electric and magnetic field intensities, respectively. To determine the relation between various plane wave parameters, we substitute (1.20) and (1.21) into the time-harmonic Maxwell equations (1.7)–(1.10) and obtain

$$|\mathbf{k}| = k = \omega \sqrt{\mu_0 \varepsilon_0} \tag{1.22}$$

and

$$\mathbf{H}_0 = \frac{1}{\eta_0} \hat{\mathbf{k}} \times \mathbf{E}_0 \tag{1.23}$$

In (1.23), $\eta_0 = \sqrt{\mu_0/\varepsilon_0}$ is the *intrinsic impedance* of free space.

The *Poynting vector* is

$$\mathbf{S} = \frac{1}{2} \text{Re}[\mathbf{E} \times \mathbf{H}^*] = \frac{|\mathbf{E}_0|^2}{2\eta_0} \hat{\mathbf{k}} \tag{1.24}$$

where $*$ stands for the complex conjugation of a complex quantity.

For an isotropic medium with a refractive index n, the wave vector and the intrinsic impedance are $n\mathbf{k}$ and η_0/n, respectively. In lieu of (1.23) and (1.24), the field vectors and Poynting vector in the medium with index n are related through the following relations:

$$\mathbf{H}_0 = \frac{n}{\eta_0} \hat{\mathbf{k}} \times \mathbf{E}_0 \tag{1.25}$$

$$\mathbf{S} = \frac{n|E_0|^2}{2\eta_0} \hat{\mathbf{k}} \tag{1.26}$$

In linear isotropic media, $\mathbf{E}$ and $\mathbf{D}$ are in parallel. So are $\mathbf{B}$ and $\mathbf{H}$. It is also clear from (1.23) to (1.26) that $\mathbf{E}$ and $\mathbf{D}$ are perpendicular to $\mathbf{B}$ and $\mathbf{H}$. These field vectors are also perpendicular to $\mathbf{k}$ and $\mathbf{S}$ as depicted in Figure 1.2(a). Since electric and magnetic field vectors are transverse to the direction of propagation, uniform plane waves in isotropic media are *transverse electromagnetic* (TEM) *waves*. These remarks hold for isotropic media. But for anisotropic media, many statements have to be modified [see Fig. 1.2(b)] [2, 4].

1.4 STATE OF POLARIZATION

In the last section, we consider fields in the frequency domain. It is often instructive to examine the fields in the time domain as well. In the time-domain description, we can "visualize" the motion of a field vector as a function of time. We refer the evolution of the field vector in time as the *state of polarization*. Consider the electric field at a certain point. We suppose that the electric field is confined in a plane,

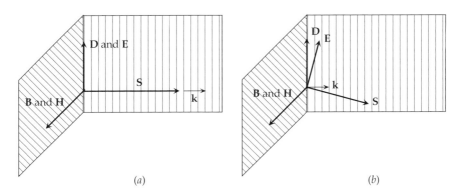

Figure 1.2   **E, D, B, H, S,** and **k** vectors of uniform plane waves in (*a*) isotropic and (*b*) anisotropic dielectric media.

which is taken as the *xy* plane for convenience. In the time-domain representation, the electric field at this point is

$$\mathcal{E} = \hat{\mathbf{x}}\mathcal{E}_x + \hat{\mathbf{y}}\mathcal{E}_y = \hat{\mathbf{x}}E_{x0}\cos(\omega t + \phi_x) + \hat{\mathbf{y}}E_{y0}\cos(\omega t + \phi_y) \qquad (1.27)$$

where E_{x0} and E_{y0} are amplitudes of the two components, and they are real and positive quantities. ϕ_x and ϕ_y are the phases of the two components relative to an arbitrary time reference. $\hat{\mathbf{x}}$ and $\hat{\mathbf{y}}$ are unit vectors in the x and y directions. The corresponding frequency-domain representation is

$$\mathbf{E} = \hat{\mathbf{x}}E_x + \hat{\mathbf{y}}E_y = \hat{\mathbf{x}}E_{x0}e^{j\phi_x} + \hat{\mathbf{y}}E_{y0}e^{j\phi_x} \qquad (1.28)$$

Depending on the amplitude ratio E_{y0}/E_{x0} and the phase difference $\Delta\phi = \phi_y - \phi_x$, the "tip" of the field vector may trace a linear, circular, or an elliptical trajectory in a left-hand or right-hand sense. If the two components are in time phase, that is, $\phi_x = \phi_y = \phi$, (1.27) can be simplified to

$$\mathcal{E} = (\hat{\mathbf{x}}E_{x0} + \hat{\mathbf{y}}E_{y0})\cos(\omega t + \phi)$$

While the length of the field vector changes as a cosine function, the field vector points to a fixed direction $\hat{\mathbf{x}}E_{x0} + \hat{\mathbf{y}}E_{y0}$. In other words, the tip of $\mathcal{E}$ moves along a straight line as time advances. We refer fields with $\phi_x = \phi_y$ as *linearly polarized* or *plane polarized fields* or *waves*.

If the two field components have the same amplitudes and are in time quadrature, that is, $E_{x0} = E_{y0}$ and $\Delta\phi = \phi_y - \phi_x = -\pi/2$, (1.27) becomes

$$\mathcal{E} = E_{x0}[\hat{\mathbf{x}}\cos(\omega t + \phi_x) + \hat{\mathbf{y}}\sin(\omega t + \phi_x)]$$

As time advances, the tip of $\mathcal{E}$ traces a circular path. If the wave under consideration moves in the $+z$ direction, then the $\mathcal{E}$ vector rotates in the counterclockwise sense for observers looking toward the source. If our right thumb points to the direction of propagation, that is, the $+z$ direction, our right-hand fingers would curl in the same sense as the motion of the tip of the electric field vector. Thus waves having field components specified by $E_{x0} = E_{y0}$ and $\phi_y - \phi_x = -\pi/2$ and moving in the $+z$ direction are *right-hand circularly polarized waves* [3, 5–7].

Similarly, if $E_{x0} = E_{y0}$ and $\Delta\phi = \phi_y - \phi_x = +\pi/2$, the field in the time-domain representation is

$$\mathcal{E} = E_{x0}[\hat{\mathbf{x}} \cos(\omega t + \phi_x) - \hat{\mathbf{y}} \sin(\omega t + \phi_x)]$$

For waves propagating in the $+z$ direction, the tip of the field vector traces a circle in the clockwise direction to observers facing the approaching waves. In other words, the field vector rotates in the left-hand sense. The fields with $E_{x0} = E_{y0}$ and $\Delta\phi = \phi_y - \phi_x = +\pi/2$ are *left-hand circularly polarized waves*. This is the Institute of Electrical and Electronics Engineers' (IEEE) definition for the right-handedness or left-handedness of the waves. The terminology used in the physics and optics literature is exactly the opposite. In many books on physics and optics, waves having $\Delta\phi = -\pi/2$ and $\Delta\phi = +\pi/2$ are identified, respectively, as the left- and right-hand circularly polarized waves [3, 5–7].

In general, (1.27), or (1.28), describes *elliptically polarized fields*. To elaborate this point further, we combine (1.27) and (1.28) and obtain (Problem 1)

$$\frac{\mathcal{E}_x^2}{E_{x0}^2} + \frac{\mathcal{E}_y^2}{E_{y0}^2} - 2\frac{\mathcal{E}_x}{E_{x0}}\frac{\mathcal{E}_y}{E_{y0}} \cos \Delta\phi = \sin^2 \Delta\phi \qquad (1.29)$$

The equation describes a *polarization ellipse* inscribed into a $2E_{x0} \times 2E_{y0}$ rectangle as shown in Figure 1.3. The shape and the orientation of the ellipse depend on the amplitude ratio E_{y0}/E_{x0} and the phase difference $\Delta\phi$. The sense of rotation depends only on the phase difference. The major and minor axes of a polarization ellipse do not necessarily coincide with the x and y axes. Therefore, there is no simple way to relate the major and minor axes of the ellipse specified in (1.29) to E_{x0} and E_{y0}. By rotating the coordinates, (1.29) can be transformed to a canonical form for ellipses [3, 5]. In the canonical form, the lengths of major and minor axes, $2E_{mj}$ and $2E_{mn}$, are readily identified. We take E_{mj} and E_{mn} as positive and $E_{mj} \geq E_{mn}$. The shape of the ellipse may also be quantified in terms of the *ellipticity*:

$$\text{Ellipticity} = \frac{E_{mn}}{E_{mj}} \qquad (1.30)$$

or the *visibility* (VS)

$$\text{VS} = \frac{E_{mj}^2 - E_{mn}^2}{E_{mj}^2 + E_{mn}^2} = \frac{1 - (E_{mn}^2/E_{mj}^2)}{1 + (E_{mn}^2/E_{mj}^2)} \qquad (1.31)$$

The ellipticity and visibility of a polarization ellipse are functions of E_{mn}/E_{mj}.

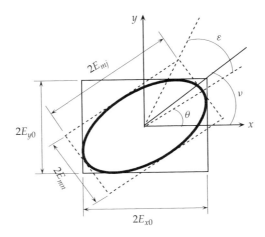

Figure 1.3 Parameters of elliptically polarized waves.

As shown in Figure 1.3, θ is the angle of the major axis relative to the x axis. We refer θ as the *azimuth* of the ellipse.

The sense of rotation of elliptically polarized field is the sense of motion of the tip of $\mathcal{E}$ as a function of time. For $0 < \Delta\phi < \pi$, the tip of $\mathcal{E}$ rotates in the same way as our left-hand fingers curl with the left thumb pointing in the direction of propagation. Thus fields with a phase difference $0 < \Delta\phi < \pi$ rotate in the left-hand sense. Fields with $\pi < \Delta\phi < 2\pi$ rotate in the right-hand direction. The limiting cases of $\Delta\phi = 0$ or π correspond to linearly polarized waves.

In summary, the state of polarization can be specified by E_{y0}/E_{x0} and $\Delta\phi$. It can also be cast in terms of $E_{mj}/E_{mn}, \theta$, and the sense of rotation. The transformation from E_{y0}/E_{x0} and $\Delta\phi$ to $E_{mn}/E_{mj}, \theta$, and the sense of rotation is facilitated by the following relations [3]:

$$\varepsilon_{\mathrm{lr}} \frac{E_{mn}}{E_{mj}} = \tan\varepsilon \qquad -\frac{\pi}{4} \le \varepsilon \le \frac{\pi}{2} \tag{1.32}$$

$$\sin 2\varepsilon = (\sin 2v)\sin\Delta\phi \tag{1.33}$$

$$\tan 2\theta = (\tan 2v)\cos\Delta\phi \qquad 0 \le \theta < \pi \tag{1.34}$$

$$\frac{E_{y0}}{E_{x0}} = \tan v, \qquad 0 \le v \le \frac{\pi}{2} \tag{1.35}$$

In (1.32), $\varepsilon_{\mathrm{lr}}$ is $+1$ for the left-handed rotation and -1 for the right-handed rotation. Detail derivation for these relations is left as an exercise for the reader (Problem 2). The physical meanings of E_{y0}/E_{x0}, E_{mn}/E_{mj}, $\Delta\phi$, and θ are easily understood from Figure 1.3. Although ε and v also have geometrical meaning of their own, as shown in Figure 1.3, we merely view ε and v as the two auxiliary variables introduced to specify the ratios E_{y0}/E_{x0} and E_{mn}/E_{mj}.

As implied in (1.28), an elliptically polarized field can be considered as the superposition of two orthogonal linearly polarized fields. An elliptically polarized field can also be viewed as the superposition of two counterrotating circularly polarized fields. To demonstrate this point, we rewrite (1.28) as

$$\mathbf{E} = \hat{\mathbf{x}}E_x + \hat{\mathbf{y}}E_y = \frac{E_x + jE_y}{\sqrt{2}}\hat{\mathbf{R}} + \frac{E_x - jE_y}{\sqrt{2}}\hat{\mathbf{L}} \tag{1.36}$$

where

$$\hat{\mathbf{R}} = \frac{\hat{\mathbf{x}} - j\hat{\mathbf{y}}}{\sqrt{2}} \quad \text{and} \quad \hat{\mathbf{L}} = \frac{\hat{\mathbf{x}} + j\hat{\mathbf{y}}}{\sqrt{2}}$$

are the *basis vectors* for right-hand and left-hand circularly polarized fields.

In isotropic media, the refractive indices "seen" or "experienced" by two orthogonal linearly polarized field components or the two counterrotating circularly polarized field components are the same. Thus the phase difference between the two field components remains unchanged as waves propagate. As a result, the two field components change at the same rate and by the same amount. Thus there is no change of the state of polarization as waves propagate in isotropic media.

The situation is quite different for waves in anisotropic media. An anisotropic medium may be birefringent, dichroic, or both. In birefringent media, different field components experience different refractive indices and travel with different phase velocities. Thus, $\Delta\phi$ changes as waves propagate in birefringent media. Because of the phase difference change, the state of polarization evolves as the waves propagate. In dichroic media, the two field components decay with different rates. Then E_{y0}/E_{x0} changes as waves propagate. Thus the state of polarization also evolves as waves propagate. In short, the state of polarization evolves as waves travel in anisotropic media. Further discussion on the subject can be found in [3, 8].

1.5 REFLECTION AND REFRACTION BY A PLANAR BOUNDARY BETWEEN TWO DIELECTRIC MEDIA

In homogeneous and isotropic media, uniform plane waves propagate along straight-line paths until they impinge upon boundaries. In inhomogeneous media, the rays turn continuously until they reach the boundary. At boundaries, waves are reflected and refracted abruptly. In this section, we consider the reflection and refraction of uniform plane waves at a planar boundary separating the two dielectric media having indices n_1 and n_2 (Fig. 1.4). Uniform plane waves propagate in $\mathbf{k}_{\text{in}}$ prior to impinging on the boundary. Note that $\mathbf{k}_{\text{in}} = k\hat{\mathbf{k}}_{\text{in}}$ and $k = \omega/c$ is the vacuum wave vector of the same angular frequency. It is convenient to refer various field components to a *plane of incidence*, which is defined by a unit vector $\hat{\mathbf{n}}_1$ normal to the boundary and the incident wave vector $n_1\mathbf{k}_{\text{in}}$. For the geometry shown in Figure 1.4, the plane of incidence is the xz plane. An arbitrary incident plane wave may be resolved into two orthogonal polarizations. One polarization has the electric field *normal* to the plane of incidence, and the other has the electric field *parallel* or *in* the plane of incidence. In the following discussion, we identify the two field components as $E_{\perp}$ and $E_{\parallel}$, respectively, and treat the two polarizations separately.

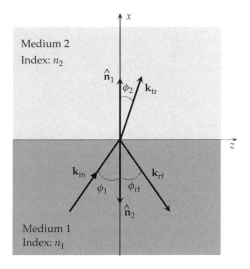

Figure 1.4 Reflection and refraction by a planar boundary.

1.5.1 Perpendicular Polarization

If the incident electric field is normal to the plane of incidence, so are the reflected and transmitted electric fields. Since all electric fields are normal to the plane of incidence, we refer this polarization as the *perpendicular polarization*. The electric fields are also perpendicular to the direction of propagation; the waves are the *transverse electric waves*, or simply *TE waves*. In some literature, they are also referred to as the *s waves* where *s* is the first letter of *senkrecht*, the German word for perpendicular.

With reference to the coordinates shown in Figure 1.4, the incident electric field is in the y direction. The incident magnetic field accompanying the incident electric field is in the direction of $\hat{\mathbf{k}}_{\text{in}} \times \hat{\mathbf{y}}$. Thus, the incident electric and magnetic fields are

$$\mathbf{E}_{\text{in}} = \hat{\mathbf{y}} E_{\text{in}0} e^{-jn_1 \mathbf{k}_{\text{in}} \cdot \mathbf{r}} \tag{1.37}$$

$$\mathbf{H}_{\text{in}} = \hat{\mathbf{k}}_{\text{in}} \times \hat{\mathbf{y}} \frac{n_1 E_{\text{in}0}}{\eta_0} e^{-jn_1 \mathbf{k}_{\text{in}} \cdot \mathbf{r}} \tag{1.38}$$

where $E_{\text{in}0}$ is the amplitude of the incident electric field. Let $n_1 \mathbf{k}_{\text{rf}}$ and $n_2 \mathbf{k}_{\text{tr}}$ be the wave vectors of the reflected and transmitted planes wave. Then, the reflected and transmitted fields may be written as, respectively,

$$\mathbf{E}_{\text{rf}} = \hat{\mathbf{y}} E_{\text{rf}0} e^{-jn_1 \mathbf{k}_{\text{rf}} \cdot \mathbf{r}} \tag{1.39}$$

$$\mathbf{H}_{\text{rf}} = \hat{\mathbf{k}}_{\text{rf}} \times \hat{\mathbf{y}} \frac{n_1 E_{\text{rf}0}}{\eta_0} e^{-jn_1 \mathbf{k}_{\text{rf}} \cdot \mathbf{r}} \tag{1.40}$$

$$\mathbf{E}_{\text{tr}} = \hat{\mathbf{y}} E_{\text{tr}0} e^{-jn_2 \mathbf{k}_{\text{tr}} \cdot \mathbf{r}} \tag{1.41}$$

$$\mathbf{H}_{\text{tr}} = \hat{\mathbf{k}}_{\text{tr}} \times \hat{\mathbf{y}} \frac{n_2 E_{\text{tr}0}}{\eta_0} e^{-jn_2 \mathbf{k}_{\text{tr}} \cdot \mathbf{r}} \tag{1.42}$$

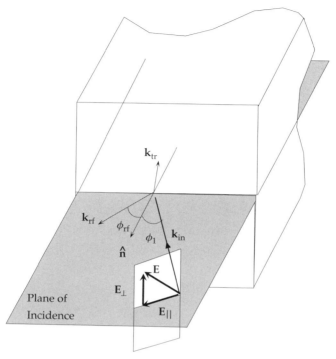

Figure 1.5 Plane of incidence and the perpendicular and parallel field components.

where $E_{\mathrm{rf}0}$ and $E_{\mathrm{tr}0}$ are the amplitudes of the reflected and transmitted electric fields. The remaining task is to determine various quantities as functions of $\mathbf{k}_{\mathrm{in}}$, $E_{\mathrm{in}0}$, n_1, and n_2.

Let ϕ_1 and ϕ_{rf} be the angles between the normal $\hat{\mathbf{n}}_1$ and the incident and reflected wave vectors as shown in Figure 1.4. Then, $\mathbf{k}_{\mathrm{in}}$ and $\mathbf{k}_{\mathrm{rf}}$ are

$$\mathbf{k}_{\mathrm{in}} = (\hat{\mathbf{x}} \cos \phi_1 + \hat{\mathbf{z}} \sin \phi_1)k$$

$$\mathbf{k}_{\mathrm{rf}} = (-\hat{\mathbf{x}} \cos \phi_{\mathrm{rf}} + \hat{\mathbf{z}} \sin \phi_{\mathrm{rf}})k$$

When the incident angle is smaller than a critical angle, a term to be introduced shortly, it is possible and meaningful to interpret $\mathbf{k}_{\mathrm{tr}}$ as a real vector having a real angle ϕ_2 relative to the normal $\hat{\mathbf{n}}_2$. Then $\mathbf{k}_{\mathrm{tr}}$ can be written as

$$\mathbf{k}_{\mathrm{tr}} = (\hat{\mathbf{x}} \cos \phi_2 + \hat{\mathbf{z}} \sin \phi_2)k$$

When the incident angle is larger than the critical angle, it is not possible to associate ϕ_2 with a real or physical angle. In the next subsection, we first discuss the cases where the incident angle is small. Then we discuss the necessary changes for a large incident angle.

1.5.1.1   Reflection and Refraction

To determine E_{rf0} and E_{tr0}, ϕ_{rf}, and ϕ_2, we make use of the boundary conditions (1.16) and (1.17). From the continuation of the tangential components of **E** and **H**, we obtain two equations:

$$E_{in0}e^{-jn_1k\sin\phi_1 z} + E_{rf0}e^{-jn_1k\sin\phi_{rf} z} = E_{tr0}e^{-jn_2k\sin\phi_2 z} \tag{1.43}$$

$$n_1\cos\phi_1 E_{in0}e^{-jn_1k\sin\phi_1 z} - n_1\cos\phi_{rf}E_{r0}e^{-jn_1k\sin\phi_{rf} z} = n_2\cos\phi_2 E_{tr0}e^{-jn_2k\sin\phi_2 z} \tag{1.44}$$

The two equations hold for all values of z if and only if all exponents are same. In other words,

$$n_1\sin\phi_1 = n_1\sin\phi_{rf} = n_2\sin\phi_2$$

We conclude immediately that the reflection angle equals to the incident angle,

$$\phi_{rf} = \phi_1 \tag{1.45}$$

This is the *law of reflection*. We also deduce

$$n_1\sin\phi_1 = n_2\sin\phi_2 \tag{1.46}$$

that is *Snell's law of refraction* or the *law of refraction*.

With ϕ_{rf} and $\sin\phi_2$ given by (1.45) and (1.46), (1.43) and (1.44) are simplified to

$$E_{in0} + E_{rf0} = E_{tr0} \tag{1.47}$$

$$n_1\cos\phi_1 E_{in0} - n_1\cos\phi_1 E_{rf0} = n_2\cos\phi_2 E_{tr0} \tag{1.48}$$

We solve for E_{rf0} and E_{tr0} in terms of E_{in0}. By defining the *reflection coefficient of the perpendicular polarization* as $\Gamma_\perp = E_{rf0}/E_{in0}$, we obtain

$$\Gamma_\perp = \frac{n_1\cos\phi_1 - n_2\cos\phi_2}{n_1\cos\phi_1 + n_2\cos\phi_2} \tag{1.49}$$

Equation (1.49) is the *Fresnel equation for the perpendicular polarization*. The equation can be cast in several equivalent forms. For example, it can be rearranged as

$$\Gamma_\perp = \frac{n_1\cos\phi_1 - \sqrt{n_2^2 - n_1^2\sin^2\phi_1}}{n_1\cos\phi_1 + \sqrt{n_2^2 - n_1^2\sin^2\phi_1}} \tag{1.50}$$

For given n_1, n_2, and ϕ_1, (1.49) or (1.50) can be used to calculate $\Gamma_\perp$. Typical plots of the reflection coefficient versus the incident angle are shown in Figures 1.6, 1.7, and 1.8. When medium 2 is denser than medium 1, that is, $n_2 > n_1$, $\Gamma_\perp$ is real

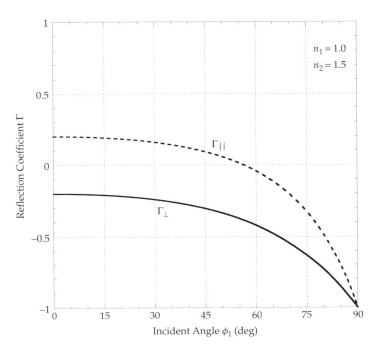

Figure 1.6 Reflection coefficient at a planar boundary between two dielectric media with $n_1 = 1.0$ and $n_2 = 1.5$.

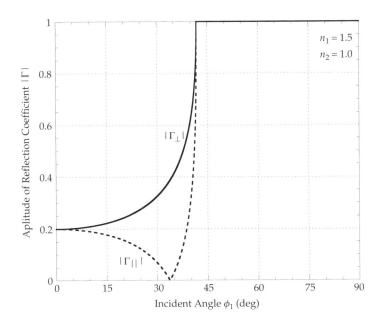

Figure 1.7 Amplitude of reflection coefficient at a planar boundary between two dielectric media with $n_1 = 1.5$ and $n_2 = 1.0$.

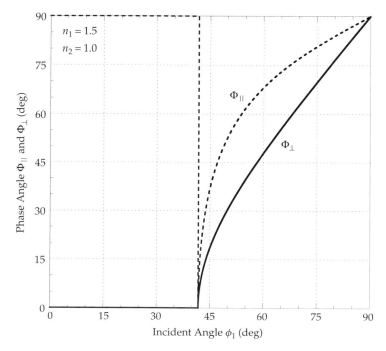

Figure 1.8 Phase of reflection coefficient at a planar boundary between two dielectric media with $n_1 = 1.5$ and $n_2 = 1.0$.

and negative for all values of ϕ_1, as shown in Figure 1.6. When medium 1 is denser than medium 2 and if the incident angle ϕ_1 is small such that $n_1 \sin \phi_1 \leq n_2$, then $\Gamma_\perp$ is real and positive. If the incident angle is large such that $n_1 \sin \phi_1 > n_2$, then $\Gamma_\perp$ becomes a complex quantity. More importantly, $|\Gamma_\perp|$ is 1. This is the case of the total internal reflection to be discussed in the next section.

The percentage of power reflected by the planar boundary is $|\Gamma_\perp|^2$, which is commonly referred to as the *power reflection coefficient* or the *reflectance*.

1.5.1.2   Total Internal Reflection

When medium 1 is denser than medium 2, there exists an incident angle such that $n_1 \sin \phi_1 = n_2$. This is the *critical angle* mentioned earlier. If the incident angle ϕ_1 is smaller than the critical angle, then $\Gamma_\perp$ as given in (1.49) and (1.50) is a real quantity. But if the incident angle ϕ_1 is greater than the critical angle, $\Gamma_\perp$ becomes a complex quantity. Although (1.46) is still valid, ϕ_2 is a complex quantity and we cannot assign a physically intuitive meaning to it. For $n_1 \sin \phi_1 > n_2, n_2 \cos \phi_2 = \pm j\sqrt{n_1^2 \sin^2 \phi_1 - n_2^2}$. To select a proper sign for $n_2 \cos \phi_2$, we return to (1.41) and (1.42) and examine the exponents as functions of x and z:

$$n_2 \mathbf{k}_{\mathrm{tr}} \cdot \mathbf{r} = xkn_2 \cos \phi_2 + zkn_2 \sin \phi_2 = \pm jxk\sqrt{n_1^2 \sin^2 \phi_2 - n_2^2} + zkn_1 \sin \phi_1$$

$$(1.51)$$

In terms of x and z explicitly, the transmitted electric field is

$$\mathbf{E}_{tr} = \hat{y} E_{tr0} e^{\pm xk \sqrt{n_1^2 \sin^2 \phi_2 - n_2^2} - jzkn_1 \sin \phi_1} \qquad (1.52)$$

A corresponding expression can be written for the transmitted magnetic fields. As shown in Figure 1.4, x is positive in region 2. Since the fields in region 2 must decay as x increases, the exponent in (1.52) must have a negative real part. This is possible if and only if we choose the minus sign in (1.51) and (1.52). In other words,

$$n_2 \cos \phi_2 = -j \sqrt{n_1^2 \sin^2 \phi_1 - n_2^2} \qquad (1.53)$$

Using this choice of $n_2 \cos \phi_2$, (1.40) becomes

$$\Gamma_\perp = \frac{n_1 \cos \phi_1 + j \sqrt{n_1^2 \sin^2 \phi_1 - n_2^2}}{n_1 \cos \phi_1 - j \sqrt{n_1^2 \sin^2 \phi_1 - n_2^2}} \qquad (1.54)$$

when $n_1 \sin \phi_1 > n_2$. Obviously, $\Gamma_\perp$ is now a complex quantity. By writing $\Gamma_\perp = |\Gamma_\perp| e^{j2\Phi_\perp}$, and we note immediately

$$|\Gamma_\perp| = 1 \qquad (1.55)$$

It means that all incident power is reflected by the planar boundary when the incident angle is greater than the critical value. This is known as the *total internal reflection* (TIR). It is the mechanism responsible for the lossless or low-loss wave-guiding in thin-film waveguides and optical fibers. We also deduce from (1.54) that the phase angle $\Phi_\perp$ is a positive angle:

$$\Phi_\perp = \tan^{-1} \frac{\sqrt{n_1^2 \sin^2 \phi_1 - n_2^2}}{n_1 \cos \phi_1} \qquad (1.56)$$

A plot of $\Phi_\perp$ as a function of ϕ_1 for the case with $n_1 = 1.5$ and $n_2 = 1.0$ is shown in Figure 1.8. We can also show analytically that $\Phi_\perp$ tends to $\pi/2$ as ϕ_1 approaches $\pi/2$. This observation is crucial in analyzing modes guided by graded-index waveguides and fibers.

1.5.2 Parallel Polarization

In the parallel polarization, the incident, reflected, and transmitted electric fields are in the plane of incidence. In other words, the electric fields have a component normal to and a component in parallel with the boundary. However, the accompanying magnetic fields are normal to the plane of incidence and the direction of propagation. The polarization is referred to as the *parallel polarization*, and the waves are the *transverse magnetic waves, TM waves,* or the *p waves* and *p* is the first letter of *parallel* in German.

For the parallel polarization, we could work with the incident electric field intensity as well. However, it is simpler to express all field quantities in terms of the incident magnetic field intensity, which is in the y direction in Figure 1.4. Thus, we write

$$\mathbf{H}_{\text{in}} = \hat{\mathbf{y}} H_{\text{in}\,0}\, e^{-jn_1 \mathbf{k}_{\text{in}} \cdot \mathbf{r}} \tag{1.57}$$

$$\mathbf{E}_{\text{in}} = \hat{\mathbf{y}} \times \hat{\mathbf{k}}_{\text{in}} \frac{\eta_0 H_{\text{in}\,0}}{n_1} e^{-jn_1 \mathbf{k}_{\text{in}} \cdot \mathbf{r}} \tag{1.58}$$

where $H_{\text{in}\,0}$ is the amplitude of the incident magnetic field intensity. We express the reflected and transmitted fields as

$$\mathbf{H}_{\text{rf}} = \hat{\mathbf{y}} H_{\text{rf}\,0} e^{-jn_1 \mathbf{k}_{\text{rf}} \cdot \mathbf{r}} \tag{1.59}$$

$$\mathbf{E}_{\text{rf}} = \hat{\mathbf{y}} \times \hat{\mathbf{k}}_{\text{rf}} \frac{\eta_0 H_{\text{rf}\,0}}{n_1} e^{-jn_1 \mathbf{k}_{\text{rf}} \cdot \mathbf{r}} \tag{1.60}$$

$$\mathbf{H}_{\text{tr}} = \hat{\mathbf{y}} H_{\text{tr}\,0} e^{-jn_2 \mathbf{k}_{\text{tr}} \cdot \mathbf{r}} \tag{1.61}$$

$$\mathbf{E}_{\text{tr}} = \hat{\mathbf{y}} \times \hat{\mathbf{k}}_{\text{tr}} \frac{\eta_0 H_{\text{tr}\,0}}{n_2} e^{-jn_2 \mathbf{k}_{\text{tr}} \cdot \mathbf{r}} \tag{1.62}$$

where $H_{\text{rf}\,0}$ and $H_{\text{tr}\,0}$ are the amplitudes of the reflected and transmitted magnetic field intensities. To solve for $\mathbf{k}_{\text{rf}}$, $\mathbf{k}_{\text{tr}}$, $H_{\text{rf}\,0}$, and $H_{\text{tr}\,0}$, we again make use of the boundary conditions (1.16) and (1.17) to rederive (1.45), (1.46). Then we obtain

$$H_{\text{in}\,0} + H_{\text{rf}\,0} = H_{\text{tr}\,0} \tag{1.63}$$

$$\frac{1}{n_1} \cos \phi_1 H_{\text{in}\,0} - \frac{1}{n_1} \cos \phi_1 H_{\text{rf}\,0} = \frac{1}{n_2} \cos \phi_2 H_{\text{tr}\,0} \tag{1.64}$$

From these equations, we solve for $H_{\text{rf}\,0}$ and $H_{\text{tr}\,0}$ in terms of $H_{\text{in}\,0}$.

1.5.2.1 Reflection and Refraction

By defining the *reflection coefficient of the parallel polarization* as $\Gamma_{\|} = H_{\text{rf}\,0}/H_{\text{in}\,0}$, we obtain

$$\Gamma_{\|} = \frac{n_2 \cos \phi_1 - n_1 \cos \phi_2}{n_2 \cos \phi_1 + n_1 \cos \phi_2} \tag{1.65}$$

This is the *Fresnel equation for the parallel polarization*. $\Gamma_{\|}$ can also be written in several equivalent forms, one of which is

$$\Gamma_{\|} = \frac{n_2^2 \cos \phi_1 - n_1 \sqrt{n_2^2 - n_1^2 \sin^2 \phi_1}}{n_2^2 \cos \phi_1 + n_1 \sqrt{n_2^2 - n_1^2 \sin^2 \phi_1}} \tag{1.66}$$

The critical angle introduced in connection with the perpendicular polarization plays an equally important role in the parallel polarization. $\Gamma_{\|}$ as given in (1.66) is real when ϕ_1 is smaller than the critical angle, and $\Gamma_\parallel$ can be positive or negative depending on the incidence angle as shown in Figures 1.6 and 1.7. As shown in these curves, $\Gamma_\parallel$ vanishes if $n_1 > n_2$ and if the incidence angle is

$$\tan \phi_1 = \frac{n_2}{n_1} \tag{1.67}$$

The incidence angle is commonly known as the *polarizing* or *Brewster angle*.

1.5.2.2 Total Internal Reflection

Following the same reasoning discussed in the perpendicular polarization, we choose $n_2 \cos \phi_2 = -j\sqrt{n_1^2 \sin^2 \phi_1 - n_2^2}$ when $n_1 \sin \phi_1 > n_2$. With this choice of $n_2 \cos \phi_2$, we obtain from (1.66)

$$\Gamma_\parallel = \frac{n_2^2 \cos \phi_1 + jn_1\sqrt{n_1^2 \sin^2 \phi_1 - n_2^2}}{n_2^2 \cos \phi_1 - jn_1\sqrt{n_1^2 \sin^2 \phi_1 - n_2^2}} \tag{1.68}$$

when $n_1 \sin \phi_1 > n_2$. Again, we note that $\Gamma_\parallel$ is a complex quantity for ϕ_1 greater than the critical angle. By writing $\Gamma_\parallel = |\Gamma_\parallel|e^{j2\Phi_\parallel}$, we obtain

$$|\Gamma_\parallel| = 1 \tag{1.69}$$

It means that the incident power is totally reflected when the incident angle is greater than the critical angle. The phase term $\Phi_\parallel$ is a positive angle:

$$\Phi_\parallel = \tan^{-1} \frac{n_1\sqrt{n_1^2 \sin^2 \phi_1 - n_2^2}}{n_2^2 \cos \phi_1} \tag{1.70}$$

As shown in Figure 1.8, $\Phi_\parallel$ also approaches $\pi/2$ in the limit of $\phi_1 = \pi/2$.

1.6 GUIDED WAVES

In this book, we are interested in waves guided by waveguides and fibers. We take the direction of wave propagation as the z axis. The waveguides and fibers have a constant cross section in planes transverse to the z axis. The index profile is independent of z as well. However, the index n may vary in the transverse directions. To consider waves propagating in the $+z$ direction, we separate the fields into the transverse and longitudinal components and write

$$\mathbf{E}(\mathbf{r}; \omega) = [\mathbf{e}_t(x, y) + \hat{\mathbf{z}}e_z(x, y)]e^{-j\beta z} \tag{1.71}$$

$$\mathbf{H}(\mathbf{r}; \omega) = [\mathbf{h}_t(x, y) + \hat{\mathbf{z}}h_z(x, y)]e^{-j\beta z} \tag{1.72}$$

where β is the propagation constant yet to be determined. The field quantities $\mathbf{e}_t(x, y)$, $\mathbf{h}_t(x, y)$, $e_z(x, y)$, and $h_z(x, y)$ may be functions of x and y as indicated explicitly. But they are independent of z. To proceed, we write $\nabla = \nabla_t - j\beta\hat{\mathbf{z}}$. Substituting these expressions in (1.7) and (1.8), we obtain

$$-j\beta\hat{\mathbf{z}} \times \mathbf{e}_t(x, y) - \hat{\mathbf{z}} \times \nabla_t e_z(x, y) = -j\omega\mu_0\mathbf{h}_t(x, y) \tag{1.73}$$

$$\nabla_t \times \mathbf{e}_t(x, y) = -j\omega\mu_0 h_z(x, y)\hat{\mathbf{z}} \tag{1.74}$$

$$-j\beta\hat{\mathbf{z}} \times \mathbf{h}_t(x, y) - \hat{\mathbf{z}} \times \nabla_t h_z(x, y) = j\omega\varepsilon_0 n^2(x, y)\mathbf{e}_t(x, y) \tag{1.75}$$

$$\nabla_t \times \mathbf{h}_t(x, y) = j\omega\varepsilon_0 n^2(x, y)e_z(x, y)\hat{\mathbf{z}} \tag{1.76}$$

Upon eliminating $\mathbf{h}_t(x, y)$ from (1.73) and (1.75), we obtain

$$\mathbf{e}_t(x, y) = \frac{j[\beta\nabla_t e_z(x, y) - \omega\mu_0\hat{\mathbf{z}} \times \nabla_t h_z(x, y)]}{\beta^2 - \omega^2\mu_0\varepsilon_0 n^2(x, y)} \tag{1.77}$$

Similarly, by eliminating $\mathbf{e}_t(x, y)$ from the two equations, we obtain

$$\mathbf{h}_t(x, y) = \frac{j[\beta\nabla_t h_z(x, y) + \omega\varepsilon_0 n^2(x, y)\hat{\mathbf{z}} \times \nabla_t e_z(x, y)]}{\beta^2 - \omega^2\mu_0\varepsilon_0 n^2(x, y)} \tag{1.78}$$

In principle, all transverse field components can be determined once the longitudinal field components are known. But it is rather difficult to determine $e_z(x, y)$ and $h_z(x, y)$ if the index in each region is a function of x and y. The differential equations can be obtained by substituting (1.77) and (1.78) into (1.74) and (1.76). But the differential equations contain $e_z(x, y)$ and $h_z(x, y)$. In other words, $e_z(x, y)$ and $h_z(x, y)$ are coupled. In addition, the boundary conditions (1.16) and (1.17) can be met only when $e_z(x, y)$ and $h_z(x, y)$ are considered simultaneously. In short, it is not a trivial matter to solve for $e_z(x, y)$ and $h_z(x, y)$ if the index in each region varies with x and y. The same conclusion can be reached by considering the wave equations obtained directly from Maxwell's equations (1.7) and (1.8):

$$\nabla^2\mathbf{E}(\mathbf{r}; \omega) + k^2 n^2(x, y)\mathbf{E}(\mathbf{r}; \omega) + \nabla\left[\frac{\mathbf{E}(\mathbf{r}; \omega) \cdot \nabla n^2(x, y)}{n^2(x, y)}\right] = 0 \tag{1.79}$$

$$\nabla \times \left[\frac{\nabla \times \mathbf{H}(\mathbf{r}; \omega)}{n^2(x, y)}\right] + k^2\mathbf{H}(\mathbf{r}; \omega) = 0 \tag{1.80}$$

If, however, under certain conditions, the differential equations for $e_z(x, y)$ and $h_z(x, y)$ are decoupled, and boundary conditions can be satisfied by considering $e_z(x, y)$ and its derivative alone, or by considering $h_z(x, y)$ and its derivative alone, then we have the fields and propagation constant of *TM* and *TE modes*, respectively. Two special cases are considered below.

1.6.1 Transverse Electric Modes

Consider waveguides that extend indefinitely in the y direction and the index in each region is a constant independent of y and z. In other words, the index in each region is $n(x)$ instead of $n(x, y)$. All field quantities are also independent of y. Taking advantage of these properties, we drop all partial derivatives with respect to y and obtain from (1.73)–(1.76):

$$j\beta e_y(x) = -j\omega\mu_0 h_x(x) \tag{1.81}$$

$$-j\beta e_x(x) - \frac{de_z(x)}{dx} = -j\omega\mu_0 h_y(x) \tag{1.82}$$

$$\frac{de_y(x)}{dx} = -j\omega\mu_0 h_z(x) \tag{1.83}$$

$$j\beta h_y(x) = j\omega\varepsilon_0 n^2(x) e_x(x) \tag{1.84}$$

$$-j\beta h_x(x) - \frac{dh_z(x)}{dx} = j\omega\varepsilon_0 n^2(x) e_y(x) \tag{1.85}$$

$$\frac{dh_y(x)}{dx} = j\omega\varepsilon_0 n^2(x) e_z(x) \tag{1.86}$$

Note that (1.81), (1.83), and (1.85) involve only $e_y(x)$, $h_x(x)$, and $h_z(x)$. Note in particular that the electric field intensity has just one Cartesian component, $e_y(x)$. We express the two magnetic field components in terms of $e_y(x)$ and obtain

$$h_x(x) = -\frac{\beta}{\omega\mu_0} e_y(x) \tag{1.87}$$

$$h_z(x) = \frac{j}{\omega\mu_0} \frac{de_y(x)}{dx} \tag{1.88}$$

Using these two expressions in conjunction with (1.85), we obtain a differential equation for $e_y(x)$:

$$\frac{d^2 e_y(x)}{dx^2} + [k^2 n^2(x) - \beta^2] e_y(x) = 0 \tag{1.89}$$

By solving $e_y(x)$ from the above equation and choosing constants to satisfy the boundary conditions, we have the fields and propagation constant of modes guided by the waveguide. Since the electric field is normal to the direction of propagation, the modes are classified as the *TE modes*.

1.6.2 Transverse Magnetic Modes

On the other hand, $e_x(x)$, $e_z(x)$, and $h_y(x)$ alone are involved in (1.82), (1.84), and (1.86). Since there is just one magnetic field component $h_y(x)$, we express

$e_x(x)$ and $e_z(x)$ in terms of $h_y(x)$, and obtain

$$e_x(x) = \frac{\beta}{\omega\varepsilon_0 n^2(x)} h_y(x) \tag{1.90}$$

$$e_z(x) = -\frac{j}{\omega\varepsilon_0 n^2(x)} \frac{dh_y(x)}{dx} \tag{1.91}$$

and

$$\frac{d}{dx}\left[\frac{1}{n^2(x)}\frac{dh_y(x)}{dx}\right] + \left[k^2 - \frac{\beta^2}{n^2(x)}\right]h_y(x) = 0 \tag{1.92}$$

By solving $h_y(x)$, we obtain the fields and propagation constant of the *TM modes*.

1.6.3   Waveguides with Constant Index in Each Region

In the case where n in each region is independent of y, z, *and* x, (1.87), (1.88), (1.90), and (1.91) remain valid. However, the two differential equations (1.89) and (1.92) are simplified further:

$$\frac{d^2 e_y(x)}{dx^2} + (k^2 n^2 - \beta^2)e_y(x) = 0 \tag{1.93}$$

$$\frac{d^2 h_y(x)}{dx^2} + (k^2 n^2 - \beta^2)h_y(x) = 0 \tag{1.94}$$

Equations (1.93) and (1.94) are two ordinary differential equations with constant coefficients and they can be solved readily. Detail study of TE and TM modes for dielectric waveguides with constant index in each region are discussed in Chapter 2. In fact, (1.90), (1.91), (1.92), and (1.93) are the starting point for the discussions in Chapter 2.

PROBLEMS

---

1. Show that from (1.27) and (1.28) that

$$\frac{\mathcal{E}_x}{E_{x0}} \sin \Delta\phi = \cos(\omega t + \phi_x) \sin \Delta\phi$$

$$\frac{\mathcal{E}_x}{E_{x0}} \cos \Delta\phi - \frac{\mathcal{E}_y}{E_{y0}} = \sin(\omega t + \phi_x) \sin \Delta\phi$$

Then, (1.29) follows immediately.

2. Derive (1.32) to (1.35).

3. Derive (1.79) and (1.80).

REFERENCES

1. R. E. Collin, *Field Theory of Guided Waves*, 2nd ed., IEEE Press, New York, 1991.
2. S. Ramo, J. R. Whinnery, and T. Van Duzer, *Fields and Waves in Communication Electronics*, 3rd ed., Wiley, New York, 1994.
3. M. Born and E. Wolf, *Principles of Optics*, 6th ed., Pergamon, Oxford, 1980.
4. D. B. Melrose and R. C. McPhedran, *Electromagnetic Processes in Dispersive Media*, Cambridge University Press, Cambridge, 1991.
5. E. Hecht, *Optics*, 3rd ed., Addison Wesley, Reading, MA, 1998.
6. John Krause and D. Fleisch, *Electromagnetics with Applications*, 5th ed., McGraw-Hill, New York, 1998.
7. U. S. Inan and A. S. Inan, *Electromagnetic Waves*, Prentice Hall, Upper Saddle River, NJ, 2000.
8. C. L. Chen, *Elements of Optoelectronics and Fiber Optics*, R. D. Irwin, Chicago, 1996.

2

STEP-INDEX THIN-FILM
WAVEGUIDES

2.1 INTRODUCTION

In the simplest form, an optical thin-film waveguide is a long structure having three dielectric regions as shown in Figure 2.1. The three dielectric regions are a thick region with an index n_s, a thin layer of an index n_f, and a thick region with an index n_c. The thin layer has the largest index and is referred to as the *film region*. The thick region with the smallest index is the *cover region*. The region having the second lowest index is the *substrate region*. Throughout our discussions, we assume that $n_f > n_s$ and n_c. Since the film region has the largest index n_f, fields are mainly confined in this region. The film thickness h is comparable to the operating wavelength λ. In contrast, the cover and substrate regions are much thicker than λ. We take the two thick regions as infinitely thick as an approximation.

Before presenting the thin-film waveguide theory, we describe briefly how thin-film waveguides are made. Probably, the simplest method to make a thin-film waveguide is to dip a plate in, or to spin-coat a plate with, polymer or photoresist. By dipping or coating, a thin layer of polymer or photoresist is formed on the plate and a thin-film waveguide is thereby formed. The plate mentioned above can be, and usually is, a glass slide. Thin-film waveguides can also be made by

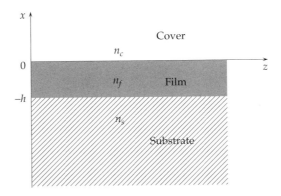

Figure 2.1 Step-index thin-film waveguide.

sputtering one type of glass on another type of glass or by dipping a glass slide into molten $AgNO_3$, thereby forming a high-index layer via the Ag ion exchange processes. Numerous techniques exist to form high-index layers on Si, GaAs, InP, or other semiconductor substrates. Although Si is lossy at the visible spectra, it is transparent at 1.3 μm or longer wavelengths. While many processes may be used to form waveguide layers on lithium niobate ($LiNbO_3$) and lithium tantalate ($LiTaO_3$) substrates, the in-diffusion process is the most popular method of making $LiNbO_3$ and $LiTaO_3$ waveguides. At present, most passive waveguide structures are based on glass, lithium niobate, lithium tantalate, Si, GaAs, and InP. The main attribute of lithium niobate, lithium tantalate, GaAs, or InP waveguides is that these materials are electrooptic, and the waveguide characteristics can be tuned by applying static, radio-frequency, or microwave electric fields to the waveguide material. Si wafers and glass slides are widely available and at a reasonable cost. Besides, Si and glass technologies are two matured and advanced technologies.

The index profile of the photoresist–glass or polymer–glass waveguides can be represented by three straight-line segments with abrupt index change at the boundaries as shown in Figure 2.2(*a*). Waveguides with such an index profile are known as *step-index waveguides*. The index of $LiNbO_3$, $LiTaO_3$, semiconductor waveguides, or ion-exchanged glass waveguides changes gradually in the film and substrate regions, as shown in Figure 2.2(*b*). In fact, the film and substrate regions may merge into one region and the index changes continuously as a function of position. The precise index profile would depend on the waveguide materials and the fabrication processes involved. For example, the index profile of many in-diffused $LiNbO_3$ or $LiTaO_3$ waveguides can be approximated by an exponential function. These waveguides are generally referred to as the *graded-index waveguides*. We restrict our discussions in this chapter to the step-index waveguides only. Discussions on graded-index waveguides are deferred until Chapter 3.

The cross sections of several waveguides are shown in Figure 2.3. In Figure 2.3(*a*), the waveguide dimension in the *y* direction is much larger than the film thickness *h* in the *x* direction and the operating wavelength λ. In studying the wave propagation in the *z* direction, we may ignore the field variation in the *y* direction

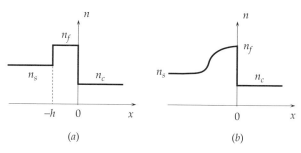

Figure 2.2 Two index profiles: (a) step-index and (b) graded-index profiles.

as an approximation. These waveguides are *two-dimensional waveguides*. They are also referred to as *thin-film waveguides, dielectric slab waveguides*, or *planar waveguides*. For waveguides shown in Figure 2.3(*b*), the channel, ridge, or strip width in the *y* direction is comparable to *h* and λ. They are *three-dimensional waveguides*. Depending on the waveguide geometry and the fabrication processes, they are known as channel waveguides, ridge or rib waveguides, and embedded strip waveguides. We will discuss three-dimensional waveguides in Chapter 5.

2.2   DISPERSION OF STEP-INDEX WAVEGUIDES

To study waves guided by two-dimensional step-index waveguides [1–4], we begin with the time-harmonic ($e^{j\omega t}$) Maxwell equations (1.7)–(1.10) for dielectric media. For a dielectric medium with an index n, the permittivity is $\varepsilon = n^2\varepsilon_0$ and the permeability is $\mu = \mu_0$. Different regions have different indices of refraction. The basic waveguide geometry is depicted in Figures 2.1 and 2.3(*a*). The film region is $-h \leq x \leq 0$ and it has the largest index n_f. The cover region is $x > 0$ and it has the smallest index n_c. The substrate region, $x \leq -h$, has an index between n_f and n_c. We choose the coordinate system such that waves propagate in the z direction. Since waves propagate in the z direction, a phase factor $e^{-j\beta z}$ is present in all terms. Accordingly, we write **E** and **H** as

$$\mathbf{E} = [\hat{\mathbf{x}}e_x(x, y) + \hat{\mathbf{y}}e_y(x, y) + \hat{\mathbf{z}}e_z(x, y)]e^{-j\beta z}$$

$$\mathbf{H} = [\hat{\mathbf{x}}h_x(x, y) + \hat{\mathbf{y}}h_y(x, y) + \hat{\mathbf{z}}h_z(x, y)]e^{-j\beta z}$$

Clearly, $\partial/\partial z = -j\beta$. Since the material properties and the waveguide geometry are independent of y, all field components are also independent of y. In other words, e_x and h_y and so forth are functions of x only. Then Maxwell's equations, (1.7) and (1.8), in the component form can be written as

$$j\beta e_y(x) = -j\omega\mu_0 h_x(x) \tag{2.1}$$

$$-j\beta e_x(x) - \frac{de_z(x)}{dx} = -j\omega\mu_0 h_y(x) \tag{2.2}$$

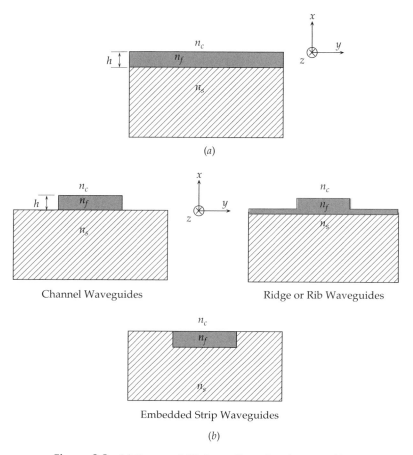

Figure 2.3   (a) Two- and (b) three-dimensional waveguides.

$$\frac{de_y(x)}{dx} = -j\omega\mu_0 h_z(x) \tag{2.3}$$

$$j\beta h_y(x) = j\omega\varepsilon_0 n^2 e_x(x) \tag{2.4}$$

$$-j\beta h_x(x) - \frac{dh_z(x)}{dx} = j\omega\varepsilon_0 n^2 e_y(x) \tag{2.5}$$

$$\frac{dh_y(x)}{dx} = j\omega\varepsilon_0 n^2 e_z(x) \tag{2.6}$$

These equations have been derived previously in Chapter 1. They are repeated here for convenience. Equations (2.1)–(2.6) are naturally divided into two groups: The first group—(2.1), (2.3), and (2.5)—involves e_y, h_x, and h_z only. Since the electric field is in the y direction that is perpendicular to the direction of propagation, this group of fields is referred to as the *transverse electric* (TE) *modes*. The second group—(2.2), (2.4), and (2.6)—involves h_y, e_x, and e_z. They form the *transverse magnetic* (TM) *modes*. We will study each group separately.

2.2.1 Transverse Electric Modes

A TE mode [3, 4] has an electric field component, e_y, and two magnetic field components, h_x and h_z. In particular, the electric field component is in parallel with the waveguide surface and perpendicular to the direction of propagation. The two magnetic field components can be expressed in terms of the electric field component. More explicitly,

$$h_x(x) = -\frac{\beta}{\omega\mu_0}e_y(x) \tag{2.7}$$

$$h_z(x) = j\frac{1}{\omega\mu_0}\frac{de_y(x)}{dx} \tag{2.8}$$

Substituting the two expressions in (2.5), we obtain a wave equation for e_y:

$$\frac{d^2e_y(x)}{dx^2} + (k^2n^2 - \beta^2)e_y(x) = 0 \tag{2.9}$$

The boundary conditions are the continuation of e_y, h_z, and b_x at $x = 0$ and $x = -h$. In view of (2.7), it is clear that b_x is continuous when e_y is continuous. It also follows from (2.8) that h_z is continuous at the boundary if de_y/dx is continuous there. In short, all boundary conditions are met if the electric field component e_y and its normal derivative de_y/dx are continuous at the boundaries. In the following sections, we solve e_y from (2.9) subject to the continuity of e_y and de_y/dx at the two interfaces.

In the film region ($-h \leq x \leq 0$, $n = n_f$), we expect e_y to be an oscillatory function of x. This is possible only if $k^2n_f^2 - \beta^2$ is positive. Upon introducing $\kappa_f = \sqrt{k^2n_f^2 - \beta^2}$, we write

$$e_y(x) = E_f \cos(\kappa_f x + \phi) \tag{2.10}$$

where E_f and ϕ are two constants yet undetermined. For future use, we also note

$$\frac{de_y(x)}{dx} = -\kappa_f E_f \sin(\kappa_f x + \phi) \tag{2.11}$$

For the substrate region ($x \leq -h$, $n = n_s$), we expect e_y to decay as x becomes more negative. This can be true only if $k^2n_s^2 - \beta^2$ is negative. Thus we introduce $\gamma_s = \sqrt{\beta^2 - k^2n_s^2}$ and write

$$e_y(x) = E_s e^{\gamma_s(x+h)} \tag{2.12}$$

$$\frac{de_y(x)}{dx} = \gamma_s E_s e^{\gamma_s(x+h)} \tag{2.13}$$

Similarly, we expect $k^2 n_c^2 - \beta^2$ to be negative for the cover region $(x > 0,$ $n = n_c)$. We define $\gamma_c = \sqrt{\beta^2 - k^2 n_c^2}$ and write

$$e_y(x) = E_c e^{-\gamma_c x} \tag{2.14}$$

$$\frac{de_y(x)}{dx} = -\gamma_c E_c e^{-\gamma_c x} \tag{2.15}$$

The constants E_f, E_s, E_c, and the phase ϕ in (2.10)–(2.15) and the propagation constant β are chosen to satisfy the boundary conditions noted earlier. The continuation of e_y and de_y/dx at $x = 0$ leads to

$$E_f \cos \phi = E_c \tag{2.16}$$

$$\kappa_f E_f \sin \phi = \gamma_c E_c \tag{2.17}$$

From the two equations, we obtain an expression for ϕ:

$$\tan \phi = \frac{\gamma_c}{\kappa_f} \tag{2.18}$$

Similarly, from the boundary conditions at $x = -h$, we have

$$E_f \cos(-\kappa_f h + \phi) = E_s \tag{2.19}$$

$$-E_f \kappa_f \sin(-\kappa_f h + \phi) = \gamma_f E_s \tag{2.20}$$

When the two equations are combined, we obtain a second expression for ϕ:

$$\tan(-\kappa_f h + \phi) = -\frac{\gamma_s}{\kappa_f} \tag{2.21}$$

Eliminating ϕ from (2.18) and (2.21), we obtain the *dispersion relation* or *characteristic equation for TE modes*:

$$\kappa_f h = \tan^{-1} \frac{\gamma_c}{\kappa_f} + \tan^{-1} \frac{\gamma_s}{\kappa_f} + m\pi \tag{2.22}$$

where $m = 0, 1, 2, 3, \ldots$ is an integer, and m is known as the *mode number*. The solution of (2.22) with a specific value of m gives the propagation constant of TE$_m$ mode. Once β is known, three of the four constants can be determined. The fourth constant, say E_f or E_c, represents the amplitude of the guided mode. Thus the TE modes guided by a thin-film waveguide are completely determined. Note that E_c is the electric field intensity at the cover–film boundary and E_f is the peak electric field intensity of the TE mode. Therefore E_f and E_c are of interest.

2.2.2 Transverse Magnetic Modes

A TM mode [3, 4] has a magnetic field component, h_y, and two electric field components, e_x and e_z. The transverse electric field component e_x is normal to the waveguide surface and the direction of propagation. In addition, the two electric field components can be expressed in terms of h_y. Specifically, we obtain from (2.4) and (2.6)

$$e_x(x) = \frac{\beta}{\omega \varepsilon_0 n^2} h_y(x) \tag{2.23}$$

$$e_z(x) = -j \frac{1}{\omega \varepsilon_0 n^2} \frac{dh_y(x)}{dx} \tag{2.24}$$

The boundary conditions for h_y, e_z, and d_x are met if h_y and $(1/n^2)/(dh_y/dx)$ are continuous at the boundaries. Following the same procedure used in the last subsection in analyzing TE modes, we write h_y in the cover, film, and substrate regions as

$$h_y(x) = \begin{cases} H_c e^{-\gamma_c x} & (2.25) \\ H_f \cos(\kappa_f x + \phi') & (2.26) \\ H_s e^{\gamma_s (x+h)} & (2.27) \end{cases}$$

where H_c, H_f, H_s, and ϕ' are constants to be determined. By matching the boundary values, we obtain the *dispersion relation* or *characteristic equation for TM modes*:

$$\kappa_f h = \tan^{-1} \frac{n_f^2}{n_c^2} \frac{\gamma_c}{\kappa_f} + \tan^{-1} \frac{n_f^2}{n_s^2} \frac{\gamma_s}{\kappa_f} + m'\pi \tag{2.28}$$

The mode number m' is also an integer. When β and three of the four constants are determined, the TM problem is solved. The fourth constant, say H_f or H_c, represents the amplitude of the TM mode. In particular, H_c is the magnetic field intensity at the cover–film boundary and H_f is the peak magnetic field intensity of the TM mode in question.

For a given waveguide operating at a given wavelength, n_f, n_s, n_c, h, and λ are known. β can be determined from the dispersion relation (2.22) for TE modes and (2.28) for TM modes. No analytic solution of the two transcendental equations is known. Numerical techniques are applied to determine β. But it is not easy or trivial to evaluate β numerically either. To appreciate the complication involved, we introduce the *effective guide index N* such that $\beta = kN$. For a guided mode, N is between n_f and n_s. For most waveguides of practical interest, n_f and n_s are numerically close and the index difference $(n_f - n_s)/n_f$ is less than a few percent. Thus N differs from n_f or n_s very slightly. Therefore, an accurate determination of N is difficult. Besides, the numerical results are valid for a specific set of waveguide parameters only. To circumvent these difficulties, Kogelnik and Ramaswany [5] introduced the generalized parameters. An alternate set of parameters has been introduced by Pandraud and Parriaux [6]. It is customary to use Kogelnik and Ramaswany's a, b, c, d, and V as the generalized parameters. One of the advantages of using their parameters is that b and V are also useful to characterize step-index fibers.

2.3   GENERALIZED PARAMETERS

As explained earlier, it is rather difficult to evaluate β and N numerically. Therefore, it is desirable to have a set of universal curves that are applicable to most waveguides of interest. These curves have been constructed by Kogelnik and Ramaswamy [5]. This is the subject of this section.

2.3.1   The a, b, c, d, and V Parameters

Three generalized parameters [3, 5, 7] are sufficient to describe all TE modes guided by three-layer step-index waveguides [3, 5]. The three generalized parameters are
(a) the *asymmetry measure*

$$a = \frac{n_s^2 - n_c^2}{n_f^2 - n_s^2} \tag{2.29}$$

(b) the *generalized frequency*, also known as the *generalized film thickness*,

$$V = kh\sqrt{n_f^2 - n_s^2} \tag{2.30}$$

and (c) the *generalized guide index*

$$b = \frac{N^2 - n_s^2}{n_f^2 - n_s^2} \tag{2.31}$$

The generalized parameters a and b are the differences $n_s^2 - n_c^2$ and $N^2 - n_s^2$ normalized with respect to $n_f^2 - n_s^2$. In other words, these generalized parameters are in terms of the differences of index squares rather than the indices themselves.
A little manipulation will show that

$$\kappa_f h = kh\sqrt{n_f^2 - N^2} = V\sqrt{1-b} \tag{2.32}$$

$$\gamma_s h = kh\sqrt{N^2 - n_s^2} = V\sqrt{b} \tag{2.33}$$

and

$$\gamma_c h = kh\sqrt{N^2 - n_c^2} = V\sqrt{a+b} \tag{2.34}$$

Using these expressions in (2.22), we cast the dispersion relation for TE modes in terms of generalized parameters:

$$V\sqrt{1-b} = \tan^{-1}\sqrt{\frac{a+b}{1-b}} + \tan^{-1}\sqrt{\frac{b}{1-b}} + m\pi \qquad (2.35)$$

For TM modes, we need one more parameter [5, 7] in addition to a, b, and V defined above. The extra parameter can be either

$$c = \frac{n_s^2}{n_f^2} \qquad (2.36)$$

or

$$d = \frac{n_c^2}{n_f^2} = c - a(1-c) \qquad (2.37)$$

In terms of generalized parameters, the dispersion relation for TM modes is

$$V\sqrt{1-b} = \tan^{-1}\frac{1}{d}\sqrt{\frac{a+b}{1-b}} + \tan^{-1}\frac{1}{c}\sqrt{\frac{b}{1-b}} + m'\pi \qquad (2.38)$$

Except for the presence of $1/c$ and $1/d$, (2.38) is similar to (2.35). In the two transcendental equations, b spans from 0 and 1. More importantly, the solutions in terms of b are well apart and distinct. In other words, (2.35) and (2.38) can be solved easily by numerical methods.

2.3.2 The bV Diagram

For a given waveguide operating at a specific wavelength, n_f, n_s, n_c, h, and λ are known. The values of a, c, d, and V may be calculated from the waveguide parameters. For each set of a, V, and c, we determine b numerically from (2.35) for TE modes and (2.38) for TM modes. There may be one or more solutions for b, depending on V, a, and c. Each solution of b corresponds to a guided mode. The largest b corresponds to $m = 0$. Knowing b, the effective index N, the propagation constant β, three amplitude constants, and ϕ can be evaluated. Plots of b versus V for TE modes with four values of a are depicted in Figure 2.4. Similar plots for TM modes for $c = 0.98$, 0.94, and 0.90 are shown in Figure 2.5. The bV curves of TE and TM modes are really very close. To compare curves of TE and TM modes, we superimpose plots of TE and TM modes for two values of a and for $c = 0.94$ in Figure 2.6. The bV curves of TE modes are shown as thick lines and dots and that of TM modes as thin lines and dots. Obviously, the two sets of curves are really very close.

As noted earlier, each solution of b corresponds to a guided mode. As the film thickness gets thinner, corresponding to a smaller V, b becomes smaller. As b of a given mode approaches 0, the mode approaches its *cutoff*. All modes, except the TE_0

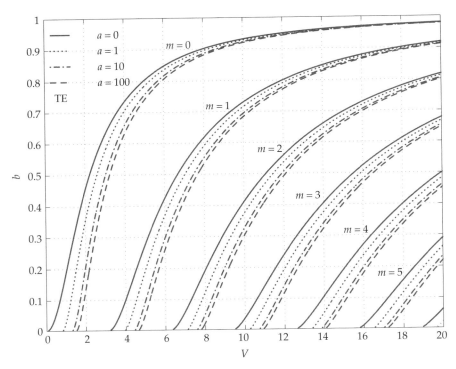

Figure 2.4 *bV* curves of TE modes guided by step-index thin-film waveguides.

and TM_0 modes guided by symmetric waveguides, are cutoff if the film region is sufficiently thin. In fact, if the film thickness is sufficiently thin, no mode is supported by the asymmetric waveguides. A waveguide is symmetric if the cover and substrate indices are the same. A waveguide having a thin glass slide surrounded by air on both sides is a good example of symmetric waveguides. The fundamental mode of *symmetric waveguides* has no cutoff. In other words, a symmetric waveguide supports at least a TE and a TM mode each no matter how thin the film region is. The TE and TM modes are the TE_0 and TM_0 modes.

We can visualize the shapes of the *bV* curves by recalling the modes guided by parallel-plate waveguides. A parallel-plate waveguide is simply two large conducting plates separated by a dielectric material of index n and thickness h. Modes guided by parallel-plate waveguides are discussed in many texts on electromagnetism [8–11]. By defining $V = knh$ and $b = N^2/n^2$, the dispersion of TE modes guided by a parallel-plate waveguide can be written as

$$b = 1 - \left[\frac{(m+1)\pi}{V} \right]^2 \tag{2.39}$$

The *bV* characteristic of parallel-plate waveguides as given in (2.39) is plotted as dots in Figure 2.7. For comparison, the *bV* curves of TE modes guided by a

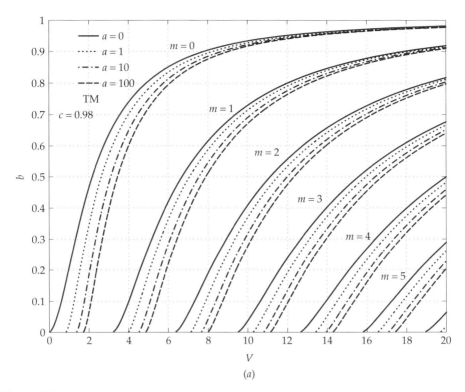

Figure 2.5 *bV* curves of TM modes guided by step-index thin-film waveguides: (*a*) $c = 0.98$, (*b*) $c = 0.94$, and (*c*) $c = 0.90$.

symmetric thin-film waveguide are depicted in the same figure as solid lines. The similarity between the two sets of curves is obvious.

To illustrate the use of the bV diagram, we consider a numerical example with $n_f = 1.500$, $n_s = n_c = 1.300$, and $h/\lambda = 3.000$. Since the cover and substrate indices are equal, the waveguide is a symmetric waveguide. For a symmetric waveguide with $V = 14.11$, we read from Figure 2.4 that b of TE$_0$ mode is about 0.96 (with a little imagination?). Using this value of b, we obtain from (2.31) that $N \sim 1.493$. The next mode is TE$_1$ mode for which b and N are 0.85 and 1.472, respectively; b and N for higher-order modes can be obtained in the same manner.

2.3.3 Cutoff Thicknesses and Cutoff Frequencies

As noted previously, the cutoff condition is $b = 0$. By setting b to zero, we have, from (2.35), the cutoff V value for TE$_m$ mode:

$$V_m = m\pi + \tan^{-1}\sqrt{a} \qquad (2.40)$$

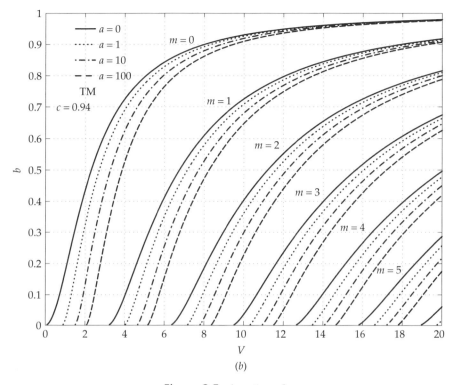

Figure 2.5 (*continued*).

In other words, a TE_m mode is supported by a thin-film waveguide if the film thickness is

$$h = \frac{m\pi + \tan^{-1}\sqrt{a}}{2\pi\sqrt{n_f^2 - n_s^2}}\lambda \tag{2.41}$$

or thicker.

The cutoff V for a $TM_{m'}$ mode is, from (2.38),

$$V_{m'} = m'\pi + \tan^{-1}\left(\frac{\sqrt{a}}{d}\right) = m'\pi + \tan^{-1}\left(\frac{n_f^2}{n_c^2}\sqrt{a}\right) \tag{2.42}$$

Since n_c is smaller than n_f, the cutoff V for TE_m mode is slightly smaller than the cutoff V of TM_m mode with the same mode number m.

2.3.4 Number of Guided Modes

It is clear from (2.40), that m TE modes may be supported by a thin-film waveguide if V is between $m\pi + \tan^{-1}\sqrt{a}$ and $(m + 1)\pi + \tan^{-1}\sqrt{a}$. Similarly, there are m TM modes if V is between $m\pi + \tan^{-1}(\sqrt{a}/d)$ and $(m + 1)\pi + \tan^{-1}(\sqrt{a}/d)$.

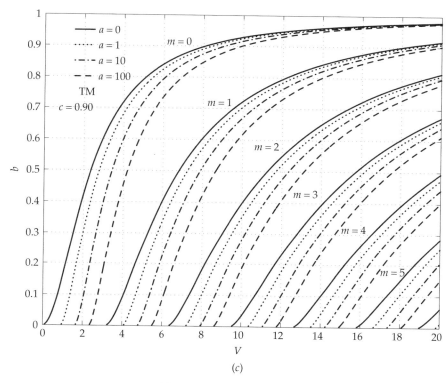

Figure 2.5 (continued).

We also note that for symmetric waveguides, the cutoff V values for TE_m and TM_m modes are simply $m\pi$. For each increment of V by π, there is an additional TE mode and an additional TM mode. We can also determine the number of modes guided from the bV plots if V is less than 20. For symmetric thin-film waveguides with V greater than 20, the number of TE and TM modes supported by the waveguide is approximately $2[1 + (V/\pi)]$. The numbers of TE and TM modes guided by asymmetric waveguides are approximately the same as that guided by symmetric waveguides. In short, the number of TE and TM modes guided by step-index thin-film waveguides is approximately $2[1 + (V/\pi)]$.

2.3.5 Birefringence in Thin-Film Waveguides

As shown in Figures 2.4, 2.5, and 2.6, the bV curves of TE_m and TM_m modes of the same waveguide having the same mode number may be close. But the bV curves do not cross or intersect. For the same waveguide and for the same mode number m, the b value of TM_m mode is always smaller than that of TE_m mode as shown in Figure 2.6. Thus, the propagation constant β of a TE mode is slightly different from that of a TM mode with the same mode number. In short, thin-film waveguides made of isotropic dielectric materials are *birefringent*. The birefringence of thin-film waveguides may be problematic in many applications.

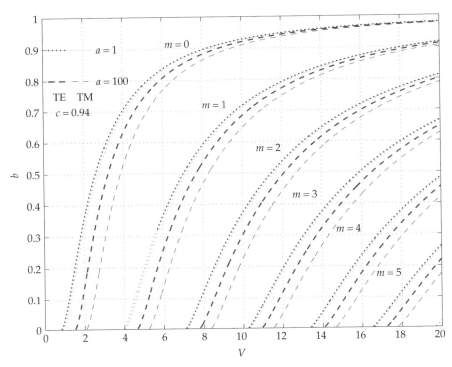

Figure 2.6   *bV* curves of TE and TM modes guided by step-index thin-film waveguides.

2.4   FIELDS OF STEP-INDEX WAVEGUIDES

2.4.1   Transverse Electric Modes

By requiring the continuation at the film–cover and film–substrate boundaries and making use of the dispersion relation (2.22), we can eliminate three of the four constants E_c, E_f, E_s, and ϕ. Then $e_y(x)$ given in (2.10), (2.12), and (2.14) may be expressed in terms of a single amplitude constant. If we keep E_c as the amplitude constant, then $e_y(x)$ is

$$
e_y(x) = \begin{cases} E_c e^{-\gamma_c x} & x \geq 0 \\ E_c \left[\cos \kappa_f x - \dfrac{\gamma_c}{\kappa_f} \sin \kappa_f x \right] & -h \leq x \leq 0 \\ E_c \left[\cos \kappa_f h + \dfrac{\gamma_c}{\kappa_c} \sin \kappa_f h \right] e^{\gamma_s (x+h)} & x \leq -h \end{cases} \tag{2.43}
$$

When written in terms of the generalized parameters, $e_y(x)$ becomes

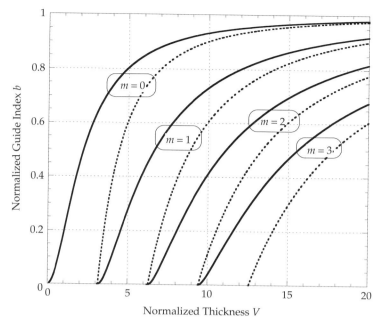

Figure 2.7 Comparison of the bV characteristic of TE modes guided by a symmetric dielectric waveguide (solid lines) with that of a parallel plate waveguide (dotted lines).

$$e_y(x) = \begin{cases} E_c e^{-V\sqrt{a+b}\,x/h} & x \geq 0 \\[2mm] E_c \left[\cos\left(\dfrac{V\sqrt{1-b}\,x}{h} \right) - \sqrt{\dfrac{a+b}{1-b}} \right. \\ \left. \times \sin\left(\dfrac{V\sqrt{1-b}\,x}{h} \right) \right] & -h \leq x \leq 0 \\[2mm] E_c \left[\cos(V\sqrt{1-b}) + \sqrt{\dfrac{a+b}{1-b}} \right. \\ \left. \times \sin(V\sqrt{1-b}) \right] e^{V\sqrt{b}[1+(x/h)]} & x \leq -h \end{cases} \qquad (2.44)$$

Typical field distributions of $e_y(x)$ of TE_0, TE_1, and TE_2 modes of asymmetric waveguides are presented in Figure 2.8. The vertical dash lines in the figures mark the film–substrate and film–cover boundaries. In the film region, the field distribution of a guided mode is an oscillatory function of x, as expected. In the two outer regions, the field decays exponentially from the film–cover and film–substrate boundaries. For all guided modes, fields in the cover and substrate regions are confined mainly in the thin layers near the film–cover or film–substrate boundaries. Beyond the thin layers immediately adjacent to the two boundaries, fields are very weak unless the mode is close to cutoff. A careful examination of these plots also

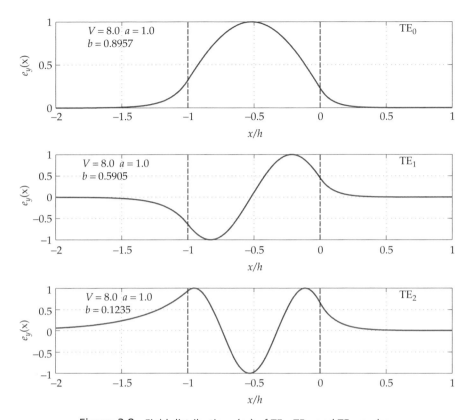

Figure 2.8 Field distributions (e_y) of TE_0, TE_1, and TE_2 modes.

reveals that fields in the substrate region are stronger than those in the cover region. This is also expected since we have assumed that $n_s \geq n_c$.

Expressions for $h_x(x)$ and $h_z(x)$ can be obtained by substituting $e_y(x)$ into (2.7) and (2.8). If desired, expressions can also be written in terms of E_f or E_s (Problem 5). Most quantities of interest, the time-average power, for example, can also be expressed in terms of the generalized parameters. We will discuss the time-average power in Section 2.6.

2.4.2 Transverse Magnetic Modes

In terms of H_c, $h_y(x)$ is

$$
h_y(x) = \begin{cases}
H_c e^{-\gamma_c x} & x \geq 0 \\[2ex]
H_c \left(\cos \kappa_f x - \dfrac{n_f^2 \gamma_c}{n_c^2 \kappa_f} \sin \kappa_f x \right) & -h \leq x \leq 0 \\[2ex]
H_c \left(\cos \kappa_f h + \dfrac{n_f^2 \gamma_c}{n_c^2 \kappa_f} \sin \kappa_f h \right) e^{\gamma_s (x+h)} & x \leq -h
\end{cases}
\tag{2.45}
$$

When expressed in terms of generalized parameters, $h_y(x)$ is

$$
h_y(x) = \begin{cases}
H_c e^{-V\sqrt{a+bx}/h} & x \geq 0 \\
H_c\left[\cos\left(\dfrac{V\sqrt{1-b}x}{h}\right) - \dfrac{1}{d}\sqrt{\dfrac{a+b}{1-b}}\right. \\
\qquad \left.\times \sin\left(\dfrac{V\sqrt{1-b}x}{h}\right)\right] & -h \leq x \leq 0 \\
H_c\left[\cos(V\sqrt{1-b}) + \dfrac{1}{d}\sqrt{\dfrac{a+b}{1-b}}\right. \\
\qquad \left.\times \sin(V\sqrt{1-b})\right]e^{V\sqrt{b}[1+(x/h)]} & x \leq -h
\end{cases}
\tag{2.46}
$$

The other field components may be obtained by substituting (2.45) or (2.46) into (2.4) and (2.6).

2.5 COVER AND SUBSTRATE MODES

In addition to a finite number of guided modes, a thin-film waveguide also supports a continuum of radiation modes. Radiation modes are not confined in a region nor guided by the dielectric boundaries. They are referred to as the *cover* (or *air*) and *substrate modes*, respectively. For a substrate mode, the total internal reflection occurs at the cover–film interface, but not the film–substrate boundary, recalling that $n_s > n_c$. In the cover region, the field of a substrate mode decreases exponentially from the film–cover boundary. In the substrate region, the field is the superposition of, and the interference between, the incident and reflected plane waves. Thus, the field in the substrate region varies in an oscillatory manner instead decaying exponentially from the film–substrate boundary. For the cover mode, there is no total internal reflection at either boundary. Thus a cover mode is a plane wave in one region and crests and troughs in the other region. In other words, a cover mode penetrates into the substrate and cover regions. This is depicted schematically in Figure 2.9.

2.6 TIME-AVERAGE POWER AND CONFINEMENT FACTORS

2.6.1 Time-Average Power Transported by TE Modes

The time-average power transported by a TE mode in a thin-film waveguide of unit width in the y direction is

$$
P = \iint \frac{1}{2}\text{Re}[\mathbf{E} \times \mathbf{H}^*]\cdot\hat{\mathbf{z}}\,dx\,dy = \frac{\beta}{2\omega\mu_o}\iint |e_y(x)|^2\,dx\,dy
$$
$$
= \frac{\beta}{2\omega\mu_o}1\left[\int_{-\infty}^{-h} E_s^2 e^{2\gamma_s(x+h)}\,dx + \int_{-h}^{0} E_f^2\cos^2(\kappa_f x + \phi)\,dx + \int_0^{\infty} E_c^2 e^{-2\gamma_c x}\,dx\right]
$$

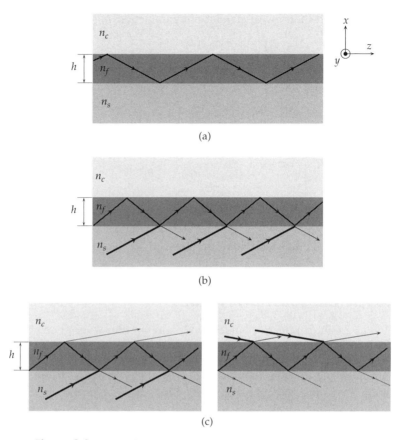

Figure 2.9 Rays of (*a*) guided, (*b*) substrate, and (*c*) cover modes.

where 1 signifies a unit width in the *y* direction. All integrals can be integrated in closed form and the result is

$$P = \frac{\beta}{4\omega\mu_0}\left[\frac{E_s^2}{\gamma_s} + E_f^2\left(h + \frac{\sin 2\phi - \sin 2(-\kappa_f h + \phi)}{2\kappa_f}\right) + \frac{E_c^2}{\gamma_c}\right] \qquad (2.47)$$

The three terms represent the time-average power transported in the substrate, film, and cover regions. By making use of (2.16)–(2.18) and (2.19)–(2.21), we can show

$$\frac{E_c^2}{\gamma_c} + \frac{E_f^2 \sin 2\phi}{2\kappa_f} = \frac{E_f^2}{\gamma_c} \qquad (2.48)$$

and

$$\frac{E_s^2}{\gamma_s} - \frac{E_f^2 \sin 2(-\kappa_f h + \phi)}{2\kappa_f} = \frac{E_f^2}{\gamma_s} \qquad (2.49)$$

Making use of these relations, we can express E_c and E_s in terms of E_f. Finally, we simplify (2.47) and obtain

$$P = \frac{\beta}{4\omega\mu_o} E_f^2 \left[h + \frac{1}{\gamma_s} + \frac{1}{\gamma_c} \right] \qquad (2.50)$$

To appreciate the significance of the bracketed term in the above expression, we consider TE modes guided by a parallel-plate waveguide with a plate separation of h. Suppose that the peak field intensity is E_f. Then the time-average power carried by TE modes of a parallel-plate waveguide of unit width is $P = (\beta/4\omega\mu_o)E_f^2 h$ [8–11]. Recall that h is the spacing between the two conducting plates. This expression is very similar to (2.50). By analogy, we define the *effective waveguide thickness* of a thin-film waveguide operating in a TE mode as

$$h_{\text{eff}} = h + \frac{1}{\gamma_s} + \frac{1}{\gamma_c} \qquad (2.51)$$

We also define the *normalized effective film thickness* as $H = kh_{\text{eff}}\sqrt{n_f^2 - n_s^2}$. In terms of the generalized parameters,

$$H = V + \frac{1}{\sqrt{a+b}} + \frac{1}{\sqrt{b}} \qquad (2.52)$$

Figure 2.10 is a plot of H versus V for the first four TE modes. In all cases, H decreases from a large value near the cutoff to a minimum, and then it increases linearly with V. This can be understood as follows. When a mode is near the cutoff, fields are not confined in the film region. Therefore, the effective film thickness is rather large. In other words, a large fraction of time-average power resides outside the film region if a mode is close to the cutoff. Far above the cutoff, however, H approaches an asymptote:

$$H \longrightarrow V + 1 + \frac{1}{\sqrt{1+a}} \qquad (2.53)$$

It is evident from Figure 2.10 that H is only slightly larger then V for large V. It means that only a small fraction of time-average power is outside the film region when a mode is far above the cutoff. In short, fields and time-average power are indeed confined in the film region for modes far above the cutoff.

The minimum value of H depends on the asymmetry measure a and the mode number m. For example, the minima of H of TE_0 mode are 4.93, 4.52, and 4.40 for $a = 0$, 1, and 10, respectively. For higher-order TE modes, the minima of H are larger, as expected.

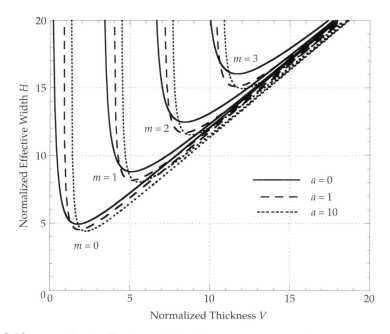

Figure 2.10 Normalized effective widths of TE modes guided by step-index thin-film waveguides.

2.6.2 Confinement Factor of TE Modes

It is instructive to study the percentage of time-average power contained in each region. To quantify the fractional power within the film region, we define the *confinement factor* Γ_f as

$$\Gamma_f = \frac{\text{Time-average power transported in the film region}}{\text{Total time-average power transported by the waveguide}}$$

From the physical meaning of terms in (2.47), we readily deduce that

$$\Gamma_f = \frac{h + \dfrac{\gamma_c}{\kappa_f^2 + \gamma_c^2} + \dfrac{\gamma_s}{\kappa_f^2 + \gamma_s^2}}{h + \dfrac{1}{\gamma_c} + \dfrac{1}{\gamma_s}} \qquad (2.54)$$

The confinement factor can also be expressed directly in terms of the generalized parameters

$$\Gamma_f = \frac{V + \sqrt{b} + \dfrac{\sqrt{a+b}}{1+a}}{V + \dfrac{1}{\sqrt{b}} + \dfrac{1}{\sqrt{a+b}}} \qquad (2.55)$$

Similarly, we introduce Γ_s and Γ_c to quantify the fractional power contained in the substrate and cover regions:

$$\Gamma_s = \frac{\text{Time-average power transported in the substrate region}}{\text{Total time-average power transported by the waveguide}}$$

$$\Gamma_c = \frac{\text{Time-average power transported in the cover region}}{\text{Total time-average power transported by the waveguide}}$$

Following the same procedure, we obtain

$$\Gamma_s = \frac{1-b}{\sqrt{b}\left[V + \dfrac{1}{\sqrt{b}} + \dfrac{1}{\sqrt{a+b}}\right]} \tag{2.56}$$

$$\Gamma_c = \frac{1-b}{(1+a)\sqrt{a+b}\left[V + \dfrac{1}{\sqrt{b}} + \dfrac{1}{\sqrt{a+b}}\right]} \tag{2.57}$$

It is simple to show indeed that $\Gamma_f + \Gamma_s + \Gamma_c = 1$. In Figure 2.11 we plot Γ_f, Γ_s, and Γ_c of TE_0 mode as functions of V. Except for thin-film waveguides operating near the cutoff, Γ_f is much larger than Γ_s and Γ_c. For waveguides operating far above the cutoff, Γ_f approaches 1, and Γ_s and Γ_c are very small. For all asymmetric waveguides, Γ_s is greater than Γ_c. It means that the field and power density are stronger in the substrate region than that in the cover region. This is understandable since $n_s > n_c$. For symmetric waveguides ($n_s = n_c$), Γ_s and Γ_c are the same. Plots of Γ_f, Γ_s, and Γ_c for higher-order modes are similar to that shown in Figure 2.11 except the cutoff is shifted to a larger V value.

2.6.3 Time-Average Power Transported by TM Modes

The time-average power transported by a TM mode is

$$P = \iint \frac{1}{2}\text{Re}[\mathbf{E} \times \mathbf{H}^*]\cdot\hat{\mathbf{z}}\,dx\,dy = \frac{\beta}{2\omega\varepsilon_0}\iint \frac{1}{n^2}|h_y(x)|^2\,dx\,dy$$

$$= \frac{\beta}{2\omega\varepsilon_0}1\left[\int_{-\infty}^{-h}\frac{1}{n_s^2}H_s^2 e^{2\gamma_s(x+h)}\,dx + \int_{-h}^{0}\frac{1}{n_f^2}H_f^2\cos^2(\kappa_f x + \phi)\,dx\right.$$

$$\left. + \int_{0}^{\infty}\frac{1}{n_c^2}H_c^2 e^{-2\gamma_c x}\,dx\right]$$

When the integrals are evaluated, we obtain

$$P = \frac{\beta}{4\omega\varepsilon_0}\left[\frac{H_s^2}{\gamma_s n_s^2} + \frac{H_f^2}{n_f^2}\left(h + \frac{\sin 2\phi' - \sin 2(-\kappa_f h + \phi')}{2\kappa_f}\right) + \frac{H_c^2}{\gamma_c n_c^2}\right] \tag{2.58}$$

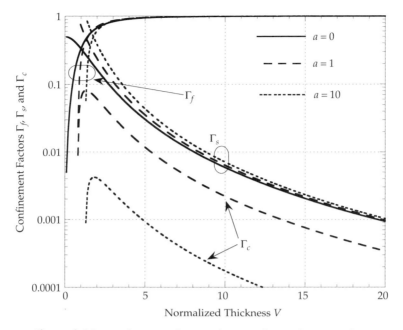

Figure 2.11   Confinement factors of TE_0 mode as a function of V.

The equation may be simplified to read

$$P = \frac{\beta}{4\omega\varepsilon_0} \frac{H_f^2}{n_f^2} \left[h + \frac{1}{\gamma_s q_s} + \frac{1}{\gamma_c q_c} \right] \qquad (2.59)$$

where

$$q_s = \frac{N^2}{n_s^2} + \frac{N^2}{n_f^2} - 1 \qquad (2.60)$$

$$q_c = \frac{N^2}{n_c^2} + \frac{N^2}{n_f^2} - 1 \qquad (2.61)$$

The effective film thickness of TM modes is

$$h_{\text{eff}} = h + \frac{1}{\gamma_s q_s} + \frac{1}{\gamma_c q_c} \qquad (2.62)$$

We can define and evaluate the normalized effective film thickness and confinement factors for TM modes in the same manner as we did for the TE modes.

2.7   PHASE AND GROUP VELOCITIES

In terms of β, the phase and group velocities are

$$v_{\text{ph}} = \frac{\omega}{\beta} = \frac{ck}{\beta} = \frac{c}{N} \quad \text{and} \quad v_{\text{gr}} = \frac{d\omega}{d\beta} = \frac{c\, dk}{d\beta}$$

respectively. Usually b and N are known numerically. Thus it is easy to compute the phase velocity numerically. But it is relatively difficult to evaluate the group velocity exactly. However, it is possible to estimate the group velocity approximately if we ignore the material dispersion. For this purpose, we note from the expressions for v_{ph} and v_{gr} that

$$\frac{c^2}{v_{\text{ph}}v_{\text{gr}}} = \frac{1}{2k}\frac{d(\beta^2)}{dk} = \frac{1}{2k}\frac{d}{dk}(kN)^2$$

If we assume that n_f, n_s, and n_c are independent of λ, then the last differentiation can be carried out and we obtain

$$\frac{c^2}{v_{\text{ph}}v_{\text{gr}}} = N^2 + \frac{k}{2}\frac{dN^2}{dk} = n_f^2\left(b + \frac{V}{2}\frac{db}{dV}\right) + n_s^2\left(1 - b - \frac{V}{2}\frac{db}{dV}\right) \qquad (2.63)$$

To find an expression for db/dV of TE modes, we differentiate both sides of the dispersion relation of TE modes (2.35) and obtain

$$\frac{db}{dV} = \frac{2(1-b)}{V + \dfrac{1}{\sqrt{b}} + \dfrac{1}{\sqrt{a+b}}} \qquad (2.64)$$

Using this expression in conjunction with the expressions for the confinement factors Γ_f, Γ_s, and Γ_c of TE modes, we obtain

$$\frac{c^2}{v_{\text{ph}}v_{\text{gr}}} = n_f^2\Gamma_f + n_s^2\Gamma_s + n_c^2\Gamma_c \qquad (2.65)$$

This is an exact relationship provided the material dispersion is ignored. It shows explicitly that the phase and group velocities and the three confinement factors of TE modes are related. For most thin-film waveguides, Γ_c is much smaller than Γ_f and Γ_s. In addition, $n_f \approx n_s$. Thus, $\Gamma_s + \Gamma_f \sim 1$. We can approximate (2.65) as

$$\frac{c^2}{v_{\text{ph}}v_{\text{gr}}} \approx n_f^2 \qquad (2.66)$$

If the phase velocity is known, the group velocity can be estimated readily. Of course, the estimate is rather poor for modes near the cutoff or if the waveguide materials are highly dispersive. The material dispersion can be accounted for by using the group indices of the three regions, as indicated by Buus [12].

PROBLEMS

1. Starting from Maxwell's equations, derive the characteristic equation for TM modes guided by a step-index thin-film waveguide. In other words, derive Eq. (2.28).

2. Show that (2.28) can be reduced to (2.38).

3. Based on the equations obtained in Problem 1, derive (2.59) for the time-average power carried by a TM mode.

4. Show that the effective waveguide thickness of TM modes is given by (2.62).

5. Rewrite the expression for $e_y(x)$ and use E_f as the amplitude constant.

6. At optical frequencies, no material is a "good" conductor, much less a "perfect" conductor. For this problem, we suppose a fictitious and "perfectly conducting" material exists at optical frequencies.

   Figure 2.12. depicts a structure that is very long in the z direction and very wide in the y direction. It consists of a perfect conductor and a dielectric slab. The perfect conductor extends from $x = 0$ to $-\infty$. The dielectric slab has a thickness of h and an index n_f and it sits at a distance d from the conducting boundary. The medium surrounding the slab has an index n_s, and n_s is smaller than n_f.

   a. Derive a dispersion relation for TE modes guided by the structure propagating in the z direction.

   b. Give a expression for E_y of TE modes. Leave an amplitude constant undetermined or unspecified.

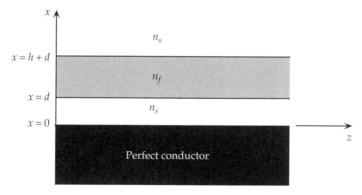

Figure 2.12 Thin-film waveguide with a dielectric slab and a "perfect" conductor. The y axis is perpendicular to the paper.

7. This problem introduces an experimental procedure for determining the asymmetry factor a [13].

   Consider a step-index thin-film waveguide. Suppose the cutoff wavelengths, λ_{c1} and λ_{c2}, of TE_1 and TE_2 modes of a thin-film waveguide are measured and the ratio

$\lambda_{c2}/\lambda_{c1}$ is R. Assume that indices n_f, n_s, and n_c are independent of λ. Show that

$$a = \tan^2 \left[\frac{(2R - 1)\pi}{1 - R} \right]$$

REFERENCES

1. P. K. Tien, "Light waves in thin films and integrated optics," *App. Opt.*, Vol. 10, pp. 2395–2413 (1971).

2. P. K. Tien, "Integrated optics and new wave phenomena in optical waveguides," *Rev. Mod. Phys.*, Vol. 49, No. 2, pp. 361–420 (April, 1977).

3. H. Kogelnik, "Theory of dielectric waveguides," in *Integrated Optics*, T. Tamir (ed.), Springer, New York, Heidelberg, Berlin, 1982, Chapter 2; and in *Guided-Wave Optoelectronics*, 2nd ed., T. Tamir, (ed.), Springer, New York, Heidelberg, Berlin, 1990, Chapter 2.

4. N. Nishihara, M. Harana, and T. Suhara, *Optical Integrated Circuits*, McGraw-Hill, New York, 1989.

5. H. Kogelnik and V. Ramaswamy, "Scaling rules for thin film optical waveguides," *Appl. Opt.*, Vol. 13, pp. 1857–1862 (1974).

6. G. Pandraud and O. Parriaux, "Zero dispersion in step index slab waveguides," *IEEE J. Lightwave Techno.*, Vol. 17, No. 11, pp. 2336–2341 (Nov. 1999).

7. G. A. Bennett and C. L. Chen, "Wavelength dispersion of optical waveguides," *Appl. Opt.*, Vol. 19, pp. 1990–1995 (1980).

8. R. E. Collin, *Field Theory of Guided Waves*, 2nd ed., IEEE Press, New York, 1991.

9. S. Ramo, J. R. Whinnery and T. Van Duzer, *Fields and Waves in Communication Electronics*, 3rd ed., Wiley, New York, 1994.

10. John Krause and D. Fleisch, *Electromagnetics with Applications*, 5th ed., McGraw-Hill, New York, 1998.

11. U. S. Inan and A. S. Inan, *Electromagnetic Waves*, Prentice Hall, Upper Saddle River, NJ, 2000.

12. J. Buus, "Dispersion of TE modes in slab waveguides with reference to double heterostructure semiconductor lasers," *App. Opt.*, Vol. 19, No. 12, pp. 1987–1989 (June 15, 1980).

13. S. I. Hosain and J. P. Meunier, "Characterization of asymmetric graded-index planar optical waveguides from the knowledge of TE_0-TE_1 mode cutoff wavelengths," *IEEE Photonics Technol. Lett.*, Vol. 3, No. 9, pp. 801–803 (Sept. 1991).

BIBLIOGRAPHY

1. A. Ghatak, *Optics*, Tata McGraw-Hill, New York, 1977.

2. H. A. Haus, *Waves and Fields in Optoelectronics*, Prentice Hall, Englewood Cliffs, NJ, 1984.

3. R. G. Hunsperger, *Integrated Optics: Theory and Technology*, 5th ed., Springer, Berlin, 2002.

4. W. B. Jones, Jr., *Introduction to Optical Fiber Communication Systems*, Holt, Reinhardt and Winston, New York, 1988.

5. B. E. A. Saleh and M. C. Teich, *Fundamentals of Photonics,* Wiley, New York, 1991.

6. J. Wilson and J. F. B. Hawkes, *Optoelectronics, An Introduction*, 2nd ed., Prentice Hall, Englewood Cliffs, NJ, 1989.

7. A. Yariv, *Introduction to Optical Electronics,* 4th ed., Saunders College, Philadelphia, 1991.

3

GRADED-INDEX THIN-FILM WAVEGUIDES

3.1 INTRODUCTION

The theory of step-index dielectric waveguides discussed in the last chapter is relatively simple and quite complete. Most quantities of interest are expressed in elementary functions. This helps greatly in understanding the properties of step-index dielectric waveguides. But most practical planar waveguides have a graded-index profile. While basic properties of graded-index waveguides are similar to that of step-index waveguides, there are subtle differences. In this chapter, we study graded-index thin-film waveguides.

Whenever possible, we use the same symbols to describe the waveguide geometry and the index profile of the graded-index waveguides as those for step-index waveguides. For many graded-index waveguides, the film and substrate regions may coalesce smoothly into a contiguous and continuous region. As a result, the index profiles in the two regions can be combined and represented by a continuous function $n(x)$ that varies continuously from n_f to n_s (Fig. 3.1). Here n_s is the index deep in the substrate region and n_f is the index at the film–cover boundary. The index in the cover region is n_c. The same set of generalized parameters, V, a, b, c, and d, introduced in the last chapter for step-index waveguides, are also used for graded-index

Foundations for Guided-Wave Optics, by Chin-Lin Chen
Copyright © 2007 John Wiley & Sons, Inc.

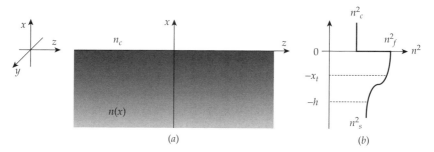

Figure 3.1   Thin-film waveguide with graded-index profile.

waveguides. The five generalized parameters are repeated here for convenience:

$$V = kh\sqrt{n_f^2 - n_s^2} \tag{3.1}$$

$$a = \frac{n_s^2 - n_c^2}{n_f^2 - n_s^2} \tag{3.2}$$

$$b = \frac{N^2 - n_s^2}{n_f^2 - n_s^2} \tag{3.3}$$

$$c = \frac{n_s^2}{n_f^2} \tag{3.4}$$

$$d = \frac{n_c^2}{n_f^2} = c - a(1 - c) \tag{3.5}$$

In Sections 3.2 and 3.3, we consider waveguides with linearly or exponentially graded-index profiles [1–5]. For the two index profiles, closed-form expressions for fields and the dispersion relation are derived. Additional examples of graded-index waveguides can be found in Smithgall and Dabby [6], Kogelnik [7], and Savatinova and Nadjakov [8]. Nevertheless, only a limited number of graded-index dielectric waveguides can be studied analytically. To study waveguides with an arbitrary graded-index index profile, we have to resort to approximate or numerical methods. In Section 3.4, we apply the WKB method to derive an approximate dispersion relation for graded-index waveguides with an arbitrary index profile. In Section 3.5, we use a different method to derive the dispersion equation for graded-index waveguides. In the last section, we compare the essential features of graded-index waveguides with those of the step-index waveguides.

3.2   TRANSVERSE ELECTRIC MODES GUIDED BY LINEARLY GRADED DIELECTRIC WAVEGUIDES

In this section, we consider TE modes guided by linearly graded waveguides [1]. In the cover and substrate regions, the indices n_c and n_s are constants. In the film

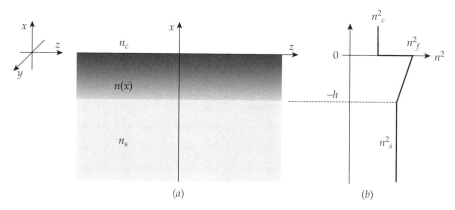

Figure 3.2   Thin-film waveguide with a linearly graded index profile.

region, $n^2(x)$ changes linearly from n_f^2 at $x = 0$ to n_s^2 at $x = -h$, as depicted schematically in Figure 3.2:

$$n^2(x) = n_f^2 + (n_f^2 - n_s^2)\frac{x}{h} \qquad -h \le x \le 0 \qquad (3.6)$$

As noted in Sections 1.5.1 and 2.2.1, all field components of TE modes can be expressed in terms of $e_y(x)$. The relations have been given in (1.87) and (1.88) and in (2.7) and (2.8). The differential equation for $e_y(x)$ has been given by (1.89) and (2.9). Since the indices in the cover and substrate regions are constants, $e_y(x)$ in the two regions are the same as that given in Section 2.2.1:

$$e_y(x) = C_1 e^{-V\sqrt{a+b}\,x/h} \qquad x \ge 0 \qquad (3.7)$$

and

$$e_y(x) = C_4 e^{V\sqrt{b}(x+h)/h} \qquad x \le -h \qquad (3.8)$$

where C_1 and C_4 are two constants yet to be determined. We save C_2 and C_3 for fields in the film region.

To determine $e_y(x)$ for the film region, we substitute (3.6) into (2.9), cast the equation in terms of the generalized parameters, and obtain

$$\frac{d^2 e_y(x)}{dx^2} + \frac{V^2}{h^3}[x + h(1-b)]e_y(x) = 0 \qquad (3.9)$$

This is a second-order differential equation with a variable coefficient. To convert the equation into a standard and easily recognizable form, we follow Marcuse [1] and define a new variable:

$$\varsigma = -\frac{V^{2/3}}{h}[x + h(1-b)]$$

In terms of the newly defined variable, (3.9) becomes

$$\frac{d^2 e_y(\varsigma)}{d\varsigma^2} - \varsigma e_y(\varsigma) = 0 \qquad (3.10)$$

This is a Bessel differential equation of order $\frac{1}{3}$. It has Airy functions $\text{Ai}(\varsigma)$ and $\text{Bi}(\varsigma)$ as the two linearly independent solutions [9]. In Figure 3.3, we plot $\text{Ai}(\varsigma)$ and $\text{Bi}(\varsigma)$ as functions of ς. The two Airy functions are oscillatory functions when ς is negative. But $\text{Ai}(\varsigma)$ decays exponentially and $\text{Bi}(\varsigma)$ grows exponentially for positive ς. In any event, $e_y(\varsigma)$ is a linear combination of the two Airy functions:

$$e_y(\varsigma) = C_2 \text{Ai}(\varsigma) + C_3 \text{Bi}(\varsigma)$$

where C_2 and C_3 are two unknown constants. When expressed in terms of x, $e_y(x)$ becomes

$$e_y(x) = C_2 \text{Ai}\left\{-\frac{V^{2/3}}{h}[x + h(1 - b)]\right\} + C_3 \text{Bi}\left\{-\frac{V^{2/3}}{h}[x + h(1 - b)]\right\} \quad (3.11)$$

To determine the constants and b, we make use of the boundary conditions that $e_y(x)$ and $[de_y(x)]/dx$ are continuous at $x = 0$ and $x = -h$. From the boundary conditions at $x = 0$, we obtain

$$\frac{C_3}{C_2} = -\frac{V^{1/3}\sqrt{a + b}\,\text{Ai}[-V^{2/3}(1 - b)] - \text{Ai}'[-V^{2/3}(1 - b)]}{V^{1/3}\sqrt{a + b}\,\text{Bi}[-V^{2/3}(1 - b)] - \text{Bi}'[-V^{2/3}(1 - b)]} \qquad (3.12)$$

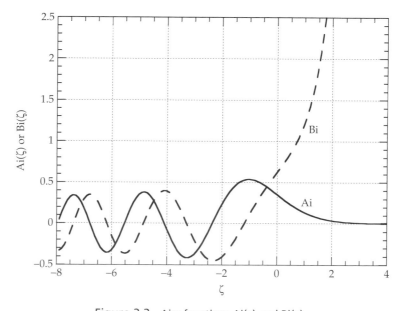

Figure 3.3   Airy functions Ai(ς) and Bi(ς).

where a prime signifies the differentiation with respect to the argument. Similarly, the boundary conditions at $x = -h$ lead to

$$\frac{C_3}{C_2} = -\frac{V^{1/3}\sqrt{b}\,\text{Ai}(V^{2/3}b) + \text{Ai}'(V^{2/3}b)}{V^{1/3}\sqrt{b}\,\text{Bi}(V^{2/3}b) + \text{Bi}'(V^{2/3}b)} \qquad (3.13)$$

Combining (3.12) and (3.13), we obtain an exact *dispersion relation* for TE modes guided by linearly graded planar waveguides:

$$\frac{V^{1/3}\sqrt{b}\,\text{Ai}(V^{2/3}b) + \text{Ai}'(V^{2/3}b)}{V^{1/3}\sqrt{b}\,\text{Bi}(V^{2/3}b) + \text{Bi}'(V^{2/3}b)}$$
$$= \frac{V^{1/3}\sqrt{a+b}\,\text{Ai}[-V^{2/3}(1-b)] - \text{Ai}'[-V^{2/3}(1-b)]}{V^{1/3}\sqrt{a+b}\,\text{Bi}[-V^{2/3}(1-b)] - \text{Bi}'[-V^{2/3}(1-b)]} \qquad (3.14)$$

By solving the dispersion equation numerically for a given a and V, we obtain b. Figure 3.4 depicts the bV curves of linearly graded waveguides with $a = 10$ and 100. Once b is known, the field $e_y(x)$ in the three regions can be calculated

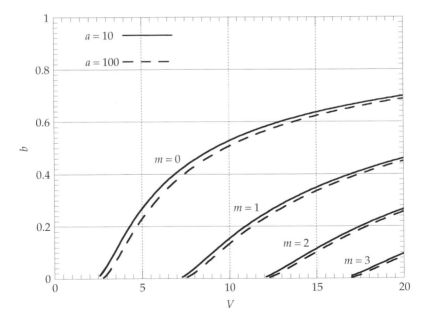

Figure 3.4 bV diagram of TE modes guided by waveguides with a linearly tapered index profile.

from (3.7), (3.8), and (3.11):

$$
e_y(x) = \begin{cases}
\begin{aligned}
& C_2\{\mathrm{Ai}[-V^{2/3}(1-b)] \\
& \quad + \frac{C_3}{C_2}\mathrm{Bi}[-V^{2/3}(1-b)]\}e^{-V\sqrt{a+bx}/h}
\end{aligned} & x \geq 0 & \quad (3.15) \\[2em]
\begin{aligned}
& C_2\left(\mathrm{Ai}\left\{-\frac{V^{2/3}[x+h(1-b)]}{h}\right\}\right. \\
& \quad \left.+\frac{C_3}{C_2}\mathrm{Bi}\left\{-\frac{V^{2/3}[x+h(1-b)]}{h}\right\}\right)
\end{aligned} & -h \leq x \leq 0 & \quad (3.16) \\[2em]
C_2\left[\mathrm{Ai}(V^{2/3}b) + \frac{C_3}{C_2}\mathrm{Bi}\,(V^{2/3}b)\right]e^{-V\sqrt{b}(x+h)/h} & x \leq -h & \quad (3.17)
\end{cases}
$$

In the above equations, C_2 is the amplitude constant and C_3/C_2 is given in (3.12) or (3.13). In Figure 3.5, we plot the field distributions of the first three TE modes with $V = 15$ and $a = 10$. For comparison, we normalize the electric field of each mode such that the peak electric field intensity is 1. As expected, $e_y(x)$ varies in an oscillatory manner in the film region and decays exponentially in the cover and substrate regions.

The field components of TM modes are $e_x(x)$, $e_z(x)$, and $h_y(x)$. As noted in Section 1.5.2, $e_x(x)$ and $e_z(x)$ can be expressed in terms of $h_y(x)$. However, the differential equation (1.92) for $h_y(x)$ has additional terms. Thus the analysis for the TM modes is more complicated, although it follows the general procedure used in this section for examining the TE modes [2, 3]. The TM mode problem is left as an exercise for the readers (Problem 1).

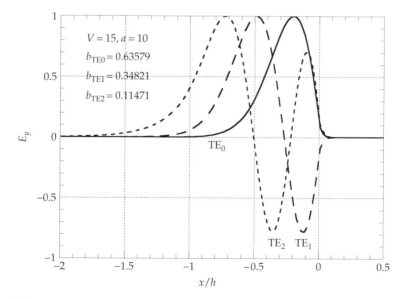

Figure 3.5 Electric fields of the first three TE modes guided by waveguides with a linearly tapered index profile.

3.3 EXPONENTIALLY GRADED DIELECTRIC WAVEGUIDES

For many dielectric waveguides, in-diffused lithium niobate, and lithium tantalate waveguides, for example, the index distribution varies smoothly from the film region to the substrate region and can be approximated by an exponential function:

$$n(x) = n_s + (n_f - n_s)e^{2x/h} = n_s + \Delta n\, e^{2x/h} \qquad x \leq 0 \qquad (3.18)$$

The index difference $\Delta n = n_f - n_s$ is much smaller than n_s and n_f. Equation (3.18) can also be described by an equivalent form:

$$n^2(x) \approx n_s^2 + 2n_s\,\Delta n\, e^{2x/h} \qquad (3.19)$$

In the two equations, h is a *characteristic distance* of the exponentially graded waveguide. It is chosen such that the index change $n(x) - n_s$ at $x = -h$ is $\Delta n/e^2$. We will need the characteristic distance later to define the normalized film thickness V.

3.3.1 Transverse Electric Modes

By substituting (3.19) into (1.87) or (2.9), we obtain a differential equation for $e_y(x)$:

$$\frac{d^2 e_y(x)}{dx^2} + k^2[2n_s\,\Delta n\, e^{2x/h} - (N^2 - n_s^2)]e_y(x) = 0 \qquad (3.20)$$

Equation (3.20) is also a second-order differential equation with a variable coefficient. To cast the equation in a canonical form, we follow Conwell's work [4, 5] and introduce a new variable $\xi = Ve^{x/h}$. Then we apply the chain rule of differentiation to (3.20) and obtain

$$\frac{d^2 e_y(\xi)}{d\xi^2} + \frac{1}{\xi}\frac{d e_y(\xi)}{d\xi} + \left(1 - \frac{V^2 b}{\xi^2}\right)e_y(\xi) = 0 \qquad (3.21)$$

This is the Bessel differential equation of order $V\sqrt{b}$ [9]. It has two linearly independent solutions $J_{V\sqrt{b}}(\xi)$ and $Y_{V\sqrt{b}}(\xi)$. In other words, e_y is a linear combination of $J_{V\sqrt{b}}(\xi)$ and $Y_{V\sqrt{b}}(\xi)$. Note that the Neumann function $Y_{V\sqrt{b}}(\xi)$ is singular at $\xi = 0$, and the origin of ξ corresponds to a point physically deep in the substrate region. Thus we have to discard $Y_{V\sqrt{b}}(\xi)$. For the film and substrate regions ($x \leq 0$), we write $e_y(x)$ as

$$e_y(x) = E_f J_{V\sqrt{b}}(Ve^{x/h}) \qquad (3.22)$$

where E_f is an amplitude constant yet to be determined.
For the cover region ($x \geq 0$), $e_y(x)$ is

$$e_y(x) = E_c\, e^{-V\sqrt{a+b}\,x/h} \qquad (3.23)$$

From the boundary conditions on $e_y(x)$ and $de_y(x)/dx$ at $x = 0$, we obtain

$$\frac{J'_{V\sqrt{b}}(V)}{J_{V\sqrt{b}}(V)} = -\sqrt{a+b} \tag{3.24}$$

where a prime indicates the differentiation with respect to the argument of the Bessel function. With the help of the recurrent relations of Bessel functions [9], the above expression can be written as

$$\frac{J_{V\sqrt{b}+1}(V) - J_{V\sqrt{b}-1}(V)}{J_{V\sqrt{b}}(V)} = 2\sqrt{a+b} \tag{3.25}$$

Equation (3.24), or equivalently (3.25), is the dispersion relation for TE modes guided by exponentially graded waveguides [4, 5]. For a given waveguide, a and V are known, b can be found numerically from the dispersion relation. Figure 3.6 depicts the bV plots of exponentially graded waveguides with $a = 10$ and 100. A comparison of Figure 3.6 and Figure 2.4 shows that the normalized guide index of exponentially graded waveguides is smaller than that of step-index waveguides having the same V. We also observe that the cutoff conditions ($b = 0$) are near the roots of $J_{V\sqrt{b}}(V)$ for highly asymmetric waveguides.

Once V, a, and b are known, fields of TE modes can be calculated from (3.22) and (3.23). Field distributions for TE$_0$, TE$_1$, and TE$_2$ guided by an exponentially graded waveguide with $V = 10$ and $a = 10$ are depicted in Figure 3.7. Fields of TE modes are oscillatory in the region near the film–cover boundary. They decay

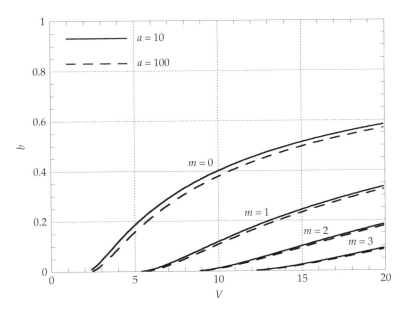

Figure 3.6   *bV* diagram of TE modes guided by waveguides with a exponential index profile.

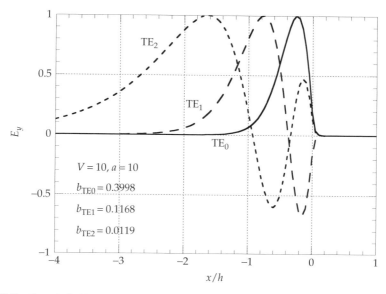

Figure 3.7 Electric fields of the first three TE modes guided by waveguides with a exponential index profile.

exponentially deep into the substrate region. But the transition from an oscillatory field to an exponentially decaying field is gradual and smooth. This is understandable since the index profile changes smoothly from the film region to the substrate region. We also note that fields of higher-order modes extend deeper into the substrate region.

3.3.2 Transverse Magnetic Modes [4, 5]

As noted in Section 1.6.2, the differential equation for $h_y(x)$ in an inhomogeneous medium is rather complicated as given in (1.92). To solve for $h_y(x)$ for the film and substrate regions, we again follow Conwell's work [5] and introduce $g(x)$ such that

$$h_y(x) = n(x)g(x) \tag{3.26}$$

Using this relation in (1.92), we obtain a differential equation for $g(x)$ for $x \leq 0$:

$$\frac{n''(x)}{n(x)}g(x) - 2\frac{[n'(x)]^2}{n^2(x)}g(x) + g''(x) + [k^2n^2(x) - \beta^2]g(x) = 0 \tag{3.27}$$

where a prime signifies the differentiation with respect to x. Substituting (3.19) in (3.27) and ignoring terms on orders of $(\Delta n)^2$ or smaller, we simplify the differential equation for $g(x)$:

$$g''(x) + \left[\left(1 + \frac{2}{k^2h^2n_s^2}\right)2k^2n_s\,\Delta n\,e^{2x/h} + (k^2n_s^2 - \beta^2)\right]g(x) = 0 \tag{3.28}$$

In terms of the normalized parameters V and b, the above equation becomes

$$g''(x) + \frac{V^2}{h^2}\left[\left(1 + \frac{2}{k^2 h^2 n_s^2}\right) e^{2x/h} - b\right] g(x) = 0 \qquad (3.29)$$

The differential equation can also be converted to a Bessel differential equation by introducing a new variable $\xi = V' e^{x/h}$ and a new parameter

$$V' = \sqrt{1 + \frac{2}{k^2 h^2 n_s^2}}\, V \qquad (3.30)$$

Clearly $g(x)$ can be expressed in terms of a Bessel function of order $V\sqrt{b}$. It follows that $h_y(x)$ can be written as

$$h_y(x) = H_f n(x) J_{V\sqrt{b}}(V' e^{x/h}) \qquad x \le 0 \qquad (3.31)$$

where H_f is an amplitude constant to be determined.

In the cover region ($x > 0$), the index is a constant and $h_y(x)$ is simply

$$h_y(x) = H_c e^{-V\sqrt{a+b}\,x/h} \qquad (3.32)$$

To eliminate one of the two constants, H_f and H_c, and to determine b, we make use of the continuation conditions of $h_y(x)$ and $(1/n^2)(dh_y(x)/dx)$ at $x = 0$. Thus, we obtain a dispersion relation for the TM modes [4, 5]:

$$\frac{J'_{V\sqrt{b}}(V')}{J_{V\sqrt{b}}(V')} = -\frac{n_f^2}{n_c^2}\frac{V}{V'}\sqrt{a+b} - 2\frac{\Delta n}{V' n_f} \qquad (3.33)$$

For most in-diffused LiNbO$_3$ and LiTaO$_3$ waveguides of interest, khn_s is rather large. Then $V' \sim V$. It is also convenient to express $(n_f/n_s)^2$ and $\Delta n/n_f$ in terms of c and d defined in (3.4) and (3.5). In terms of these parameters, the dispersion relation (3.33) for TM modes becomes

$$\frac{J_{V\sqrt{b}+1}(V) - J_{V\sqrt{b}-1}(V)}{J_{V\sqrt{b}}(V)} = \frac{2}{d}\sqrt{a+b} + 2\frac{1-c}{Vc} \qquad (3.34)$$

The above equation may be used to determine b for a given a, c, and V. The magnetic field, $h_y(x)$, in terms of generalized parameters, is given in (3.31) and (3.32). The electric field components, $e_x(x)$ and $e_z(x)$ can be deduced from (1.90) and (1.91). This completes the analysis for the TM modes guided by exponentially graded dielectric waveguides.

3.4   THE WKB METHOD

As noted earlier, only a few graded-index waveguides can be analyzed exactly. Since an exact analysis of waveguides with an arbitrary index profile is unlikely, we settle for an approximate dispersion relation. In this section, we apply a method commonly known as the *WKB* or *WKBJ method* to establish the dispersion relation of graded-index dielectric waveguides [10, 11]. The method was developed initially by G. Wentzel, H. A. Kramers, L. Brillouin, and H. Jeffreys in their study of physics problems [10]. D. Gloge applied the method to examine the optical waveguide and fiber problems [11].

We suppose that the index distribution in the film and substrate regions is given by a continuous function $f(x)$ (Fig. 3.1):

$$n^2(x) = n_s^2 + (n_f^2 - n_s^2)f(x) \qquad x \leq 0 \tag{3.35}$$

We further assume that $f(x)$ increases monotonically and continuous from points deep in the substrate, $f(-\infty) = 0$, to the boundary between the cover and film region, $f(0) = 1$. As noted earlier, we need a characteristic distance h to define the generalized film thickness V. We choose h such that the change of n^2 at $x = -h$ is $2/e$ of the total change of n^2, that is, $n^2(0) - n^2(-h) = 2(n_f^2 - n_s^2)/e$.

The WKB method consists of several major steps [10, 11]. One of the crucial steps is the identification of the turning point of the guided mode in question. Although b has yet been determined, we pretend that b is known. Then a point $(x = -x_t)$ can be identified that $f(-x_t) = b$. This is the *turning point*. A different mode has a different b and therefore a different turning point. For a continuous and monotonic $f(x)$, there exists one and only one turning point. We confine our discussions to waveguides having a single and isolated turning point only. In addition, we assume that the turning point is far from the region where the index changes abruptly or discontinuously.

Having identified the turning point, we divide the film and substrate regions into three zones. We refer the zones to the right and to the left of the turning point as the R and L *zones*, respectively (Fig. 3.8). In addition, we label the zone in the immediate vicinity of the turning point as the *transition zone*. In other words, the R zone is the region between the turning point and the cover–film boundary. The L zone refers to the region between the turning point and points deep in the substrate region. The transition zone is a narrow region near the turning point and between the R and L zones.

The next step is to derive simple and approximate expressions for $e_y(x)$ for the R and L zones. Unfortunately, the two approximate equations are too simple that they become singular at the turning point. Therefore, we cannot use the two approximate equations to enforce the boundary conditions at the turning point. It is necessary to derive another equation specifically for the transition zone. As expected, the third expression is very complicated. Since the third equation is accurate near and at the turning point, we use this complicated expression to connect the equations for the R and L zones. Thus, the third or complicated equation is often referred to as the *connecting equation*.

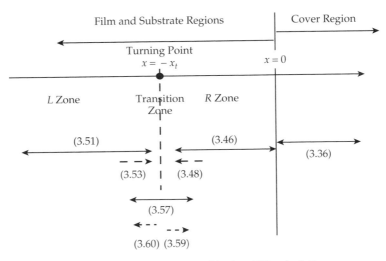

Figure 3.8 Three zones used in the WKB calculation.

Since the cover region has a constant index, $e_y(x)$ is simply

$$e_y(x) = C_1 e^{-V\sqrt{a+b}\,x/h} \qquad x \geq 0 \tag{3.36}$$

To find $e_y(x)$ for the film and substrate regions, we substitute (3.34) into the differential equation (1.87) or (2.9), and obtain for $x < 0$

$$\frac{d^2 e_y(x)}{dx^2} + V^2 \frac{f(x) - b}{h^2} e_y(x) = 0 \tag{3.37}$$

The remaining task is to solve (3.37) for $e_y(x)$ for $x < 0$.

3.4.1 Auxiliary Function

For most graded-index waveguides of interest, the waveguide is thick and V is greater than 1. In the following discussions, we treat V as a large parameter. We introduce an *auxiliary function* $S(x)$ such that $e_y(x)$ is

$$e_y(x) = e^{jVS(x)} \tag{3.38}$$

Substituting (3.38) in (3.37), we obtain a differential equation for $S(x)$:

$$jVS''(x) - V^2 S'^2(x) + V^2 \frac{f(x) - b}{h^2} = 0 \tag{3.39}$$

where $S'(x)$ and $S''(x)$ are the two derivatives of $S(x)$ with respect to x. An exact solution for $S(x)$ is not expected or is too complicated. We seek an approximate expression for $S(x)$ instead. Toward this end, we expand $S(x)$ as an asymptotic series in V:

$$S(x) = S_0(x) + \frac{1}{V} S_1(x) + \frac{1}{V^2} S_2(x) + \cdots \tag{3.40}$$

Upon substituting (3.40) into (3.39) and arranging the terms in the descending order of V, we obtain

$$\left[S_0'^2(x) - \frac{f(x) - b}{h^2} \right] V^2 - [j S_0''(x) - 2 S_0'(x) S_1'(x)] V$$

$$- [j S_1''(x) - S_1'^2(x) - 2 S_0'(x) S_2'(x)] + \cdots = 0 \tag{3.41}$$

For (3.41) to hold for an arbitrary V, the coefficient of each order of V must vanish individually. Since we are content with an approximate solution for $S(x)$, we consider the two leading terms of (3.41) only. The coefficient of the V^2 term vanishes if

$$S_0'^2(x) = \frac{f(x) - b}{h^2} \tag{3.42}$$

The coefficient of the V^1 term is zero if

$$S_1'(x) = \frac{j S_0''}{2 S_0'} \tag{3.43}$$

In the following subsections, we determine $S_0(x)$ and $S_1(x)$ and, therefore, $S(x)$. For different zones, $f(x) - b$ has a different sign. Therefore, $S_0(x)$ and $S_1(x)$ have different expressions for different zones.

3.4.2 Fields in the R Zone

To the right side of the turning point, $-x_t < x < 0$ and $f(x) - b$ is positive. Thus, we obtain from (3.42)

$$S_0(x) = \pm \frac{1}{h} \int_{-x_t}^{x} \sqrt{f(x) - b}\, dx \tag{3.44}$$

A constant of integration will be specified later. In fact, it will be absorbed by a constant yet to be introduced.

To find S_1, we integrate (3.43) and obtain

$$S_1(x) = \frac{j}{2} \ln[S_0'(x)] = \frac{j}{4} [\ln(f(x) - b) - 2 \ln h] \tag{3.45}$$

A constant of integration is also left unspecified. By substituting S_0 and S_1 so obtained in (3.38), we have an approximate expression for $e_y(x)$ in the R zone.

There are two independent solutions as signified by the plus and minus signs in (3.44). Thus $e_y(x)$ is a linear combination of

$$\frac{h^{1/2}}{[f(x) - b]^{1/4}} \exp\left[\pm j\frac{V}{h}\int_{-x_t}^{x}\sqrt{f(x) - b}\,dx\right]$$

It is convenient to express $e_y(x)$ as a sine function with an arbitrary phase instead of the two complex exponential functions. Thus we write

$$e_y(x) = C_2\frac{h^{1/2}}{[f(x) - b]^{1/4}}\sin\left[\frac{V}{h}\int_{-x_t}^{x}\sqrt{f(x) - b}\,dx + \phi\right] \qquad -x_t < x \le 0$$

$$(3.46)$$

The constants C_2 and ϕ are yet undetermined. As noted earlier, $f(x)$ at the turning point is b. For points in the immediate vicinity of the turning point, $f(x) \approx b + (x + x_t)f_1$ where f_1 stands for df/dx at the turning point. When this approximation is used in the integrand of (3.46), we can evaluate the integral analytically and obtain

$$\int_{-x_t}^{x}\sqrt{f(x) - b}\,dx \approx \int_{-x_t}^{x}\sqrt{(x + x_t)f_1}\,dx = \tfrac{2}{3}(x + x_t)^{3/2}f_1^{1/2} \qquad (3.47)$$

Thus for the region to the right of, and near, the turning point, $e_y(x)$ is approximately

$$e_y(x) \approx C_2\frac{h^{1/2}}{[(x + x_t)f_1]^{1/4}}\sin\left[\frac{2V}{3h}(x + x_t)^{3/2}f_1^{1/2} + \phi\right] \qquad -x_t \lesssim x \qquad (3.48)$$

Although (3.46) and (3.48) are useful for points to the right of the turning point, they are not accurate enough for points very close to the turning point. In fact, (3.46) and (3.48) are singular at the turning point. Therefore, we cannot use (3.46) and (3.48) directly to evaluate C_2 and ϕ.

3.4.3 Fields in the L Zone

In the L zone, $f(x) - b$ is negative. We obtain from (3.42) and (3.43)

$$S_0(x) = \pm\frac{j}{h}\int_{x}^{-x_t}\sqrt{b - f(x)}\,dx \qquad (3.49)$$

and

$$S_1(x) = \frac{j}{2}\ln[S_0'(x)] = \frac{j}{2}\ln\left[\pm\frac{j}{h}\sqrt{b - f(x)}\right] = \frac{j}{2}\left\{\ln\left[\frac{\sqrt{b - f(x)}}{h}\right] \pm j\frac{\pi}{2}\right\}$$

$$(3.50)$$

Thus, $e_y(x)$ may be written as a linear combination of two exponential functions with real exponents:

$$\frac{h^{1/2}}{[b - f(x)]^{1/4}} \exp\left[\mp \frac{V}{h} \int_x^{-x_t} \sqrt{b - f(x)}\, dx\right]$$

From the physical consideration, we expect $e_y(x)$ to decay exponentially as x recedes into the substrate. Thus we drop the term with a positive exponent and keep the term with a negative exponent:

$$e_y(x) = C_5 \frac{h^{1/2}}{[b - f(x)]^{1/4}} \exp\left[-\frac{V}{h} \int_x^{-x_t} \sqrt{b - f(x)}\, dx\right] \qquad x < -x_t \quad (3.51)$$

where C_5 is also a constant to be determined.

Again we are interested in $e_y(x)$ at points near the turning point. For points near and to the left of the turning point, x is slightly less than $-x_t$ and

$$b - f(x) \approx -(x + x_t) f_1 = |x + x_t| f_1 \qquad (3.52)$$

Then

$$\int_x^{-x_t} \sqrt{b - f(x)}\, dx \approx \int_x^{-x_t} \sqrt{|x + x_t| f_1}\, dx = \frac{2}{3}(|x + x_t|)^{3/2} f_1^{1/2} \qquad (3.53)$$

Thus for points in the immediate vicinity of and to the left of the turning point, $e_y(x)$ is, approximately,

$$e_y(x) \approx C_5 \frac{h^{1/2}}{[(|x + x_t| f_1)]^{1/4}} \exp\left[-\frac{2V}{3h}(|x + x_t|)^{3/2} f_1^{1/2}\right] \qquad x \lesssim -x_t \quad (3.54)$$

Again, we note that (3.54) is singular at the turning point.

3.4.4 Fields in the Transition Zone

Obviously, fields must be finite everywhere in the film and substrate regions. An accurate expression for $e_y(x)$ must be regular at the turning point. Yet the approximate expressions (3.46), (3.48), (3.51), and (3.54) are singular at the turning point. In other words, these expressions are not accurate enough for the turning point. It is necessary to derive an alternate and more accurate expression for $e_y(x)$ explicitly for the transition zone. This is done in this section. For points in the transition zone, $f(x) \approx b + (x + x_t) f_1$. Using this approximation in the differential equation (3.36), we have

$$\frac{d^2 e_y(x)}{dx^2} + \frac{V^2 f_1}{h^2}(x + x_t) e_y(x) = 0 \qquad (3.55)$$

In other words, the index profile in the transition zone is approximately linearly graded. Following the procedure discussed in Section 3.2, we define $\varsigma = -(V^2 f_1/h^2)^{1/3}(x + x_t)$. Then (3.55) becomes

$$\frac{d^2 e_y(\varsigma)}{d\varsigma^2} - \varsigma e_y(\varsigma) = 0 \tag{3.56}$$

Therefore $e_y(\varsigma)$ in the transition zone is the linear combination of the two Airy functions:

$$e_y(\varsigma) = C_3 \text{Ai}(\varsigma) + C_4 \text{Bi}(\varsigma) \tag{3.57}$$

The two constants, C_3 and C_4, are to be determined. For $x \lesssim -x_t$, ς is positive. As x moves away from the turning point and farther into the substrate, ς increases and $\text{Bi}(\varsigma)$ grows exponentially. But $e_y(x)$ cannot increase exponentially as x decreases farther into the substrate. Therefore, we drop $\text{Bi}(\varsigma)$ by setting C_4 to zero and obtain

$$e_y(\varsigma) = C_3 \text{Ai}(\varsigma) \tag{3.58}$$

For $x \gtrsim -x_t$, ς is negative. As x moves away from the turning point toward the cover–film boundary, $|\varsigma|$ increases and $\text{Ai}(\varsigma)$ is an oscillatory function of ς. At the turning point, ς vanishes and

$$e_y(-x_t) = C_3 \text{Ai}(0) = C_3 3^{-2/3} / \Gamma \left(\tfrac{2}{3}\right) = 0.35502 C_3 \tag{3.59}$$

We note in particular that $e_y(x)$ as given by (3.58) is finite at the turning point so long as C_3 is finite.

Next we consider points in the vicinity of the turning point. As V is a large parameter, $|\varsigma|$ can be quite large even for points very close to the turning point. Thus, we are justified using the asymptotic expression of $\text{Ai}(\varsigma)$ for a large $|\varsigma|$ [9].

For $x \gtrsim -x_t$, ς is negative. The asymptotic expression for $\text{Ai}(\varsigma)$ for a large and negative ς is

$$\text{Ai}(\varsigma) \approx \frac{1}{\sqrt{\pi}(|\varsigma|)^{1/4}} \sin \left(\frac{2}{3}|\varsigma|^{3/2} + \frac{\pi}{4}\right)$$

Thus $e_y(x)$ is

$$e_y(x) \approx \frac{C_3 h^{1/6}}{\sqrt{\pi}(V^2 f_1)^{1/12}(|x_t + x|)^{1/4}} \sin \left[\frac{2V f_1^{1/2}}{3h}(|x_t + x|)^{3/2} + \frac{\pi}{4}\right] \tag{3.60}$$

On the other hand, for a large and positive ς, the asymptotic expression for $\text{Ai}(\varsigma)$ is

$$Ai(\varsigma) \approx \frac{1}{2\sqrt{\pi}\,\varsigma^{1/4}} \exp \left(-\frac{2}{3}\varsigma^{3/2}\right)$$

Thus for $x \lesssim -x_t$, $e_y(x)$ is approximately

$$e_y(x) \approx \frac{C_3 h^{1/6}}{2\sqrt{\pi}(V^2 f_1)^{1/12}(|x + x_t|)^{1/4}} \exp\left[-\frac{2V f_1^{1/2}}{3h}(|x + x_t|)^{3/2}\right] \quad (3.61)$$

We use (3.60) and (3.61) in conjunction with (3.48) and (3.54) to determine the constants.

3.4.5 The Constants

Instead of matching $e_y(x)$ and $de_y(x)/dx$ at discrete points, we compare the functional form of $e_y(x)$ in a small and finite region near the turning point. Equations (3.48) and (3.61) are valid for a finite region in the vicinity of, and to the right of, the turning point. The two equations are identical if

$$C_3 = C_2 \frac{\pi^{1/2} V^{1/6} h^{1/3}}{f_1^{1/6}} \quad (3.62)$$

and $\phi = \pi/4$.

On the other hand, (3.54) and (3.60) hold for all points in the immediate neighborhood of, and to the left of, the turning point. The two expressions are identical if

$$C_5 = C_3 \frac{f_1^{1/6}}{2\sqrt{\pi} V^{1/6} h^{1/3}} \quad (3.63)$$

and $\phi = \pi/4$. From (3.62) and (3.63), we finally connect C_2 with C_5:

$$C_5 = \frac{C_2}{2} \quad (3.64)$$

To summarize the results of this section, we list expressions for e_y for the cover region and the three zones of the film and substrate regions:

$$e_y(x) = \begin{cases} C_1 e^{-V\sqrt{a+bx}/h}, & x \geq 0 \\ C_2 \dfrac{h^{1/2}}{[f(x) - b]^{1/4}} \sin\left\{\dfrac{V}{h}\displaystyle\int_{-x_t}^{x} [f(x) - b]^{1/2}\, dx + \dfrac{\pi}{4}\right\}, & -x_t < x \leq 0 \\ C_2 \dfrac{\pi^{1/2} V^{1/6} h^{1/3}}{f_1^{1/6}} \mathrm{Ai}\left[-\left(\dfrac{V^2 f_1}{h^2}\right)^{1/3}(x + x_t)\right], & x \sim -x_t \\ C_2 \dfrac{1}{2} \dfrac{h^{1/2}}{[b - f(x)]^{1/4}} \exp\left\{-\dfrac{V}{h}\displaystyle\int_{x}^{-x_t} [b - f(x)]^{1/2}\, dx\right\}, & x < -x_t \end{cases}$$

$$(3.65)$$

The constants C_1 and C_2 are related through the boundary conditions at the cover–film interface. This is discussed in the next subsection.

3.4.6 Dispersion Relation

To determine the dispersion relation, we enforce the boundary conditions at $x = 0$. Since $e_y(x)$ is continuous at $x = 0$, we have

$$C_1 = C_2 \frac{h^{1/2}}{(1-b)^{1/4}} \sin\left\{ \frac{V}{h} \int_{-x_t}^{0} [f(x) - b]^{1/2}\, dx + \frac{\pi}{4} \right\} \qquad (3.66)$$

From the continuation of $de_y(x)/dx$ at $x = 0$, we deduce

$$-C_1 \frac{V\sqrt{a+b}}{h} = C_2 \frac{V(1-b)^{1/4}}{h^{1/2}} \cos\left\{ \frac{V}{h} \int_{-x_t}^{0} [f(x) - b]^{1/2}\, dx + \frac{\pi}{4} \right\}$$

$$- \frac{1}{4} C_2 \frac{h^{1/2}}{(1-b)^{5/4}} f'(0) \sin\left\{ \frac{V}{h} \int_{-x_t}^{0} [f(x) - b]^{1/2}\, dx + \frac{\pi}{4} \right\} \qquad (3.67)$$

By combining the two equations, we obtain an *approximate dispersion relation for TE modes*:

$$-\sqrt{\frac{1-b}{a+b}} \left(1 - \frac{h}{4V} \frac{f'(0)}{\sqrt{1-b}} \tan\left\{ \frac{V}{h} \int_{-x_t}^{0} [f(x) - b]^{1/2}\, dx + \frac{\pi}{4} \right\} \right)$$

$$= \tan\left\{ \frac{V}{h} \int_{-x_t}^{0} [f(x) - b]^{1/2}\, dx + \frac{\pi}{4} \right\} \qquad (3.68)$$

Since V is usually a large parameter, we can simplify the equation further by dropping the $h/(4V)$ term on the left-hand side and obtain

$$\tan\left\{ \frac{V}{h} \int_{-x_t}^{0} [f(x) - b]^{1/2}\, dx + \frac{\pi}{4} \right\} = -\sqrt{\frac{1-b}{a+b}} \qquad (3.69)$$

We rearrange the equation as

$$\frac{V}{h} \int_{-x_t}^{0} [f(x) - b]^{1/2}\, dx = m\pi - \frac{\pi}{4} - \tan^{-1}\sqrt{\frac{1-b}{a+b}} \qquad (3.70)$$

This is the approximate dispersion relation for TE modes guided by graded-index waveguides having a single and isolated turning point.

3.4.7 An Example

To assess the accuracy of (3.70), we use it to study exponentially graded waveguides and to compare the WKB result with that based on the exact dispersion relation (3.24). For exponentially graded dielectric waveguides, the index profile in the film and substrate regions is given in (3.19). By comparing (3.19) with (3.35), we see that $f(x) = e^{2x/h}$. Thus, the turning point is simply $x_t = -(h/2)\ln b$. Since

$f(x)$ is a simple exponential function, the integral in (3.70) can be evaluated in close form. Thus from (3.70) we obtain an approximate dispersion relation for the exponentially graded waveguides:

$$V\left(\sqrt{1-b} - \sqrt{b}\tan^{-1}\sqrt{\frac{1-b}{b}}\right) = m\pi - \frac{\pi}{4} - \tan^{-1}\sqrt{\frac{1-b}{a+b}} \qquad (3.71)$$

Equations (3.25) and (3.71) look quite different analytically. To see if they are equivalent numerically, we resort to a numerical comparison. For this purpose, we evaluate (3.71) numerically and plot the bV curves in Figure 3.9 for the first four modes with $a = 10$. When the curves of Figure 3.9 are compared with those based on (3.25), the two sets of plots are indistinguishable. In other words, (3.25) and (3.71) are numerically equivalent even though they appear different analytically.

Recall that the dispersion relations, (3.70) and (3.71), are predicated on the assumption that V is large. Numerical calculation shows that (3.71) are accurate even for V as small as 3. The accuracy improves as V increases further. Nevertheless, we note that (3.70) is valid only for graded-index waveguides with a single isolated turning point. If $f(x)$ is not a monotonic function, there may be two or more turning points. Or, there is just one turning point. But the turning point is near the index discontinuities. In these cases, (3.70) would fail. Although the WKB method can be applied to waveguides having two or more turning points, or having a turning point near the index discontinuities, the procedure is quite tedious and the results are very complicated. We refer readers to the literature [11–16] for details.

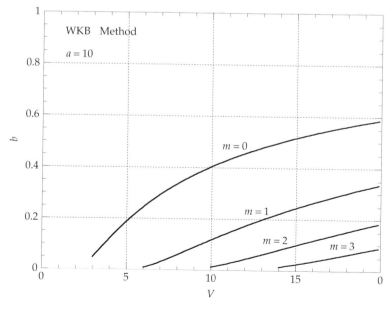

Figure 3.9 bV diagram of TE modes guided by waveguides with an exponentially graded-index profile as determined by WKB method.

In principle, we could use (3.65) to approximate field expressions in various regions once b is known. While the dispersion relation (3.70) based on the WKB method is reasonably accurate, the WKB expression (3.65) for fields may not be as accurate as we would wish. This is a shortcoming of the WKB method in general [10, 16, 17].

3.5   HOCKER AND BURNS' NUMERICAL METHOD

The most versatile and accurate methods for studying waveguide problems are those based on numerical techniques. Many numerical techniques are available for analyzing graded-index dielectric waveguides [18]. Here we discuss a method formulated by Hocker and Burns [19]. In their method, the motivation for each step is clear, the physical meaning of each term is evident, and the resulting equation is amenable to numerical calculations. However, like most numerical methods for solving the dispersion relation, it does not provide any information on the field distribution of the guided modes.

Conceptually, we can view waveguide modes as fields bouncing back and forth in the thin-film region. A dispersion relation is simply a mathematical expression equating the *total transverse phase change* to an integer multiple of 2π. In step-index waveguides, rays follow straight-line paths in the film region and bend abruptly at the film–cover and film–substrate boundaries, as shown in Figure 3.10(a). The total transverse phase change consists of three parts, the phase delays Φ_{cf} and Φ_{fs} due to the total internal reflection at the two interfaces, and the phase delay Φ_{OPL} due to the zigzagging optical path in the film region. Modes in graded-index waveguides can be understood in the same manner. At the film–cover boundary, a ray turns sharply. We use Φ_{cf} to denote the phase delay at the cover–film boundary. Since the index in the graded-index region changes continuously, the ray follows a curved trajectory and turns continuously [Fig. 3.10(b)]. We designate Φ_{OPL} as the phase delay of the curved ray trajectory. For a given guided mode and at a certain point in the graded-index region, the ray trajectory is exactly in parallel with the z axis. This point is the turning point of the particular trajectory or mode. Let Φ_t be the phase delay at the turning point. The total transverse phase change is the sum of Φ_{cf}, Φ_{OPL}, and Φ_t.

To calculate the phase delay of the curved optical path, we model the graded-index region as a stack of thin layers [Fig. 3.10(b)]. Each layer is very thin that the index variation in the layer can be ignored. In other words, each layer is treated as a region with a constant index. Within each thin layer, a ray follows a straight-line path. But the indices of neighboring layers differ only slightly. As a result, the ray direction changes very slightly at the boundary between the neighboring layers. Thus the ray turns gradually as it moves from layer to layer. In the limit of infinitesimally thin layers, the optical path becomes a smooth and continuously turning trajectory. Suppose the jth layer centers at x_j and has a thickness of Δx_j. Then, the index of the jth layer is $n(x_j)$. Let θ_j be the ray angle in the jth layer relative to the z axis. Then the transverse phase change in the jth layer is

$$\Delta \Phi_j = -kn(x_j)\sin\,\theta_j\,\Delta x_j$$

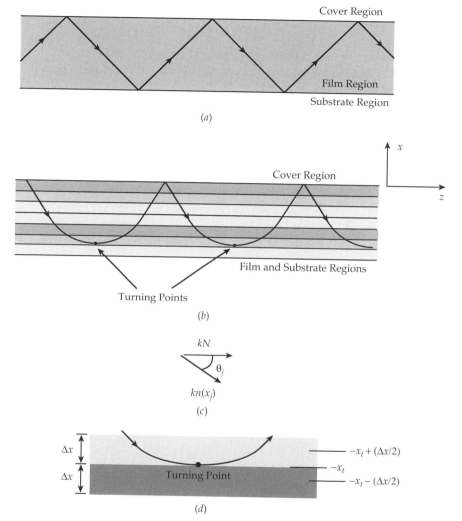

Figure 3.10 Rays in (a) step-index and (b) graded-index waveguides. (c) Ray and angle in jth layer and (d) ray at the turning point.

The minus sign in the above equation can be understood as follows. Throughout this book, we consider time-harmonic fields of the form of $e^{j\omega t}$. Thus a wave propagates in $+z$ is given by $e^{j(\omega t - \beta z)}$. The phase delay due to a path length increase of z is $-\beta z$. Note the minus sign.

Let the effective index of the guided mode be N. Then θ_j is given by $\cos \theta_j = N/n(x_j)$. It follows that $\sin \theta_j = \sqrt{1 - [N/n(x_j)]^2}$. Therefore, the transverse phase change in the jth layer is

$$\Delta \Phi_j = -k \sqrt{n^2(x_j) - N^2}\, \Delta x_j \qquad (3.72)$$

Starting from an arbitrary point at the cover–film boundary, we trace a ray through all layers to the turning points and back. By summing the contribution to the phase change from all layers from the cover–film boundary ($x = 0$) to the turning point where $n(-x_t) = N$ and back to the cover–film boundary, we obtain the round-trip transverse phase change associated with the curved optical path length. In the limit of infinitesimal layer thickness, the summation becomes an integral:

$$\Phi_{\mathrm{OPL}} = -2 \sum_j \Delta\Phi_j \longrightarrow -2k \int_{-x_t}^{0} \sqrt{n^2(x) - N^2} \, dx \qquad (3.73)$$

It is simple to show, from the index profile (3.35) and the definition for b, that

$$n^2(x) - N^2 = (n_f^2 - n_s^2)[f(x) - b] \qquad (3.74)$$

Thus the round-trip transverse phase delay associated with the curved optical path is

$$\Phi_{\mathrm{OPL}} = -2k \int_{-x_t}^{0} \sqrt{n^2(x) - N^2} \, dx = -\frac{2V}{h} \int_{-x_t}^{0} \sqrt{f(x) - b} \, dx \qquad (3.75)$$

3.5.1   Transverse Electric Modes

The phase change of TE modes due to the total internal reflection at a dielectric boundary is given in (1.56). In (1.56), ϕ is the angle between the ray vector and the normal to the boundary. In terms of the angle θ relative to the z axis, Φ_{cf} is

$$\Phi_{\mathrm{cf}} = \tan^{-1} \frac{\sqrt{n_f^2 \cos^2 \theta - n_c^2}}{n_f \sin \theta} \qquad (3.76)$$

When written in terms of the normalized parameters a and b, the above equation becomes

$$\Phi_{\mathrm{cf}} = \tan^{-1} \sqrt{\frac{a + b}{1 - b}} \qquad (3.77)$$

To determine the phase delay Φ_t at the turning point, we consider the two thin layers, one on each side of the turning point, as depicted in Figure 3.10(d). At the turning point, $n(-x_t) = N$. For convenience, we assume that the two layers have the same thickness Δx. Then the indices of the two layers in question are $n[-x_t \pm (\Delta x)/2]$. For TE waves incident from a layer with an index $n[-x_t + (\Delta x/2)]$ upon a layer with an index $n[-x_t - (\Delta x/2)]$, the phase change due to the total internal reflection at the boundary is

$$\Phi_t = \tan^{-1} \frac{\sqrt{n^2[-x_t + (\Delta x/2)] \cos^2 \theta - n^2[-x_t - (\Delta x/2)]}}{n[-x_t + (\Delta x/2)] \sin \theta} \qquad (3.78)$$

As mentioned earlier, $N = n(x_j)\cos\theta_j$, we can rewrite the above equation as

$$\Phi_t = \tan^{-1}\frac{\sqrt{N^2 - n^2[-x_t - (\Delta x/2)]}}{\sqrt{n^2[-x_t + (\Delta x/2)] - N^2}} \qquad (3.79)$$

Since the two layers are very thin and Δx is very small, we approximate $n[-x_t \pm (\Delta x/2)]$ as

$$n[-x_t \pm (\Delta x/2)] \approx n(-x_t) \pm \frac{\Delta x}{2}\frac{dn}{dx} = N \pm \delta$$

where

$$\delta = \frac{\Delta x}{2}\frac{dn}{dx}$$

Then

$$\sqrt{N^2 - n^2[-x_t - (\Delta x/2)]} \approx \sqrt{N^2 - (N - \delta)^2} = \sqrt{2N\delta - \delta^2} \qquad (3.80)$$

$$\sqrt{n^2[-x_t + (\Delta x/2)] - N^2} \approx \sqrt{(N + \delta)^2 - N^2} = \sqrt{2N\delta + \delta^2} \qquad (3.81)$$

Using these approximations in (3.78), we obtain

$$\Phi_t \approx \tan^{-1}\frac{\sqrt{2N - \delta}}{\sqrt{2N + \delta}} \qquad (3.82)$$

In the limit of vanishing layer thickness, Δx and δ tend zero. Then Φ_t given in the above equation approaches $\pi/4$. Combining (3.75), (3.77), and (3.82), we have the dispersion relation of TE modes:

$$\frac{2V}{h}\int_{-x_t}^{0}\sqrt{f(x) - b}\,dx = 2m''\pi + \frac{\pi}{2} + 2\tan^{-1}\sqrt{\frac{a + b}{1 - b}} \qquad (3.83)$$

where m'' is an integer. The equation can be reduced to

$$\frac{V}{h}\int_{-x_t}^{0}\sqrt{f(x) - b}\,dx = m''\pi + \frac{\pi}{4} + \tan^{-1}\sqrt{\frac{a + b}{1 - b}} \qquad (3.84)$$

By making use of an identify

$$\tan^{-1}x + \tan^{-1}\frac{1}{x} = \frac{\pi}{2}$$

we rewrite (3.84) as

$$\frac{V}{h}\int_{-x_t}^{0}\sqrt{f(x) - b}\,dx = m\pi - \frac{\pi}{4} - \tan^{-1}\sqrt{\frac{1 - b}{a + b}} \qquad (3.85)$$

where $m = m'' + 1$.

Equation (3.84), or equivalently (3.85), is the dispersion relation for TE modes guided by dielectric waveguides with an arbitrary graded-index profile $f(x)$ [19]. It is interesting to note that (3.85) is exactly the same as (3.70) derived in the last section by the WKB method.

3.5.2 Transverse Magnetic Modes

For the TM modes, Φ_{cf} and Φ_t are, from (1.70),

$$\Phi_{cf} = \tan^{-1} \frac{n_f \sqrt{n_f^2 \cos^2 \theta - n_c^2}}{n_c^2 \sin \theta} = \tan^{-1} \frac{1}{d}\sqrt{\frac{a+b}{1-b}} \tag{3.86}$$

$$\Phi_t = \tan^{-1} \frac{n[-x_t + (\Delta x/2)]\sqrt{n^2[-x_t + (\Delta x/2)]\cos^2 \theta - n^2[-x_t - (\Delta x/2)]}}{n^2[-x_t - (\Delta x/2)]\sin \theta}$$

$$\approx \tan^{-1} \frac{(N+\delta)^2 \sqrt{2N-\delta}}{(N-\delta)^2 \sqrt{2N+\delta}} \tag{3.87}$$

In the limit of a vanishing layer thickness, Δx and δ end to zero, and Φ_t approaches $\pi/4$. Thus the *dispersion relation for TM modes* is

$$\frac{V}{h} \int_{-x_t}^{0} \sqrt{f(x) - b}\, dx = m'\pi + \frac{\pi}{4} + \tan^{-1} \frac{1}{d}\sqrt{\frac{a+b}{1-b}} \tag{3.88}$$

This completes the Hocker and Burns analysis for TM modes [19].

3.6 STEP-INDEX THIN-FILM WAVEGUIDES VERSUS GRADED-INDEX DIELECTRIC WAVEGUIDES

In this section we compare the essential features of modes guided by graded-index waveguides with that of step-index waveguides. Typical plots of the field distributions and the dispersion of TE modes guided by step-index waveguides are shown in Figures 2.4 and 2.8. The bV characteristics of TE modes guided by linearly and exponentially graded waveguides are shown in Figures 3.4 and 3.6. The field distributions of TE modes guided by these waveguides are depicted in Figures 3.5 and 3.7. The dispersion of TE modes guided by waveguides with complimentary error function, Gaussian function, and parabolic index profiles can be found in the literature [2, 6–8]. Additional field plots can be found in [6]. Based on these figures, we make five observations.

1. In general, the bV characteristics of step-index waveguides rise rapidly. But, the bV curves of graded-index waveguides increase rather slowly. A comparison of Figures 3.4 and 3.6 shows that a steeper index tapering leads to a slower rate of rising in the bV curves.

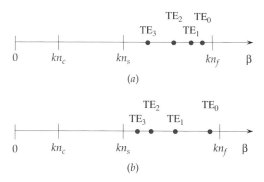

Figure 3.11   Mode Spectra of (a) step-index and (b) graded-index waveguides.

2. In all waveguides studied, the b value of step-index waveguides is the largest of all waveguides having the same V, a, and m.

3. An examination of the bV curves also reveals that the difference between of b of two neighboring modes, that is, $b_{m+1} - b_m$, increases with m for step-index waveguides. In contrast, $b_{m+1} - b_m$ of graded-index waveguides decrease as m increases. In other words, the mode spectrum of step-index waveguides is quite different from that of the graded-index waveguides. This is depicted schematically in Figure 3.11.

4. The field distributions of the fundamental modes of most dielectric waveguides are essentially the same irrespective of the index profiles.

5. The field distributions of higher-order modes are profile dependent. In the case of step-index waveguides, all field maxima and minima are confined in the film region. Furthermore the field strength of all field maxima is essentially the same. In graded-index waveguides, however, the field maxima of higher-order modes shift gradually deeper into the substrate region. As they shift, the strength of the field maxima also increases.

PROBLEMS

1. Consider TM modes guided by thin-film waveguides with a linearly tapered index profile. Derive the dispersion relation and an expression for $h_y(x)$ of TM modes.

2. Consider TE modes guided by dielectric waveguides with an index profile given by (3.35) with (a) $f(x) = e^{x/h}$ and (b) $f(x) = [1 + (x/h)^2]^{-1}$. Apply the Hocker and Burns method to set up a dispersion relation for each index profile. Then solve the dispersion relation numerically. The final results for each index profile should be a set of bV curves for $a = 1, 10$, and 100 for V ranging from 0 to 20.

3. Starting from the Fresnel equation for plane waves with the electric fields in the plane of incidence, derive (3.86) and (3.87) and finally (3.88), the dispersion relation for TM modes.

REFERENCES

1. D. Marcuse, "TE modes of graded index slab waveguides," *IEEE J. Quantum Electronics*, Vol. QE-9, pp. 1000–1006 (1973).

2. J. D. Love and A. K. Ghatak, "Exact solution for TM modes in graded index slab waveguides," *IEEE J. Quantum Electron.*, Vol. QE-15, pp. 14–16 (Jan. 1979).

3. K. Ogusu, "Piecewise linear approximation for analyzing TM modes in graded index planar waveguides," *Opt. Comm.*, Vol. 57, No. 4, pp. 274–278 (Mar. 15, 1986).

4. E. M. Conwell, "Modes in optical waveguides formed by diffusion," *Appl. Phys. Lett.*, Vol. 23, No. 6, pp. 328–329 (1973).

5. E. M. Conwell, "Modes in anisotropic optical waveguides formed by diffusion," *IEEE J. Quantum Electron.*, Vol. QE-10, No. 8, pp. 608–612 (1974).

6. D. H. Smithgall and F. W. Dabby, "Graded-index planar dielectric waveguides," *IEEE J. Quantum Electron.*, Vol. QE-9, pp. 1023–1028 (Oct. 1973).

7. H. Kogelnik, "Theory of dielectric waveguides," in *Integrated Optics*, T. Tamir (ed.), 1975; and *Guided-Wave Optoelectronics*, T. Tamir (ed.), Springer, Berlin, 1988 and 1990, Chapter 2.

8. I. Savatinova and E. Nadjakov, "Modes in diffused optical waveguides (parabolic and Gaussian models)," *Appl. Phys.*, Vol. 8, pp. 245–250 (1975).

9. M. Abramowitz and I. A. Stegun, *Handbook of Mathematical Functions with Formulas, Graphs and Mathematical Tables*, Dover, New York, 1972.

10. P. M. Morse and H. Feshbach, *Methods of Theoretical Physics*, McGraw-Hill, New York, 1953, Chapter 9.

11. D. Gloge, "Propagation effects in optical fibers," *IEEE Trans. Microwave Theory Techniques*, Vol. MTT-23, pp. 102–120 (Jan. 1975).

12. R. Srivastava, C. K. Kao, and R. V. Ramaswamy, "WKB analysis of planar surface waveguides with truncated index profiles," *IEEE J. Lightwave Techno.*, Vol. LT-5, No. 11, pp. 1605–1609 (1987).

13. J. Wang and L. Qiao, "A refined WKB method for symmetric planar waveguides with truncated-index profiles and graded-index profiles," *IEEE J. Quantum Electron.*, Vol. 27, No. 4, pp. 878–883 (Apr. 1991).

14. L. Qiao and J. Wang, "A modified ray-optic method for arbitrary dielectric waveguides," *IEEE J. Quantum Electron.*, Vol. 28, No. 12, pp. 2721–2727 (1992).

15. Z. Cao, Q. Liu, Y. Jiang, Q. Shen, and X. Dou, "Phase shift at a turning point in a planar optical waveguide," *J. Opt. Soc. Am., A*, Vol. 18, No. 9, pp. 2161–2163 (Sept. 2001).

16. A. Gedeon, "Comparison between rigorous theory and WKB-analysis of modes in graded-index waveguides," *Opt. Comm.*, Vol. 12, No. 3, pp. 329–332 (Nov. 1974).

17. J. Janta and J. Ctyroky, "On the accuracy of WKB analysis of TE and TM modes in planar graded-index waveguides," *Opt. Comm.*, Vol. 25, No. 1, pp. 49–52 (Apr. 1978).

18. K. S. Chiang, "Review of numerical and approximate methods for the modal analysis of general dielectric waveguides," *Opt. Quantum Electron*, Vol. 29, pp. s113–s134 (1994).

19. G. B. Hocker and W. K. Burns, "Modes in diffused optical; waveguides of arbitrary index profile," *IEEE J. Quantum Electron.*, Vol. QE-11, pp. 270–276 (1975).

4

PROPAGATION LOSS IN THIN-FILM WAVEGUIDES

4.1 INTRODUCTION

In the last two chapters, we considered ideal waveguides in which the waveguide materials are loss free and the waveguide geometries and index profiles are perfect. Therefore, waves propagate without attenuation. But no waveguide geometry or index profile is perfect. Loss in real dielectric materials, while small, is not zero. As a result, waves in real waveguides decay as they propagate. There are two basic loss mechanisms: the *absorption* in the waveguide materials and the *scattering* by geometry irregularities, material defect, and inhomogeneities. In waveguides based on glass and other amorphous materials, the grain size and composition fluctuations would lead to the scattering loss as well. By material absorption, we mean the attenuation of light due to the interaction with atoms or molecules of the media. As a result of the interaction, light is converted to heat or acoustic waves. The material absorption is a basic and inherent loss mechanism present in all materials. By scattering, we mean the partial conversion of a guided mode to the radiation mode, other guided modes, or the same guided mode propagating in the opposite direction. Scattering loss depends on the distribution of the material defect, surface roughness, and waveguide geometry irregularities [1, 2]. It is a complicated subject

Foundations for Guided-Wave Optics, by Chin-Lin Chen
Copyright © 2007 John Wiley & Sons, Inc.

and we will not discuss it further. Furthermore, the scattering loss depends critically on the material, fabrication processes, preparation, and surface treatment and may be reduced or minimized by improving the fabrication processes and treatment.

$LiNbO_3$, $LiTaO_3$, $BaTiO_3$, GaAs, and so forth are electrooptic materials in that the refractive indices can be varied by applying electric fields to them. Many optical devices, optical switches, and modulators, for example, are designed to take advantage of the electrooptic properties of these materials. To facilitate the modulation or to switch optical beams electrically, electric fields are established by applying voltages to metallic electrodes. The electrodes are either directly over or in the vicinity of the optical waveguides. But, all metals are lossy at the optical wavelengths. Their presence causes waves to attenuate. In general, if a metallic layer is present, either as a part of the waveguide or near the waveguide structure, then the propagation loss due to metallic materials is the dominant cause of loss.

But metallic layers also lead to new waveguide effects that form the basis of new optical devices. We will show shortly that the loss of TM modes of metal-clad thin-film waveguides is much greater than that of TE modes. Loss of higher-order modes is also much greater than that of lower-order modes. Thus, metal-clad waveguides can serve as waveguide polarizers and mode filters [3, 4] to discriminate against TM modes or higher-order modes. In other words, metallic layers may be useful in many applications. This is also a motivation for studying effects of metallic layers.

Depending on the position of electrodes relative to the optical waveguide, the electric field lines may be primarily in parallel or perpendicular to the waveguides. Figure 4.1 depicts two possible electrode configurations and the electric fields established by the electrodes [5, 6]. In Figure 4.1(a), the applied electric fields in the waveguide region are mainly in parallel with the waveguide surface, that is, the y direction. In Figure 4.1(b), the metallic films are directly over the optical waveguide and the applied electric fields are mainly normal to the waveguide surface. In other words, the electric fields are mainly in the x direction.

4.2   COMPLEX RELATIVE DIELECTRIC CONSTANT AND COMPLEX REFRACTIVE INDEX

In engineering terminology, we characterize dielectrics and conductors in terms of the permittivity ε and the conductivity σ [7–9]. By combining ε and σ, we define the *complex relative dielectric constant* $\tilde{\varepsilon}_{rel}$ as

$$\tilde{\varepsilon}_{rel} = \frac{\varepsilon}{\varepsilon_0} - j\frac{\sigma}{\omega\varepsilon_0} \tag{4.1}$$

In terms of ε and σ, or $\tilde{\varepsilon}_{rel}$, one of the time-harmonic ($e^{j\omega t}$) Maxwell equations, (1.8), becomes

$$\nabla \times \mathbf{H} = (j\omega\varepsilon + \sigma)\mathbf{E} = j\omega\varepsilon_0\tilde{\varepsilon}_{rel}\mathbf{E} \tag{4.2}$$

The real and imaginary parts of the complex relative dielectric constant can be written in terms of the *plasma frequency* ω_p and *collision frequency* ω_c of the

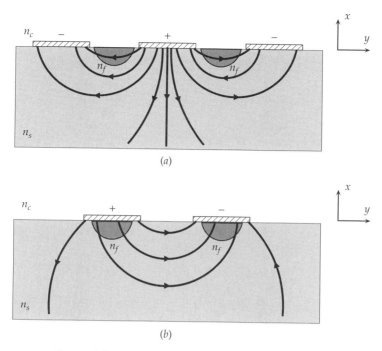

Figure 4.1   Electric field lines produced by electrodes.

material [9]:

$$\tilde{\varepsilon}_{\text{rel}} = \varepsilon_r - j\varepsilon_i = 1 - \frac{\omega_p^2}{\omega^2 + \omega_c^2} - j\frac{\omega_c\omega_p^2}{\omega(\omega^2 + \omega_c^2)} \tag{4.3}$$

In the optics literature, material properties are often expressed in terms of the *refractive index n* and *extinction coefficient κ* [9–12]. In terms of n and κ, (4.2) becomes

$$\nabla \times \mathbf{H} = j\omega\varepsilon_0(n - j\kappa)^2\,\mathbf{E} \tag{4.4}$$

Thus, n and κ are the real and imaginary parts of a *complex refractive index* $\tilde{n} = n - j\kappa$. In summary, the dielectric and loss properties of materials can be described in terms of $\tilde{\varepsilon}_{\text{rel}}$ or $\tilde{n}$. In lossless media, the relative dielectric constant and the refractive index are real quantities, and $\varepsilon_{\text{rel}} = n^2$. In real dielectric media, $\tilde{\varepsilon}_{\text{rel}}$ and $\tilde{n}$ are complex quantities and $\tilde{\varepsilon}_{\text{rel}} = \tilde{n}^2$. Many real dielectric materials are weakly absorbing in that the extinction coefficient is small in comparison with the refractive index. For example, the extinction coefficients of the ordinary waves and extraordinary waves in barium titanate (BaTiO$_3$) at 0.65 μm are 6.2×10^{-7} and 1.03×10^{-6}, and the ordinary and extraordinary indices of the material are, respectively, 2.416 and 2.364 [13]. On the other hand, n and κ of metals and semiconductors are of the same orders of magnitude. Table 4.1 lists the values of n, κ, ε_r, and ε_i of several metals and semiconductors. An extensive tabulation

TABLE 4.1   Refractive indices and extinction coefficients of selected metals and semiconductors

Material	λ (μm)	n	κ	$\varepsilon_r - j\varepsilon_i = (n - j\kappa)^2$	References
Au	0.633	0.17	3.0	$-8.97-j1.02$	14
	0.653	0.166	3.15	$-9.89-j1.05$	12
	1.55	0.550	11.5	$-132-j12.6$	12
Ag	0.633	0.065	3.9	$-15.2-j0.507$	14
	0.653	0.140	4.15	$-17.2-j1.16$	12
	1.55	0.514	10.8	$-116-j11.1$	12
Cu	0.633	0.14	3.15	$-9.91-j0.88$	14
	0.653	0.214	3.67	$-13.4-j1.57$	12
	1.55	0.606	8.26	$-67.9-j10.0$	12
Al	0.633	1.2	7	$-47.56-j16.8$	14
	0.653	1.49	7.82	$-58.9-j23.3$	12
	1.55	1.44	16.0	$-254-j46.1$	12
Cr	0.633	3.19	2.26	$+5.07-j14.4$	14
	1.590	4.13	5.03	$-8.24-j41.5$	13
Ge	0.633	4.5	1.7	$+17.4-j15.3$	14
	0.653	5.294	0.638	$+27.6-j6.76$	12
	1.55	4.275	0.00567	$+18.3-j0.049$	12
GaAs	0.633	3.856	0.196	$+14.8-j1.51$	12
	0.653	3.826	0.179	$+14.6-j1.37$	12
	1.55	3.3737	—	$+11.4$	12
Si	0.633	3.882	0.019	$+15.07-j0.148$	12
	0.653	3.847	0.016	$+15.0-j0.123$	12
	1.532	3.4784	—	$+12.1$	12

Source: From Refs.[12–14].

of material properties can be found in [11–13]. For most metals, ε_r is negative and $|\varepsilon_r|$ is larger than ε_i at optical frequencies. There are a few exceptions. For example, ε_r of chromium in the visible spectra is positive and it is smaller than ε_i [14]. For semiconductors, ε_r is also greater than ε_i. Figure 4.2 depicts the real and imaginary parts of the complex dielectric constants of many materials at 0.8 and 1.3 μm [15].

4.3   PROPAGATION LOSS IN STEP-INDEX WAVEGUIDES

4.3.1   Waveguides Having Weakly Absorbing Materials

As indicated previously, many dielectric materials are weakly absorbing and have very small extinction coefficients. Kane and Osterberg [16] showed that for waveguides made of weakly absorbing materials, the field distributions and the propagation constants are essentially the same as those of waveguides made of lossless materials. The main effect of a small material absorption is to attenuate the guided modes.

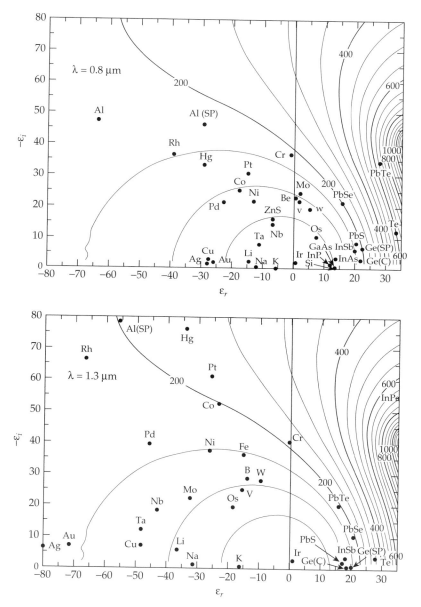

Figure 4.2 Real and imaginary parts of the complex dielectric constants of various materials at 0.8 μm (*top*) and 1.3 μm (*bottom*). The letters SP denote sputtered films and Ge(C) represents bulk crystal Ge. Quality factor contours are drawn with an interval of 50. (from [15].)

Effects of a weak material absorption on the field distribution and propagation constant are of the second order. But effects on the attenuation constant are of the first order. We can estimate the waveguide loss due to a small material absorption by using the perturbation formula [8, 17]. For TE modes, the attenuation constant is approximately

$$\alpha_{TE} \approx \frac{k}{N} \frac{n_s\kappa_s \int_{-\infty}^{-h} |e_y(x)|^2 \, dx + n_f\kappa_f \int_{-h}^{0} |e_y(x)|^2 \, dx + n_c\kappa_c \int_{0}^{\infty} |e_y(x)|^2 \, dx}{\int_{-\infty}^{\infty} |e_y(x)|^2 \, dx}$$

(4.5)

In the above expression, n's and κ's are the refractive indices and extinction coefficients of the three regions, and $e_y(x)$ and N are the field and the effective index of refraction of the mode guided by a lossless waveguide. Similar and slightly more complicated equation can be established for TM modes.

4.3.2   Metal-Clad Waveguides

If a metallic layer, or layers, is a part of the waveguide structure, the situation changes completely. The perturbation formula mentioned in the last subsection is not applicable since metals are very lossy at optical frequencies. To study waves in planar waveguides containing lossy materials, we begin with Maxwell equation (4.2) or (4.4). We use the complex refractive indices $\tilde{n}_f$, $\tilde{n}_s$, $\tilde{n}_c$ in lieu of n_f, n_s, and n_c to account for the loss in various regions. The rest of the analysis is exactly the same as that discussed in Chapter 2. The resulting dispersion relation is formally identical to (2.22) for TE modes and (2.28) for TM modes. However, the dispersion relations are complex transcendental equations since $\tilde{n}_f$, $\tilde{n}_s$, and $\tilde{n}_c$ are complex quantities. Numerical methods are needed to determine the complex propagation constant $\tilde{\beta} = \beta - j\alpha$, where α is the propagation loss of the waveguide. Once $\tilde{\beta}$ is known, fields can be calculated [17–19].

As an example, we consider the propagation, attenuation, and fields of waves guided by a silver–polymer–glass waveguide (Fig. 4.3) [19]. The polymer film and the glass substrate are assumed to be lossless and have indices of 1.5884 and 1.5133, respectively. The silver cover has a complex refractive index of $0.0065 - j4$. The plots of b and α/k as functions of V for TE and TM modes are shown in Figure 4.4. The field distributions of TE and TM modes are shown in Figures 4.5 and 4.6.

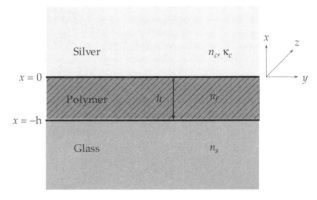

Figure 4.3   Metal-clad waveguide (silver–polymers–glass waveguide).

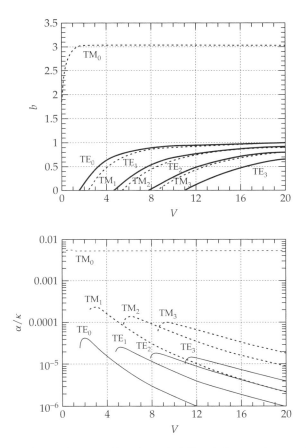

Figure 4.4   Propagation and attenuation characteristics of silver–polymer–glass waveguide.

For all modes, expect TM_0 mode, the bV curves have essentially the same shape as that of thin-film waveguides made of lossless materials. For all modes, other than TM_0 mode, α increases from a relatively small value near the cutoff to a large peak, and then it decreases monotonically as V increases. The peak loss depends on the mode and mode number. Also note that α/k may change by 2 orders of magnitude. In contrast, the attenuation of TM_0 mode changes very little as V increases. The loss in thick waveguides can be understood in the following manner. For higher-order modes, a large fraction of time-average power resides in the metallic cladding and this leads to a large attenuation constant. Thus α is larger for higher-order modes. For the same mode number, the loss in TM modes is much larger than that in TE modes. A simple and qualitative explanation is as follows. We assume that modes are far from the cutoff. Refer to Figure 4.3. TE modes have an electric field component only, the y component, and it is in parallel with the metal–dielectric interface. The boundary condition requires a vanishing tangential electric field component at a perfectly conducting boundary, and a very small tangential electric field component at a good conducting surface. Thus TE waves with the electric fields parallel to

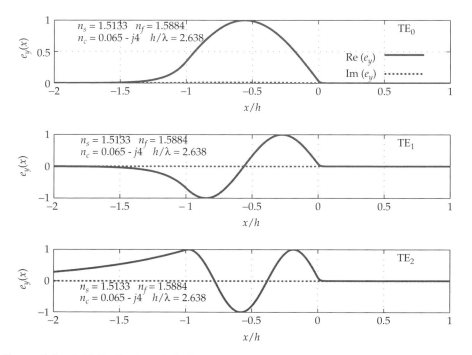

Figure 4.5 Field distributions (e_y) of TE$_0$, TE$_1$, and TE$_2$ modes. The real and imaginary parts of e_y are depicted as solid and dashed lines.

the conducting surface are "reflected" or "repelled" by the conducting boundary. In other words, the penetration of TE modes into the metallic cover region is very small. As a result, the ohmic loss of TE modes in the metallic region is small. On the other hand, TM modes have a magnetic field component and two electric field components. The magnetic field component is in parallel with the metal–dielectric boundary. The two electric components are e_x and e_z. The transverse electric field component, e_x, is the dominating electric field component and it is normal to the metal–dielectric boundary. The pertinent boundary condition is the continuation of the normal component of the electric flux density at the boundary. Thus the dominant electric field component is not "pushed away" by the conducting boundary. As a result, fields of TM modes penetrate deep into the conductors and thus experience significant ohmic loss. In short, for waveguides with a metal cladding, α_{TM} is much greater than α_{TE} for the same mode number.

Figure 4.5 depicts e_y of the first three TE modes guided by a silver-clad thin-film waveguide with $h/\lambda = 2.638$. The solid curves show the distribution of $\text{Re}(e_y)$ and dashed curves for $\text{Im}(e_y)$. A comparison of Figures 4.5 and 2.8 reveals that fields of TE mode of metal-clad waveguides are essentially the same as that of dielectric waveguides with three lossless regions. However, fields in the silver cover, that is, $x > 0$, are very small as discussed previously. For the waveguide under consideration, TE$_2$ mode is very close to the cutoff. Thus, the field of TE$_2$ mode is quite strong in and extends deep into the glass substrate ($x < -h$).

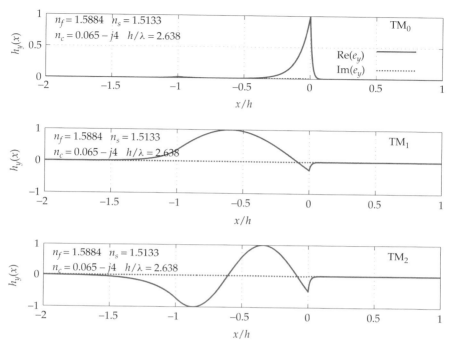

Figure 4.6 Field distributions (h_y) of TM_0, TM_1, and TM_2 modes. The real and imaginary parts of h_y are depicted as solid and dashed lines.

Figure 4.6 shows h_y of the first three TM modes for a silver-clad thin-film waveguide. Note in particular the sharp spikes at the silver–polymer boundary. The rest of the field distributions of TM_1 and TM_2 modes are similar to the TM modes guided by lossless thin-film waveguides. But the field distribution of TM_0 mode is quite different in that fields are mainly confined to a thin layer near the metal–dielectric boundary. In fact TM_0 mode is a surface wave mode. We will discuss TM_0 mode further in Section 4.6.

4.4 ATTENUATION IN THICK WAVEGUIDES WITH STEP-INDEX PROFILES

As noted earlier, a quantitative study of the attenuation of guided modes would require a detailed numerical study of the complex dispersion relation with complex $\tilde{n}_f$, $\tilde{n}_s$, and $\tilde{n}_c$. However, a simple asymptotic approximation for α for thick waveguides exists. In this section, we derive an *asymptotic expression* for α for thin-film waveguides with a large kh. The expression is valid for all modes, except TM_0 mode, guided by metal-clad dielectric waveguides. It leads a better understanding of effects of material absorption on the waveguide loss.

As shown in Figure 4.4, b and α/k vary in a simple manner if the waveguide is thick. In particular, β approaches kn_f and α decreases as the waveguide gets thicker.

In addition, the real part of $\tilde{\beta}$ is much larger than the imaginary part. We will make use of these observations in approximating and simplifying various expressions. Since $\tilde{\beta}$ approaches kn_f when a waveguide is thick, $|\tilde{\gamma}_c|$ and $|\tilde{\gamma}_s|$ are much larger than $|\tilde{\kappa}_f|$, that is, $|\tilde{\kappa}_f| \ll |\tilde{\gamma}_c|$ and $|\tilde{\kappa}_f| \ll |\tilde{\gamma}_s|$. Taking advantage of this fact, we approximate the inverse tangent function of (2.22) as

$$\tan^{-1}\frac{\tilde{\gamma}_s}{\tilde{\kappa}_f} = \frac{\pi}{2} - \tan^{-1}\frac{\tilde{\kappa}_f}{\tilde{\gamma}_s} \approx \frac{\pi}{2} - \frac{\tilde{\kappa}_f}{\tilde{\gamma}_s} \tag{4.6}$$

A similar approximation can be made for $\tan^{-1}\tilde{\gamma}_c/\tilde{\kappa}_f$. Therefore, we approximate the dispersion relation for TE modes, (2.22), as

$$\tilde{\kappa}_f h \approx (m+1)\pi - \frac{\tilde{\kappa}_f}{\tilde{\gamma}_c} - \frac{\tilde{\kappa}_f}{\tilde{\gamma}_s} \tag{4.7}$$

Solving for $\tilde{\kappa}_f$, we obtain

$$\tilde{\kappa}_f \approx \frac{(m+1)\pi}{h}\left(1 + \frac{1}{\tilde{\gamma}_c h} + \frac{1}{\tilde{\gamma}_s h}\right)^{-1} \tag{4.8}$$

From the above equation, we find $\tilde{\beta}_{\text{TE}m}$ of TE$_m$ mode and obtain

$$\tilde{\beta}_{\text{TE}m} \approx k\tilde{n}_f - \frac{(m+1)^2\pi^2}{2k\tilde{n}_f h^2}\left(1 + \frac{1}{\tilde{\gamma}_c h} + \frac{1}{\tilde{\gamma}_s h}\right)^{-2} \tag{4.9}$$

To examine effects of the loss due to metallic cover and/or metallic substrate, we assume that the film region is lossless. Thus, we revert the complex $\tilde{n}_f$ back to a real n_f while keeping $\tilde{n}_s$ and $\tilde{n}_c$ as complex quantities. For modes far from cutoff, β may be approximated by kn_f. Then $\tilde{\gamma}_s$ and $\tilde{\gamma}_c$ in the right-hand side of (4.9) may be approximated by $k\sqrt{n_f^2 - \tilde{n}_s^2}$ and $k\sqrt{n_f^2 - \tilde{n}_c^2}$, respectively. Thus, we obtain from (4.9)

$$\frac{\tilde{\beta}_{\text{TE}m}}{k} \approx n_f - \frac{(m+1)^2\pi^2}{2n_f(kh)^2}\left(1 + \frac{1}{kh\sqrt{n_f^2 - \tilde{n}_c^2}} + \frac{1}{kh\sqrt{n_f^2 - \tilde{n}_s^2}}\right)^{-2} \tag{4.10}$$

By equating real and imaginary parts of both sides of the equation, we obtain expressions for β and α. It is simple to see that the change in β due to the absorption in the cover and substrate regions is minimal. But the effect of the material absorption on α is considerable. Explicitly,

$$\frac{\alpha_{\text{TE}m}}{k} \approx \frac{(m+1)^2\pi^2}{2n_f(kh)^2}\,\text{Im}\left(1 + \frac{1}{kh\sqrt{n_f^2 - \tilde{n}_c^2}} + \frac{1}{kh\sqrt{n_f^2 - \tilde{n}_s^2}}\right)^{-2} \tag{4.11}$$

As shown in Table 4.1, $|\varepsilon_r|$ is much larger than ε_i for most metals. To take advantage of this fact, we rewrite the above equation in terms of the complex dielectric constant. For the cover and substrate regions, we write $\tilde{n}_c^2 = \tilde{\varepsilon}_{rel\ c} = \varepsilon_{cr} - j\varepsilon_{ci}$ and $\tilde{n}_s^2 = \tilde{\varepsilon}_{rel\ s} = \varepsilon_{sr} - j\varepsilon_{si}$. By noting that $|\varepsilon_{cr}| \gg \varepsilon_{ci}$, and $|\varepsilon_{sr}| \gg \varepsilon_{si}$, we obtain

$$\frac{\alpha_{TE\,m}}{k} \approx \frac{(m+1)^2\pi^2}{2n_f(kh)^3}\left[\frac{\varepsilon_{ci}}{(n_f^2 - \varepsilon_{cr})^{3/2}} + \frac{\varepsilon_{si}}{(n_f^2 - \varepsilon_{sr})^{3/2}}\right] \tag{4.12}$$

Similar results are obtained from the dispersion relation for TM modes. Corresponding to (4.10) and (4.12) we have $\tilde{\beta}_{m'}$ for $TM_{m'}$ modes:

$$\frac{\tilde{\beta}_{TM\,m'}}{k} \approx n_f - \frac{(m'+1)^2\pi^2}{2n_f(kh)^2}\left(1 + \frac{\tilde{n}_c^2}{n_f^2}\frac{1}{kh\sqrt{n_f^2 - \tilde{n}_c^2}} + \frac{\tilde{n}_s^2}{n_f^2}\frac{1}{kh\sqrt{n_f^2 - \tilde{n}_s^2}}\right)^{-2} \tag{4.13}$$

The loss in the TM modes is (Problem 1)

$$\frac{\alpha_{TM\,m'}}{k} \approx \frac{(m'+1)^2\pi^2}{2n_f(kh)^3}\left[\frac{\varepsilon_{ci}(2n_f^2 - \varepsilon_{cr})}{n_f^2(n_f^2 - \varepsilon_{cr})^{3/2}} + \frac{\varepsilon_{si}(2n_f^2 - \varepsilon_{sr})}{n_f^2(n_f^2 - \varepsilon_{sr})^{3/2}}\right] \tag{4.14}$$

From (4.12) and (4.14) we note that for thick waveguides, $\alpha_{TE\,m}$ and $\alpha_{TM\,m}$ approach asymptotically:

$$\alpha_{TE\,m,\,TM\,m} \approx C_{TE,\,TM}\frac{(m+1)^2}{k^2h^3} \tag{4.15}$$

where C_{TE} and C_{TM} are two mode-specific constants. Clearly, $\alpha_{TE\,m}$ and $\alpha_{TM\,m}$ increase as $(m+1)^2$ and decreases as h^{-3}. It is also evident that $\alpha_{TE1} \sim 4\alpha_{TE0}$, $\alpha_{TE2} \sim 9\alpha_{TE0}$ and $\alpha_{TM2} \sim (\frac{9}{4})\alpha_{TM1}$, and so forth.

For the same waveguide structure and for the same mode number m, we obtain from (4.12) and (4.14)

$$\frac{\alpha_{TE\,m}}{\alpha_{TM\,m}} \approx \left[\frac{\varepsilon_{ci}}{(n_f^2 - \varepsilon_{cr})^{3/2}} + \frac{\varepsilon_{si}}{(n_f^2 - \varepsilon_{sr})^{3/2}}\right]\left[\frac{\varepsilon_{ci}(2n_f^2 - \varepsilon_{cr})}{n_f^2(n_f^2 - \varepsilon_{cr})^{3/2}} + \frac{\varepsilon_{si}(2n_f^2 - \varepsilon_{sr})}{n_f^2(n_f^2 - \varepsilon_{sr})^{3/2}}\right]^{-1} \tag{4.16}$$

Recall that ε_r is negative for most metals as shown in Table 4.1. Then $\alpha_{TM\,m}$ is greater than $\alpha_{TE\,m}$ for the same mode number. Take the silver–polymer–glass waveguide as an example. We take glass as a lossless substrate material and set ε_{si} to zero. For the polymer film, $n_f = 1.5884$. For the silver cover, $\varepsilon_{cr} = -15.99$ and $\varepsilon_{ci} = 0.52$ at 0.633 μm. Thus $\alpha_{TM\,m}/\alpha_{TE\,m}$ is about 8.34. It clearly shows that the attenuation of TM modes is much greater than that of the TE modes.

4.5   LOSS IN TM$_0$ MODE

As seen in Figure 4.4, β and α of TM$_0$ mode are rather insensitive to change of the film thickness. Figure 4.6 reveals that the field distribution of the TM$_0$ mode is quite different from that of other modes. We will show in this section that a TM$_0$ mode is really a surface wave mode guided by the metal–dielectric boundary only. The other boundary, that is, the film–substrate boundary, and the film thickness have little or no effect on the properties of the TM$_0$ mode. It would be straightforward to set up equations describing surface waves guided by metal–dielectric boundary directly from Maxwell's equations [8]. Here, we take advantage of the dispersion relation (2.28) for TM modes to arrive at the same result.

To reduce the complexity, we consider a symmetric waveguide where the cover and substrate regions are made of the same metallic material (Fig. 4.7). To allow for loss in the metallic regions, we use $\tilde{n}_s$ and $\tilde{n}_c$ in lieu of n_s and n_c. Obviously, $\tilde{n}_s = \tilde{n}_c$. Recall that $\tilde{\kappa}_f = \sqrt{k^2 n_f^2 - \tilde{\beta}^2}$, $\tilde{\gamma}_s = \sqrt{\tilde{\beta}^2 - k^2 \tilde{n}_s^2}$. Thus, the dispersion relation (2.28) for TM modes becomes

$$\sqrt{k^2 n_f^2 - \tilde{\beta}^2}\, h = m'\pi + 2\,\tan^{-1} \frac{n_f^2}{\tilde{n}_s^2} \frac{\sqrt{\tilde{\beta}^2 - k^2 \tilde{n}_s^2}}{\sqrt{k^2 n_f^2 - \tilde{\beta}^2}} \qquad (4.17)$$

As discussed in Chapter 2, the effective refractive index of a guided mode of an ideal waveguide is between n_f and n_s. That is, $k n_s < \mathrm{Re}\,(\tilde{\beta}) < k n_f$. This is also true for all modes, except for the TM$_0$ mode, of waveguides having a metallic substrate. For TM$_0$ mode guided by a metal-clad waveguide, Re $(\tilde{\beta})$ is greater than $k n_f$. We focus our attention to the special case where Re$(\tilde{\beta}) > k n_f$, and write $\tilde{\kappa}_f$ as

$$\tilde{\kappa}_f = \sqrt{k^2 n_f^2 - \tilde{\beta}^2} = j\sqrt{\tilde{\beta}^2 - k^2 n_f^2}$$

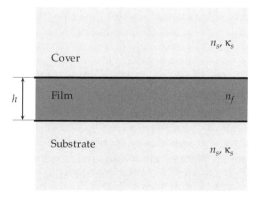

Figure 4.7   Symmetric waveguide with a metallic cover and substrate.

In addition, we make use of a trigonometric identity

$$\tan^{-1}(jz) = j \tanh^{-1} z$$

Then (4.17) becomes

$$j\sqrt{\tilde{\beta}^2 - k^2 n_f^2}\, h = m'\pi - j2 \tanh^{-1} \frac{n_f^2}{\tilde{n}_s^2} \frac{\sqrt{\tilde{\beta}^2 - k^2 \tilde{n}_s^2}}{\sqrt{\tilde{\beta}^2 - k^2 n_f^2}} \qquad (4.18)$$

As indicated previously, for most metals, $\tilde{n}_s^2$ has a large and negative real part. To facilitate discussion, we consider a fictitious material having a purely real $\tilde{n}_s^2$. To stress the fact we are discussing a fictitious waveguide with a fictitious cover and substrate materials, we write $\mathrm{Re}(\tilde{n}_s^2)$ in lieu of $\tilde{n}_s^2$. Then the above equation becomes

$$j\sqrt{\tilde{\beta}^2 - k^2 n_f^2}\, h = m'\pi - j2 \tanh^{-1} \frac{n_f^2}{\mathrm{Re}(\tilde{n}_s^2)} \frac{\sqrt{\tilde{\beta}^2 - k^2 \mathrm{Re}(\tilde{n}_s^2)}}{\sqrt{\tilde{\beta}^2 - k^2 n_f^2}} \qquad (4.19)$$

We search for a possible solution with a purely real $\tilde{\beta}$. First, consider the cases of $m' \neq 0$. If $\tilde{\beta}$ were purely real, the left-hand side of (4.19) would be purely imaginary while the right-hand side would be complex. But a purely imaginary quantity cannot possibly equal to a complex quantity. In other words, no solution with a purely real $\tilde{\beta}$ exists if m' is not zero. The case of $m' = 0$ is a special case and has to be treated separately. For the special case, (4.19) becomes

$$j\sqrt{\tilde{\beta}^2 - k^2 n_f^2}\, h = -2j \tanh^{-1} \frac{n_f^2}{\mathrm{Re}(\tilde{n}_s^2)} \frac{\sqrt{\tilde{\beta}^2 - k^2 \mathrm{Re}(\tilde{n}_s^2)}}{\sqrt{\tilde{\beta}^2 - k^2 n_f^2}} \qquad (4.20)$$

Although both sides are imaginary, they have opposite signs if $\mathrm{Re}(\tilde{n}_s^2)$ is positive. In other words, even for $m' = 0$, no solution with a real $\tilde{\beta}$ exists if $\mathrm{Re}(\tilde{n}_s^2)$ is positive. A solution is possible only if $\mathrm{Re}(\tilde{n}_s^2)$ is negative. We would arrive at the same conclusion if $\tilde{n}_s^2$ is a complex quantity and has a negative real part.

Having arrived at the desired conclusion, we go back to the dispersion relation (4.18). In terms of a complex $\tilde{n}_s^2$, (4.18) with $m' = 0$ can be written as

$$\tanh\left(\frac{\sqrt{\tilde{\beta}^2 - k^2 n_f^2}\, h}{2}\right) = -\frac{n_f^2}{\tilde{n}_s^2} \frac{\sqrt{\tilde{\beta}^2 - k^2 \tilde{n}_s^2}}{\sqrt{\tilde{\beta}^2 - k^2 n_f^2}} \qquad (4.21)$$

In general, we can solve for $\tilde{\beta}$ numerically. But a simple approximate expression for $\tilde{\beta}$ exists if the waveguide is thick. To derive the analytic expression, we note that $\tanh z \to 1$ as $z \to \infty$. Thus, in the limit of a very thick film region, (4.21) becomes

$$\frac{n_f^2}{\tilde{n}_s^2} \frac{\sqrt{\tilde{\beta}^2 - k^2\tilde{n}_s^2}}{\sqrt{\tilde{\beta}^2 - k^2 n_f^2}} \approx -1 \tag{4.22}$$

It is now trivial to solve for $\tilde{\beta}$ and we obtain

$$\frac{\tilde{\beta}}{k} \approx \sqrt{\frac{n_f^2 \tilde{n}_s^2}{n_f^2 + \tilde{n}_s^2}} \tag{4.23}$$

As given by (4.23) $\tilde{\beta}$ is independent of the film thickness h. It means that the TM_0 mode under investigation is a surface wave. This is the *Zenneck wave* guided by the metal–dielectric boundary [9]. It is also known as the *surface plasma wave* guided by the metal–dielectric boundary. The key feature of TM_0 mode is that the propagation and attenuation constants are insensitive to the dielectric film thickness, as shown in Figure 4.4.

4.6   METAL-CLAD WAVEGUIDES WITH GRADED-INDEX PROFILES

In discussions so far, we are mainly concerned with the loss in step-index waveguides. The attenuation in graded-index waveguides can be treated in the same manner. Of particular interest is the graded-index waveguides with a metal cladding. Metal-clad waveguides with linearly, parabolically, and exponentially graded-index profiles have been studied by many investigators [20–23]. In all waveguides mentioned above, the loss of TM modes is about an order of magnitude greater than that of TE modes. As discussed in Section 4.4, the attenuation constant of modes guided by metal-clad waveguides with step-index profile decreases as h^{-3} if the film region is thickness. The situation is different for graded-index waveguides. In particular, α of metal-clad waveguides with linearly and parabolically graded-index profiles decreases as h^{-1} and $h^{-1.5}$, respectively [24].

PROBLEM

1. Derive (4.14) for the TM modes.

REFERENCES

1. P. K. Tien, "Light waves in thin films and integrated optics," *Appl. Opt.*, Vol. 10, pp. 2395–2413, (Nov., 1971)
2. D. Marcuse, "Mode conversion caused by surface imperfections of a dielectric slab waveguide" *B. S. T. J.*, Vol. 48, pp. 3187–3215, (1969)

3. Y. Suematsu, M. Hakuta, K. Furuya, K. Chiba, and R. Hasumi, "Fundamental transverse electric field (TE$_0$) mode selection for thin film asymmetric light guides," *Appl. Phys. Lett.*, Vol. 21, No. 6, pp. 291–293, (Sept. 15, 1972)

4. N. Polky and G. L. Mitchell, "Metal-clad planar dielectric waveguides for integrated optics," *J. Opt. Soc. Am.*, Vol. 64, pp. 274–279, (1974)

5. D. Marcuse, "Optical electrode design for integrated optics modulators," *J. IEEE of Quantum Electron.*, Vol. QE-18, pp. 393–398, (March 1982)

6. C. M. Kim and R. V. Ramaswamy, "Overlap integral factors in integrated optic modulators and switches," *J. IEEE of Lightwave Technol.*, Vol. 7, pp. 1063–1070, (June 1989)

7. S. Ramo, J. R. Whinnery and T. Van Duzer, *Fields and Waves in Communication Electronics*, Third Ed., John Wiley and Sons, New York, (1994)

8. R. E. Collin, *Field Theory of Guided Waves*, Second Ed., IEEE Press, New York, (1991)

9. M. Born and E. Wolf, *Principles of Optics*, Sixth Ed., Pergamon Press, Oxford, (1980)

10. J. I. Pankove, *Optical processes in semiconductors*, Dover Publications, Inc., New York, NY, (1971)

11. G. Hass and L. Hadley, "Optical properties of metals," Chapter 6g of *American Institute of Physics Handbook*, edited by D. E. Gray, McGraw-Hill Book Company, Third ed., (1972)

12. E. D. Palik, *Handbook of Optical Constants of Solids*, New York, Academic Press, pp. 275–602, (1985)

13. E. D. Palik, *Handbook of Optical Constants of Solids II*, New York, Academic Press, pp. 341–747, (1991)

14. T. E. Batchman and K. A. McMillan, "Measurement on positive-permittivity metal-clad waveguides," *J. IEEE of Quantum Electron.*, Vol. QE-13, pp. 187–192, (April, 1977)

15. K. Shiraishi, H. Hatakeyama, H. Matsumoto and K. Matsumura, "Laminated polarizers exhibiting high performance over a wide range of wavelength," *J. IEEE of Lightwave Technol.*, Vol. 15, pp. 1042–1050, (June 1997)

16. J. Kane and H. Osterberg, "Optical characteristics of planar guided modes," *J. of Opt. Soc. Am.*, Vol. 54, pp. 347–352, (March 1964)

17. A. Reisinger, "Characteristics of optical guided modes in lossy waveguides," *Appl. Opt.*, Vol. 12, pp. 1015–1023, (1973)

18. T. Takano, J. Hamasaki, "Propagation modes of a metal-clad-dielectric-slab waveguide for integrated optics," *J. IEEE of Quantum Electron.*, Vol. QE-8, pp. 206–212, (Feb., 1972)

19. I. P. Kaminow, W. L. Mammel and H. P. Weber, "Metal-clad optical waveguide: analytical and experimental study", *Appl. Opt.*, Vol. 13, No. 2, pp. 396–405, (1974)

20. M. Masuda, A. Tanji, Y. Ando and J. Koyama, "Propagation losses of guided modes in an optical graded-index slab waveguide with metal cladding," *IEEE Trans. Microwave Theory and Technology*, Vol. MTT-25, pp. 774–776, (Sept. 1977).

21. T. Findaklay and C. L. Chen, "Diffused optical waveguides with exponential profile: effects of metal-clad and dielectric overlay," *Appl. Opt.*, Vol. 17, pp. 469–474, (Feb. 1978)

22. S. J. Al-Bader, "Ohmic loss in metal-clad graded-index optical waveguides," *IEEE J. Quant. Electron.*, Vol. 22, (Jan. 1986)

23. W. Y. Lee and S. Y. Wang, "Guided-wave characteristics of optical graded-index planar waveguides with metal cladding: a simple analysis method," *J. of Lightwave Technology.*, Vol. 13, pp. 416–421, (March 1995)

24. S. J. Al-Bader and H. A. Jamid, "Comparison of absorption loss in metal-clad optical waveguides," *IEEE Trans. Microwave Theory and Tech.*, Vol. MTT-34, pp. 310–314, (1986)

THREE-DIMENSIONAL WAVEGUIDES WITH RECTANGULAR BOUNDARIES

5.1   INTRODUCTION

The geometries, index profiles, and fields of thin-film waveguides discussed in the last three chapters are independent of one of the transverse directions, the y direction in Figure 5.1(a). These waveguides are often referred to as *two-dimensional waveguides*. The properties of two-dimensional (2D) waveguides are relatively simple and expressions may be written in closed forms. This is particularly true for step-index waveguides. Therefore, 2D waveguides are "welcome" from the theoretical point of view. But fields guided by these waveguides are not confined in the y direction and, therefore, 2D waveguides are not suitable for many applications. For example, we may wish to reduce the component size or to pack as many components as possible in a given space so as to make the most economical use of the available "real estate." In densely packed systems, the cross talk, or unwanted interaction, between components has to be minimized. Thus, fields in the y direction must be confined somehow. To restrict fields in the transverse directions, geometric boundaries and/or index discontinuities are introduced purposely in the two transverse directions. This leads to *three-dimensional waveguides*. Examples of three-dimensional (3D) waveguides are shown in Figures 5.1(b)–5.1(e). To describe the 3D waveguides, we begin with the

Foundations for Guided-Wave Optics, by Chin-Lin Chen
Copyright © 2007 John Wiley & Sons, Inc.

2D waveguide shown in Figure 5.1(*a*). If we remove the film material in the outer regions by some means, while keeping the film layer in the central portion intact, then we have the *raised stripe* or *channel waveguides* [Fig. 5.1(*b*)]. *Ridge* or *rib waveguides* [Fig. 5.1(*c*)] are similar to the raised strip waveguides except that the film layer on the two sides is partially removed. If we place a dielectric strip on the top of the film layer, as shown schematically in Figure 5.1(*d*), we have the *strip-loaded waveguides*. By embedding a high-index bar in the substrate region, we have the *buried* or *embedded strip waveguides* [Fig. 5.1(*e*)]. Channel, ridge, strip-loaded, and buried strip waveguides are 3D waveguides with rectangular boundaries. Circular and elliptical fibers, discussed in Chapters 9 and 12, are 3D waveguides with curved boundaries.

In this chapter, we analyze modes guided by 3D waveguides with rectangular geometries. The chapter consists of five sections. Since fields of 3D waveguides are complicated and difficult to analyze, we begin with a qualitative description

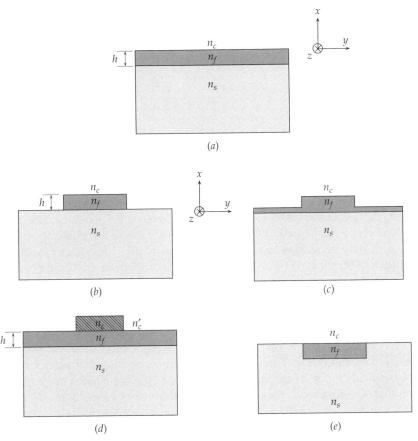

Figure 5.1   Five types of dielectric waveguides: (*a*) Thin-film waveguide, (*b*) raised strip or channel waveguide, (*c*) ridge or rib waveguide, (*d*) strip-loaded waveguide, and (*e*) buried or embedded strip waveguide.

of fields guided by *weakly guiding 3D waveguides*. This is done in Section 5.2. In Section 5.3, we present a self-consistent, order-of-magnitude, estimate of various field components. Through the order-of-magnitude analysis, we demonstrate that for weakly guiding rectangular waveguides, one of the transverse electric field components is much weaker than the other electric field components. When the weakest transverse field component is ignored as an approximation, the rectangular waveguide problem becomes manageable. Results of the order-of-magnitude analysis are the basis for approximating fields of the 3D waveguides. Two approximate methods are discussed in details. They are the Marcatili method (Section 5.4) and the effective index method (Section 5.5). In the last section, we compare the accuracy of the two methods. We also comment briefly on variations of the two methods.

Throughout this chapter, we consider time-harmonic ($e^{j\omega t}$) fields propagating in the z direction and denote the propagation constant as β. In other words, a factor $e^{-j\beta z}$ is present in all field components. We also assume that all regions are made of lossless, nonmagnetic dielectric materials.

5.2 FIELDS AND MODES GUIDED BY RECTANGULAR WAVEGUIDES

In 2D waveguides, one of the dimensions transverse to the direction of propagation is very large in comparison to the operating wavelength. This is the y direction in Figure 5.1(a). The waveguide width in this direction is treated as infinitely large. As a result, fields guided by 2D dielectric waveguides can be classified as *transverse electric* (TE) or *transverse magnetic* (TM) modes as discussed in Chapters 1, 2, and 3. For TE modes, the longitudinal electric field component, E_z, is zero, and all other field components can be expressed in terms of H_z. For TM modes, H_z vanishes and all other field components can be expressed in terms of E_z. In 3D dielectric waveguides, the waveguide width and height are comparable to the operating wavelength. Neither the width nor height can be treated as infinitely large. Neither E_z nor H_z vanishes, except for special cases. As a result, modes guided by 3D dielectric waveguides are neither TE nor TM modes except for the special cases. In general, they are *hybrid modes*. A complicated scheme is needed to designate the hybrid modes. Since all field components are present, the analysis for hybrid modes is very complicated. Intensive numerical computations are often required [1].

In many dielectric waveguide structures, the index difference is small. As a result, one of the transverse electric field components is much stronger than the other transverse electric field component. Goell has suggested a physically intuitive scheme to describe hybrid modes [2]. In Goell's scheme, a hybrid mode is labeled by the direction and distribution of the strong transverse electric field component. If the dominant electric field component is in the x (or y) direction and if the electric field distribution has $p-1$ nulls in the x direction and $q-1$ nulls in the y direction, then the hybrid mode is identified as E^x_{pq} (or E^y_{pq}) modes. The superscript denotes the direction of the *dominant transverse electric*

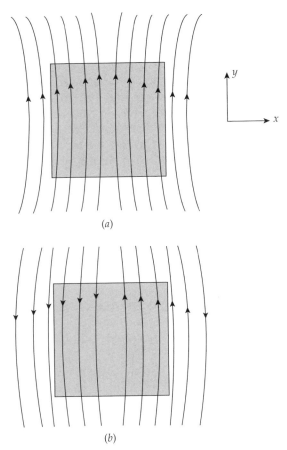

Figure 5.2 Two lowest order E^y modes guided by a weakly guiding square waveguide. Electric fields of (a) E_{11}^y and (b) E_{21}^y modes.

field component. The subscripts indicate the presence of $p - 1$ and $q - 1$ nulls in the x and y directions, respectively. In Figure 5.2, we sketch schematically the electric field lines of the two lowest order E^y modes guided by a square dielectric waveguide. Fields depicted in Figures 5.2(a) and 5.2(b) are E_{11}^y and E_{21}^y modes, respectively. An accurate plot of fields guided by a square dielectric waveguide with $n_1/n_2 = 1.5$ can be found in [1]. In the limit of an infinitesimal index difference, E_{11}^x and E_{11}^y modes become to the x-polarized and y-polarized uniform plane waves, respectively.

5.3 ORDERS OF MAGNITUDE OF FIELDS

In an unbounded, homogeneous medium, all propagating waves can be viewed as the superposition of uniform plane waves. Uniform plane waves are *transverse electromagnetic* (TEM) waves. The electric and magnetic field intensities are transverse to the direction of propagation and they are also mutually perpendicular. The propagation constant β of uniform plane waves is kn.

If the medium is inhomogeneous, very little can be said about fields in general. In many waveguides of practical interest, however, fields may be approximated by TEM waves. Well-guided modes guided by weakly guiding waveguides or fibers are good examples. For a well-guided mode, the propagation constant is very close to kn, and fields vary slowly in the transverse directions. If we use the transverse wave vectors κ_x and κ_y to describe field variations in x and y directions, then κ_x and κ_y are much smaller than kn. Take the step-index thin-film waveguides discussed in Chapter 2 as an example. The transverse wave vector κ_f of the film region has been given in (2.32) and it may be written as

$$\kappa_f^2 = k^2 n_f (n_f + n_s)(1 - b)\,\Delta \approx 2k^2 n_f^2 (1 - b)\,\Delta$$

where Δ is $(n_f - n_s)/n_f$. Obviously, $\kappa_f/(kn_f)$ is on the order of $\sqrt{(1-b)\Delta}$. For modes far from cutoff, b is close to 1. Then $\sqrt{(1-b)\Delta}$ is smaller than $\sqrt{\Delta}$. Clearly, $\kappa_f/(kn_f)$ of well-guided modes is of the first order of smallness, and we refer to it as an order of δ. Also $\kappa_x/(kn_f)$ and $\kappa_y/(kn_f)$ of well-guided modes of weakly guiding 3D waveguides are of the same order of magnitude. In the following discussions, we will make repeated use of the fact that $\kappa_x/(kn_f)$, $\kappa_y/(kn_f)$, and Δ are terms on the order of δ.

Based our discussions in Chapters 2 and 3, we expect that a Cartesian component of a propagating field may be approximated by the product of sinusoidal or exponential functions of x and y and an exponential factor $e^{j\beta z}$. Let the expression of a Cartesian field component be F. Then

$$\frac{\partial F}{\partial z} = -j\beta F$$

But $\partial F/\partial x$ and $\partial F/\partial y$ are not known precisely unless F is specified explicitly. However, the derivative of $\sin \kappa_x x$, $\cos \kappa_x x$, and $e^{\pm \kappa_x x}$ are simple, and they are $\kappa_x \cos \kappa_x x$, $-\kappa_x \sin \kappa_x x$, and $\pm \kappa_x e^{\pm \kappa_x x}$, respectively. Therefore $|\partial F/\partial x|$ and $|\partial F/\partial y|$ are on the order of $|\kappa_x F|$ and $|\kappa_y F|$, respectively. In the next subsection, we use this type of argument to estimate orders of magnitude of various field components. It should be stressed that we are interested in the order of magnitude of each term only. We are not, and cannot be, concerned with the detail variation of each term. When we compare magnitudes of electric and magnetic fields, we compare $n|E|/\eta_0$, in lieu of $|E|$, with $|H|$, since $n|E|/\eta_0$ and $|H|$ have the same unit.

5.3.1 The E^y Modes

Intuitively, we expect fields of E^y modes to behave like the y-polarized uniform plane waves in the limit of an infinitesimal index difference. Accordingly, we use the expression for y-polarized uniform plane waves as the starting point in our consideration for E^y modes. Then we proceed to show that the final results are consistent with the initial assumption. For y-polarized uniform plane waves, E_x, E_z, H_y, and H_z are zero. Thus, we expect E_x and H_y of E^y modes of weakly guiding 3D waveguides to be negligibly small. Therefore, we set H_y to zero and wish to demonstrate that

1. $|H_x|$ and $n|E_y|/\eta_0$ are of the same order of magnitude. We take them as the order of 1.
2. $|H_z|$ and $n|E_z|/\eta_0$ are smaller than $|H_x|$ and $n|E_y|/\eta_0$ by an order of magnitude. In other words, $|H_z|$ and $n|E_z|/\eta_0$ are of the first order of δ.
3. $|H_y|$ and $n|E_x|/\eta_0$ are smaller than $|H_x|$ and $n|E_y|/\eta_0$ by 2 orders of magnitude. That is, $|H_y|$ and $n|E_x|/\eta_0$ are of the order of δ^2.

If we are able to demonstrate the three items listed above, then we have begun by assuming that H_y is weak and proceeded to show that H_y is indeed small. This is the self-consistent analysis of fields. It is unlikely that we will be able to find exact expressions of the propagation constant and fields. We must content ourselves with approximate expressions accurate to the first order of δ. Thus, we ignore all terms on the order of, and smaller than, δ^2. Accordingly, we drop the smallest field components, H_y and nE_x/η_0, completely.

When we ignore H_y as an approximation, we obtain from (1.9)

$$H_z \approx -\frac{j}{\beta}\frac{\partial H_x}{\partial x} \tag{5.1}$$

Since $|(1/\beta)(\partial H_x/\partial x)|$ is on the same order as $|(\kappa_x/nk)H_x|$, $|H_z|$ is smaller than $|H_x|$ by a factor of κ_x/nk. In other words, $|H_z|$ is on the first order of δ.

We rearrange a component of Maxwell's equations, (1.8), as

$$\frac{n}{\eta_0}E_y = \frac{j}{kn}\frac{\partial H_z}{\partial x} - \frac{\beta}{kn}H_x \tag{5.2}$$

In view of (5.1), we conclude that

$$\left|\frac{1}{kn}\frac{\partial H_z}{\partial x}\right| \approx \left|\frac{1}{kn\beta}\frac{\partial^2 H_x}{\partial x^2}\right|$$

and that is on the order of $\left|(\kappa_x^2/k^2n^2)H_x\right|$. Then the second term on the right-hand side of (5.2) is the dominant term. By dropping the first term of the right-hand side of (5.2), we obtain

$$\frac{n}{\eta_0}E_y \approx -\frac{\beta}{kn}H_x \tag{5.3}$$

It is also convenient to write E_y in a slightly different form. For this purpose, we substitute (5.1) into (5.2) and rewrite the equation as

$$\frac{n}{\eta_0} E_y \approx \frac{1}{kn\beta} \left(\frac{\partial^2 H_x}{\partial x^2} - \beta^2 H_x \right)$$

Making use of the wave equation for H_x, we reduce the above equation further:

$$\frac{n}{\eta_0} E_y \approx -\frac{1}{kn\beta} \left(\frac{\partial^2 H_x}{\partial y^2} + k^2 n^2 H_x \right)$$

Clearly the first term on the right-hand side is smaller than the second term by 2 orders of magnitude. This is due to the factor $\kappa_y^2 / k^2 n^2$. By ignoring the first term, we obtain

$$\frac{n}{\eta_0} E_y \approx -\frac{kn}{\beta} H_x \tag{5.4}$$

From either (5.3) or (5.4), it is clear that $n|E_y|/\eta_0$ and $|H_x|$ are on the same order of magnitude. Similarly, $n|E_z|/\eta_0$ and $n|E_x|/\eta_0$ are smaller than H_x by factors on the order of κ_y/kn and $\kappa_x\kappa_y/(kn)^2$, respectively.

If we take $|H_x|$ as a term on the order of 1, then $|H_z|$ is on the order of δ. We have assumed that $|H_y|$ is very small and ignore it completely. Similarly, $n|E_y|/\eta_0$, $n|E_z|/\eta_0$, and $n|E_x|/\eta_0$ are on the orders of 1, δ, and δ^2, respectively.

For future reference, we list all field components of E^y modes below:

$$\frac{n}{\eta_0} E_x = O(\delta^2) \tag{5.5}$$

$$\frac{n}{\eta_0} E_y = -\frac{\beta}{kn} H_x + O(\delta^2) = -\frac{kn}{\beta} H_x + O(\delta^2) \tag{5.6}$$

$$\frac{n}{\eta_0} E_z = \frac{j}{kn} \frac{\partial H_x}{\partial y} + O(\delta^2) \tag{5.7}$$

$$H_y = O(\delta^2) \tag{5.8}$$

$$H_z = -\frac{j}{\beta} \frac{\partial H_x}{\partial x} + O(\delta^2) \tag{5.9}$$

In (5.5)–(5.9), all field components are expressed in terms of H_x and its derivatives. All equations are accurate to the first order of δ.

5.3.2 The E^x Modes

Similar considerations can be applied to E^x modes. By setting H_x to zero as an approximation and taking $|H_y|$ as a term on the order of 1, we can demonstrate that $|H_z|$ is on the order of δ. We also show that $n|E_x|/\eta_0$, $n|E_z|/\eta_0$, and $n|E_y|/\eta_0$ are on the order of 1, δ, and δ^2, respectively. In addition, all field components may be expressed in terms of H_y and expressions are accurate to the first order of δ:

$$\frac{n}{\eta_0} E_x = \frac{\beta}{kn} H_y + O(\delta^2) = \frac{kn}{\beta} H_y + O(\delta^2) \tag{5.10}$$

$$\frac{n}{\eta_0} E_y = O(\delta^2) \tag{5.11}$$

$$\frac{n}{\eta_0} E_z = -\frac{j}{kn} \frac{\partial H_y}{\partial x} + O(\delta^2) \tag{5.12}$$

$$H_x = O(\delta^2) \tag{5.13}$$

$$H_z = -\frac{j}{\beta} \frac{\partial H_y}{\partial y} + O(\delta^2) \tag{5.14}$$

This completes the order-of-magnitude estimate for E^x and E^y modes. These equations are the basis of the Marcatili and the effective index methods for analyzing modes of rectangular dielectric waveguides.

5.4   MARCATILI METHOD

Now we consider a weakly guiding rectangular waveguide with a core of index n_1 surrounded by regions with smaller indices n_j, where j is 2, 3, 4, or 5. The waveguide cross section is shown in Figure 5.3(a). By *weakly guiding*, we mean that the index differences $n_1 - n_j$ is much smaller than n_1 and n_j. In the following discussions, we ignore fields in the corner regions completely. Therefore, there is no need to specify indices in the corner regions, except that these indices are smaller than n_1. The cross section covers many 3D waveguides of interest. For example, if indices of the two upper corners are the same as n_2, if indices of the two lower corners equal to n_3, and if $n_3 = n_4 = n_5$, then the rectangular waveguide shown in Figure 5.3(a) becomes an embedded strip waveguide shown in Figure 5.1(e). A channel waveguide [Fig. 5.1(b)] corresponds to a structure where regions 2, 3, 5, and two upper corners have the same index n_2 and the two lower corners have an index n_4.

In the Marcatili method [3], the 3D waveguide shown in Figure 5.3(a) is replaced by two 2D waveguides. Based on our knowledge gained from preceding chapters, we expect fields to be strong in region 1 and weak in surrounding regions. Furthermore, fields in region 1 vary in an oscillatory manner in x and y directions, while fields in outer regions decay exponentially from boundaries. Fields in the four corners are very weak and are ignored completely in our consideration. Thus, we ignore the boundary conditions along the dash lines of Figure 5.3(a) all together.

5.4.1   The E^y Modes

To study E^y modes, we begin by writing an expression for H_x in each region in terms of trigonometric and exponential functions with unknown amplitude constants,

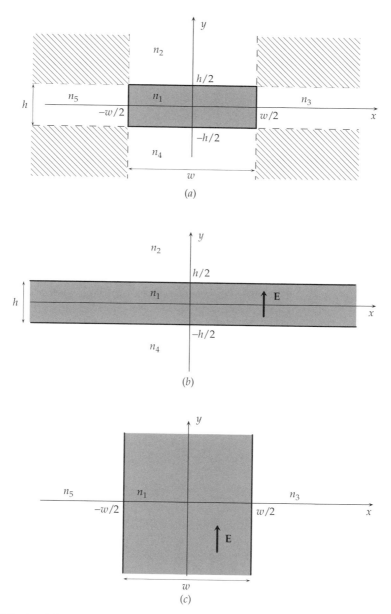

Figure 5.3 Model used to analyze E^y modes of a (a) rectangular waveguide, (b) waveguide H, and (c) waveguide W.

phases, and transverse wave vectors. Then we apply the boundary conditions along the solid lines shown in Figure 5.3(a) to determine the transverse wave vectors. Finally, we use the transverse wave vectors so obtained to evaluate the propagation constant of E^y modes.

5.4.1.1 Expressions for H_x

We write H_x in the five regions as

$$H_{x1} = C_1 \cos(\kappa_{x1} x + \phi_{x1}) \cos(\kappa_{y1} y + \phi_{y1}) e^{-j\beta z} \qquad \text{(region 1)}$$

$$H_{x2} = C_2 \cos(\kappa_{x2} x + \phi_{x2}) e^{-j\kappa_{y2} y} e^{-j\beta z} \qquad \text{(region 2)}$$

$$H_{x3} = C_3 e^{-j\kappa_{x3} x} \cos(\kappa_{y3} y + \phi_{y3}) e^{-j\beta z} \qquad \text{(region 3)} \qquad (5.15)$$

$$H_{x4} = C_4 \cos(\kappa_{x4} x + \phi_{x4}) e^{+j\kappa_{y4} y} e^{-j\beta z} \qquad \text{(region 4)}$$

$$H_{x5} = C_5 e^{+j\kappa_{x5} x} \cos(\kappa_{y5} y + \phi_{y5}) e^{-j\beta z} \qquad \text{(region 5)}$$

where C_j, ϕ_{xj}, and ϕ_{yj} are constants to be determined, κ_{xj} and κ_{yj} are components of the transverse wave vectors. For each region, κ_{xj}, κ_{yj}, and β are related as required by Maxwell's equations:

$$\kappa_{xj}^2 + \kappa_{yj}^2 + \beta^2 = k^2 n_j^2 \qquad j = 1, 2, 3, 4, \text{ or } 5 \qquad (5.16)$$

There is no additional constraint on the transverse wave vectors. In fact κ_{y2}, κ_{x3}, κ_{y4}, and κ_{x5} are imaginary. Now we consider the boundary conditions. A quick glance at Figure 5.3(a) reveals that H_x is tangent to the horizontal boundaries and B_x is normal to the vertical boundaries. Recall that B_x is simply $\mu_0 H_x$. Thus H_x must be continuous at all points along the solid-line boundaries shown in Figure 5.3(a). The continuation condition can be met if and only if $\kappa_{x1} = \kappa_{x2} = \kappa_{x4}$, $\kappa_{y1} = \kappa_{y3} = \kappa_{y5}$, $\phi_{x1} = \phi_{x2} = \phi_{x4}$, and $\phi_{y1} = \phi_{y3} = \phi_{y5}$. Therefore, we write κ_x in lieu of κ_{x1}, κ_{x2} and κ_{x4}, κ_y in the place of κ_{y1}, κ_{y3}, and κ_{y5}. Similarly, we write ϕ_x and ϕ_y in lieu of ϕ_{xj} and ϕ_{yj}. When expressed in terms of κ_x, κ_y, ϕ_x, and ϕ_y, Eq. (5.15) is greatly simplified:

$$H_{x1} = C_1 \cos(\kappa_x x + \phi_x) \cos(\kappa_y y + \phi_y) e^{-j\beta z} \qquad \text{(region 1)}$$

$$H_{x2} = C_2 \cos(\kappa_x x + \phi_x) e^{-j\kappa_{y2} y} e^{-j\beta z} \qquad \text{(region 2)}$$

$$H_{x3} = C_3 e^{-j\kappa_{x3} x} \cos(\kappa_y y + \phi_y) e^{-j\beta z} \qquad \text{(region 3)} \qquad (5.17)$$

$$H_{x4} = C_4 \cos(\kappa_x x + \phi_x) e^{+j\kappa_{y4} y} e^{-j\beta z} \qquad \text{(region 4)}$$

$$H_{x5} = C_5 e^{+j\kappa_{x5} x} \cos(\kappa_y y + \phi_y) e^{-j\beta z} \qquad \text{(region 5)}$$

By substituting H_x given in (5.17) into (5.5)–(5.9), we obtain explicit expressions for various field components. As indicated in (5.5)–(5.9), the five equations are accurate to the first order of δ. Consistent with this order of accuracy, we ignore all terms on the order of δ^2 and smaller in the boundary condition considerations.

5.4.1.2 Boundary Conditions along Horizontal Boundaries, $y = \pm h/2$, $|x| < w/2$

Along the horizontal boundaries, the tangential field components are E_x, E_z, H_x, and H_z. We ignore E_x since it is on the order of δ^2, as indicated in (5.5).

1. It is clear from (5.7) that E_z is continuous at the horizontal boundary if $(1/n_j^2)(\partial H_x/\partial y)$ is continuous there.

2. Being tangential to the horizontal lines, H_x must be continuous everywhere along the horizontal lines. If H_x is continuous everywhere along the horizontal lines, the tangential derivative $\partial H_x/\partial x$ and therefore H_z must also be continuous on the horizontal lines, as indicated in (5.9). In other words, if H_x is continuous at the horizontal lines, so is H_z.

On the horizontal boundaries, the normal components are B_y and D_y. Note that $D_y = \varepsilon_0 n^2 E_y$. We conclude from (5.6) that D_y is continuous at the horizontal boundaries if H_x is continuous there. It would be pointless to consider B_y since it is on the order of δ^2.

In short, at the horizontal boundaries, $y = \pm h/2$, $|x| < w/2$, all boundary conditions are met when H_x and $(1/n_j^2)(\partial H_x/\partial y)$ are continuous at these boundaries. These equations are accurate to the first order of δ.

5.4.1.3 Boundary Conditions along Vertical Boundaries, $x = \pm w/2$, $|y| < h/2$

On the vertical boundaries, the tangential components are the y and z components and the normal component is the x component. There is no need to consider H_y and D_x since they are on the order of δ^2. From (5.6), we note that E_y is continuous if H_x is continuous. Similarly, from (5.9), we observe that requiring the continuation of H_z amounts to requiring the continuation of $\partial H_x/\partial x$. The only field component left to be considered is E_z. Consider the boundary conditions at $x = w/2$. The vertical boundary is between regions 1 and 3. To simplify equations, we write $E_{z1} \equiv E_z(w^-/2, y, z)$ and $E_{z3} \equiv E_z(w^+/2, y, z)$. And H_{x1} and H_{x3} are defined in the same manner. Then we obtain from (5.7)

$$
\begin{aligned}
E_{z1} - E_{z3} &= \frac{j\eta_0}{k}\left(\frac{1}{n_1^2}\frac{\partial H_{x1}}{\partial y} - \frac{1}{n_3^2}\frac{\partial H_{x3}}{\partial y}\right) + O(\delta^2) \\
&= \frac{j\eta_0}{k}\frac{1}{n_1^2}\frac{\partial}{\partial y}(H_{x1} - H_{x3}) - \frac{j\eta_0}{n_3}\left[\left(\frac{n_1^2 - n_3^2}{n_1^2}\right)\left(\frac{1}{kn_3}\frac{\partial H_{x3}}{\partial y}\right)\right] \\
&\quad + O(\delta^2)
\end{aligned}
\tag{5.18}
$$

Since $(1/kn_3)(\partial H_{x3}/\partial y)$ and $n_1^2 - n_3^2/n_1^2$ are of the first order of δ, the product of the two terms, that is, the term in the square bracket, is of the order of δ^2 or smaller. So far, we have kept terms of orders of 1 and δ and have ignored all terms on the order of δ^2 and smaller. It would be meaningless to include terms of the order of δ^2 and smaller in considering the boundary condition at the vertical boundaries. Thus we ignore the bracketed term and we obtain

$$
E_{z1} - E_{z3} = \frac{j\eta_0}{k}\frac{1}{n_1^2}\frac{\partial}{\partial y}(H_{x1} - H_{x3}) + O(\delta^2)
\tag{5.19}
$$

In other words, E_z is continuous at the vertical boundary if H_x is continuous there. Similar argument applies to the left boundary at $x = -w/2$. In summary, all boundary conditions are met when H_x and $\partial H_x/\partial x$ are continuous at the vertical lines of $x = \pm w/2$, $|y| < h/2$.

5.4.1.4 Transverse Wave Vector κ_x

Now we apply the boundary conditions discussed above to determine the transverse wave vectors. From the continuation of H_x and $(1/n_j^2)(\partial H_x/\partial y)$ at $y = h/2$, we obtain

$$C_1 \cos\left(\tfrac{1}{2}\kappa_y h + \phi_y\right) = C_2 e^{-j\kappa_{y2}h/2} \tag{5.20}$$

$$-\frac{\kappa_y}{n_1^2} C_1 \sin\left(\frac{1}{2}\kappa_y h + \phi_y\right) = -\frac{j\kappa_{y2}}{n_2^2} C_2 e^{-j\kappa_{y2}h/2} \tag{5.21}$$

By combining the two equations, we obtain

$$\tan\left(\frac{1}{2}\kappa_y h + \phi_y\right) = \frac{j\kappa_{y2} n_1^2}{\kappa_y n_2^2} \tag{5.22}$$

We also deduce from (5.16) that

$$j\kappa_{y2} = \sqrt{k^2(n_1^2 - n_2^2) - \kappa_y^2}. \tag{5.23}$$

Thus we can express the above equation in terms of κ_y:

$$\tan\left(\frac{1}{2}\kappa_y h + \phi_y\right) = \frac{n_1^2 \sqrt{k^2(n_1^2 - n_2^2) - \kappa_y^2}}{n_2^2 \kappa_y} \tag{5.24}$$

that may be rewritten as

$$\tfrac{1}{2}\kappa_y h + \phi_y = q'\pi + \tan^{-1}\left[\frac{n_1^2 \sqrt{k^2(n_1^2 - n_2^2) - \kappa_y^2}}{n_2^2 \kappa_y}\right] \tag{5.25}$$

where q' is an integer.

Similarly, we deduce from the boundary conditions at $y = -h/2$ that

$$\tfrac{1}{2}\kappa_y h - \phi_y = q''\pi + \tan^{-1}\left[\frac{n_1^2 \sqrt{k^2(n_1^2 - n_4^2) - \kappa_y^2}}{n_4^2 \kappa_y}\right] \tag{5.26}$$

where q'' is also an integer.

Eliminating ϕ_y from (5.25) and (5.26), and identifying q as $q' + q''$, we obtain

$$\kappa_y h = q\pi + \tan^{-1}\left[\frac{n_1^2\sqrt{k^2(n_1^2 - n_2^2) - \kappa_y^2}}{n_2^2\kappa_y}\right] + \tan^{-1}\left[\frac{n_1^2\sqrt{k^2(n_1^2 - n_4^2) - \kappa_y^2}}{n_4^2\kappa_y}\right]$$

(5.27)

This equation is equivalent to the dispersion relation (2.26) for TM modes guided by a thin-film waveguide with a film layer of index n_1 and thickness h bounded by regions with indices n_2 and n_4. Such a 2D waveguide is shown in Figure 5.3(b). In the following discussions, we refer to the horizontal 2D waveguide as *waveguide H*. We can be understood (5.27) in the following manner. For E^y modes, the dominant electric field component is E_y and it is normal to the boundaries of waveguide H. Therefore, we would expect fields to be similar to that of TM modes guided by a thin-film waveguide. Thus the equation for κ_y must be the dispersion relation of TM modes.

5.4.1.5 *Transverse Wave Vector* κ_x

Next, we consider κ_x. If the original 3D waveguide were infinitely wide in the x direction, the analysis leading to (5.27) would be exact and κ_x would be zero. But the width w of the 3D waveguide shown in Figure 5.3(a) is finite. We have to take the boundary conditions at $x = \pm w/2$ into account. From the boundary conditions on H_x and $\partial H_x/\partial x$ at the vertical boundaries, we obtain an equation for κ_x:

$$\kappa_x w = p\pi + \tan^{-1}\left[\frac{\sqrt{k^2(n_1^2 - n_3^2) - \kappa_x^2}}{\kappa_x}\right] + \tan^{-1}\left[\frac{\sqrt{k^2(n_1^2 - n_5^2) - \kappa_x^2}}{\kappa_x}\right]$$

(5.28)

Note that the dominate electric field component of E^y modes is in parallel with the vertical boundaries. Thus, modes being considered must be TE modes guided by a vertical thin-film waveguide. Simple manipulation reveals that (5.28) is indeed equivalent to the dispersion relation (2.22) of TE modes guided by a thin-film waveguide with a film index n_1 and thickness w bounded by regions with indices n_3 and n_5 [Fig. 5.3(c)]. We identify the 2D waveguide shown in Figure 5.3(c) as *waveguide W*.

5.4.1.6 *Approximate Dispersion Relation*

Having determined κ_y from (5.27) for a given set of n_1, n_2, n_4, and h, and κ_x from (5.28) for a given set of n_1, n_3, n_5, and w, we use (5.16) to evaluate the propagation constant. More specifically,

$$\beta^2 = k^2 n_1^2 - \kappa_x^2 - \kappa_y^2$$

(5.29)

This completes the study of the dispersion of E^y modes.

5.4.2 The E^x Modes

To consider E^x modes, we express H_y in the form of (5.17). On the horizontal boundaries, the boundary conditions are the continuation of H_y and $\partial H_y/\partial y$ at $y = \pm h/2$. The equation for κ_y is

$$\kappa_y h = q\pi + \tan^{-1}\left[\frac{\sqrt{k^2(n_1^2 - n_2^2) - \kappa_y^2}}{\kappa_y}\right] + \tan^{-1}\left[\frac{\sqrt{k^2(n_1^2 - n_4^2) - \kappa_y^2}}{\kappa_y}\right] \tag{5.30}$$

It corresponds to the dispersion relation (2.22) for TE modes guided by waveguide H [Fig. 5.4(b)].

On the vertical boundaries, the boundary conditions are the continuation of H_y and $(1/n_j^2)(\partial H_y/\partial x)$ at $x = \pm w/2$. The resulting equation for κ_x is

$$\kappa_x w = p\pi + \tan^{-1}\left[\frac{n_1^2\sqrt{k^2(n_1^2 - n_3^2) - \kappa_x^2}}{n_3^2\kappa_x}\right] + \tan^{-1}\left[\frac{n_1^2\sqrt{k^2(n_1^2 - n_5^2) - \kappa_x^2}}{n_5^2\kappa_x}\right] \tag{5.31}$$

It corresponds to the dispersion relation (2.26) for TM modes guided by *waveguide W* [Fig. 5.4(c)]. We can understand (5.30) and (5.31) in the following manner. The dominant electric field of E^x modes is in parallel with the horizontal boundaries. Thus, we use the dispersion equation of TE modes guided by waveguide H to determine κ_y. The dominant electric field component of E^x modes is perpendicular to the vertical boundaries of waveguide W. Therefore, we use the dispersion for TM modes guided by the 2D waveguide to evaluate κ_x. With κ_x and κ_y known, the propagation constant can be determined from (5.29).

5.4.3 Discussions

In the Marcatili method, a rectangular waveguide is modeled by two 2D waveguides: waveguides H and W. Waveguide H is obtained by extending the waveguide width w to infinity. Waveguide W is obtained by extending the waveguide height h to infinity. For E^y modes, we determine κ_y from the dispersion relation for TM mode of waveguide H [Fig. 5.3(b)], and κ_x from that for TE modes of waveguide W [Fig. 5.3(c)]. Then we determine β from (5.29). With κ_y known, we use (5.25) to determine the phase angle ϕ_y. Amplitude constants C_1, C_2, and C_4 are related through the continuation conditions on H_x. Phase angle ϕ_x and constants C_1, C_3, and C_5 are obtained in the same manner. An explicit expression for H_x, except for one of the amplitude constants, can be written down immediately. Then, we use (5.5)–(5.9) to determine other field components. This completes the analysis for fields of E^y modes.

In the case of E^x modes, we determine κ_y from TE modes guided by waveguide H [Fig. 5.4(b)] and κ_x from TM modes guided by waveguide W [Fig. 5.4(c)]. Having determined κ_x and κ_y, we use (5.29) to evaluate β. Then, we determine expressions for field components.

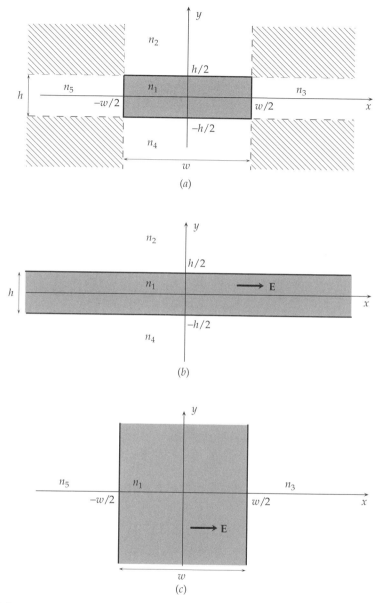

Figure 5.4   Model used to analyze E^x modes of a (a) rectangular waveguide, (b) waveguide H, and (c) waveguide W.

5.4.4   Generalized Guide Index

It is possible to express the propagation constant in terms of the generalized guide index as well. In the following discussions, we express the generalized guide index

b of E^y modes of a rectangular waveguide in terms of the generalized parameters of the two 2D waveguides. For convenience, we assume that $n_5 > n_4 > n_2$ and $n_5 > n_3$. We denote the generalized parameters of waveguide H with a subscript H:

$$V_{\mathrm{H}} = kh\sqrt{n_1^2 - n_4^2} \tag{5.32}$$

$$a_{\mathrm{H}} = \frac{n_4^2 - n_2^2}{n_1^2 - n_4^2} \tag{5.33}$$

$$b_{\mathrm{H}} = \frac{\beta_H^2 - k^2 n_4^2}{k^2(n_1^2 - n_4^2)} \tag{5.34}$$

We also define $c_{\mathrm{H}} = n_4^2/n_1^2$ and $d_{\mathrm{H}} = c_{\mathrm{H}} - a_{\mathrm{H}}(1 - c_{\mathrm{H}})$. With V_{H}, a_{H}, c_{H}, and d_{H} known, we use Figure 2.5 for TM modes to estimate b_{H}. From b_{H}, we evaluate β_{H} and κ_y.

Similarly, we label all terms related to waveguide W with a subscript W:

$$V_{\mathrm{W}} = kw\sqrt{n_1^2 - n_5^2} \tag{5.35}$$

$$a_{\mathrm{W}} = \frac{n_5^2 - n_3^2}{n_1^2 - n_5^2} \tag{5.36}$$

$$b_{\mathrm{W}} = \frac{\beta_W^2 - k^2 n_5^2}{k^2(n_1^2 - n_5^2)} \tag{5.37}$$

Based on V_{W} and a_{W}, we use Figure 2.4 for TE modes to estimate b_{W}. Then we determine β_{W} and κ_x.

With κ_x and κ_y known, we obtain from (5.29), (5.34), and (5.37)

$$\beta^2 = k^2(n_4^2 + n_5^2 - n_1^2) + b_{\mathrm{W}}k^2(n_1^2 - n_5^2) + b_{\mathrm{H}}k^2(n_1^2 - n_4^2) \tag{5.38}$$

The generalized guide index for the 3D waveguide, in terms of n_1 and n_5, is

$$b_M = \frac{\beta^2 - k^2 n_5^2}{k^2(n_1^2 - n_5^2)} \tag{5.39}$$

Upon substituting (5.38) into (5.39), we obtain the normalized guide index for E^y modes guided by a rectangular waveguide:

$$b_M = b_{\mathrm{W}} + \frac{n_1^2 - n_4^2}{n_1^2 - n_5^2}(b_{\mathrm{H}} - 1) \tag{5.40}$$

In (5.39) and (5.40), we add a subscript M to stress the fact that we have applied the Marcatili method to analyze the waveguide problem.

5.5 EFFECTIVE INDEX METHOD

Like the Marcatili method discussed in the last section, the *effective index method* is also an approximate method for analyzing rectangular waveguides. In the Marcatili method, a 3D waveguide [Fig. 5.3(*a*)] is replaced by two 2D waveguides: waveguides H and W depicted in Figures 5.3(*b*) and 5.3(*c*). The two 2D waveguides are *mutually independent* in that the waveguide parameters of the two 2D waveguides come directly from the original 3D waveguide. Furthermore, the two 2D waveguides are solved independently. In the effective index method, advanced by Knox and Toulios [4], a 3D waveguide is replaced by two *related* 2D waveguides. The first 2D waveguide is exactly the same as waveguide H or waveguide W introduced in the Marcatili method. We identify the first waveguide as *waveguide I or I'*. The effective index n_{eff} of waveguide I or I' is then used to define the second 2D waveguide that is labeled as *waveguide II or II'*. The propagation constant of waveguide II or II' is an approximation for that of the original 3D waveguide. This is the essence of the effective index method [4]. Clearly, it is necessary to solve waveguide I or I' before waveguide II or II' is attempted. In short, waveguide II or II' is dependent on waveguide I or I'.

To provide a theoretical basis for the effective index method, and to understand errors incurred in the approximation, we adapt an approach reported by Kumar et al. [5, 6]. In lieu of the original 3D waveguide, we consider a *pseudowaveguide* that can be resolved into aforementioned waveguides I and II or I' and II'. The pseudowaveguide is chosen such that waveguides I and II, or I' and II', can be easily identified and analyzed. The dispersion of waveguide II, or II', is used as an approximation for β of the original 3D waveguide.

Consider E^y modes guided by a 3D waveguide shown in Figure 5.5(*a*). As discussed in Section 5.3, all field components of E^y modes can be expressed in terms of H_x. We write H_x as $h_x(x, y) e^{-j\beta z}$ where β is the propagation constant and $h_x(x, y)$ satisfies the wave equation:

$$\left[\frac{\partial^2}{\partial x^2} + \frac{\partial^2}{\partial y^2} + k^2 n^2(x, y) - \beta^2 \right] h_x(x, y) = 0 \qquad (5.41)$$

where $n(x, y)$ is the index profile of the original 3D waveguide. For region j, the index is

$$n(x, y) = n_j \qquad j = 1, 2, \ldots 5 \qquad (5.42)$$

Instead of considering the original 3D waveguide itself, we examine a pseudowaveguide with a modified index profile $n_{ps}(x, y)$. The modified index profile is chosen so that $n_{ps}^2(x,y)$ can be expressed as the sum of two parts, one part being independent of x and the other independent of y:

$$n_{ps}^2(x, y) = n_x^2(x) + n_y^2(y) \qquad (5.43)$$

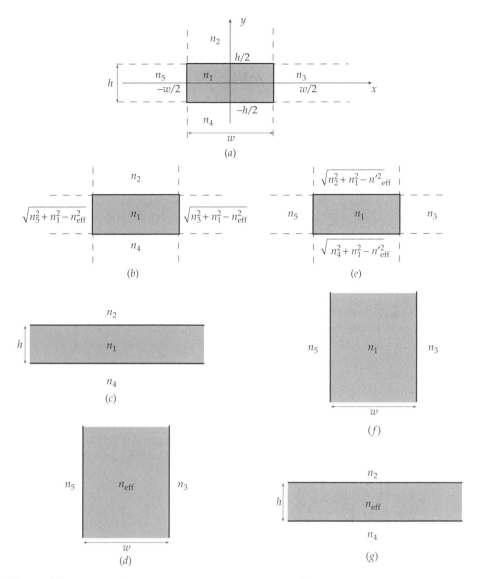

Figure 5.5 Model of pseudowaveguides used in the effective index method (a) Rectangular waveguide, (b) pseudowaveguide, (c) waveguide I, (d) waveguide II, (e) alternate pseudowaveguide, (f) waveguide I′, and (g) waveguide II′.

Then we determine $h_x(x, y)$ by the usual method of separation of variables. We suppose that $h_x(x, y)$ can be expressed as the product of two functions, $X(x)$ and $Y(y)$. In terms of $X(x)$ and $Y(y)$, (5.41) becomes

$$\frac{1}{X(x)}\frac{d^2 X(x)}{dx^2} + \frac{1}{Y(y)}\frac{d^2 Y(y)}{dy^2} + \left[k^2 n_x^2(x) + k^2 n_y^2(y) - \beta^2\right] = 0 \qquad (5.44)$$

By rearranging the terms, the above equation becomes

$$\frac{1}{Y(y)}\frac{d^2Y(y)}{dy^2} + k^2 n_y^2(y) = -\frac{1}{X(x)}\frac{d^2X(x)}{dx^2} - \left[k^2 n_x^2(x) - \beta^2\right] \tag{5.45}$$

Terms on the left-hand side of (5.45) are independent of x, and terms on the right-hand side are independent of y. Yet, the two sides are equal for all values of x and y. This is possible if and only if each side is a constant. Let the constant be $(kn_{\text{eff}})^2$. Then we obtain two equations from (5.45):

$$\frac{1}{Y(y)}\frac{d^2Y(y)}{dy^2} + k^2 \left[n_y^2(y) - n_{\text{eff}}^2\right] = 0 \tag{5.46}$$

$$\frac{1}{X(x)}\frac{d^2X(x)}{dx^2} + k^2 \left[n_x^2(x) + n_{\text{eff}}^2\right] - \beta^2 = 0 \tag{5.47}$$

When the two differential equations are solved subject to appropriate boundary conditions, the pseudowaveguide problem is solved completely. As noted earlier, we use β of waveguide II as an approximation for β of the original 3D waveguide, $H_x(x, y, z)$ of the original 3D waveguide is $X(x)Y(y)e^{-j\beta z}$.

An alternate and equally valid arrangement of terms is to write

$$\frac{1}{X(x)}\frac{d^2X(x)}{dx^2} + k^2 n_x^2(x) = -\frac{1}{Y(y)}\frac{d^2Y(y)}{dy^2} - \left[k^2 n_y^2(y) - \beta^2\right] \tag{5.48}$$

In lieu of (5.44) and (5.45), we obtain two alternate differential equations:

$$\frac{1}{X(x)}\frac{d^2X(x)}{dx^2} + k^2 \left[n_x^2(x) - n_{\text{eff}}'^2\right] = 0 \tag{5.49}$$

$$\frac{1}{Y(y)}\frac{d^2Y(y)}{dy^2} + k^2 \left[n_y^2(y) + n_{\text{eff}}'^2\right] - \beta^2 = 0 \tag{5.50}$$

that can be used to determine a new and different set of $X(x)$ and $Y(y)$.

It should be stressed that the constant $(kn_{\text{eff}})^2$ introduced in (5.46) and (5.47) is different from $(kn_{\text{eff}}')^2$ used in (5.49) and (5.50), n_{eff} is the effective refractive index of waveguide I and n_{eff}' is that of waveguide I'. As there are two possible arrangements of terms, there are two possible pseudowaveguides, and two sets of β $X(x)$ and $Y(y)$. Either set may be used as the approximations for the propagation constant and fields of the original 3D waveguide. Depending on the index profiles and geometries of the original 3D waveguides, one pseudowaveguide may be more accurate than the other. This will be discussed later.

5.5.1   A Pseudowaveguide

For the 3D waveguide shown in Figure 5.5(a), one possible choice of $n_{ps}^2(x, y)$ is

$$
n_{ps}^2(x, y) = \begin{cases} n_1^2 & \text{region 1} \\ n_2^2 & \text{region 2} \\ n_3^2 + n_1^2 - n_{eff}^2 & \text{region 3} \\ n_4^2 & \text{region 4} \\ n_5^2 + n_1^2 - n_{eff}^2 & \text{region 5} \end{cases} \tag{5.51}
$$

where n_{eff} is yet unspecified. The index profile $n(x, y)$ of the original 3D waveguide and $n_{ps}(x, y)$ of the pseudowaveguide are depicted in Figures 5.5(a) and 5.5(b) for comparison. The indices in regions 1,2 and 4 of the two waveguides are the same. But the indices in regions 3 and 5 of the pseudowaveguide differ from that of regions 3 and 5 of the original 3D waveguide. However, the differences are rather small since n_{eff} differs only slightly from n_1.

Given in (5.51), $n_{ps}^2(x, y)$ is the sum of two parts:

$$
n_y^2(y) = \begin{cases} n_2^2 & y > h/2 \\ n_1^2 & -h/2 \leqslant y \leqslant h/2 \\ n_4^2 & y < -h/2 \end{cases} \tag{5.52}
$$

$$
n_x^2(x) = \begin{cases} n_3^2 - n_{eff}^2 & x > w/2 \\ 0 & -w/2 \leqslant x \leqslant w/2 \\ n_5^2 - n_{eff}^2 & x < -w/2 \end{cases} \tag{5.53}
$$

The differential equation (5.46) with $n_y(y)$ given by (5.52) describes a 2D waveguide with a film thickness h and indices n_2, n_1, and n_4 [Fig. 5.5(c)]. This is the waveguide I mentioned earlier. Waveguide I is realized by making the waveguide infinitely wide. It is the same waveguide H introduced in the Marcatili method and shown in Figure 5.3(b). When (5.46) is solved subject to the continuation of Y and $(1/n_y^2)(dY/dy)$ at $y = \pm h/2$, we obtain the dispersion and fields of TM modes guided by waveguide I. Let the propagation constant be kn_{eff}, then the characteristic equation is

$$
kh\sqrt{n_1^2 - n_{eff}^2} = q\pi + \tan^{-1}\left[\frac{n_1^2}{n_2^2} \frac{\sqrt{n_{eff}^2 - n_2^2}}{\sqrt{n_1^2 - n_{eff}^2}} \right] + \tan^{-1}\left[\frac{n_1^2}{n_4^2} \frac{\sqrt{n_{eff}^2 - n_4^2}}{\sqrt{n_1^2 - n_{eff}^2}} \right] \tag{5.54}
$$

Equation (5.54) is equivalent to the dispersion relation (2.26) for TM modes. It also corresponds to (5.27). An expression for $Y(y)$ can be written down readily except for an amplitude constant.

Equations (5.47) and (5.53) pertain to a 2D waveguide having an index profile $\sqrt{n_x^2(x) + n_{eff}^2}$ where $n_x^2(x)$ is given in (5.53). The 2D waveguide is waveguide

II. The indices in the three regions of waveguide II are, respectively, n_3, n_{eff}, and n_5 as shown in Figure 5.5(d). By solving (5.47), subject to the continuation conditions of X and dX/dx at $x = \pm w/2$, we obtain the propagation constant and fields of TE modes guided by waveguide II. Let the effective index of TE modes guided by waveguide II be N, then the dispersion relation in terms of N is

$$kw\sqrt{n_{\text{eff}}^2 - N} = p\pi + \tan^{-1}\frac{\sqrt{N^2 - n_3^2}}{\sqrt{n_{\text{eff}}^2 - N^2}} + \tan^{-1}\frac{\sqrt{N^2 - n_5^2}}{\sqrt{n_{\text{eff}}^2 - N^2}} \tag{5.55}$$

The propagation constant of the pseudowaveguide is $\beta = kN$. We also determine $X(x)$ except for an amplitude constant. Finally, we assemble all terms and write $H_x(x, y, z) = X(x)Y(y)e^{-j\beta z}$. This completes the analysis of the pseudowaveguide. As noted earlier, we use β and $H_x(x, y, z)$ so obtained as approximations for the corresponding terms of the original 3D waveguide.

5.5.2 Alternate Pseudowaveguide

An alternate choice is to select $n_{\text{ps}}'^2(x, y)$ to accompany (5.49) and (5.50):

$$n_{\text{ps}}'^2(x, y) = \begin{cases} n_1^2 & \text{region 1} \\ n_2^2 + n_1^2 - n_{\text{eff}}'^2 & \text{region 2} \\ n_3^2 & \text{region 3} \\ n_4^2 + n_1^2 - n_{\text{eff}}'^2 & \text{region 4} \\ n_5^2 & \text{region 5} \end{cases} \tag{5.56}$$

$n_{\text{ps}}'^2(x, y)$ is the sum of two parts:

$$n_x^2(x) = \begin{cases} n_3^2 & x > w/2 \\ n_1^2 & -w/2 \leqslant x \leqslant w/2 \\ n_5^2 & x < -w/2 \end{cases} \tag{5.57}$$

$$n_y^2(y) = \begin{cases} n_2^2 - n_{\text{eff}}'^2 & y > h/2 \\ 0 & -h/2 \leqslant y \leqslant h/2 \\ n_4^2 - n_{\text{eff}}'^2 & y < -h/2 \end{cases} \tag{5.58}$$

The index profile of the alternate pseudowaveguide is shown in Figure 5.5(e). Waveguide I', defined by (5.49) and (5.57), has indices n_3, n_1, and n_5 in the three regions depicted in Figure 5.5(f). The film width is w. The boundary conditions are the continuation of X and dX/dx at $x = \pm w/2$. From these boundary conditions, we determine the dispersion relation of TE modes. In particular, the effective index n'_{eff} is determined from

$$kw\sqrt{n_1^2 - n'^2_{\text{eff}}} = p\pi + \tan^{-1}\frac{\sqrt{n'^2_{\text{eff}} - n_3^2}}{\sqrt{n_1^2 - n'^2_{\text{eff}}}} + \tan^{-1}\frac{\sqrt{n'^2_{\text{eff}} - n_5^2}}{\sqrt{n_1^2 - n'^2_{\text{eff}}}} \qquad (5.59)$$

The new n'_{eff} is then used in conjunction with (5.50) and (5.58) to specify waveguide II$'$, which is shown in Figure 5.5(g). The indices in the three regions of waveguide II$'$ are n_2, n'_{eff}, and n_4. The boundary conditions are the continuation of Y and $(1/n_y^2)(dY/dy)$ at $y = \pm h/2$. The resulting dispersion relation of waveguide II$'$, in terms of effective index N, is

$$kh\sqrt{n'^2_{\text{eff}} - N^2} = q\pi + \tan^{-1}\left[\frac{n'^2_{\text{eff}}}{n_2^2}\frac{\sqrt{N^2 - n_2^2}}{\sqrt{n'^2_{\text{eff}} - N^2}}\right] + \tan^{-1}\left[\frac{n'^2_{\text{eff}}}{n_4^2}\frac{\sqrt{N^2 - n_4^2}}{\sqrt{n'^2_{\text{eff}} - N^2}}\right]$$

(5.60)

where N is the approximate effective refractive index of the original 3D waveguide under investigation.

5.5.3 Generalized Guide Index

It is also possible to cast the dispersion relation obtained by the effective index method in terms of generalized parameters. Suppose we use waveguides I and II in considering E^y mode guided by the 3D waveguide. Again, we assume that $n_5 > n_4 > n_2$ and $n_5 > n_3$. For waveguide I, we calculate the usual generalized parameters V_I, a_I, c_I, and d_I. Then we estimate b_I from Figure 2.5 for TM modes. In terms of these parameters, we have the effective index n_{eff} of waveguide I:

$$n_{\text{eff}}^2 = n_4^2 + b_I(n_1^2 - n_4^2) \qquad (5.61)$$

For waveguide II, we use n_3, n_{eff}, and n_5 to define V_{II} and a_{II}. Then we use Figure 2.4 for TE modes to estimate b_{II}. Then

$$N^2 = n_5^2 + b_{II}(n_{\text{eff}}^2 - n_5^2) = n_5^2 + b_{II}[n_4^2 - n_5^2 + b_I(n_1^2 - n_4^2)] \qquad (5.62)$$

Thus the generalized guide index b of the original 3D waveguide based on effective index method is

$$b_{KT} = b_{II} + \frac{n_1^2 - n_4^2}{n_1^2 - n_5^2}b_{II}(b_I - 1) \qquad (5.63)$$

The subscripts KT are inserted to honor the two researchers (Knox and Toulios) who proposed the effective index method [4]. A similar equation for b_{KT} can also be written in terms of generalized parameters of waveguides I$'$ and II$'$. It is left as an exercise for readers.

5.6 COMPARISON OF METHODS

In the last two sections, we discussed two approximate methods for analyzing weakly guiding rectangular waveguides. In the Marcatili method, a rectangular waveguide is replaced by two *mutually independent* thin-film waveguides. The two 2D waveguides are independent in that they are specified directly from the 3D waveguide. In the effective index method, a 3D waveguide is replaced by two *related* 2D waveguides. The two 2D waveguides are related in that the effective index of waveguide I (or I′) is used as the *film index* of waveguide II (or II′). In both methods, the dispersion relations for TE and TM modes of 2D waveguides discussed in Chapter 2 are used to estimate the propagation constant of the original 3D waveguide. In this section, we estimate the accuracy of the two methods. Since an exact expression of β of the rectangular waveguides is not known, we take the value determined by various numerical techniques as *numerically exact* and use it in the comparison [6–8].

As an example, we consider a dielectric bar of index n_1 immersed in a medium with index n_2 [Fig. 5.6(a)]. The waveguide corresponds to the structure shown in Figure 5.3(a) with $n_2 = n_3 = n_4 = n_5$ and $w = 2h$. To facilitate comparison, we define the generalized guide thickness V and normalized guide index b in terms of

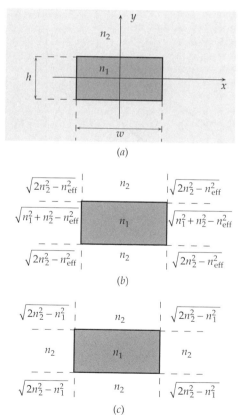

Figure 5.6 (a) Rectangular waveguide, (b) pseudowaveguide for effective index method, and (c) equivalent waveguide for Marcatili method.

n_1, n_2, and h:

$$V = kh\sqrt{n_1^2 - n_2^2}$$

$$b = \frac{\beta^2 - k^2 n_2^2}{k^2(n_1^2 - n_2^2)}$$

Figure 5.7 depicts the bV curves of the first four E^y modes [7]. Results based on a finite-element method, the Marcatili method, and the effective index method are shown as dots, dot–dash, and dashed lines, respectively. As noted earlier, the numerical results obtained by a finite-element method are taken as "numerically exact." Clearly the two approximate methods are accurate for modes far above cutoff. But discrepancy appears when modes are near the cutoff. The error can be traced to the basic assumption. Both approximate methods are predicated on the assumption that fields are confined mainly in the core region. In fact, we have ignored fields in the corner regions completely in both methods. The assumption holds for modes far above the cutoff. But the assumption breaks down for modes near the cutoff.

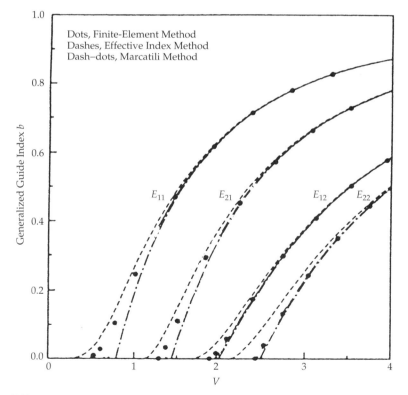

Figure 5.7 Comparison of three methods applied to a rectangular dielectric waveguide ($w = 2\,h$). (After Chaing [7]).

For the example under consideration, the normalized guide index b obtained by the effective index method is larger, and that obtained by the Marcatili method is smaller, than the *numerically exact* value. In other words, one method *overestimates* while the other method *underestimates* the normalized guide index. To identify the source of error, we note that the perturbation on β^2 by a small change of n^2 is

$$\delta(\beta^2) = \frac{k^2 \int \int |E(x, y, z)|^2 \Delta[n^2(x, y)] \, dx \, dy}{\int \int |E(x, y, z)|^2 \, dx \, dy} \tag{5.64}$$

In the above equation, $\delta(\beta^2)$ is the change or perturbation of β^2, and $\Delta(n^2)$ is the change of n^2 or the relative dielectric constant ε_r. To apply the perturbation equation [9] to the waveguide under consideration, we compare the index profile of the original 3D waveguide with that of the pseudowaveguide. For the pseudowaveguide, the index distribution is given by

$$n_{ps}^2(x, y) = \begin{cases} n_1^2 & \text{region 1} \\ n_2^2 & \text{region 2} \\ n_2^2 + n_1^2 - n_{eff}^2 & \text{region 3} \\ n_2^2 & \text{region 4} \\ n_2^2 + n_1^2 - n_{eff}^2 & \text{region 5} \end{cases} \tag{5.65}$$

And $n_x^2(x)$ and $n_y^2(y)$ are

$$n_y^2(y) = \begin{cases} n_2^2 & y > h/2 \\ n_1^2 & -h/2 \leqslant y \leqslant h/2 \\ n_2^2 & y < -h/2 \end{cases} \tag{5.66}$$

$$n_x^2(x) = \begin{cases} n_2^2 - n_{eff}^2 & x > w/2 \\ 0 & -w/2 \leqslant x \leqslant w/2 \\ n_2^2 - n_{eff}^2 & x < -w/2 \end{cases} \tag{5.67}$$

The index distribution of the pseudowaveguide, including the corner regions, is depicted in Figure 5.6(b). Note that indices in regions 3 and 5 are $\sqrt{n_2^2 + (n_1^2 - n_{eff}^2)}$, which is slightly larger than n_2. On the other hand, indices of the corner regions are $\sqrt{n_2^2 - (n_{eff}^2 - n_2^2)}$, which are slightly smaller than n_2 of the original 3D waveguide. Since fields in regions 3 and 5 are stronger than that in the corner regions, $\Delta(n^2)$ in the side regions:

$$\Delta(n^2) = [n_2^2 + (n_1^2 - n_{eff}^2)] - n_2^2 = n_1^2 - n_{eff}^2$$

is the main source of perturbation. Since $n_1^2 - n_{eff}^2$ is positive, $\delta(\beta^2)$ as indicated in (5.64) is positive. Thus, the effective index method *over estimates* the propagation constant.

It has been shown by Kumar et al.[5] that the Marcatili method amounts to solving an *equivalent* waveguide exactly. Corresponding to the waveguide shown in Figure 5.6(a), the equivalent waveguide has an index distribution $n_{eq}(x, y)$ given by

$$n_{eq}^2(x, y) = [n'(x)]^2 + [n''(y)]^2 \tag{5.68}$$

where $[n'(x)]^2$ and $[n''(y)]^2$ are given by

$$[n'(x)]^2 = \begin{cases} n_2^2 - (n_1^2/2) & x > w/2 \\ n_1^2/2 & -w/2 \leqslant x \leqslant w/2 \\ n_2^2 - (n_1^2/2) & x < -w/2 \end{cases} \tag{5.69}$$

$$[n''(y)]^2 = \begin{cases} n_2^2 - (n_1^2/2) & y > h/2 \\ n_1^2/2 & -h/2 \leqslant y \leqslant h/2 \\ n_2^2 - (n_1^2/2) & y < -h/2 \end{cases} \tag{5.70}$$

If we apply (5.68)–(5.70) to all regions, including the corner regions, we obtain an index distribution shown in Figure 5.6(c). Indices in regions 1–5 are the same as that in the corresponding regions in the original 3D waveguide. But the index of the corner regions of the equivalent waveguide is $\sqrt{n_2^2 - (n_1^2 - n_2^2)}$, which is slightly smaller than n_2. Thus the perturbation is

$$\Delta(n^2) = (2n_2^2 - n_1^2) - n_2^2 = n_2^2 - n_1^2$$

is negative. From (5.64), we see that the Marcatili method *underestimates* the propagation constant.

This example shows that the numerically exact result is bounded on one side by the result obtained by the Marcatili method and on the other side by that based on the effective index method. But this is just an example. It would be incorrect to conclude that the Marcatili method would always underestimate, and the effective index method would always overestimate, the propagation constant. It would be best to examine each case individually. For a detailed examination of errors involved in the two approximate methods, readers are referred to [12].

Study also shows that the error increases as the aspect ratio of the 3D waveguide approaches 1. Thus, the error is worst for square waveguides. This is true for both methods. The effect of the aspect ratio on the accuracy of the two methods can be understood as follows. A waveguide with a large aspect ratio, that is, $h >> w$ or $h << w$, corresponds to a thin or a slender rectangular waveguide. It is reasonable to approximate a thin or slender rectangular waveguide by a 2D waveguide as the first step. Thus, the error introduced by the first 2D waveguide is smaller for waveguides with a larger aspect ratio. The final result would be more accurate too.

Several methods have been proposed to improve the accuracy of the two methods. They include the dual effective index method [10] and several perturbation methods [5, 8, 11, 12]. While errors are reduced in the improved methods, they are much more complicated. Besides, the root cause of the inaccuracy for modes near the cutoff remains, albeit smaller. Although the effective index method was proposed originally for waveguides with rectangular boundaries, it has been applied to waveguides with nonplanar boundaries, circular fibers, and coupled waveguides [13–17].

PROBLEMS

1. Consider E^x modes guided by the rectangular waveguide shown in Figure 5.3(a). Apply the Marcatili method to the problem and express the generalized guide index in terms of the generalized parameters of waveguides H and W.

2. Consider E^y modes guided by the rectangular waveguide shown in Figure 5.3(a). Use the effective index method to analyze the problem and express the generalized guide index in terms of the generalized parameters of waveguides I' and II'.

3. Use the effective index method to estimate the effective refractive index N of the lowest E^y mode guided by the channel waveguide shown in Figure 5.8. A numerical answer is expected.

4. Shown in Figure 5.9 are two coupled rectangular dielectric waveguides. The waveguide thickness h is very small compared to the width and separation w. The index difference is small too, that is, $|n_1 - n_2| << n_1$. We are interested in E^x modes guided by the coupled waveguides and propagating in the $+z$ direction.
   (a) Identify the boundary conditions at the vertical and horizontal boundaries.
   (b) Use the effective index method to determine β of E^x modes.

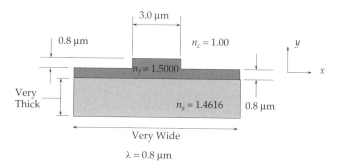

Figure 5.8   Channel waveguide.

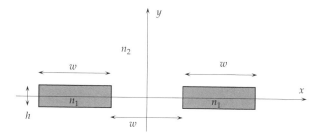

Figure 5.9   Two coupled dielectric waveguides.

REFERENCES

1. K. Ogusu, "Numerical analysis of the rectangular dielectric waveguide and its modifications," *IEEE Trans. Microwave Theory Technol.*, Vol. MTT-25, No. 11, pp. 874–885 (1977).

2. J. E. Goell, "A circular-harmonic computer analysis of rectangular dielectric waveguide," *B. S. T. J.*, Vol. 48, pp. 2133–2160 (1969).

3. E. A. J. Marcatili, "Dielectric rectangular waveguide and directional coupler for integrated optics," *B. S. T. J.*, Vol. 48, pp. 2079–2102 (1969).

4. R. M. Knox, and P. P. Toulios, "Integrated circuits for the millimeter through optical frequency range," *Proceedings of MRI Symposium on Submillimeter Waves,* J. Fox (ed), Polytechnic Press, Brooklyn, New York, 1970; pp. 497–516.

5. A. Kumar, K. Thyagarajan, and A. K. Ghatak, "Analysis of rectangular-core waveguides: An accurate perturbation approach," *Opt. Lett.*, Vol. 8, pp. 63–65 (1983).

6. A. Kumar, D. F. Clark, and B. Culshaw, "Explanation of errors inherent in the effective index method for analyzing rectangular-core waveguides," *Opt. Lett.*, Vol. 13, pp. 1129–1131 (1988).

7. K. S Chiang, "Analysis of rectangular dielectric waveguides: effective index method with built-in perturbation correction," *Electron. Lett.*, Vol. 28, No. 4, pp. 388–390 (Feb. 1992).

8. K. S. Chiang, K. M. Lo, and K. S. Kwok, "Effective-index method with built-in perturbation correction for integrated optical waveguides," *J. IEEE Lightwave Technol.*, Vol. 14, No. 2, pp. 223–228 (1996).

9. H. A. Haus, *Waves and Fields in Optoelectronics*, Prentice-Hall, Englewood Cliffs, NJ, 1984.

10. K. S. Chiang, "Dual effective-index method for the analysis of rectangular dielectric waveguides," *Appl. Opt.*, Vol. 13, pp. 2169–2174 (1986).

11. C. M. Kim, B. G. Jung, and C. W. Lee, "Analysis of dielectric rectangular waveguide by modified effective index method," *Electron. Lett.*, Vol. 22, No. 6, pp. 296–298 (1986).

12. K. S. Chiang, "Performance of the effective-index method for the analysis of dielectric waveguides," *Opt. Lett.*, Vol. 16, pp. 714–718 (1991).

13. K. S. Chiang, "Analysis of optical fibers by the effective index method," *Appl. Opt.*, Vol. 25, pp. 348–354 (1986).

14. J. Buus, "Application of the effective index method to nonplanar structures," *J. IEEE Quantum Electron.,* Vol. QE-20, No. 10, pp. 1106–1109 (1984).

15. K. S. Chiang, "Effective-index method for the analysis of optical waveguide couplers and arrays: An asymptotic theory," *J. IEEE Lightwave Technol.*, Vol. 9, No. 2, pp. 62–72 (1991).

16. K. S. Chiang, "Review of numerical and approximate methods for the modal analysis of general optical dielectric waveguides," *Opt. Quantum Electron.*, Vol. 26, pp. S113–S134 (1994).

17. K. Van de Velde, H. Thienpont, and R. Van Geen, "Extending the effective index method for arbitrarily shaped inhomogeneous optical waveguides," *IEEE J. Lightwave Technol.*, Vol. 6, pp. 1153–1159 (1988).

6

OPTICAL DIRECTIONAL COUPLERS AND THEIR APPLICATIONS

6.1 INTRODUCTION

In the preceding chapters, we have studied modes of isolated waveguides. In this chapter, we consider waves guided by two or more coupled waveguides. When several optical waveguides sit side by side, significant interaction between waveguides arises that may lead to a significant power exchange between them. Power exchange is substantial if the interacting waveguides are in parallel and the guided modes propagate with the same phase velocities. Under the ideal conditions, power transferred from one waveguide to another accumulates coherently. As a result, a significant fraction of power, or the total power under certain conditions, is transferred from one waveguide to the other waveguide. Many guided wave devices have been conceived to take advantage of the coherent and directional interaction of fields. These devices are referred to as *directional couplers*. In this chapter, we study the operation principle and applications of directional couplers. Specifically, we study optical switches, power dividers, power combiners, multiplexers, demultiplexers, optical filters, and intensity modulators based on directional couplers.

We begin by describing the interaction of guided waves qualitatively. The description is couched in terms of the coupling of modes propagating in the two isolated waveguides. The interaction can also be viewed as the interference of normal

Foundations for Guided-Wave Optics, by Chin-Lin Chen
Copyright © 2007 John Wiley & Sons, Inc.

modes of the composite waveguide structure. In Section 6.3, Marcatili's improved coupled-mode equations are discussed. The essential characteristics of directional couplers are summarized in Section 6.4. Also introduced in Section 6.4 is the switching diagram of simple or conventional directional couplers. In Section 6.5, we analyze the switched $\Delta\beta$ directional couplers. The use of switched $\Delta\beta$ directional couplers as optical filters is examined in Section 6.6. The operation of intensity modulators based on electrooptic directional couplers is studied in Section 6.7.

In Section 6.8, we return to the basic theory of directional couplers. However, we view the directional interaction as the interference between two propagating normal modes. Although directional couplers may have two, three, or more waveguides, much of our discussions on directional couplers pertain to two interacting waveguides. In the last section, we consider couplers with three waveguides. The applications of directional couplers with three or more waveguides are briefly mentioned.

6.2   QUALITATIVE DESCRIPTION OF THE OPERATION OF DIRECTIONAL COUPLERS

To describe the directional coupler operation qualitatively, we consider two waveguides in close proximity. The two waveguides may be two dielectric slabs having one slab stacked on top of the other as shown schematically in Figure 6.1(a), or two channel waveguides placed side by side as shown in Figure 6.1(b). In either configuration, we begin by considering two isolated waveguides. When the two waveguides are far apart and isolated, each waveguide supports a number of guided modes and possibly a continuum of a radiation mode. Experimental investigations and numerical simulations show that only modes propagating with identical, or nearly identical, phase velocities and in the same direction interact strongly [1]. For simplicity, we concentrate on the two guided modes, one from each waveguide. We write the modes guided by the two isolated waveguides as $a_{10}\mathbf{e}_1(x, y)e^{-j\beta_1 z}$ and $a_{20}\mathbf{e}_2(x, y)e^{-j\beta_2 z}$ where a_{10} and a_{20} are the amplitude constants, $\mathbf{e}_1(x, y)$ and $\mathbf{e}_2(x, y)$ are the fields, and β_1 and β_2 are the propagation constants, of the two guided modes. When the two waveguides are close physically, $\mathbf{e}_1(x, y)e^{-j\beta_1 z}$ extends to waveguide 2 and $\mathbf{e}_2(x, y)e^{-j\beta_2 z}$ to waveguide 1. But $\mathbf{e}_1(x, y)e^{-j\beta_1 z}$ does not satisfy the boundary conditions at boundaries of waveguide 2, nor does $\mathbf{e}_2(x, y)e^{-j\beta_2 z}$ meet the boundary conditions at interfaces of waveguide 1. In other words, waveguide 2 perturbs $\mathbf{e}_1(x, y)e^{-j\beta_1 z}$ and waveguide 1 disturbs $\mathbf{e}_2(x, y)e^{-j\beta_2 z}$. Unless the two waveguides are very close, the perturbation would be small. Under the assumption of a weak perturbation, fields guided by the two coupled waveguides may be approximated by a linear combination of $\mathbf{e}_1(x, y)e^{-j\beta_1 z}$ and $\mathbf{e}_2(x, y)e^{-j\beta_2 z}$. Thus we write the field of the coupled waveguides as

$$\mathbf{E}(x, y, z) \approx a_1(z)\mathbf{e}_1(x, y)e^{-j\beta_1 z} + a_2(z)\mathbf{e}_2(x, y)e^{-j\beta_2 z} \qquad (6.1)$$

where $a_1(z)$ and $a_2(z)$ are the two amplitude functions.

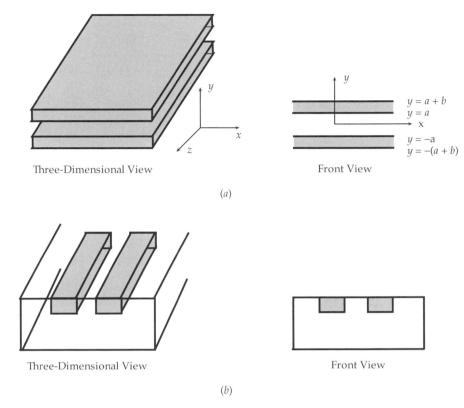

Figure 6.1 Directional couplers with two coupled waveguides: (a) two dielectric slab waveguides and (b) two channel waveguides.

Alternately, we treat the coupled waveguides as a composite waveguide structure. We could in principle, and indeed we can in many cases, analyze modes guided by the composite waveguide rigorously. The field expressions would satisfy Maxwell's equations and meet all boundary conditions of the composite structure. We refer to these modes as the *normal modes* of the composite waveguide. Then fields guided by the composite waveguide are the superposition of the normal mode fields. In the structures where the two coupled waveguides are identical, the normal mode fields are either *symmetric* or *antisymmetric* with respect to the centerline of the composite structure. In Figure 6.2, we sketch the field distributions of the two lowest normal modes of a composite structure having two identical waveguides sitting side by side. Let the two lowest normal mode fields and propagation constants be $\mathbf{e}_s(x, y)$, $\mathbf{e}_a(x, y)$, β_s, and β_a. If the two waveguides are not identical, fields are neither symmetric nor antisymmetric. In short, subscripts s and a imply no physical meaning unless the two interacting waveguides are identical. In terms of $\mathbf{e}_s(x, y)$ and $\mathbf{e}_a(x, y)$, the field guided by the composite waveguide can be written as

$$\mathbf{E}(x, y, z) = a_{s0}\mathbf{e}_s(x, y)e^{-j\beta_s z} + a_{a0}\mathbf{e}_a(x, y)e^{-j\beta_a z} \qquad (6.2)$$

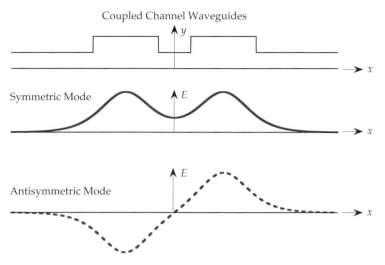

Figure 6.2 Symmetric and antisymmetric modes.

where a_{s0} and a_{a0} are the normal mode amplitudes. As the two normal modes propagate in the composite waveguide structure, they interfere continuously. The continuous and periodic power exchange between the two waveguides is due to the interference of the two normal modes.

In summary, we may treat the field of two coupled waveguides either as the superposition of $\mathbf{e}_1(x, y)e^{-j\beta_1 z}$ and $\mathbf{e}_2(x, y)e^{-j\beta_2 z}$ guided by two isolated waveguides as indicated by (6.1) or the interference of $\mathbf{e}_s(x, y)e^{-j\beta_s z}$ and $\mathbf{e}_a(x, y)e^{-j\beta_a z}$ of the two normal modes of the composite waveguide as represented by (6.2). If we view the field as the superposition of $\mathbf{e}_1(x, y)e^{-j\beta_1 z}$ and $\mathbf{e}_2(x, y)e^{-j\beta_2 z}$, we need to determine the amplitude functions $a_1(z)$ and $a_2(z)$. Marcatili derived two coupled differential equations governing the evolution of $a_1(z)$ and $a_2(z)$ [2]. His theory is discussed in Section 6.3. If we consider the field as the interference of the two normal modes, we have to determine fields $\mathbf{e}_s(x, y)$ and $\mathbf{e}_a(x, y)$ and the propagation constants β_s and β_a of the two normal modes. This is discussed in Section 6.8.

6.3 MARCATILI'S IMPROVED COUPLED-MODE EQUATIONS

We consider a structure consisting of two dielectric waveguides of arbitrary cross sections. For example, one waveguide may be a rectangular waveguide and the other an elliptical waveguide, as shown in Figure 6.3. We label the two waveguides as waveguides 1 and 2. Each waveguide is a small index variation in a medium of index n. The index variation $\delta n_i(x, y)$ vanishes everywhere except where waveguide i is. Even at waveguide i, $\delta n_i(x, y)$ is much smaller than n. As discussed in Chapter 5, fields guided by a weakly guiding structure are approximately TEM waves. This is also true for coupled waveguides. In other words, fields guided by the coupled waveguides are also approximately TEM waves since $\delta n_i(x, y) \ll n$.

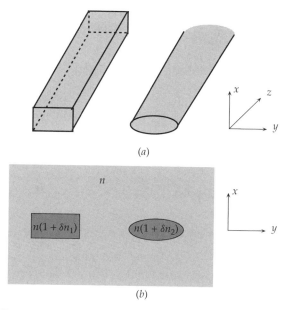

Figure 6.3   Coupling of two arbitrary waveguides: (a) two coupled waveguides and (b) index perturbation.

6.3.1   Fields of Isolated Waveguides

Let the electric field guided by waveguide 1 in the absence of waveguide 2 be $\mathbf{e}_1(x, y)e^{-j\beta_1 z}$. The expressions for $\mathbf{e}_1(x, y)$ and β_1 are determined from the wave equation and the associated boundary conditions:

$$\{\nabla_t^2 + k^2 n^2 [1 + \delta n_1(x, y)]^2 - \beta_1^2\}\mathbf{e}_1(x, y) = 0 \tag{6.3}$$

where

$$\nabla_t^2 = \frac{\partial}{\partial x^2} + \frac{\partial}{\partial y^2}$$

is the transverse Laplacian operator.

Since $\delta n_1(x, y)$ is small everywhere, β_1 is approximately kn. Then $[\delta n_1(x, y)]^2$ and $(kn - \beta_1)^2$ are terms of the second-order smallness. When terms of second- and higher-order smallness are ignored, we have

$$k^2 n^2 [1 + \delta n_1(x, y)]^2 - \beta_1^2 \approx 2kn(kn - \beta_1) + 2k^2 n^2 \delta n_1(x, y)$$

Then (6.3) becomes approximately

$$[\nabla_t^2 + 2k^2 n^2 \delta n_1(x, y) + 2kn(kn - \beta_1)]\mathbf{e}_1(x, y) \approx 0 \tag{6.4}$$

Similarly, fields and propagation constants of modes guided by the isolated waveguide 2 may be determined from

$$[\nabla_t^2 + 2k^2n^2\delta n_2(x, y) + 2kn(kn - \beta_2)]\mathbf{e}_2(x, y) \approx 0 \qquad (6.5)$$

By solving (6.4) and (6.5) subjected to appropriate boundary conditions, we obtain the expressions for $\mathbf{e}_1(x, y)$, $\mathbf{e}_2(x, y)$, β_1, and β_2. In the following discussions, we assume that fields and propagation constants of isolated waveguides are known. In addition, we suppose that fields are normalized in that

$$\int_S \mathbf{e}_i(x, y) \cdot \mathbf{e}_i(x, y)\, ds = 1 \quad i = 1 \text{ and } 2 \qquad (6.6)$$

6.3.2 Normal Mode Fields of the Composite Waveguide

For the composite structure, $\delta n_1(x, y)$ and $\delta n_2(x, y)$ are present simultaneously. Let the field guided by the composite structure be $\mathbf{e}(x, y)$ and the propagation constant be β. The two quantities must satisfy the wave equation

$$\{\nabla_t^2 + 2k^2n^2[\delta n_1(x, y) + \delta n_2(x, y)] + 2kn(kn - \beta)\}\mathbf{e}(x, y) \approx 0 \qquad (6.7)$$

and associated boundary conditions. Usually the composite waveguide structure has a complicated geometry, it would be difficult, if not impossible, to solve (6.7) exactly. We look for ways to approximate $\mathbf{e}(x, y)$ and β without actually solving (6.7). This is the main thrust of the coupled-mode formulation.

6.3.3 Marcatili Relation

By making the scalar product of $\mathbf{e}_2(x, y)$ with the wave equation (6.4) and $\mathbf{e}_1(x, y)$ with (6.5), subtracting the two resulting expressions and integrating the combined expression over the entire cross section S of the composite waveguide, we obtain

$$\int_S [\mathbf{e}_2(x, y) \cdot \nabla_t^2 \mathbf{e}_1(x, y) - \mathbf{e}_1(x, y) \cdot \nabla_t^2 \mathbf{e}_2(x, y)]\, ds$$

$$= -2k^2n^2 \int_S [\delta n_1(x, y) - \delta n_2(x, y)]\mathbf{e}_1(x, y) \cdot \mathbf{e}_2(x, y)\, ds \qquad (6.8)$$

$$+ 2kn(\beta_1 - \beta_2) \int_S \mathbf{e}_1(x, y) \cdot \mathbf{e}_2(x, y)\, ds$$

Making use of Green's theorem [(B.10) of Appendix B], we show that the integral on the left-hand side of (6.8) is zero. Thus we obtain the *Marcatili relation* [2, 3]

$$\kappa_1 - \kappa_2 = c(\beta_1 - \beta_2) \qquad (6.9)$$

where κ_1 and κ_2 are the two *coupling constants*:

$$\kappa_1 = kn \int_S [\delta n_1(x, y)] \mathbf{e}_1(x, y) \cdot \mathbf{e}_2(x, y) \, ds \qquad (6.10)$$

$$\kappa_2 = kn \int_S [\delta n_2(x, y)] \mathbf{e}_1(x, y) \cdot \mathbf{e}_2(x, y) \, ds \qquad (6.11)$$

and c is the *overlap integral* or the *butt coupling*:

$$c = \int_S \mathbf{e}_1(x, y) \cdot \mathbf{e}_2(x, y) \, ds \qquad (6.12)$$

The coupling constant κ_1 quantifies the coupling of the field guided by waveguide 2 to waveguide 1, and κ_2 the coupling of the field of guided by waveguide 1 to waveguide 2. The overlap integral c is a measure of the closeness of the two waveguides and the similarity of $\mathbf{e}_1(x, y)$ and $\mathbf{e}_2(x, y)$.

The *propagation constant mismatch* $\beta_1 - \beta_2$ and the coupling constants κ_1 and κ_2 are related as indicated by the Marcatili relation (6.9). To examine the relationship further, we define $\Delta\beta$ and δ such that

$$\delta \equiv \frac{\Delta\beta}{2} \equiv \frac{\beta_1 - \beta_2}{2} \qquad (6.13)$$

In addition, we normalize δ with respect to the geometrical mean of the two coupling constants and label the normalized propagation constant mismatch as the *mismatch parameter d*:

$$d \equiv \frac{\delta}{\sqrt{\kappa_1\kappa_2}} \equiv \frac{\Delta\beta}{2\sqrt{\kappa_1\kappa_2}} \equiv \frac{\beta_1 - \beta_2}{2\sqrt{\kappa_1\kappa_2}} \qquad (6.14)$$

In terms of newly defined d, the Marcatili relation (6.9) becomes

$$\sqrt{\frac{\kappa_1}{\kappa_2}} - \sqrt{\frac{\kappa_2}{k_1}} = 2cd \qquad (6.15)$$

Solving (6.15) for κ_1/κ_2, we obtain

$$\frac{\kappa_1}{\kappa_2} = \left(cd + \sqrt{1 + c^2d^2} \right)^2 = e^{\sinh^{-1} cd} \qquad (6.16)$$

Clearly, the ratio of the two coupling constants depends on the product of the overlap integral c and the mismatch parameter d only. For many directional couplers of practical interest, the two interacting modes are either synchronized ($\beta_1 = \beta_2$), nearly synchronized ($\beta_1 \approx \beta_2$), or the two waveguides are far apart so that the overlap integral c is very small. Then cd is either zero or very small. As a result, κ_1 and κ_2 are equal or nearly equal.

6.3.4 Approximate Normal Mode Fields

From (6.4), (6.5), and (6.7), we also obtain

$$kn \int_S [\delta n_2(x, y)] \mathbf{e}(x, y) \cdot \mathbf{e}_1(x, y) \, ds = (\beta - \beta_1) \int_S \mathbf{e}(x, y) \cdot \mathbf{e}_1(x, y) \, ds \quad (6.17)$$

$$kn \int_S [\delta n_1(x, y)] \mathbf{e}(x, y) \cdot \mathbf{e}_2(x, y) \, ds = (\beta - \beta_2) \int_S \mathbf{e}(x, y) \cdot \mathbf{e}_2(x, y) \, ds \quad (6.18)$$

As noted earlier, we look for ways to approximate $\mathbf{e}(x, y)$ without actually solving the wave equation (6.7) exactly. To obtain an approximate expression for $\mathbf{e}(x, y)$, we suppose that the two waveguides are not too close. Then the field guided by the composite structure may be approximated by a linear combination of $\mathbf{e}_1(x, y)$ and $\mathbf{e}_2(x, y)$:

$$\mathbf{e}(x, y) \approx \mathbf{e}_1(x, y) + r \mathbf{e}_2(x, y) \quad (6.19)$$

where r is a yet undetermined constant. Substituting (6.19) into (6.17) and (6.18), and expressing the equations in terms of κ_1, κ_2, and c, we obtain

$$\frac{\rho_2 + r\kappa_2}{1 + rc} = \beta - \beta_1 \quad (6.20)$$

$$\frac{\kappa_1 + r\rho_1}{c + r} = \beta - \beta_2 \quad (6.21)$$

where

$$\rho_1 = kn \int_S [\delta n_1(x, y)] \mathbf{e}_2(x, y) \cdot \mathbf{e}_2(x, y) \, ds \quad (6.22)$$

$$\rho_2 = kn \int_S [\delta n_2(x, y)] \mathbf{e}_1(x, y) \cdot \mathbf{e}_1(x, y) \, ds \quad (6.23)$$

By subtracting (6.21) from (6.20), we eliminate β and obtain an expression for r:

$$\frac{\kappa_1 + r\rho_1}{c + r} - \frac{\rho_2 + r\kappa_2}{1 + rc} = \beta_1 - \beta_2 \quad (6.24)$$

Note that κ_1, κ_2, ρ_1, ρ_2, and c are integrals of fields and the index variations of the two isolated waveguides. Thus, $\kappa_1, \kappa_2, \rho_1, \rho_2$, and c can be calculated once the waveguide geometries and fields of isolated waveguides are known. Then we evaluate r from (6.24). With r known, we determine β and $\mathbf{e}(x, y)$ of the composite waveguide from (6.19) and (6.21). Exact expressions for r and β have been derived by Marcatili [2]. But the expressions are very complicated. Further simplication is desirable. From definitions of κ's, ρ's, and c and under the condition of weakly guiding, we can show that $c \ll 1$ and $\rho_i \ll \kappa_i$. Then we reduce (6.24) to a quadratic equation [2]:

$$\kappa_2 r^2 + (\beta_1 - \beta_2)r - \kappa_1 \approx 0 \quad (6.25)$$

Let the two roots of the quadratic equation be r_s and r_a, then

$$r_{s,a} \approx \frac{1}{2\kappa_2}\left[-(\beta_1 - \beta_2) \pm \sqrt{(\beta_1 - \beta_2)^2 + 4\kappa_1\kappa_2}\right] \tag{6.26}$$

Making use (6.14), we express the two roots in terms of c and d:

$$r_{s,a} \approx \sqrt{\frac{\kappa_1}{\kappa_2}}\left(-d \pm \sqrt{1+d^2}\right) = \left(cd + \sqrt{1+c^2d^2}\right)\left(-d \pm \sqrt{1+d^2}\right) \tag{6.27}$$

Upon substituting r_s and r_a back to (6.20) or (6.21), we obtain

$$\beta_{s,a} \approx \frac{\beta_1 + \beta_2}{2} \pm \sqrt{\kappa_1\kappa_2(1+d^2)} \tag{6.28}$$

It is also straightforward to establish that

$$\int_S [\mathbf{e}_1(x, y) + r_s\mathbf{e}_2(x, y)] \cdot [\mathbf{e}_1(x, y) + r_a\mathbf{e}_2(x, y)]\, ds$$
$$= 1 + r_s r_a + c(r_s + r_a) \approx -4\, dc[dc + \sqrt{1 + (dc)^2}] \tag{6.29}$$

In the limit of an infinitesimal c, the right-hand side of (6.29) vanishes. Then the integral in the left-hand side of (6.29) must be zero. In other words, fields of the two normal modes, $\mathbf{e}_1(x, y) + r_s\mathbf{e}_2(x, y)$ and $\mathbf{e}_1(x, y) + r_a\mathbf{e}_2(x, y)$, are *mutually orthogonal* in the limit of infinitesimal c even if the waveguides are not identical.

For directional couplers with two identical and uniformly spaced waveguides, the results are simple and intuitive: the normal mode fields are symmetric and antisymmetric with respect to the centerline. To elaborate this observation further, we note $\beta_1 = \beta_2, \kappa_1 = \kappa_2$, and $d = 0$ since the two isolated waveguides are identical. Then we obtain from (6.27) and (6.28) that $r_s = 1$ and $r_a = -1$. Thus the two normal mode fields are $\mathbf{e}_1(x, y) \pm \mathbf{e}_2(x, y)$. It clearly shows that the two normal mode fields are symmetric and antisymmetric with respect to the centerline of the composite structure (Fig. 6.2). The propagation constants β_s and β_a of the normal modes are $\beta_1 \pm \kappa_1$. Note that β_s and β_a differ from β_1 even for directional couplers having two identical and parallel waveguides.

6.3.5 Improved Coupled-Mode Equations

As indicated in Section 6.3.4, fields guided by the two coupled waveguides can be viewed as the superposition of two normal modes with r_s and r_a given in (6.27) and β_s and β_a given in (6.28). Then the field guided by the composite waveguide structure can be expressed as

$$\mathbf{E}(x, y, z) \approx a_{s0}[\mathbf{e}_1(x, y) + r_s\mathbf{e}_2(x, y)]e^{-j\beta_s z} + a_{a0}[\mathbf{e}_1(x, y) + r_a\mathbf{e}_2(x, y)]e^{-j\beta_a z} \tag{6.30}$$

Terms in the above expression can be grouped in various ways. The amplitude functions can also be defined in a different manner. By grouping terms and defining amplitude functions differently, we obtain different sets of coupled-mode equations. One possible choice is to collect all terms having $\mathbf{e}_1(x, y)$ as a group and those having $\mathbf{e}_2(x, y)$ as the other group as indicated in (6.1). Then

$$\mathbf{E}(x, y, z) \approx a_1(z)\mathbf{e}_1(x, y)e^{-j\beta_1 z} + a_2(z)\mathbf{e}_2(x, y)e^{-j\beta_2 z} \tag{6.31}$$

The amplitude functions are

$$a_1(z) = (a_{s0}e^{-j\sigma z} + a_{a0}e^{j\sigma z})e^{j\delta z} \tag{6.32}$$

$$a_2(z) = (a_{s0}r_s e^{-j\sigma z} + a_{a0}r_a e^{j\sigma z})e^{-j\delta z} \tag{6.33}$$

where

$$\sigma = \sqrt{\kappa_1\kappa_2(1 + d^2)} = \sqrt{\kappa_1\kappa_2 + \delta^2} \tag{6.34}$$

In (6.32) and (6.33), we cast the *amplitude functions* $a_1(z)$ and $a_2(z)$ of the composite waveguide in terms of the normal mode amplitudes a_{s0} and a_{a0}.

It is often useful to describe the evolution of $a_1(z)$ and $a_2(z)$ in terms of $a_1(z)$ and $a_2(z)$ themselves. For this purpose, we differentiate (6.32) and (6.33) with respect to z and make use of (6.15) to obtain

$$\frac{da_1(z)}{dz} = -j\kappa_2 a_2(z)e^{j2\delta z} \tag{6.35}$$

$$\frac{da_2(z)}{dz} = -j\kappa_1 a_1(z)e^{-j2\delta z} \tag{6.36}$$

The two coupled differential equations, (6.35) and (6.36), are commonly known as the *coupled-mode equations* [2]. An alternative derivation of the coupled-mode equations can be found in [4, 5]. Most basic characteristics of directional couplers are derived from (6.35) and (6.36). Note the presence of the two coupling constants, κ_1 and κ_2, in the two equations. The two coupling constants are different in general, as indicated in (6.16). We also deduce from (6.35) and (6.36) that

$$\frac{d}{dz}(|a_1(z)|^2 + |a_2(z)|^2) = j(\kappa_1^* - \kappa_2)a_1^*(z)a_2(z)e^{j2\delta z} - j(\kappa_1 - \kappa_2^*)a_1(z)a_2^*(z)e^{-j2\delta z}$$

The term on the left-hand side is the rate of change of the total power transported by the composite waveguide structure. It depends in a complicated manner on the amplitude functions, coupling constants and the propagation constant mismatch. The right-hand side vanishes completely when $\kappa_1 = \kappa_2^*$. Thus, the *total power is conserved* in the coupled waveguides if $\kappa_1 = \kappa_2^*$.

6.3.6 Coupled-Mode Equations in an Equivalent Form

It is also instructive to rewrite the coupled-mode equations (6.35) and (6.36) in a different and yet equivalent form. For this purpose, we introduce $A_1(z)$ and $A_2(z)$

such that $A_1(z) = a_1(z)e^{-j\beta_1 z}$ and $A_2(z) = a_2(z)e^{-j\beta_2 z}$. In terms of $A_1(z)$ and $A_2(z)$, (6.35) and (6.36) become

$$\frac{dA_1(z)}{dz} = -j\beta_1 A_1(z) - j\kappa_2 A_2(z) \qquad (6.37)$$

$$\frac{dA_2(z)}{dz} = -j\beta_2 A_2(z) - j\kappa_1 A_1(z) \qquad (6.38)$$

Equations (6.37) and (6.38) are also known as the coupled-mode equations [4, 5]. We can interpret these equations intuitively. The left-hand side of each equation is the rate of change or evolution of A_1 or A_2 in each waveguide. The first term on the right-hand side of each equation is the change of A_1 or A_2 due to the mode propagation in the isolated waveguide. The second term on the right-hand side gives the perturbation effect or the correction due to the presence of the neighboring waveguide. It is straightforward to generalize (6.37) and (6.38) to coupled waveguides having three or more coupled waveguides. We will come back to the equations in Section 6.9.

6.3.7 Coupled-Mode Equations in an Alternate Form

As mentioned earlier, we can define the amplitude functions differently. An alternate set of the amplitude functions are

$$f(z) = (a_{s0}\,e^{-j\beta_s z} + a_{a0}\,e^{-j\beta_a z})e^{+j(\beta_s+\beta_a)z/2} \qquad (6.39)$$

$$g(z) = (a_{s0}\,r_s e^{-j\beta_s z} + a_{a0}r_a\,e^{-j\beta_a z})e^{+j(\beta_s+\beta_a)z/2} \qquad (6.40)$$

In terms of $f(z)$ and $g(z)$, the field is

$$\mathbf{E}(x, y, z) \approx [f(z)\mathbf{e}_1(x, y) + g(z)\mathbf{e}_2(x, y)]e^{-j(\beta_1+\beta_2)z/2} \qquad (6.41)$$

In (6.41), $(\beta_1 + \beta_2)/2$ is the average propagation constant. It is left as an exercise for the readers (Problem 4) to show that the coupled-mode equations [2] in terms of $f(z)$ and $g(z)$ are

$$\frac{df(z)}{dz} = -j\frac{\beta_1 - \beta_2}{2}f(z) - j\sqrt{\kappa_1\kappa_2}\,e^{-\sinh^{-1}dc}g(z) \qquad (6.42)$$

$$\frac{dg(z)}{dz} = -j\sqrt{\kappa_1\kappa_2}\,e^{\sinh^{-1}dc}f(z) - j\frac{\beta_2 - \beta_1}{2}g(z) \qquad (6.43)$$

As indicated in (6.24), $\beta_1 - \beta_2$ relates to κ_1, κ_2, ρ_1, ρ_2, and c. In other words, all coefficients in (6.42) and (6.43) are directly in terms of integrals of fields and index variations.

6.4 DIRECTIONAL COUPLERS WITH UNIFORM CROSS SECTION AND CONSTANT SPACING

6.4.1 Transfer Matrix

To study the essential properties of optical directional couplers, we consider a directional coupler having a uniform cross section and a constant waveguide spacing as shown schematically in Figure 6.4(a). We assume that significant interaction takes place only in an interaction region of length L and ignore the interaction outside this region completely. For such a directional coupler, $\beta_1, \beta_2, \kappa_1, \kappa_2$, and therefore

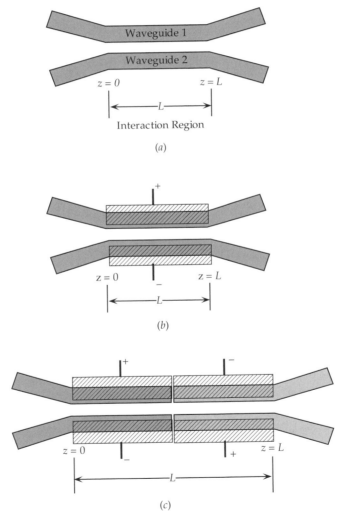

Figure 6.4   (a) Conventional directional couplers, (b) conventional directional coupler with electrodes for tuning β_1 and β_2 and (c) switched $\Delta\beta$ directional couplers.

δ are constants independent of z. Then (6.35) and (6.36) become two simple differential equations except for the presence of two exponential functions. To eliminate the exponential functions, we introduce two *auxiliary functions*, $\mathcal{R}(z)$ and $\mathcal{S}(z)$, such that $\mathcal{R}(z) = a_1(z)e^{-j\delta z}$ and $\mathcal{S}(z) = a_2(z)e^{+j\delta z}$. In terms of $\mathcal{R}(z)$ and $\mathcal{S}(z)$, the coupled-mode equations become

$$\frac{d\mathcal{R}(z)}{dz} + j\delta\mathcal{R}(z) = -j\kappa_2\mathcal{S}(z) \tag{6.44}$$

$$\frac{d\mathcal{S}(z)}{dz} - j\delta\mathcal{S}(z) = -j\kappa_1\mathcal{R}(z) \tag{6.45}$$

By differentiating (6.44) and (6.45) with respect to z and making use of (6.45) and (6.44), we obtain two second-order ordinary differential equations with constant coefficients:

$$\frac{d^2\mathcal{R}(z)}{dz^2} + (\kappa_1\kappa_2 + \delta^2)\mathcal{R}(z) = 0 \tag{6.46}$$

$$\frac{d^2\mathcal{S}(z)}{dz^2} + (\kappa_1\kappa_2 + \delta^2)\mathcal{S}(z) = 0 \tag{6.47}$$

Solutions of the two ordinary differential equations are

$$\mathcal{R}(z) = C_1 \cos \sigma z + C_2 \sin \sigma z \tag{6.48}$$

$$\mathcal{S}(z) = \frac{j}{\kappa_2}\left[(\sigma C_2 + j\delta C_1)\cos \sigma z + (j\delta C_2 - \sigma C_1)\sin \sigma z\right] \tag{6.49}$$

where C_1 and C_2 are constants determined by the boundary values $\mathcal{R}(0)$ and $\mathcal{S}(0)$. In terms of $\mathcal{R}(0)$ and $\mathcal{S}(0)$ and in a matrix form, $\mathcal{R}(z)$ and $\mathcal{S}(z)$ are

$$\begin{bmatrix} \mathcal{R}(z) \\ \mathcal{S}(z) \end{bmatrix} = \begin{bmatrix} \cos \sigma z - \dfrac{j\delta}{\sigma}\sin \sigma z & -\dfrac{j\kappa_2}{\sigma}\sin \sigma z \\ -\dfrac{j\kappa_1}{\sigma}\sin \sigma z & \cos \sigma z + \dfrac{j\delta}{\sigma}\sin \sigma z \end{bmatrix} \begin{bmatrix} \mathcal{R}(0) \\ \mathcal{S}(0) \end{bmatrix} \tag{6.50}$$

The 2×2 matrix in (6.50) is commonly referred to as the *transfer matrix*. It is tempting to treat κ_1, κ_2, and δ as three independent parameters. Actually, κ_1 and κ_2 are not mutually independent since κ_1/κ_2 is a function of cd as given in (6.16). In other words, it is more meaningful to treat c and d as two independent parameters, as noted by Marcatili [2]. In terms of c and d, (6.50) becomes

$$\begin{bmatrix} \mathcal{R}(z) \\ \mathcal{S}(z) \end{bmatrix} = \begin{bmatrix} \cos \sigma z - \dfrac{j\,d}{\sqrt{1+d^2}}\sin \sigma z & -j\dfrac{e^{-(\sinh^{-1} cd)/2}}{\sqrt{1+d^2}}\sin \sigma z \\ -j\dfrac{e^{(\sinh^{-1} cd)/2}}{\sqrt{1+d^2}}\sin \sigma z & \cos \sigma z + \dfrac{j\,d}{\sqrt{1+d^2}}\sin \sigma z \end{bmatrix} \begin{bmatrix} \mathcal{R}(0) \\ \mathcal{S}(0) \end{bmatrix} \tag{6.51}$$

In any event, $\mathcal{R}(z)$ and $\mathcal{S}(z)$ are readily calculated once the transfer matrix and the initial values, $\mathcal{R}(0)$ and $\mathcal{S}(0)$, are known. If desired, we can recover $a_1(z)$ and $a_2(z)$ from $\mathcal{R}(z)$ and $\mathcal{S}(z)$. Power carried by the two waveguides are $|a_1(z)|^2$ and $|a_2(z)|^2$ that are, respectively, $|\mathcal{R}(z)|^2$ and $|\mathcal{S}(z)|^2$. This completes the basic theory of directional couplers with a uniform cross section and a constant waveguide spacing.

6.4.2 Essential Characteristics of Couplers with $\kappa_1 = \kappa_2 = \kappa$

For brevity, we write $\kappa = \sqrt{\kappa_1\kappa_2}$. For many directional couplers of practical interests, the two waveguides are identical, similar, or far apart. Then $\kappa \approx \kappa_1 \approx \kappa_2$ and $c \approx 0$. We suppose the input power is fed to one waveguide only, say waveguide 1. By setting $\mathcal{R}(0) = 1$ and $\mathcal{S}(0) = 0$, we obtain from (6.50)

$$\mathcal{R}(z) = \cos \sigma z - \frac{j\delta}{\sigma} \sin \sigma z \tag{6.52}$$

$$\mathcal{S}(z) = -\frac{j\kappa}{\sigma} \sin \sigma z \tag{6.53}$$

Recall that $\sigma = \sqrt{\kappa_1\kappa_2 + \delta^2} = \sqrt{\kappa^2 + \delta^2}$. In Figure 6.5, we plot $|a_1(z)|^2$ and $|a_2(z)|^2$ as functions of $\kappa z/\pi$ for three values of δ/κ. First, we examine power carried by waveguide 1. At $z = 0$, $|a_1(z)|^2$ is a maximum and the maximum value is 1. This is expected since all power resides initially in waveguide 1. As z increases,

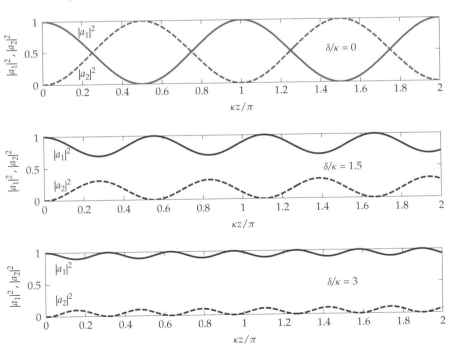

Figure 6.5 Power distribution in a directional coupler.

power in waveguide 1 decreases and shifts gradually to waveguide 2. At $z = \pi/2\sigma$, power in waveguide 1 drops to a minimum. The minimum value is

$$|a_1(z)|^2_{min} = \frac{\delta^2/\kappa^2}{1 + (\delta^2/\kappa^2)} \tag{6.54}$$

Beyond this point, power returns gradually to waveguide 1. The periodic spatial vacillation of power continues as z increases.

Next, we examine power carried by waveguide 2. At $z = 0$, no power is carried by waveguide 2. As z increases, power in waveguide 2 increases until it reaches a peak:

$$|a_2(z)|^2_{max} = \frac{1}{1 + (\delta^2/\kappa^2)} \tag{6.55}$$

It then returns to zero gradually. Again, the periodic power swing continues as the wave moves in the interaction region.

Based on these equations and plots similar to that shown in Figure 6.5, we arrive at the following conclusions.

1. The total power in a directional coupler with a uniform cross section and a constant spacing is conserved since $|a_1(z)|^2 + |a_2(z)|^2 = |\mathcal{R}(z)|^2 + |\mathcal{S}(z)|^2 = 1$. This is expected since we have assumed that the two waveguides are loss free and that the two coupling constants are real and equal.

2. A complete power transfer from one waveguide to another is possible only if δ vanishes completely. In other words, power crossover is total only if the two waveguides are *phase velocity matched*.

3. In addition, a complete power crossover occurs only at discrete locations where $z = \pi/2\kappa, 3\pi/2\kappa, 5\pi/2\kappa, \ldots$ as shown in Figure 6.5(a).

4. It is customary to use the minimum interaction length required for a complete power crossover as a measure of the interaction or coupling strength. We refer such a length, $l_c = \pi/2\kappa$, as the *coupling length* of the directional coupler. In summary, the power transfer is 100% if and only if the two waveguides are synchronized and the interaction length is an odd-integer multiple of the coupling length.

5. Irrespective of the propagation constant mismatch, all power returns to the original waveguide at $z = \pi/\sigma, 2\pi/\sigma, 3\pi/\sigma, \ldots$.

6. As shown in Figure 6.5, power vacillates back and forth between the two waveguides. The rate of power swing is the slowest when the two waveguides are synchronized. The rate of power exchange increases with δ/κ. A rapid and periodic power swing is detrimental to most applications.

6.4.3 3-dB Directional Couplers

In this subsection, we consider a directional coupler having two parallel and identical waveguides ($\delta = 0$). We also suppose the coupler length is an odd multiple of half coupling length $[L = (m + \frac{1}{2})l_c]$. Under this condition, we obtain, from (6.50) or (6.51)

$$\begin{bmatrix} \mathcal{R}(L) \\ \mathcal{S}(L) \end{bmatrix} = \frac{1}{\sqrt{2}} \begin{bmatrix} 1 & -j \\ -j & 1 \end{bmatrix} \begin{bmatrix} \mathcal{R}(0) \\ \mathcal{S}(0) \end{bmatrix} \qquad (6.56)$$

If an input is fed to a waveguide, say waveguide 1, then $\mathcal{R}(0) = 1$, $\mathcal{S}(0) = 0$. At the output, $\mathcal{R}(L) = 1/\sqrt{2}$, $\mathcal{S}(L) = -j/\sqrt{2}$, and $|\mathcal{R}(L)|^2 = |\mathcal{S}(L)|^2 = \frac{1}{2}$. In other words, the input power splits evenly between two output waveguides. The coupler is often referred to as a *3-dB directional coupler*. Also note that the two output fields are at phase quadrature. These features are the ideal properties of *ideal optical beam splitters or combiners*. Indeed, 3-dB directional couplers are often used as power dividers or combiners (Fig. 6.6).

From the above discussion, it is clear that optical signals may be switched from one waveguide to another if the propagation constant mismatch can be tuned some how. Indeed, the propagation constants can be varied through the *linear electrooptic effects* in electrooptic crystals [6], the charge carrier injection in semiconductors [7, 8], or *magnetooptic effects* in garnets [9, 10]. Thus optical signals can be switched electrically or magnetically. Optical switches based on this directional coupler configuration are discussed in the next subsection.

6.4.4 Directional Couplers as Electrically Controlled Optical Switches

In dielectric materials, the electric polarizability, and therefore the refractive index, may be varied by applying an electric field. For most materials, the index change is small and quadratic in the applied electric field. The effect is known as the *static*

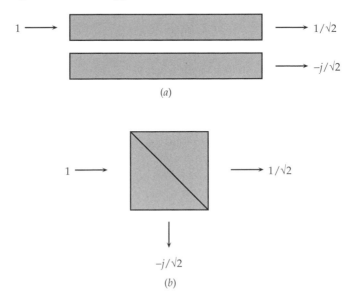

Figure 6.6   (a) 3-dB directional coupler and (b) cubic beam splitter.

Kerr effect or the *Kerr effect* for short. For crystals without a center of inversion symmetry, LiNbO$_3$, LiTaO$_3$, BaTiO$_3$, quartz, GaAs, and InP, for example, the index change Δn varies linearly with the applied electric field. More importantly, the index change is much larger than that due to the static Kerr effect. This is the linear electrooptic effect or *Pockels effect* [6].

In semiconductors, in addition to the Kerr and Pockels effects mentioned above, the refractive index also changes linearly with the injected current density. The index change depends on the doping in the semiconductor [7]. In some semiconductors, the index change is quite large. For example, in InGaAsP/InP, the index change may be as large as 0.01 for an injected current density of 4.8 kA/cm^2 [8]. Such an index change is 2 orders of magnitude larger than that due to the Pockels effect in the semiconductor. But the injected current density required for the index tuning is quite large. A semiconductor junction is usually required to facilitate the current injection.

The dielectric tensor of nonmagnetic materials such as glass can also be perturbed by constant or low-frequency magnetic fields. The effects may be understood in terms of the equation of motion of electrons in electric and magnetic fields. These effects are generally referred to as the magnetooptic effects. The magnetooptic effects in paramagnetic and diamagnetic materials are quite small. But the magnetooptic effects in ferromagnetic materials can be much larger [9, 10].

To take advantage of these effects, waveguides are built on electrooptic (EO), semiconducting, or magnetooptic materials, electrodes are placed over or in the vicinity of the optical waveguides [Fig. 6.4(b)]. When radio-frequency, microwave, or millimeter wave voltages are applied to the electrodes, the propagation constants and/or the coupling constants are affected by the applied fields. In this manner, the propagation constant mismatch between two interacting waveguides or the coupling constants can be varied electrically or magnetically. Then the optical beam is switched from one waveguide to another. For fast responses, ferroelectric or semiconducting materials are often used. These directional couplers have been used as electrically controlled optical switches, routers, lateral shifters, and intensity modulators [11–20]. In the following, we provide a qualitative description of the operation of these directional coupler devices without specifying the mechanism responsible for the changing of propagation constant mismatch. A detailed and quantitative discussion of these directional coupler devices is postponed until Sections 6.6 and 6.7.

To illustrate the optical switching in directional couplers made of EO materials, we consider a directional coupler having two identical waveguides and a constant waveguide spacing [Fig. 6.4(b)]. We further assume that the interaction length L is exactly a coupling length l_c. Recall that κ is $\sqrt{\kappa_1 \kappa_2}$. Suppose the input power is fed to waveguide 1 at $z = 0$. In the absence of an applied voltage, the two waveguides are synchronized, that is, $\delta = 0$. In Figure 6.7, we plot the power distribution in the adjacent waveguide as a function of z/l_c. The solid curve depicts the power distribution $|a_2(z)|^2$ in waveguide 2 in the absence of the applied voltage. Note that all power emerges from the output end of waveguide 2. When a voltage is applied to the electrodes, the two waveguides are no longer synchro-nized. Suppose the applied voltage is to increase β_1 by $\sqrt{3}\kappa$ and to decrease β_2 by $\sqrt{3}\kappa$. Then the propagation constant mismatch δ is $\sqrt{3}\kappa$. The resulting power distribution in the adjacent waveguide is shown as the dashed curve in Figure 6.7. Now all power emerges from the output end of waveguide 1 in the presence of an applied voltage. Clearly these directional couplers built on electrooptic or semicon-ducting materials can work as electrically controlled optical switches or intensity modulators.

In the directional coupler switch considered above, the coupler length L is exactly a coupling length l_c. If the coupler length is not exactly l_c due to a design error or fabrication tolerance, we can force all power to emerge from the output end of the input waveguide by adjusting the applied voltage. For example, if L is off by 1%, that is, L is $0.99l_c$, then all power emerges from the output end of waveguide 1 when the applied voltage increases slightly that $\delta = \sqrt{3.08}\kappa$. In other words, it is possible to adjust the applied voltage to compensate for the design error or fabrication tolerance. But power cross over the adjacent waveguide cannot be complete or 100% if L is not exactly an odd integer multiple of l_c. This is the main shortcoming of using simple or conventional directional couplers as electrically controlled switches or intensity modulators. Ways to circumvent the difficulty will be discussed in Section 6.5.

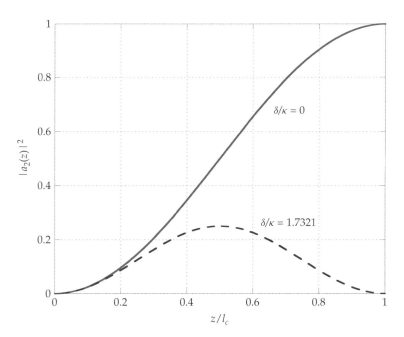

Figure 6.7 Power distribution in waveguide 2 in a directional coupler having a constant spacing and a length of one coupling length. Input power to waveguide 1 is one unit.

6.4.5 Switching Diagram

To facilitate future discussions of the directional coupler operation, we follow Schmidt and Kogelnik and introduce two terms: bar and cross states [21, 22]. Suppose the input power is fed to one and only one waveguide. If for a certain design and/or under certain conditions, all power emerges from the distal end of the same waveguide [Fig. 6.8(a)], the directional coupler is in a *bar state*. On the other hand, if all power exits from the far end of the adjacent waveguide [Fig. 6.8(b)], the directional coupler is in a *cross state*. Take the directional coupler with a uniform cross section and a constant spacing discussed in the last subsection as an example. If L is exactly l_c and if no voltage is applied to the electrodes, then $\delta = 0$ and all power fed to the input waveguide crosses over to the adjacent waveguide at the output end. The directional coupler is in a cross state. On the other hand, if a voltage is applied to the electrodes such that $\delta = \sqrt{3}\kappa$, all power emerges from the output end of the input waveguide. The directional coupler is in a bar state.

To illustrate the directional coupler operation graphically, we use (6.53) to deduce the condition for a bar state. A bar state is reached if $\sigma L = m\pi$ where m is an integer. In terms of κ and δ, the condition is

$$\left(\frac{\delta L}{\pi}\right)^2 + \left(\frac{\kappa L}{\pi}\right)^2 = m^2 \qquad (6.57)$$

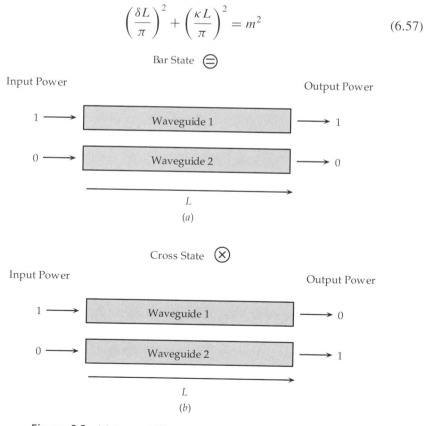

Figure 6.8   (a) Bar and (b) cross states of directional couplers.

Condition (6.57) corresponds to circular loci on the $(\kappa L/\pi)(\delta L/\pi)$ plane (Fig. 6.9). The existence of continuous loci is crucial in many practical devices. For a directional coupler of an arbitrary length built on electrooptic or semiconducting material, a voltage can always be applied, or a current can be injected, to tune the directional coupler to a bar state, assuming the applied voltage is not too high to cause the electrical breakdown or the injected current too intense to introduce the thermal instability.

On the other hand, a cross state is possible only if the two interacting waveguides are perfectly synchronized and if the coupler length is precisely an odd-integer multiple of l_c. The conditions correspond to isolated points on the abscissa on the $(\kappa L/\pi)(\delta L/\pi)$ plane. No voltage or current can be applied to tune a directional coupler to a cross state if the coupler length is not an odd integer multiple of l_c.

The use of loci in the $(\kappa L/\pi)(\delta L/\pi)$ plane to visualize the directional coupler operations and the concept of the cross and bar states were first introduced by Schmidt and Kogelnik [20, 21] in their study of directional coupler switches. Plots on the $(\kappa L/\pi)(\delta L/\pi)$ plane are the *switching diagrams*. On the switching diagrams, $\ominus$ and $\otimes$ denote the bar and cross states, respectively.

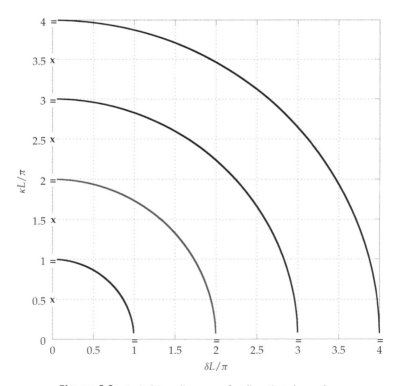

Figure 6.9   Switching diagram of a directional coupler.

6.5 SWITCHED $\Delta\beta$ COUPLERS

As mentioned in Section 6.4.4, it is impossible to tune a directional coupler to a cross state if the coupler length is not exactly an odd-integer multiple of the coupling length. Schmidt and Kogelnik have proposed and demonstrated a new directional coupler configuration to overcome the difficulty [21, 22]. In their design, a directional coupler of arbitrary length is partitioned into two or more sections of equal lengths [Figs. 6.4(c) and 6.10] and voltages are applied to each section to tune the propagation constant mismatch. In particular, the propagation constant mismatches, $\Delta\beta$, in neighboring sections are tuned in opposite directions by applying voltages, or injecting currents, of opposite polarities. The new directional coupler is known as the *switched $\Delta\beta$ coupler*. A switched $\Delta\beta$ directional coupler can always be tuned to a bar state or a cross state even if the coupler length is not exactly an odd-integer multiple of l_c. This is the advantage of the switched $\Delta\beta$ directional couplers over the conventional directional couplers. In this section, we consider the switched $\Delta\beta$ couplers with two sections of equal lengths.

Suppose a directional coupler of length L is divided into two sections of equal lengths. Let the propagation constant mismatches of the two sections be positive and negative as shown in Figure 6.10. Again, we use κ in lieu of $\sqrt{\kappa_1\kappa_2}$ for simplicity. For the first coupler section, the transfer matrix is, from (6.50),

$$[M_1^+] = \begin{bmatrix} \mathcal{A}_1 & -j\mathcal{B}_1 \\ -j\mathcal{B}_1^* & \mathcal{A}_1^* \end{bmatrix} \tag{6.58}$$

where

$$\mathcal{A}_1 = \cos\frac{\sigma L}{2} - \frac{j\delta}{\sigma}\sin\frac{\sigma L}{2} \tag{6.59}$$

and

$$\mathcal{B}_1 = \frac{\kappa}{\sigma}\sin\frac{\sigma L}{2} \tag{6.60}$$

For the second coupler section, the polarity of $\Delta\beta$ is reversed. Thus, the transfer matrix of the second section is

$$[M_1^-] = \begin{bmatrix} \mathcal{A}_1^* & -j\mathcal{B}_1 \\ -j\mathcal{B}_1^* & \mathcal{A}_1 \end{bmatrix} \tag{6.61}$$

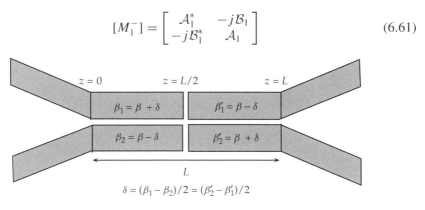

$$\delta = (\beta_1 - \beta_2)/2 = (\beta_2' - \beta_1')/2$$

Figure 6.10 Switched $\Delta\beta$ directional coupler.

The overall transfer matrix of the two-section switched $\Delta\beta$ coupler is the ordered product of $[M_1^-]$ and $[M_1^+]$:

$$[M_2] = [M_1^-][M_1^+] = \begin{bmatrix} \mathcal{A}_2 & -j\mathcal{B}_2 \\ -j\mathcal{B}_2^* & \mathcal{A}_2^* \end{bmatrix} \tag{6.62}$$

where $\mathcal{A}_2 = 1 - 2|\mathcal{B}_1|^2 = 2|\mathcal{A}_1|^2 - 1$, and $\mathcal{B}_2 = 2\mathcal{A}_1^*\mathcal{B}_1$.

A bar state of the two-section switched $\Delta\beta$ coupler requires a vanishing $\mathcal{B}_2$ that requires either $\mathcal{A}_1$ or $\mathcal{B}_1$ to be zero. $\mathcal{A}_1$ is zero only if $\delta = 0$ and if L is $2l_c, 6l_c, 10l_c, \ldots$. The conditions correspond to isolated points $\kappa L/\pi = 1, 3, 5, \ldots$ on the abscissa of the $(\kappa L/\pi)(\delta L/\pi)$ plane (Fig. 6.11). The isolated points are of no practical interest. The loci of $\mathcal{B}_1 = 0$ are circular arcs of radii $2, 4, 6, \ldots$ on the $(\kappa L/\pi)(\delta L/\pi)$ plane. They are shown as solid curves in Figure 6.11.

The condition for a cross states is

$$\mathcal{A}_2 = 1 - 2|\mathcal{B}_1|^2 = 0$$

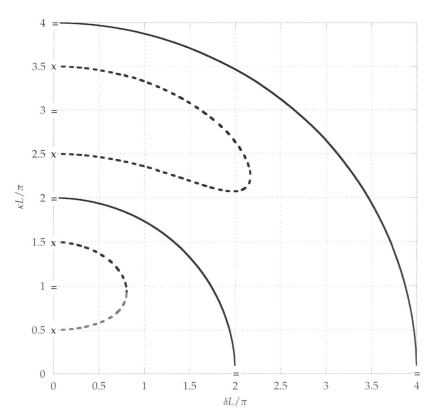

Figure 6.11   Switching diagram of a two-section switched $\Delta\beta$ directional coupler.

which can be expressed in terms of κ, δ, and L explicitly,

$$\frac{\kappa^2}{\kappa^2 + \delta^2} \sin^2 \sqrt{\kappa^2 + \delta^2} \frac{L}{2} = \frac{1}{2} \qquad (6.63)$$

The loci have to be determined numerically. In Figure 6.11, we plot the loci of cross states as dashed curves. The end points of the loci are $\kappa L/\pi = 0.5, 1.5, 2.5, \ldots$ on the vertical axis of the $(\kappa L/\pi)(\delta L/\pi)$ plane. The existence of continuous loci for the cross states is most significant since it makes the design, fabrication, and operation of switched $\Delta\beta$ couplers simple and practical.

To elaborate the operation of switched $\Delta\beta$ couplers further, we consider a two-section switched $\Delta\beta$ coupler with an overall length L of $1.800l_c$ as an example. In Figure 6.12(a) and 6.12(b), we plot $|a_1(z)|^2$ (solid curves) and $|a_2(z)|^2$ (dashed curves) under the conditions of bar and cross states. To tune the coupler into a bar state, a voltage is applied to the electrodes so that $\delta \approx 1.98\kappa$. As shown in Figure 6.12(a), $|a_1(L)|^2$ is 1 and $|a_2(L)|^2$ vanishes completely and the switched $\Delta\beta$ coupler is in a bar state. With $\delta \approx 1.98\kappa$, and $L = 1.800l_c$, each half of the switched $\Delta\beta$ coupler is really a conventional directional coupler operating at a

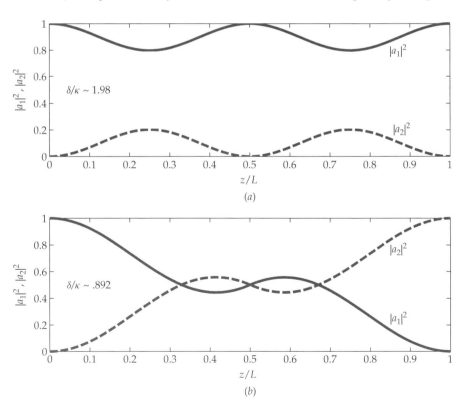

Figure 6.12 Power distribution in a two-section switched $\Delta\beta$ coupler ($L = 1.800l_c$): (a) bar state and (b) cross state.

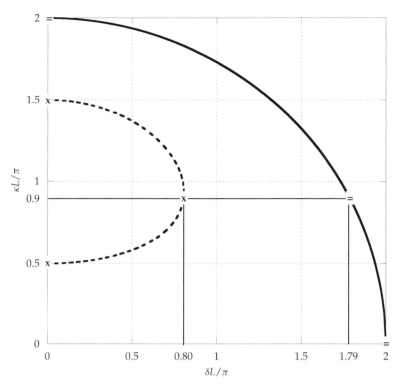

Figure 6.13 Switching diagram of a two-section switched $\Delta\beta$ directional coupler ($L = 1.800l_c$).

bar state, since $|a_1(L/2)|^2 = 1$ and $|a_2(L/2)|^2 = 0$ at the midpoint of the coupler. The two conventional directional couplers are concatenated in series. To tune the two-section switched $\Delta\beta$ coupler to a cross state, we decrease the applied voltage so that $\delta \approx 0.892\kappa$. The power distributions in the two waveguides are shown in Figure 6.12(b). With $\delta \approx 0.892\kappa$, the output power in the input waveguide vanishes completely, and the output power in the adjacent waveguide increases to 1, that is, $|a_1(L)|^2 = 0$ and $|a_2(L)|^2 = 1$. Thus, the switched $\Delta\beta$ coupler with $\delta \approx 0.892\kappa$ is indeed at a cross state. In Figure 6.13, the operation of the switched $\Delta\beta$ coupler with $L = 1.800l_c$ corresponds to two points in the switching diagram. They are (1.78, 0.9) and (0.8, 0.9) on the $(\kappa L/\pi)(\delta L/\pi)$ plane, respectively.

The operation of switched $\Delta\beta$ couplers with three or more sections can be understood in the same manner. The analysis and the switching diagrams of multiple-section switched $\Delta\beta$ couplers can be found in [21–24].

6.6 DIRECTIONAL COUPLERS AS OPTICAL FILTERS

Optical filters exist in a bulk optic or guided-wave optic form. Bulk optical filters may be based on *thin dielectric films* or *birefringent plates* [25]. The center

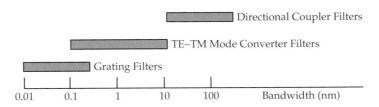

Figure 6.14 Bandwidth of various guided wave optical filters. (After [18].)

wavelength and bandwidth of thin-film optical filters depends on the film thickness, number of films, and the contrast of refractive indices. Guided-wave optical filters can be based on *fiber gratings, integrated optic gratings, TE-TM mode converters,* or directional couplers. The geometries and the operation principles of guided-wave filters are quite different, so are their useful bandwidths as shown in Figure 6.14 [19]. In this section, we discuss the directional coupler filters. Depending on the similarity of and spacing between the interacting waveguides, the filter bandwidth, side lobe levels, and tunability of the directional coupler filters can be quite different. Figure 6.15(a) shows a directional coupler with two identical waveguides and a constant waveguide spacing. In Figures 6.15(b) and 6.15(c), the two waveguides are different. A constant waveguide spacing in the interaction region is shown in Figures 6.15(b). In Figure 6.15(c), the spacing varies with z. In the following subsections, we consider the characteristics of these directional coupler filters.

For directional couplers having a uniform cross section and a constant spacing [Figs. 6.15(a) and 16.5(b)], β_1, β_2, κ_1, κ_2, and δ are constants independent of z. We suppose that an optical signal of unity power is fed to a waveguide, say waveguide 1, at the input ($z = 0$). Power emerging from the adjacent waveguide, waveguide 2, at $z = L$, is taken as the filter output. The frequency response of the directional coupler filter is, from (6.50),

$$|a_2(L)|^2 = \kappa_1^2 L^2 \left[\frac{\sin(\sqrt{\kappa_1\kappa_2 + \delta^2}\, L)}{\sqrt{\kappa_1\kappa_2 + \delta^2}\, L} \right]^2 \qquad (6.64)$$

In the preceding sections, we were interested in directional couplers operating at a single wavelength. Therefore, we were not concerned with the variation of coupling constants and the propagation constant mismatch as functions of λ. In considering the filter characteristics, we have to account for the dependence of κ_1, κ_2, and δ on λ. In fact, the dependence of κ_1, κ_2, and δ on λ is the physical origins of the frequency selectivity of the directional coupler devices. The coupling constants, and equivalently the coupling length l_c, are wavelength dependent on three accounts. As defined in Chapter 2, V is $kh\sqrt{n_f^2 - n_s^2}$. The indices n_f and n_s (and n_c if the cover region is not vacuum) are wavelength dependent. So are V and all parameters depending on V. As the wavelength increases, V decreases, guided modes becomes closer to the cutoffs, and fields of guided modes extend more into the cover and substrate regions. As a result, κ_1, κ_2, and c increase and l_c decreases as λ increases. For many directional couplers, l_c varies inversely with λ when operating in

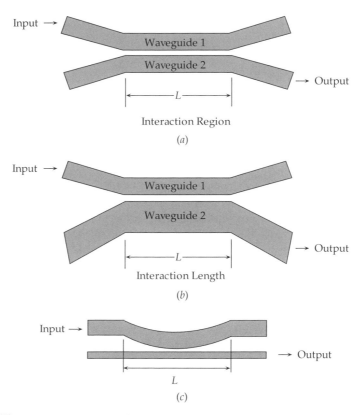

Figure 6.15 Directional coupler filters: (a) Directional coupler filter with two identical waveguides and uniform spacing, (b) directional coupler with two nonidentical waveguides and uniform spacing, and (c) directional coupler filter with two dissimilar waveguides and tapered spacing.

a narrow spectral range. Specific examples are shown in Figure 6.16 [26]. Although the film thickness h is independent of λ, kh decreases as $1/\lambda$. Similarly, the ratios of waveguide width, spacing, and the coupler length over wavelength also depend on λ.

6.6.1 Directional Coupler Filters with Identical Waveguides and Uniform Spacing

First, we consider directional coupler filters with two parallel and identical waveguides [27–29]. The waveguide spacing is a constant as shown in Figure 6.15(a). For these directional couplers, we set $\kappa_1 = \kappa_2$, $\delta = 0$ and obtain from (6.64)

$$|a_2(L)|^2 = \sin^2 \kappa_1 L \qquad (6.65)$$

Since the coupling constant κ_1 is wavelength dependent, so is the output $|a_2(L)|^2$. Suppose the coupler length L is an odd-integer multiple of the coupling length at

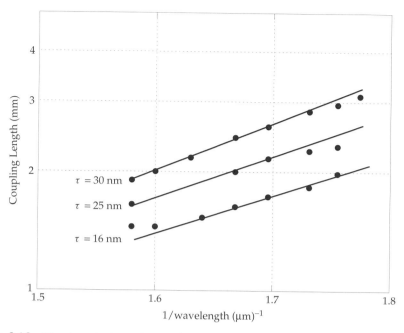

Figure 6.16 Wavelength dependence of the coupling length of directional couplers built on titanium diffused LiNbO$_3$ substrate. (After [26].)

λ_0, that is, $L = (2m + 1)l_c(\lambda)$, then

$$\kappa_1(\lambda_0)L = \left(m + \tfrac{1}{2}\right)\pi \quad m = 0, 1, 2, 3, \ldots \tag{6.66}$$

Clearly the filter response (6.65) has a peak value of 1 at λ_0. To study the bandwidth of the optical filter, we consider two wavelengths λ_1 and λ_2 on either side of λ_0 such that

$$\kappa_1(\lambda_1)L = \left(m + \tfrac{3}{4}\right)\pi \tag{6.67}$$

$$\kappa_1(\lambda_2)L = \left(m + \tfrac{1}{4}\right)\pi \tag{6.68}$$

At λ_1 and λ_2, $|a_2(L)|^2$ is $\tfrac{1}{2}$. In other words, λ_1 and λ_2 are the half-power wavelengths of the frequency response (Fig. 6.17). To arrive at an approximate expression of the half-power bandwidth, we approximate $\kappa_1(\lambda)$ and $\kappa_2(\lambda)$ by the two leading terms of the Taylor series expansion and obtain from (6.66) to (6.68),

$$\Delta\lambda \equiv \lambda_1 - \lambda_2 \approx \frac{\pi}{2L\dfrac{d\kappa_1}{d\lambda}} \tag{6.69}$$

which was first established by Digonnet and Shaw for fiber directional coupler filters [30].

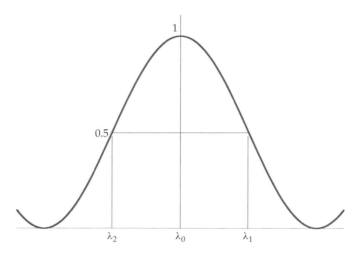

Figure 6.17   Frequency response of a directional coupler filter.

For many coupled waveguides, l_c varies inversely with λ in a narrow range as shown in Figure 6.16. In other words, $\kappa_1(\lambda)$ varies linearly with λ in the narrow spectral range. Then, (6.66) can be written as $\kappa_1(\lambda_0)L = K\lambda_0 L = (m + 1/2)\pi$ where K is proportionality constant. Similar expressions may be written for $\kappa_1(\lambda_1)L$ and $\kappa_1(\lambda_2)L$. Then (6.69) can be simplified as

$$\frac{\Delta\lambda}{\lambda_0} \approx \frac{1}{2m + 1} \tag{6.70}$$

Thus the filter bandwidth $\Delta\lambda$ of a directional coupler filter with two identical and parallel waveguides and a length of $3l_c$ is $0.33\lambda_0$. The bandwidth decreases to $0.20\lambda_0$ when the coupler length increases to $5l_c$. Even for directional couplers as long as $5l_c$, the bandwidth is still very broad. Also note that the filter response (6.65) is the square of a sine function. Thus the side lobes are as large as the main peak. This is a major drawback of directional coupler filters having two identical waveguides and constant spacing.

6.6.2   Directional Coupler Filters with Nonidentical Waveguides and Uniform Spacing

If the two interacting waveguides are different, as shown in Figures 6.15(b) and 6.15(c), the spectral width is much narrower [18, 31–34]. Consider a directional coupler with two dissimilar, uniformly spaced channel waveguides shown in Figure 6.18(a). If the wider and thicker waveguide (waveguide 2) has a smaller index difference compared to that of the thinner and narrower waveguide (waveguide 1), β_1 may equal to β_2 at a certain wavelength [18]. But the rate of change of β_1 and β_2 are different. Then δ varies with λ. We suppose that δ vanishes at λ_0. In a narrow spectral range near λ_0, we may take the refractive indices as wavelength independent. Then β_1, β_2, and therefore δ vary linearly with λ as depicted

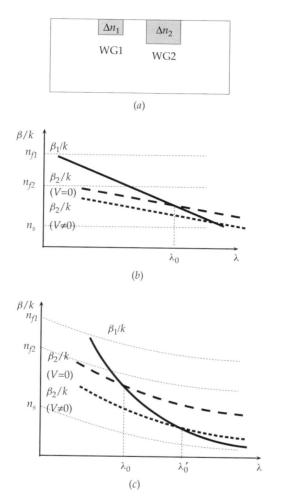

Figure 6.18 Directional coupler with two nonidentical waveguides: (*a*) Two waveguides with different width, thickness, and index difference ($\Delta n_1 > n_2$), (*b*) β_1 and β_2 as functions of λ, assuming wavelength-independent n_s and n_f, and (*c*) β_1 and β_2 as functions of λ, assuming wavelength-dependent n_s and n_f.

schematically in Figure 6.18(*b*). If the index dispersion of waveguide materials is taken into account, then δ varies nonlinearly with λ. This is shown schematically in Figure 6.18(*c*) [18]. Although κ_1, κ_2, and l_c are also wavelength dependent, the dominant term affecting the filter characteristic is the variation of δ with λ. In Figure 6.19, we plot $|a_2(L)|^2$ as a function of $\delta/\sqrt{\kappa_1\kappa_2}$ for three values of $\sqrt{\kappa_1\kappa_2}L$. If the two modes guided by the two waveguides are synchronized and L is exactly an odd-integer multiple of l_c at λ_0, then the filter response peaks at λ_0 and the peak response is κ_1/κ_2 and that is close to 1. Also note that the side lobes are smaller than the major peak.

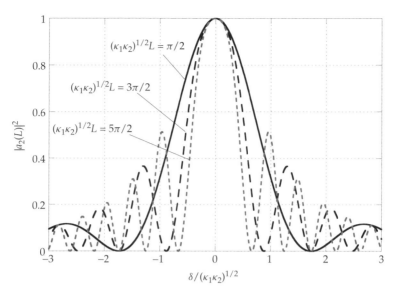

Figure 6.19   Frequency response of directional coupler filters with nonidentical waveguides.

To determine the filter spectral width, we find values of $\delta/\sqrt{\kappa_1\kappa_2}$ where the filter response drops to half of the peak value. For the three cases shown in Figure 6.19, the half-power points are $\delta/\sqrt{\kappa_1\kappa_2} \sim \pm0.798, \pm0.538$, and ±0.429, respectively. For convenience, we label the numerical values as q_0, q_1, and q_2. Then

$$\delta_{\text{HP } m} = q_m \sqrt{\kappa_1 k_2} \quad m = 0, 1, 2 \qquad (6.71)$$

To relate $\delta_{\text{HP}m}$ to the wavelength directly, we recall from (6.13) that

$$\delta(\lambda) \equiv \frac{\beta_2(\lambda) - \beta_1(\lambda)}{2} = \frac{\pi}{\lambda}[N_2(\lambda) - N_1(\lambda)] \qquad (6.72)$$

where $N_1(\lambda)$ and $N_2(\lambda)$ are the effective refractive indices of modes guided by waveguides 1 and 2. At λ_0, $N_1(\lambda_0) = N_2(\lambda_0)$ and $\delta(\lambda_0) = 0$. For wavelengths near λ_0, we again use the two leading terms of the Taylor series expansion to approximate $\delta(\lambda)$:

$$\delta(\lambda) = \delta(\lambda_0) + (\lambda - \lambda_0)\left.\frac{d\delta(\lambda)}{d\lambda}\right|_{\lambda_0} + \cdots \qquad (6.73)$$

Recalling that $\delta(\lambda_0) = 0$, the above equation becomes

$$\delta(\lambda) \approx (\lambda - \lambda_0)\frac{\pi}{\lambda_0}\left.\left(\frac{dN_2}{d\lambda} - \frac{dN_1}{d\lambda}\right)\right|_{\lambda_0} \qquad (6.74)$$

Combining (6.71), (6.72), (6.73), and (6.74), we obtain

$$\frac{\lambda_{\mathrm{HP}\,m} - \lambda_0}{\lambda_0} \approx \frac{q_m \sqrt{\kappa_1 \kappa_2}}{\pi \left(\dfrac{dN_2}{d\lambda} - \dfrac{dN_1}{d\lambda} \right)\bigg|_{\lambda_0}} \tag{6.75}$$

Making use of the fact that $\sqrt{\kappa_2 \kappa_1} L = \left(m + \frac{1}{2} \right) \pi$, we obtain

$$\frac{\lambda_{\mathrm{HP}\,m} - \lambda_0}{\lambda_0} \approx \frac{q_m \left(m + \frac{1}{2} \right)}{L \left(\dfrac{dN_2}{d\lambda} - \dfrac{dN_1}{d\lambda} \right)\bigg|_{\lambda_0}} \tag{6.76}$$

Thus the full width between half-power points is

$$\frac{\Delta\lambda}{\lambda_0} = 2\frac{\lambda_{\mathrm{HP}\,m} - \lambda_0}{\lambda_0} \approx \frac{q_m(2m+1)}{L \left(\dfrac{dN_2}{d\lambda} - \dfrac{dN_1}{d\lambda} \right)\bigg|_{\lambda_0}} \tag{6.77}$$

For $m = 0$, 1, and 2, $q_m(2m + 1)$ are 0.798, 1.614, and 2.145, respectively. For $m = 0$, (6.77) is identical to the expression derived by Marcus for fiber directional coupler filters [31]. When typical values of $(dN_1/d\lambda) - (dN_2/d\lambda)$ are used, we obtain $\Delta\lambda \sim 0.02\lambda_0$. In other words, the bandwidth of directional coupler filters with two dissimilar and uniformly spaced waveguides is much narrower than that of the directional coupler filters having two similar waveguides.

To estimate the side lobe levels, again we consider a directional coupler filter with $L = l_c$. As indicated by (6.64), the main response peak is κ_1/κ_2. Numerical calculation shows that the two largest minor peaks are $0.116\kappa_1/\kappa_2$ and they occur at $\delta/\sqrt{\kappa_1 \kappa_2} = \pm 2.83$. In other words, the largest side lobe levels are -9.4 dB below the main peak. If the coupler length is $3l_c$ or $5l_c$, the largest side lobe increases to $0.366\kappa_1/\kappa_2$ (-4.4 dB) and 0.514 (-2.9 dB), respectively.

Since the waveguide widths, thickness, and index differences of the two channel waveguides can be chosen independently, the design of directional coupler filters with dissimilar channel waveguides is more flexible than that of directional couplers with two identical waveguides. The side lobes are also smaller if the two waveguides are dissimilar. If the directional coupler filters are built on electrooptic or semiconducting substrates, δ can be tuned by applying voltages to the electrodes near the waveguide regions or by injecting current into semiconductors [32–34]. In Figure 6.18(b) and 6.18(c), we show, schematically, the effect of an applied voltage on β_2/k and thus the peak frequency response wavelength of the filter. It clearly shows that the peak transmission wavelength can be tuned electrically.

6.6.3 Tapered Directional Coupler Filters

While the side lobes of directional coupler filters with dissimilar and uniformly spaced waveguides are reduced, they are still quite large even if the two interacting waveguides are quite different. The side lobes can be reduced further if the spacing between waveguides is tapered [Figs. 6.15(c) and 6.20] [35, 36]. For directional couplers with tapered spacing, $\kappa_1(z)$ and $\kappa_2(z)$ vary with z. On the other hand, if the two waveguide cross sections do not change with z, then δ is independent of z.

By tracing the steps leading to the coupled-mode equations (6.35) and (6.36), or (6.44) and (6.45), we conclude that the coupled-mode equations remain valid even if $\kappa_1(z)$ and $\kappa_2(z)$ vary with z. But the final results, the transfer matrix (6.50) and (6.51) for example, based on the assumption of constant κ_1 and κ_2 are no longer valid. In this subsection, we begin with (6.44) and (6.45) and analyze the filter response of directional coupler filters with tapered spacing.

Again, we assume that power is fed to waveguide 1 and output is extracted from waveguide 2. For convenience, we take $z = -L/2$ as the input point and $z = L/2$ as the output point (Fig. 6.20). The boundary conditions are $\mathcal{R}(-L/2) = 1$ and $\mathcal{S}(-L/2) = 0$. To study the filter response, it is necessary to find $\mathcal{S}(L/2)$. To proceed, we follow Alferness and Cross and define [36]

$$\rho(z) = -j\frac{\mathcal{S}(z)}{\mathcal{R}(z)} \tag{6.78}$$

From the definition of $\rho(z)$ and that power is conserved in the directional couplers, that is, $|\mathcal{R}(z)|^2 + |\mathcal{S}(z)|^2 = 1$, we show

$$|\mathcal{S}(z)|^2 = \frac{|\rho(z)|^2}{1 + |\rho(z)|^2} \tag{6.79}$$

The filter output $|\mathcal{S}(L/2)|^2$ is readily obtained if $|\rho(L/2)|^2$ is determined. To solve for $\rho(z)$, we obtain from (6.44) and (6.45) a nonlinear differential equation:

$$\frac{d\rho(z)}{dz} = j2\delta\rho(z) - [\kappa_1(z) + \kappa_2(z)\rho^2(z)] \tag{6.80}$$

The boundary condition is $\rho(-L/2) = 0$. A general and analytic solution for $\rho(z)$ with an arbitrary $\kappa_1(z)$, $\kappa_2(z)$, and δ is not known. But the solution of a special

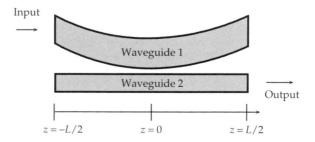

Figure 6.20   Directional coupler filter with a tapered spacing.

TABLE 6.1　Side Lobe Levels of Tapered Directional Coupler Filters [36]

Taper	Tapering Function	Side Lobe Level	
		$\kappa_0 = \pi/2$ (dB)	$\kappa_0 = 3\pi/2$ (dB)
Uniform	$\sqrt{\kappa_1\kappa_2}L = \kappa_0$	-9.4	-4.4
Hamming	$\sqrt{\kappa_1\kappa_2}L = \kappa_0[1 + 0.852\cos(2\pi z/L)]$	-29.7	-25.6
Raised cosine	$\sqrt{\kappa_1\kappa_2}L = \kappa_0[1 + \cos(2\pi z/L)]$	-24.1	-19.2
Blackman	$\sqrt{\kappa_1\kappa_2}L = \kappa_0[1 + 1.19\cos(2\pi z/L)$ $+ 0.19\cos(4\pi z/L)]$	-33.3	-28.4
Kaiser	$\sqrt{\kappa_1\kappa_2}L = \kappa_0 10\mathrm{csch}(10)I_0$ $\times (10\sqrt{1 - (2z/L)^2})$	-35.2	-30.7

case is known. The special case is $\delta = 0$ and $\kappa_1(z) = \kappa_2(z)$. Under these conditions, (6.80) becomes

$$\frac{1}{1 + \rho^2(z)}\frac{d\rho(z)}{dz} = -\kappa_1(z) \tag{6.81}$$

The solution is

$$\rho(z) = -\tan\left[\int_{-L/2}^{z}\kappa_1(z')\,dz'\right] \tag{6.82}$$

Thus, the output of the directional coupler filters with tapered spacing is

$$|S(L/2)|^2 = \sin^2\left[\int_{-L/2}^{L/2}\kappa_1(z')\,dz'\right] \tag{6.83}$$

This is the filter response of a special tapered directional coupler.

Alferness and Cross have studied the nonlinear differential equation (6.80) numerically [36]. The key results for directional couplers with Hamming, Blackman, and Kaiser and raised cosine tapering are tabulated in Table 6.1. Note that the side lobes are at least -19 dB below the main peak.

6.7　INTENSITY MODULATORS BASED ON DIRECTIONAL COUPLERS

As indicated in Section 6.4.4, the propagation constant of a guided mode may be varied electrically if the optical waveguide is made of electrooptic or semiconducting materials. The propagation constant change leads to the *phase modulation*. But an intensity modulation is often preferred. The phase modulation can be converted to the *intensity modulation* if the electrooptic or semiconducting waveguide is in one arm of a Mach–Zehnder interferometer or a part of a directional coupler [13, 14, 19, 23, 24, 37–39]. In this section, we use a lithium niobate directional coupler as an example to illustrate the operation of directional coupler modulators.

6.7.1 Electrooptic Properties of Lithium Niobate

Lithium niobate (LiNbO$_3$) is a ferroelectric crystal with a *trigonal (3m) symmetry*. A trigonal crystal has a unique axis. When the trigonal crystal is rotated by 120° with the unique axis as the axis of rotation, the rotated crystal lattice is indistinguishable from its original lattice prior to the rotation. In crystallography, the unique axis is referred to as the *c axis*. In the following discussions, we designate the crystallographic axes of LiNbO$_3$ crystal as the X, Y, and Z axes and choose the c axis as the Z axis. From an optics point of view, LiNbO$_3$ is a *uniaxial material*. The particular axis mentioned above is the *optic axis* of the uniaxial material. Electric fields polarized in directions normal to the optic axis "experience" an ordinary index of refraction n_o and those polarized in the direction of the optic axis "see" an extraordinary index of refraction n_e [6, 40].

In analyzing modes propagating in waveguides, we often choose the z direction as the direction of propagation and x and y directions as two directions normal to the direction of propagation. In the following discussions, we will retain this practice. Clearly, x, y, and z axes may or may not be the same as the crystallographic X, Y, and Z axes. In Figure 6.21(a), we show a directional coupler made of a LiNbO$_3$ crystal. Fields of guided modes are confined mainly in a thin layer near the top surface. As indicated in Figure 6.21(a), the top surface is normal to the Z axis. We refer such a LiNbO$_3$ plate or wafer as a Z cut plate or wafer.

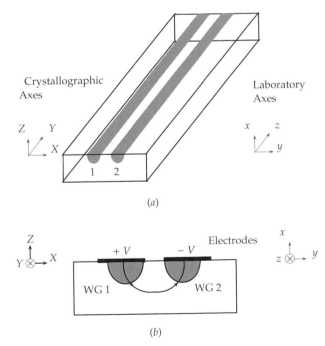

(a)

(b)

Figure 6.21 Directional coupler on a Z-cut Y-propagating LiNbO3 substrate: (a) without electrodes and (b) cross-sectional view of directional coupler with electrodes.

To describe the electrooptic effects, we have to specify fields of the guided modes and that of the modulating field. While the optical beam is in the visible or near-infrared (IR) region, the modulating field is at millimeter, microwave, radiowave, or lower frequencies. We denote the electric field intensity of the optical beam by $\mathbf{E}$ and the modulating electric field by $\mathbf{E}_m$ with a subscript m.

Because of the crystal symmetry, many linear EO coefficients of $3m$ materials are zero. The nonzero linear EO coefficients of crystals with $3m$ symmetry are r_{22}, r_{13}, r_{33}, and r_{51}. When a $3m$ material is subjected a modulating field $\mathbf{E}_m$, the change of the principal indices are [6, 39–41]

$$\Delta n_X \approx -\tfrac{1}{2} n_o^3 (-r_{22} E_{mY} + r_{13} E_{mZ}) \tag{6.84}$$

$$\Delta n_Y \approx -\tfrac{1}{2} n_o^3 (r_{22} E_{mY} + r_{13} E_{mZ}) \tag{6.85}$$

$$\Delta n_Z \approx -\tfrac{1}{2} n_e^3 r_{33} E_{mZ} \tag{6.86}$$

where E_{mX}, E_{mY}, and E_{mZ} are the X, Y, and Z components of $\mathbf{E}_m$. At 0.633 μm, the principal indices and linear EO coefficients of LiNbO$_3$ are [6, 40]

$$n_o = 2.286 \quad n_e = 2.200$$

$$r_{13} = 8.6 \times 10^{-12} \text{ m/V} \quad r_{22} = 3.4 \times 10^{-12} \text{ m/V} \tag{6.87}$$

$$r_{33} = 30.8 \times 10^{-12} \text{ m/V} \quad r_{51} = 28 \times 10^{-12} \text{ m/V}$$

Note that n_o is slightly larger than n_e. But r_{33} is much larger than r_{13} and r_{22}. And r_{51} has no effect on the principal indices directly. Although n_o, n_e, and r_{ij} vary with wavelength, r_{33} is always the largest linear EO coefficient in visible and near-IR regions. To take advantage of r_{33}, EO waveguides are often built on Z-cut LiNbO$_3$ substrates and electrodes are positioned to maximizing E_{mZ} in the waveguide region.

6.7.2 Dielectric Waveguide with an Electrooptic Layer

To discuss the EO effect in dielectric waveguides, we begin by considering a thin-film waveguide with indices n_c, n_f, and n_s and film thickness h. The effective index N of a guided mode is, assuming that $n_f - n_s \ll n_s$,

$$N \approx n_s + b(n_f - n_s) \tag{6.88}$$

If n_f changes slightly by Δn_f for some reason, N also changes. The change of the effective index is

$$\Delta N \approx b \, \Delta n_f + (n_f - n_s) \frac{db}{dn_f} \Delta n_f = \left(b + \tfrac{1}{2} V \frac{db}{dV} \right) \Delta n_f \tag{6.89}$$

Suppose the dielectric waveguide in question has a Z-cut LiNbO$_3$ layer as the thin-film region. We also assume that the modulating field is in the Z direction and

E_{mZ} is uniform in the thin-film region. Then the index changes due to the EO effect in the LiNbO$_3$ layer can be obtained by substituting (6.86) into (6.89):

$$\Delta N \approx \left(b + \frac{1}{2}V\frac{db}{dV}\right)\Delta n_f = -\frac{1}{2}\left(b + \frac{1}{2}V\frac{db}{dV}\right)n_f^3 r_{33}E_{mZ} \qquad (6.90)$$

If the waveguide is a three-dimensional waveguide and/or the modulating field E_{mZ} is not uniform in the film region, we need to account for the field distribution of the guided mode and that of the modulating field. This can be done by introducing an *overlap factor* or *overlap integral* ξ in the above expression. The overlap factor ξ is an integral of the field of the guided mode and that of the modulating field [38]. Thus, the effective index change is

$$\Delta N \approx -\frac{1}{2}\left(b + \frac{1}{2}V\frac{db}{dV}\right)\xi n^3 r_{33}E_{mZ} \qquad (6.91)$$

6.7.3   Directional Coupler Modulator Built on a Z-Cut LiNbO$_3$ Plate

Now we consider the directional coupler shown in Figure 6.21(a). The directional coupler has two identical waveguides. Electrodes are placed directly above the waveguide regions as sketched schematically in Figure 6.21(b). Two complications are obvious. First, the two interacting waveguides are three-dimensional waveguides, not two-dimensional thin-film waveguides. Second, the modulating field is not uniform in the waveguide regions and it has an X component in addition to the Z component. As noted earlier, we use an overlap integral ξ to account for these complications.

Since the two waveguides are identical, the two effective refractive indices are identical when no voltage is applied to the electrodes. Thus $\delta = 0$ when E_{mZ} is zero. When a voltage is applied to electrodes, the effective indices are perturbed by the EO effect in LiNbO$_3$. Note that E_{mZ} in the two waveguide regions points to the opposite directions. Thus ΔN_1 and ΔN_2 have the opposite polarities. In the presence of E_{mZ}, the propagation constant mismatch is, from (6.13),

$$\delta = \tfrac{1}{2}k(\Delta N_1 - \Delta N_2) = k\,\Delta N_1 = -\frac{1}{2}k\left(b + \frac{1}{2}V\frac{db}{dV}\right)\xi n^3 r_{33}E_{mZ} \qquad (6.92)$$

To consider the modulator operation, we suppose that power is fed to one waveguide only, say waveguide 1, then $|a_1(0)|^2 = 1$ and $|a_2(0)|^2 = 0$. There are two waveguides and power emerging from either waveguide may be chosen as the modulator output. If we take the distal end of the straight through waveguide as the output port, then is $|a_1(L)|^2$ is the modulator output. Then the *modulation transfer function* is, from (6.53),

$$H_{11}(E_{mZ}) = \frac{|a_1(L)|^2}{|a_1(0)|^2} = 1 - \frac{\kappa^2}{\kappa^2 + \delta^2}\sin^2\sqrt{\kappa^2 + \delta^2}L \qquad (6.93)$$

By expressing the coupling coefficient κ in terms of the coupling length l_c, and making use of (6.92), we obtain

$$H_{11}(E_{mZ}) = 1 - \frac{1}{1 + q^2 E_{mZ}^2} \sin^2\left(\frac{\pi}{2} \frac{L}{l_c} \sqrt{1 + q^2 E_{mZ}^2}\right) \tag{6.94}$$

where

$$q = 2\frac{l_c}{\lambda}\left(b + \frac{1}{2}V\frac{db}{dV}\right)\xi n^3 r_{33} \tag{6.95}$$

If we take the far end of the crossover waveguide as the output port, then the modulation transfer function is

$$H_{12}(E_{mZ}) = \frac{1}{1 + q^2 E_{mZ}^2} \sin^2\left(\frac{\pi}{2} \frac{L}{l_c} \sqrt{1 + q^2 E_{mZ}^2}\right) \tag{6.96}$$

In Figures 6.22 and 6.23, we plot H_{11} as functions of $q E_{mZ}$ for two coupler lengths. The modulation transfer function of the conventional directional couplers are shown as solid curves. Note that the modulating field E_{mZ} required to increase the modulation transfer functions to 1 is smaller for a longer directional coupler $(L = 3l_c)$ than a shorter coupler $(L = l_c)$.

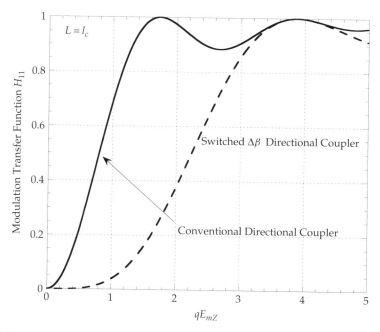

Figure 6.22 Modulation transfer function H_{11} of conventional and switched $\Delta\beta$ directional couplers with $L = l_c$.

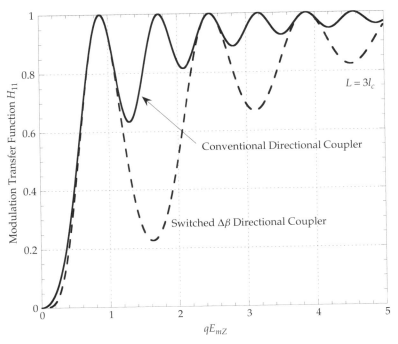

Figure 6.23 Modulation transfer function H_{11} of conventional and switched $\Delta\beta$ directional couplers with $L = 3l_c$.

We can use switched $\Delta\beta$ couplers as intensity modulators too. If we use the output of the straight through waveguide as the modulator output, then the modulation transfer function is, from (6.62),

$$H_{11}(E_{mZ}) = |\mathcal{A}_2|^2 = [1 - 2|\mathcal{B}_1|^2]^2 \tag{6.97}$$

If we use the far end of the crossover waveguide as the output port, then the modulation transfer function is

$$H_{12}(E_{MZ}) = |\mathcal{B}_2|^2 = 4|\mathcal{A}_1\mathcal{B}_1|^2 = 4(1 - |\mathcal{B}_1|^2)\,|\mathcal{B}_1|^2 \tag{6.98}$$

where

$$|\mathcal{B}_1|^2 = \frac{1}{1 + q^2 E_{mZ}^2}\sin^2\left(\frac{\pi}{4}\frac{L}{\ell_c}\sqrt{1 + q^2 E_{mZ}^2}\right) \tag{6.99}$$

Function H_{11} of switched $\Delta\beta$ directional coupler modulators is depicted in Figures 6.21 and 6.22 as dashed curves. For $L = l_c$, the conventional and switched $\Delta\beta$ directional coupler modulators have different H_{11}. The E_{mZ} required to change H_{11} to 1 is twice as large for switched $\Delta\beta$ directional couplers as that of the conventional directional couplers. For $L = 3l_c$, the modulation transfer functions of the two directional couplers are most identical for $|q E_{mZ}| < 1.2$.

As shown in Figures 6.21 and 6.22, none of the modulation transfer functions is linear in E_{mz} or the applied voltage. An intensity modulator with a nonlinear transfer function would cause signal distortion and generate harmonics and inter-modulation products. Clearly, these effects are harmful to the analog communication systems [42]. We could linearize the transfer function by shifting the operating point away from the origin and introducing a bias δ or a bias voltage. In doing so, the dynamic range suffers. Many schemes have been proposed to linearize the modulation transfer function. Details can be found in Refs. [41, 42].

6.8 NORMAL MODE THEORY OF DIRECTIONAL COUPLERS WITH TWO WAVEGUIDES

As mentioned in Section 6.1, an alternate approach to the directional coupler problems is to express the field in terms of the normal modes of the composite waveguide. For this purpose, we return to the coupled-mode equations (6.37) and (6.38). For a directional coupler with uniform cross section and constant spacing, the coupled differential equations can be written in a matrix form:

$$\frac{d}{dz}\begin{bmatrix} A_1(z) \\ A_2(z) \end{bmatrix} = -j\begin{bmatrix} \beta_1 & \kappa_1 \\ \kappa_2 & \beta_2 \end{bmatrix}\begin{bmatrix} A_1(z) \\ A_2(z) \end{bmatrix} \qquad (6.100)$$

For brevity, we introduce two matrices:

$$[Q] = \begin{bmatrix} \beta_1 & \kappa_1 \\ \kappa_2 & \beta_2 \end{bmatrix} \quad \text{and} \quad [A] = [\,A_1(z) \quad A_2(z)\,]^t$$

The superscript t stands for the matrix transposition. Then the above matrix equation becomes

$$\frac{d}{dz}[A] = -j[Q][A] \qquad (6.101)$$

Matrix $[Q]$ has two eigenvalues and they are

$$\beta_{s,a} = \frac{1}{2}\left[\beta_1 + \beta_2 \pm \sqrt{(\beta_1 - \beta_2)^2 + 4\kappa_1\kappa_2}\right]. \qquad (6.102)$$

Let the eigenvectors associated with the two eigenvalues be $[v_{s1} \quad v_{s2}]^t$ and $[v_{a1} \quad v_{a2}]^t$. Then

$$\begin{bmatrix} \beta_1 & \kappa_1 \\ \kappa_2 & \beta_2 \end{bmatrix}\begin{bmatrix} v_{s1} \\ v_{s2} \end{bmatrix} = \beta_s\begin{bmatrix} v_{s1} \\ v_{s2} \end{bmatrix} \qquad (6.103a)$$

$$\begin{bmatrix} \beta_1 & \kappa_1 \\ \kappa_2 & \beta_2 \end{bmatrix}\begin{bmatrix} v_{a1} \\ v_{a2} \end{bmatrix} = \beta_a\begin{bmatrix} v_{a1} \\ v_{a2} \end{bmatrix} \qquad (6.103b)$$

Based on the eigenvectors, we construct a new 2×2 matrix:

$$[V] = \begin{bmatrix} v_{s1} & v_{a1} \\ v_{s2} & v_{a2} \end{bmatrix} \tag{6.104}$$

that may be used to diagonalize $[Q]$. Specifically, we apply the similarity transformation [43] and obtain

$$[V]^{-1}[Q][V] = [\Lambda] \tag{6.105}$$

where

$$[\Lambda] = \begin{bmatrix} \beta_s & 0 \\ 0 & \beta_a \end{bmatrix} \tag{6.106}$$

is a diagonal matrix having β_s and β_a as the diagonal matrix elements.

By defining $[U] = [V]^{-1}[A]$ and substituting it into (6.100), we transform (6.100) to a new matrix differential equation:

$$\frac{d}{dz}[U] = -j[\Lambda][U] \tag{6.107}$$

Since $[\Lambda]$ is a diagonal matrix, the solution for $[U]$ becomes obvious

$$[U] = \begin{bmatrix} a_{s0}e^{-j\beta_s z} \\ a_{a0}e^{-j\beta_a z} \end{bmatrix} \tag{6.108}$$

In the above equation, a_{s0} and a_{a0} are amplitudes and β_s and β_a are the propagation constants of the normal modes. In other words, the two matrix elements of $[U]$ simply describe the propagation of the two normal modes. Once $[U]$ is known, we recover $[A]$ by noting $[A] = [V][U]$

$$\begin{bmatrix} A_1(z) \\ A_2(z) \end{bmatrix} = \begin{bmatrix} v_{s1}a_{s0}e^{-j\beta_s z} + v_{a1}a_{a0}e^{-j\beta_a z} \\ v_{s2}a_{s0}e^{-j\beta_s z} + v_{a2}a_{a0}e^{-j\beta_a z} \end{bmatrix} \tag{6.109}$$

This completes the normal mode theory of directional couplers having two waveguides.

For directional couplers with the two identical waveguides, the results are particularly simple and intuitive since $\beta_1 = \beta_2$ and $\kappa_1 = \kappa_2$. The two eigenvalues are $\beta_s = \beta_1 + \kappa_1$ and $\beta_a = \beta_1 - \kappa_1$. The associated eigenvectors are $[1 \quad 1]^t$ and $[1 \quad -1]^t$. Thus, we arrive at the same conclusion obtained in Section 6.3.4. Namely, the normal mode fields are symmetric and antisymmetric with respect to the centerline as shown in Figure 6.2. In terms of a_{s0}, a_{a0}, β_s, and β_a, matrix $[A]$ is

$$\begin{bmatrix} A_1(z) \\ A_2(z) \end{bmatrix} = \begin{bmatrix} a_{s0}e^{-j\beta_s z} + a_{a0}e^{-j\beta_a z} \\ a_{s0}e^{-j\beta_s z} - a_{a0}e^{-j\beta_a z} \end{bmatrix} \tag{6.110}$$

Clearly, the periodic power exchange between the two waveguides is due to the interference of the symmetric and antisymmetric normal modes. The propagation constants $\beta_1 \pm \kappa_1$ of the two normal modes are different because of coupling constant κ_1. It also reveals the existence of a simple relation between the coupling constant and the propagation constants of the normal modes: $\kappa_1 = (\beta_s - \beta_a)/2$. This fact has been used to determine κ_1 experimentally or numerically.

6.9 NORMAL MODE THEORY OF DIRECTIONAL COUPLERS WITH THREE OR MORE WAVEGUIDES

So far, we have discussed directional couplers with two waveguides. In this section, we extend our discussions to directional couplers having three or more waveguides. To illustrate the basic procedure, we consider a coupler with three waveguides. For a directional coupler with three waveguides, we generalize the coupled mode equations given in (6.99) and obtain

$$\frac{d}{dz} \begin{bmatrix} A_1(z) \\ A_2(z) \\ A_3(z) \end{bmatrix} = -j \begin{bmatrix} \beta_1 & \kappa_{12} & \kappa_{13} \\ \kappa_{21} & \beta_2 & \kappa_{23} \\ \kappa_{31} & \kappa_{32} & \beta_3 \end{bmatrix} \begin{bmatrix} A_1(z) \\ A_2(z) \\ A_3(z) \end{bmatrix} \qquad (6.111)$$

where β_i is the propagation constant of the ith isolated waveguide, and κ_{ij} is the coupling constants quantifying the interaction between waveguides i and j. To solve the coupled-mode equations, we follow the procedure described in the last section. First, we determine the eigenvalues of 3×3 matrix $[Q]$. We find the three eigenvalues and label them as β_a, β_b, and β_c. They are the three propagation constants of the three normal modes. Then the amplitudes of the field guided by waveguide i can be written as

$$A_i(z) = P_i e^{j\beta_a z} + Q_i e^{j\beta_b z} + R_i e^{j\beta_c z} \qquad (6.112)$$

The propagation constants are functions of β_i and κ_{ij} and the amplitude constants depend on the excitation. While the procedure is straightforward, the results are complicated. Some results have been reported by Iwasaki et al. [44]. To illustrate the basic procedure, we simplify the special case further.

We consider a directional coupler having three identical and equally spaced waveguides as shown in Figure 6.24. Then $\beta_1 = \beta_2 = \beta_3 = \beta$ and $\kappa_{12} = \kappa_{23} = \kappa$. Since the coupling constant decreases precipitously as the waveguide separation increases, the coupling between the two outer waveguides is much smaller than that

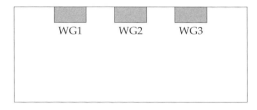

WG1   WG2   WG3

Figure 6.24   Directional coupler with three waveguides.

between nearest neighbors. As an approximation, we ignore κ_{13} and κ_{31} completely. In other words, we account for the interaction between nearest waveguides only. Then, (6.111) can be simplified as

$$\frac{d}{dz}\begin{bmatrix} A_1(z) \\ A_2(z) \\ A_3(z) \end{bmatrix} = -j\begin{bmatrix} \beta & \kappa & 0 \\ \kappa & \beta & \kappa \\ 0 & \kappa & \beta \end{bmatrix}\begin{bmatrix} A_1(z) \\ A_2(z) \\ A_3(z) \end{bmatrix} \tag{6.113}$$

The three eigenvalues of the 3×3 matrix are $\beta \pm \sqrt{2}\kappa$ and β. The eigenvectors are $1/\sqrt{2}[1 \quad \sqrt{2} \quad 1]^t$, $1/\sqrt{2}[-1 \quad \sqrt{2} \quad -1]^t$, and $1/\sqrt{2}[\sqrt{2} \quad 0 \quad -\sqrt{2}]^t$. Based on the three eigenvectors, we construct $[V]$:

$$[V] = \frac{1}{\sqrt{2}}\begin{bmatrix} 1 & -1 & \sqrt{2} \\ \sqrt{2} & \sqrt{2} & 0 \\ 1 & -1 & -\sqrt{2} \end{bmatrix} \tag{6.114}$$

and obtain

$$[V]^{-1} = \frac{1}{2\sqrt{2}}\begin{bmatrix} 1 & \sqrt{2} & 1 \\ -1 & \sqrt{2} & -1 \\ \sqrt{2} & 0 & -\sqrt{2} \end{bmatrix} \tag{6.115}$$

By defining

$$\begin{bmatrix} U_1(z) \\ U_2(z) \\ U_3(z) \end{bmatrix} = [V]^{-1}[A] = \frac{1}{2\sqrt{2}}\begin{bmatrix} 1 & \sqrt{2} & 1 \\ -1 & \sqrt{2} & -1 \\ \sqrt{2} & 0 & -\sqrt{2} \end{bmatrix}\begin{bmatrix} A_1(z) \\ A_2(z) \\ A_3(z) \end{bmatrix} \tag{6.116}$$

we transform (6.113) to a matrix differential equation having a diagonal matrix on the right-hand side [43]. The result is

$$\frac{d}{dz}\begin{bmatrix} U_1(z) \\ U_2(z) \\ U_3(z) \end{bmatrix} = -j\begin{bmatrix} \beta + \sqrt{2}\kappa & 0 & 0 \\ 0 & \beta - \sqrt{2}\kappa & 0 \\ 0 & 0 & \beta \end{bmatrix}\begin{bmatrix} U_1(z) \\ U_2(z) \\ U_3(z) \end{bmatrix} \tag{6.117}$$

Solutions of the matrix differential equation are

$$\begin{bmatrix} U_1(z) \\ U_2(z) \\ U_3(z) \end{bmatrix} = \begin{bmatrix} U_1(0)e^{-j(\beta+\sqrt{2}\kappa)z} \\ U_2(0)e^{-j(\beta-\sqrt{2}\kappa)z} \\ U_3(0)e^{-j\beta z} \end{bmatrix} \tag{6.118}$$

In the above equation, $U_1(0)$, $U_2(0)$, and $U_3(0)$ are the boundary values. We recover $[A]$ by recalling that $[A] = [V][U]$. The result is

$$
\begin{bmatrix} A_1(z) \\ A_2(z) \\ A_3(z) \end{bmatrix}
= \frac{e^{-j\beta z}}{\sqrt{2}} \left(U_1(0)e^{-j\sqrt{2}\kappa z}\begin{bmatrix} 1 \\ \sqrt{2} \\ 1 \end{bmatrix} + U_2(0)e^{+j\sqrt{2}\kappa z}\begin{bmatrix} -1 \\ \sqrt{2} \\ -1 \end{bmatrix} + U_3(0)\begin{bmatrix} \sqrt{2} \\ 0 \\ -\sqrt{2} \end{bmatrix} \right)
$$

(6.119)

To conclude this chapter, we give two examples. As the first example, we suppose that power is fed to the center waveguide (waveguide 2) only. Thus, $A_2(0) = 1$, and $A_1(0) = A_3(0) = 0$. Then $U_1(0) = U_2(0) = \frac{1}{2}$, $U_3(0) = 0$ and

$$A_1(z) = A_3(z) = -j\frac{1}{\sqrt{2}}\sin\sqrt{2}\kappa z\, e^{-j\beta z} \tag{6.120}$$

$$A_2(z) = \cos\sqrt{2}\kappa z\, e^{-j\beta z} \tag{6.121}$$

At $z = \pi/(\sqrt{2}\kappa)$ or odd multiples of $\pi/(\sqrt{2}\kappa)$, power is transferred completely to, and evenly split between, the two outer waveguides. In other words, the three-waveguide directional couplers under investigation can work as lossless power splitters when the input power is fed to the center waveguide.

As the second example, we suppose the input power is fed to one of the outer waveguides, say waveguide 3. Then, $A_1(0) = A_2(0) = 0$ and $A_3(0) = 1$. We obtain $U_1(0) = 1/(2\sqrt{2})$, $U_2(0) = -1/(2\sqrt{2})$, and $U_3(0) = -\frac{1}{2}$. Then,

$$A_1(z) = \frac{1}{2}(-1 + \cos\sqrt{2}\kappa z)e^{-j\beta z} \tag{6.122}$$

$$A_2(z) = -j\frac{1}{\sqrt{2}}(\sin\sqrt{2}\kappa z)e^{-j\beta z} \tag{6.123}$$

$$A_3(z) = \frac{1}{2}(1 + \cos\sqrt{2}\kappa z)e^{-j\beta z} \tag{6.124}$$

At $z = \pi/(\sqrt{2}\kappa), 3\pi/(\sqrt{2}\kappa), 5\pi/(\sqrt{2}\kappa)\ldots$, $A_2 = A_3 = 0$ and $|A_3|^2 = 1$. In short, power is transferred completely from waveguide 3 to waveguide 1, that is, from an outer waveguide to the waveguide on the opposite side. These directional couplers may be used as routers [45, 46].

Directional couplers with four, five, or more waveguides can be, and have been, studied in the same manner [47, 48]. These directional couplers with multiple waveguides have been proposed as power dividers, samplers, and optical filters [47–49].

PROBLEMS

1. Consider a coupler having three identical waveguides shown in Figure 6.25. The three waveguides are spaced evenly on a circle of radius R. Each waveguide, when isolated from the others, has a propagation constant β.

   (a) Write down the coupled-mode equations for the coupler with three identical waveguides. Let the coupling constant and coupler length be κ and L, respectively.

   (b) At the input ($z = 0$), a power P_{in} is fed to one of the waveguides and none to the other two waveguides. Based on the coupled-mode equations obtained in (a), find expressions for power transferred to the other two waveguides at $z = L$.

   (c) Is it possible that all power fed to the input waveguide is transferred to the other two waveguides? If it is possible, what is the condition of a complete power transfer?

2. Consider a coupler with five identical channel waveguides as shown in Figure 6.26. Clearly the spacing between waveguides 1 and 3, 2 and 4, 3 and 5 is larger than the spacing between two neighboring waveguides. In other words, the coupling constants κ_{12}, κ_{23}, κ_{34}, and κ_{45} have the same value and they are greater than κ_{13}, κ_{24}, κ_{35}, and so forth.

   (a) Write down the coupled-mode equations for the five-waveguide coupler.

   (b) Make reasonable assumption and then solve the coupled-mode equations.

   (c) At the input, power P_{in} is fed to waveguide 3 and none to the other waveguides. Determine the power transferred to waveguides 1, 2, 4, and 5 at the output.

   (d) Is it possible to transfer all power to waveguides 1, 2, 4, and 5, if so under what condition or conditions?

3. Show that (6.29) is correct.

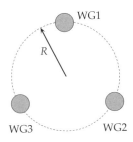

Figure 6.25　Three waveguides equally spaced on a circle of radius R.

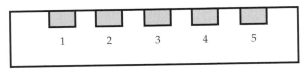

Figure 6.26　Directional coupler with five waveguides.

4. Show that the coupled-mode equations in terms of $f(z)$ and $g(z)$ are given by (6.42) and (6.43).

5. (a) Analyze a five-section switched $\Delta\beta$ directional coupler and plot the switching diagram. (Suggestion: Read Ref. [21] before attempting this problem.)
(b) Analyze a six-section switched $\Delta\beta$ directional coupler and plot the switching diagram.

6. Refer to Ref. [50] by Tietgen and Kersten. Use transfer matrices to show the three-section switched $\Delta\beta$ coupler depicted in Figure 3(a) of Tietgen and Kersten's paper indeed works as a $180°$ turn.

REFERENCES

1. J. Ctyroky and L. Thylen, "Analysis of a directional coupler by coupled modes of a single waveguide," *Opt. Lett.*, Vol. 19, pp. 1621–1623 (Oct. 15, 1994).

2. E. Marcatili, "Improved coupled-mode equations for dielectric guides," *IEEE J. Quantum Electron.*, Vol. QE-22, pp. 988–993 (1986).

3. E. A. J. Marcatili, L. L. Buhl, and R. C. Alferness, "Experimental verification of the improved coupled-mode equations," *Appl. Phys. Lett.*, Vol. 49, No. 25, pp. 1692–1693 (Dec. 22, 1986).

4. A. Yariv, "Coupled mode theory for guided-wave optics," *IEEE J. Quantum Electron.*, Vol. QE-9, pp. 919–933 (1973).

5. H. F. Taylor and A. Yariv, "Guided wave optics," *Proc. IEEE*, Vol. 62, No. 8, pp. 1044–1060 (Aug. 1974).

6. I. P. Kaminow, *An Introduction to Electrooptic Devices*, New York, Academic, 1974.

7. J. G. Mendoza-Alvarez, F. D. Nunes, and N. B. Patel, "Refractive index dependence on the free carriers for GaAs," *J. Appl. Phys.*, Vol. 51, pp. 4365–4367 (1980).

8. K. Ishida, H. Nakaruma, and H. Matsumura, "InGaAsP/InP optical switches using carrier induced index change," *Appl. Phys. Lett.*, Vol. 50, pp. 141–142 (1987).

9. H. J. Zeigler and G. W. Pratt, *Magnetic Interactions in Solids*, Oxford University Press, Oxford, UK, 1973.

10. P. K. Tien, R. J. Martin, R. Wolfe, R. C. LeCraw, and S. C. Blank, "Switching and modulation of light in megnetooptic waveguide on garnet films," *Appl. Phys. Lett.*, Vol. 21, pp. 394–396 (1972).

11. M. Papuchon, Y. Combemale, X. Mathieu, D. B. Ostrowsky, L. Reiber, A. M. Roy, B. Sejourne, and M. Weiner, "Electrical switchable optical directional coupler: COBRA," *Appl. Phys. Lett.*, Vol. 27, pp. 289–291 (1975).

12. W. E. Martin, "A new waveguide switch/modulator for integrated optics," *Appl. Phys. Lett.*, Vol. 26, pp. 562–563 (1973).

13. O. Mikami, J. Noda, and M. Fukuma, "Directional coupler type light modulator using LiNbO$_3$ waveguides," *Tran. IECE Japan*, Vol. 61E, pp. 144–147 (1978).

14. R. C. Alferness, "Titanium-diffused lithium niobate waveguide devices," in *Guided-wave Optoelectronics*, 2nd ed., T. Tamir (ed.), Springer, Berlin, 1990.

15. J. C. Campbell, F. A. Blum, D. W. Shaw, and K. C. Lawley, "GaAs electrooptic directional coupler switch," *Appl. Phys. Lett.*, Vol. 29, pp. 203–205 (1975).

16. F. J. Leonberger, J. P. Donnelly, and C. O. Bozler, "GaAs p+n-n+ directional coupler switch," *Appl. Phys. Lett.*, Vol. 29, pp. 652–654 (1975).

17. H. Kawaguchi, "GaAs rib waveguide directional coupler switch with Schottky barriers," *Electron. Lett.*, Vol. 14, pp. 387–388 (June 22, 1978).

18. B. Broberg, B. S. Lindgren, M. G. Oberg, and H. Jiang, "A novel integrated optics wavelength filter in InGaAsP-InP," *J. Lightwave Technol.*, Vol. 4, pp. 196–203 (1986).

19. R. C. Alferness, "Guided-wave devices for optical communication," *IEEE J. Quantum Electron.*, Vol. QE-17, No. 6, pp. 946–959 (July 1981).

20. W. A. Stallard, A. R. Beaumont, and R. C. Booth, "Integrated optics devices for coherent transmission," *J. Lightwave Technol.*, Vol. LT-4, No. 7, pp. 852–857 (July 1986).

21. R. V. Schmidt and H. Kogelnik, "Electro-optically switched coupler with stepped $\Delta\beta$ reversal using Ti-diffused LiNbO$_3$ waveguides," *Appl. Phys. Lett.*, Vol. 28, No. 2, pp. 503–506 (1976).

22. H. Kogelnik and R. K. Schmidt, "Switched directional coupler modulators with alternating $\Delta\beta$," *IEEE J. Quantum Electron.*, Vol. QE-12, No. 7, pp. 396–401 (1976).

23. R. V. Schmidt and R. C. Alferness, "Directional coupler switches, modulators, and filters using alternating $\Delta\beta$ techniques," *IEEE Trans. Circuits Systems*, Vol. CAS-26, No. 12, pp. 1099–1108 (Dec. 1979).

24. R. V. Schmidt and P. S. Cross, "Efficient optical waveguide switch/amplitude modulator," *Opt. Lett.*, Vol. 2, pp. 45–47 (1978).

25. H. A. MacLeod, *Thin-Film Optical Filters*, Institute of Physics Publishing, Bristol, CT, (2001).

26. J. Noda, M. Fukuma, and O. Mikami, "Design calculations for directional couplers fabricated by Ti-diffused LiNbO$_3$ waveguides," *Appl. Opt.*, Vol. 20, No. 13, pp. 2284–2290 (1981).

27. H. F. Taylor, "Frequency-selective coupling in parallel dielectric waveguides," *Opt. Commun.*, Vol. 8, pp. 421–425 (1973).

28. H. F. Taylor, "Optical switching and modulation in parallel dielectric waveguides," *J. Appl. Phys.*, Vol. 44, No. 7, pp. 3257–3262 (July, 1973).

29. O. Parriaus, F. Bernoux, and G. Chartier, "Wavelength selective distributed coupling between single-mode optical fibers for multiplexing," *J. Opt. Commun.*, Vol. 2, pp. 105–109 (1981).

30. M. Digonnet and H. J. Shaw, "Wavelength multiplexing in single-mode fiber couplers," *Appl. Opt.*, Vol. 22, No. 3, pp. 484–491 (Feb. 1983).

31. D. Marcuse, "Directional-coupler filter using dissimilar optical fiber," *Electron. Lett.*, Vol. 21, pp. 726–727 (August 15, 1985).

32. R. C. Alferness and R. V. Schmidt, "Tunable optical waveguide directional coupler filter," *App. Phys. Lett.*, Vol. 33, pp. 161–163 (1978).

33. C. Wu, C. Rolland, N. Puetz, R. Bruce, K. D. Chik, and J. M. Xu, "A vertical coupled InGaAsP/InP directional coupler filter of ultranarrow bandwidth," *IEEE Photonic Tech. Lett.*, Vol. 3, No. 6, pp. 519–521 (June, 1991).

34. B. Liu, A. Shakouri, P. Abraham, Y. J. Chiu, S. Zhang, and J. E. Bowers, "Fused InP-GaAs vertical coupled filters," *IEEE Photonic Tech. Lett.*, Vol. 11, No. 1, pp. 93–95 (Jan. 1999).

35. R. C. Alferness, "Optical directional couplers with weighted coupling," *Appl. Phys. Lett.*, Vol. 35, No. 3, pp. 260–262 (Aug. 1979).

36. R. C. Alferness and P. S. Cross, "Filter characteristics of codirectional coupled waveguides with weighted coupling," *IEEE J. Quantum Electron.*, Vol. QE-14, No. 11, pp. 843–847 (1978).

37. C. H. Bulmer and W. K. Burns, "Linear interferometric modulators in Ti:LiNbO3," *IEEE J. Lightwave Tech.*, Vol. LT-2, No. 8, pp. 512–521 (Aug. 1984).

38. R. C. Alferness, "Waveguide electrooptic modulators," *IEEE Trans. Microwave Theory Tech.*, Vol. MTT-30, No. 8, pp. 1121–1137 (Aug. 1982).

39. H. F. Schlaak, "Modulation behavior of integrated optical directional couplers," *J. Optical Commun.*, Vol. 5, No. 4, pp. 122–134 (April 1984).

40. C. L. Chen, *Elements of Optoelectronics and Fiber Optics*, R. D. Irwin, Chicago, 1996.

41. N. Dagli, "Wide-bandwidth lasers and modulators for RF photonics," *IEEE Trans. Microwave Theory Tech.*, Vol. MTT-47, No. 7, pp. 1151–1171 (July 1999).

42. W. B. Bridges and J. H. Schaffner, "Distortion in linearized electrooptic modulators," *IEEE Trans. Microwave Theory Tech.*, Vol. MTT-43, No. 9, pp. 2184–2197 (Sept. 1995).

43. I. S. Sokolnikoff and R. M. Redheffer, *Mathematics of Physics and Modern Engineering*, 2nd ed., McGraw-Hill, New York, 1966.

44. K. Iwasaki, S, Kurazono, and K. Itakura, "The coupling of modes in three dielectric slab waveguides," *Electron. Commun. Japan*, Vol. 58-C, No. 8, pp. 100–108 (Aug. 1975).

45. H. A. Haus and C. G. Fonstad, Jr., "Three-waveguide couplers for improved sampling and filtering," *IEEE J. Quantum Electron.*, Vol. QE-17, No. 12, pp. 2321–2325 (Dec. 1981).

46. J. P. Donnelly, N. L. DeMeo, Jr., and G. A. Ferrate, "Three-guide optical couplers in GaAs," *IEEE J. Lightwave Tech.*, Vol. LT-1, No. 2, pp. 417–424 (June 1983).

47. H. A. Haus and L. Molter-Orr, "Coupled multiple waveguide systems," *IEEE J. Quantum Electron.*, Vol. QE-19, No. 5, pp. 840–844 (May 1983).

48. J. P. Donnelly, H. A. Haus, and N. Whitaker, "Symmetric three-guide optical coupler with nonidentical center and outside guides," *IEEE J. Quantum Electron.*, Vol. QE-23, No. 4, pp. 401–406 (April 1987).

49. S. Somekh, E. Garmire, A. Yariv, H. L. Garvin, and R. G. Hunsperger, "Channel optical waveguide directional couplers," *Appl. Phys. Lett.*, Vol. 22, pp. 46–47 (Jan. 15, 1973).

50. K. H. Tietgen and R. Th. Kersten, "180°-turns in integrated optics," *Opt. Commu.*, Vol. 36, No. 4, pp. 281–284 (1981).

7

GUIDED-WAVE GRATINGS

7.1 INTRODUCTION

Directional couplers, *guided-wave gratings*, and *arrayed-waveguide gratings* are three basic types of passive guided-wave components used in many optical devices and systems. We have discussed directional couplers in the last chapter and we will study arrayed-waveguide gratings in the next chapter. In this chapter, we concentrate on the guided-wave gratings.

7.1.1 Types of Guided-Wave Gratings

There are three basic types of guided-wave gratings. They are the static, programmable, and moving gratings [1].

7.1.1.1 Static Gratings

Static gratings are periodic topological structures or index variations built permanently on and are integral parts of optical waveguides (Fig. 7.1). Examples include arrays of grooves, indentations, ridges, bars, or other surface corrugations built onto,

Foundations for Guided-Wave Optics, by Chin-Lin Chen
Copyright © 2007 John Wiley & Sons, Inc.

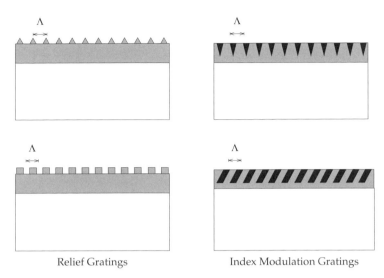

Relief Gratings   Index Modulation Gratings

Figure 7.1   Static thin-film gratings.

or etched in, optical waveguides or optical fibers. The static gratings having topological features are also known as *relief gratings* [2–5]. Gratings having index variations
are also known as *index modulation gratings*.

7.1.1.2 Programmable Gratings

As noted in the last chapter, the refractive index of electrooptic materials may change
when materials are subjected to electric fields. To make use of the *electrooptic
effects*, interdigital electrodes are deposited over or near the waveguide. An example
is shown in Figure 7.2 [6]. When a voltage is applied to the electrodes, electric fields
are established in the electrooptic material. Periodic index variations are induced
in the waveguide by the electric field and through the electrooptic effects. The
periodic index changes are the index gratings. When the voltage is turned off, the
index perturbation, and therefore the index grating, disappears. In short, the index
grating induced electrooptically may be turned on or off by controlling the voltage
to the electrodes [6–8].

Figure 7.3 depicts schematically a zigzag electrode deposited on the surface of
an iron garnet film ($Y_3Ga_{0.75}Sc_{0.5}Fe_{3.75}O_{12}$) grown on a $Gd_3Ga_5O_{12}$ substrate [9].
When a radio-frequency current passes through the serpentine electrode, it estab-
lishes a radio-frequency magnetic field. The magnetic field produces periodic magnetization variations through the magnetooptic effect of the iron garnet film [9, 10].
The magnetization grating vanishes when the applied radio-frequency current is
turned off.

Since the gratings induced in electrooptic or magnetooptic materials are controllable by electric or magnetic fields, they are the *controllable* or *programmable
gratings*.

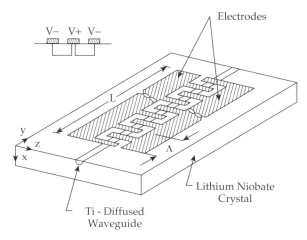

Figure 7.2 Electrooptic induced grating for TE ↔ TM mode conversion. (From [6].)

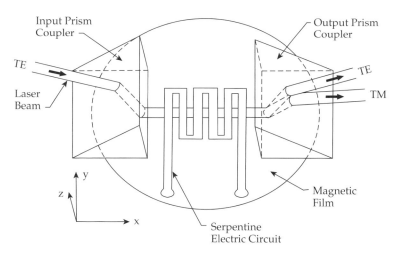

Figure 7.3 Magnetization grating on a ferromagnetic film of $Y_3Ga_{0.75}Sc_{0.5}Fe_{3.75}O_{12}$ on a substrate of $Gd_3Ga_5O_{12}$. (From [9].)

7.1.1.3 Moving Grating

By applying radio or microwave frequency voltages to interdigital transducers deposited on the surface of a piezoelectric material, surface acoustic waves may be generated near the surface of the piezoelectric material. As the surface acoustic waves propagate, they produce periodic dilation and compression in the medium, which leads to index variations in the medium through the acoustooptic effects [11]. In other words, the propagation of surface acoustic waves is accompanied by a *moving index grating*. The moving index grating diffracts the optical beams. Since the frequency of surface acoustic waves is tunable, so is the acoustooptic interaction.

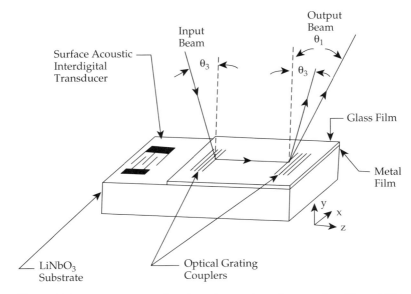

Figure 7.4   Acoustooptic-induced gratings and grating couplers. (From [12].)

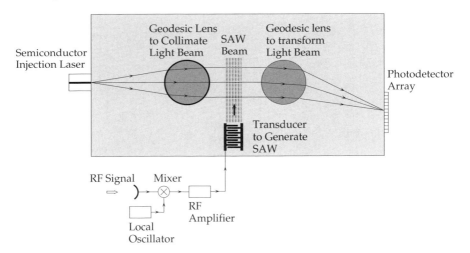

Figure 7.5   Schematic for integrated optical radio frequency spectrum analyzer. (After [15].)

Guided-wave mode converters, tunable filters (Fig. 7.4) and radio-frequency spectrum analyzers (Fig. 7.5) are but a few guided-wave devices based on the moving gratings [12–18].

7.1.2   Applications of Guided-Wave Gratings

The first known application of guided-wave gratings is to use photoresist gratings to couple light into or out of thin-film waveguides. This is shown schematically in Figure 7.4 [12, 19]. Gratings have been and are used, for example, as mode converters, polarizers [6–8, 12], narrow-band optical filters [20, 21], and distributed Bragg reflectors (DBR) [22, 23]. They are also the key components of distributed feedback (DFB) lasers [24–27]. The list of possible applications of gratings is very extensive. A partial list of guided-wave grating devices and pertinent references follow:

1. Gratings as input/output devices for guided-wave optics [12, 19]
2. Gratings for focusing or redirecting guided waves [28–32]
3. Gratings to enhance the coupling of directional couplers having nonidentical waveguides [33, 34]
4. Gratings as TE–TM mode converter filters and polarizers [6–8]
5. Gratings as $TE_m–TE_n$ mode converters [12]
6. Narrow-band optical filters [20, 21]
7. Distributed Bragg reflectors [6–8, 22, 23]
8. Distributed feedback lasers [24–28, 35]
9. Chirped gratings for dispersion compensation [36]
10. Fiber gratings for sensing stress and strain in large structures [2–5, 37]

In 1986, Ura, Suhara, Nishihara, and Koyam proposed using guided-wave gratings as part of the optic pickup units of compact disk players (Fig. 7.6) [28, 32]. It will have a tremendous impact on the integrated optics technology if guided-wave gratings can be manufactured by mass production processes and produced economically.

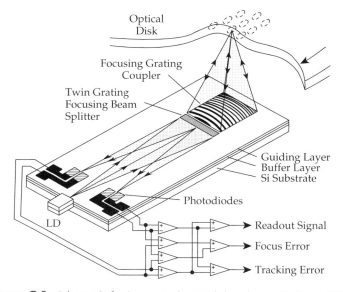

Figure 7.6 Schematic for integrated optic disk pickup unit. (From [28].)

7.1.3 Two Methods for Analyzing Guided-Wave Grating Problems

There are two basic methods for analyzing the guided-wave grating problems. Since all guided-wave gratings are periodic structures, we can apply the Floquet or Bloch theorem to analyze wave propagation in the guided wave gratings [38, 39]. While the analysis is exact, results are often complicated. Alternatively, we may treat a guided-wave grating as small perturbations to a perfect waveguide. We assume that modes guided by the perfect waveguides as known, and we study effects of the geometric or index perturbations on the propagating modes. Specifically, two coupled-mode equations [40, 41] are derived to describe the effects of gratings on the interacting modes. Although the two methods appear different, they are equivalent as shown in Ref. [42].

The perturbation method is physically intuitive and often leads to simple and closed-form expressions. We often gain physical insight to the grating problem through the perturbation analysis. Besides, we have seen the coupled-mode formulation once before in Chapter 6 in connection with our study of directional couplers. Therefore, we will concentrate on the perturbation method in this chapter. Although our discussions are couched in terms of thin-film waveguides, the results are applicable to gratings on channel and other waveguide types and optical fibers [2–5, 36, 37].

7.2 PERTURBATION THEORY

In Chapter 6, we used the coupled-mode equations to study the operation of directional couplers. A directional coupler is essentially two closely spaced waveguides. The operation of directional couplers is based on the interaction between two modes propagating in the same direction with equal, or nearly equal, phase velocities. The two modes are guided by the two waveguides, one by each waveguide.

The operation of guided-wave grating devices also involves the interaction between two or more modes. But the interacting modes are guided by the same waveguide. In addition, the modes may propagate in the same direction or in opposite directions. Thus the coupling and interaction of modes in grating devices are fundamentally different from that of directional couplers.

As noted earlier, we view a guided-wave grating as a periodic perturbation on an unperturbed waveguide. In Figure 7.7(a), we show a thin-film waveguide and denote the relative dielectric constants in various regions by a z-independent permittivity distribution function $\varepsilon_r(x, y)$. A guided-wave grating is depicted schematically in Figure 7.7(b) as a perturbed waveguide with a relative permittivity distribution function $\varepsilon_r(x, y) + \Delta\varepsilon_r(x, y, z)$. Here $\Delta\varepsilon_r(x, y, z)$ is possibly, and usually is, dependent on z.

7.2.1 Waveguide Perturbation

We suppose that all modes guided by the original or unperturbed waveguide are known and we denote the field as $\mathbf{E}_0$ and $\mathbf{H}_0$. Of course, $\mathbf{E}_0$ and $\mathbf{H}_0$ are solutions

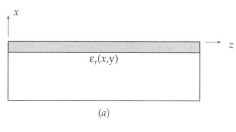

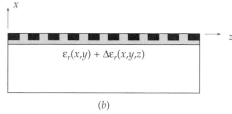

Figure 7.7 Waveguide grading as a perturbation of a thin-film waveguide: (a) original thin-film waveguide and (b) waveguide grating as a perturbation on the original waveguide.

of Maxwell's equations with a relative permittivity distribution function $\varepsilon_r(x, y)$:

$$\nabla \times \mathbf{E}_0 = -j\omega\mu_0\mathbf{H}_0 \tag{7.1}$$

$$\nabla \times \mathbf{H}_0 = +j\omega\varepsilon_0\varepsilon_r(x, y)\mathbf{E}_0 \tag{7.2}$$

subjected to the appropriate boundary conditions at interfaces between various regions.

Let the field of the perturbed waveguide be $\mathbf{E}$ and $\mathbf{H}$. The field vectors satisfy Maxwell's equations with a relative permittivity distribution function $\varepsilon_r(x,y) + \Delta\varepsilon_r(x,y,z)$:

$$\nabla \times \mathbf{E} = -j\omega\mu_0\mathbf{H} \tag{7.3}$$

$$\nabla \times \mathbf{H} = +j\omega\varepsilon_0[\varepsilon_r(x, y) + \Delta\varepsilon_r(x, y, z)]\mathbf{E} \tag{7.4}$$

and suitable boundary conditions. By forming the dot products of $\mathbf{E}_0^*$ with (7.4), $\mathbf{H}$ with the complex conjugation of (7.1), and combining the two resulting equations, we obtain

$$\nabla \cdot (\mathbf{E}_0^* \times \mathbf{H}) = j\omega\mu_0 H \cdot H_0^* - j\omega\varepsilon_0[\varepsilon_r(x, y) + \Delta\varepsilon_r(x, y, z)]\mathbf{E} \cdot \mathbf{E}_0^* \tag{7.5}$$

Similarly, from (7.2) and (7.3), we obtain

$$\nabla \cdot (\mathbf{E} \times \mathbf{H}_0^*) = -j\omega\mu_0 H \cdot H_0^* + j\omega\varepsilon_0\varepsilon_r(x, y)\mathbf{E} \cdot \mathbf{E}_0^* \tag{7.6}$$

Combining (7.5) and (7.6), we obtain

$$\nabla \cdot [\mathbf{E}_0^* \times \mathbf{H} + \mathbf{E} \times \mathbf{H}_0^*] = -j\omega\varepsilon_0\,\Delta\varepsilon_r(x, y, z)\mathbf{E} \cdot \mathbf{E}_0^* \tag{7.7}$$

Integrating (7.7) over the entire waveguide cross section S, we obtain

$$\int_S \nabla_t \cdot [\mathbf{E}_0^* \times \mathbf{H} + \mathbf{E} \times \mathbf{H}_0^*]_t \, ds + \int_S \frac{d}{dz} [(\mathbf{E}_0^* \times \mathbf{H} + \mathbf{E} \times \mathbf{H}_0^*) \cdot \hat{\mathbf{z}}] \, ds$$

$$= -j\omega\varepsilon_0 \int_S \Delta\varepsilon_r(x, y, z) \mathbf{E} \cdot \mathbf{E}_0^* \, ds \tag{7.8}$$

where the subscript t signifies the transverse components of the vector. Since $[\mathbf{E}_0^* \times \mathbf{H} + \mathbf{E} \times \mathbf{H}_0^*]_t$ is a vector in the transverse plane, we apply the two-dimensional divergence theorem discussed in Appendix B and show that the first integral on the left-hand side of (7.8) vanishes as the waveguide cross section S extends indefinitely. Thus (7.8) reduces to

$$\int_S \frac{d}{dz} [(\mathbf{E}_{0t}^* \times \mathbf{H}_t + \mathbf{E}_t \times \mathbf{H}_{0t}^*) \cdot \hat{\mathbf{z}}] \, ds = -j\omega\varepsilon_0 \int_S \Delta\varepsilon_r(x, y, z) \mathbf{E} \cdot \mathbf{E}_0^* \, ds \tag{7.9}$$

This is the desired *perturbation formula*. It relates $\mathbf{E}$, $\mathbf{H}$, and the permittivity perturbation $\Delta\varepsilon_r(x, y, z)$ of the perturbed waveguide to $\mathbf{E}_0$ and $\mathbf{H}_0$ of the unperturbed waveguide. Terms on the left-hand side of (7.9) involves only the transverse field components. But the right-hand side term involves all field components. Equation (7.9) is the starting point of the perturbation analysis.

It is instructive to compare the perturbation formula (7.9) with (C.8) of the Lorentz reciprocity discussed in Appendix C. The procedures used to establish the two formulas are very similar. In both cases, we are concerned with fields at a single and specific frequency. But the Lorentz reciprocity pertains to fields in a specific region having a specific permittivity and permeability. For example, fields under consideration may be modes guided by a certain waveguide. On the other hand, the perturbation formula (7.9) pertains to modes guided by two different waveguides, an unperturbed waveguide and a perturbed waveguide.

7.2.2 Fields of Perturbed Waveguide

The next step is to find expressions for $\mathbf{E}$ and $\mathbf{H}$ guided by the perturbed waveguide. We begin with fields $\mathbf{E}_0$ and $\mathbf{H}_0$ of the unperturbed waveguide. Let $\mathbf{E}_0$ and $\mathbf{H}_0$ be fields of mode ν guided by the unperturbed waveguide, that is, $\mathbf{E}_0(x, y, z) = \mathbf{e}_\nu(x, y)e^{-j\beta_\nu z}$ and $\mathbf{H}_0(x, y, z) = \mathbf{h}_\nu(x, y)e^{-j\beta_\nu z}$. For brevity, we drop all arguments (x, y, z) or (x, y) in the following discussions. The mode fields $\mathbf{e}_\nu$ and $\mathbf{h}_\nu$ and the propagation constant β_ν are related through Maxwell's equations:

$$\nabla \times [(\mathbf{e}_{\nu t} + \hat{\mathbf{z}}e_{\nu z})e^{-j\beta_\nu z}] = -j\omega\mu_0[(\mathbf{h}_{\nu t} + \hat{\mathbf{z}}h_{\nu z})e^{-j\beta_\nu z}] \tag{7.10}$$

$$\nabla \times [(\mathbf{h}_{\nu t} + \hat{\mathbf{z}}h_{\nu z})e^{-j\beta_\nu z}] = j\omega\varepsilon_0\varepsilon_r(x, y)[(\mathbf{e}_{\nu t} + \hat{\mathbf{z}}e_{\nu z})e^{-j\beta_\nu z}] \tag{7.11}$$

We take $\mathbf{e}_\nu$, $\mathbf{h}_\nu$, and β_ν as known quantities. We also assume that $\mathbf{e}_\nu$ and $\mathbf{h}_\nu$ are normalized in that the time-average power carried by the mode is 1 W.

Since the perturbed waveguide has a complicated geometry or index profile, it is difficult to obtain explicit expressions of the fields guided by it. On the other hand, the perturbation is small; **E** and **H** of the perturbed waveguide may be expressed in terms of $\mathbf{E}_0$ and $\mathbf{H}_0$ of unperturbed waveguide [41]. For the *transverse field components* of **E** and **H**, we write

$$\mathbf{E}_t = \sum_v a_v(z)\mathbf{e}_{vt}e^{-j\beta_v z} \tag{7.12}$$

$$\mathbf{H}_t = \sum_v a_v(z)\mathbf{h}_{vt}e^{-j\beta_v z} \tag{7.13}$$

where $a_v(z)$ is the mode amplitude function, $\mathbf{e}_{vt}$ and $\mathbf{h}_{vt}$ are the transverse field components, and e_{vz} and h_{vz} are the z component of fields of mode v guided by the unperturbed waveguide. The summation is over all possible guided modes. If desired, an integral may be inserted in (7.12) and (7.13) to account for the radiation mode explicitly.

Marcuse showed that the *longitudinal field component* of the electric field in the perturbed waveguide has to be treated differently [41]. To search for a suitable representation for E_z, we return to (7.4) and note

$$\hat{\mathbf{z}} \cdot \nabla \times \mathbf{H} = \hat{\mathbf{z}} \cdot \nabla_t \times \mathbf{H}_t = j\omega\varepsilon_0[\varepsilon_r(x, y) + \Delta\varepsilon_r(x, y, z)]E_z \tag{7.14}$$

Substituting (7.13) into (7.14) and making use of (7.11), we obtain

$$\hat{\mathbf{z}} \cdot \nabla_t \times \mathbf{H}_t = \hat{\mathbf{z}} \cdot \nabla_t \times \left(\sum_v a_v(z)\mathbf{h}_{vt}e^{-j\beta_v z}\right) = \sum_v a_v(z)(\hat{\mathbf{z}} \cdot \nabla_t \times \mathbf{h}_{vt})e^{-j\beta_v z}$$

$$= \sum_v a_v(z)(j\omega\varepsilon_0\varepsilon_r(x, y)e_{vz})e^{-j\beta_v z}$$

$$= j\omega\varepsilon_0[\varepsilon_r(x, y) + \Delta\varepsilon_r(x, y, z)]E_z \tag{7.15}$$

Thus we obtain

$$E_z = \sum_v \frac{\varepsilon_r(x, y)}{\varepsilon_r(x, y) + \Delta\varepsilon_r(x, y, z)}a_v(z)e_{vz}e^{-j\beta_v z} \tag{7.16}$$

In short, the z component of the perturbed electric field can be expressed in terms of $a_v(z)$ and $e^{-j\beta_v z}$. But the expansion coefficients are slightly different from $a_v(z)$.

Following the same procedure, we obtain an expression for H_z of the perturbed field in the presence of the dielectric perturbation:

$$H_z = \sum_v a_v(z)h_{vz}e^{-j\beta_v z} \tag{7.17}$$

In summary, fields of the perturbed waveguide may be written as [41]

$$\mathbf{E} = \sum_{v} a_v(z) \left[\mathbf{e}_{vt} + \hat{\mathbf{z}} \frac{\varepsilon_r(x, y)}{\varepsilon_r(x, y) + \Delta\varepsilon_r(x, y, z)} e_{vz} \right] e^{-j\beta_v z} \qquad (7.18)$$

$$\mathbf{H} = \sum_{v} a_v(z) [\mathbf{h}_{vt} + \hat{\mathbf{z}} h_{vz}] e^{-j\beta_v z} \qquad (7.19)$$

7.2.3 Coupled Mode Equations and Coupling Coefficients

Now we are ready to study guided-wave gratings. Specifically, we derive the coupled-mode equations and the coupling coefficients for the guided-wave gratings. Suppose we are interested in effects of a guided-wave grating on mode ℓ. We express $\mathbf{E}_0$ and $\mathbf{H}_0$ in terms of the field of mode ℓ explicitly:

$$\mathbf{E}_0 = (\mathbf{e}_{\ell t} + \hat{\mathbf{z}} e_{\ell z}) e^{-j\beta_\ell z} \qquad (7.20)$$

$$\mathbf{H}_0 = (\mathbf{h}_{\ell t} + \hat{\mathbf{z}} h_{\ell z}) e^{-j\beta_\ell z} \qquad (7.21)$$

The next step is to substitute $\mathbf{E}_0$, $\mathbf{H}_0$, $\mathbf{E}$, and $\mathbf{H}$ given in (7.18)–(7.21) into (7.9). As the first step of the substitution, we examine the integrand in the left-hand side of (7.9) and note

$$\hat{\mathbf{z}} \cdot [\mathbf{E}_{0t}^* \times \mathbf{H}_t + \mathbf{E}_t \times \mathbf{H}_{0t}^*] = \sum_{v} a_v(z) e^{j(\beta_\ell - \beta_v)z} [\mathbf{e}_{\ell t}^* \times \mathbf{h}_{vt} + \mathbf{e}_{vt} \times \mathbf{h}_{\ell t}^*] \cdot \hat{\mathbf{z}} \quad (7.22)$$

Upon integrating the above expression over the waveguide cross section S and applying the orthonormality relation Eq.(C.22) in the Appendix C5 the integral in the left-hand side of (7.9) is greatly simplified:

$$\int_S \frac{d}{dz} \left\{ \sum_{v} a_v(z) e^{j(\beta_\ell - \beta_v)z} [\mathbf{e}_{\ell t}^* \times \mathbf{h}_{vt} + \mathbf{e}_{vt} \times \mathbf{h}_{\ell t}^*] \cdot \hat{\mathbf{z}} \right\} ds$$

$$= \frac{d}{dz} \sum_{v} \{ a_v(z) e^{j(\beta_\ell - \beta_v)z} 4\delta(\ell, v) \} = \pm 4 \frac{d}{dz} a_\ell(z) \qquad (7.23)$$

where $\delta(\ell, v)$ is the Kronecker delta and the $\pm$ sign is $+$ for $\beta_\ell > 0$ and $-$ for $\beta_\ell < 0$ respectively.

In terms of $\mathbf{e}_v$ and $\mathbf{h}_v$, the right-hand side term of (7.9) is

$$- j\omega\varepsilon_0 \int_S \Delta\varepsilon_r(x, y, z) \mathbf{E} \cdot \mathbf{E}_0 \, ds = - j\omega\varepsilon_0 \sum_{v} a_v(z) e^{j(\beta_\ell - \beta_v)z} \int_S \Delta\varepsilon_r(x, y, z)$$

$$\times \left[\mathbf{e}_{vt} \cdot \mathbf{e}_{\ell t}^* + \frac{\varepsilon_r(x, y)}{\varepsilon_r(x, y) + \Delta\varepsilon_r(x, y, z)} e_{vz} e_{\ell z}^* \right] ds \qquad (7.24)$$

Substituting (7.23) and (7.24) into (7.9), we obtain the *coupled-mode equations*:

$$\pm\frac{da_\ell(z)}{dz} = -j \sum_\nu [\kappa_{\ell\nu}^t(z) + \kappa_{\ell\nu}^z(z)] e^{j(\beta_\ell - \beta_\nu)z} a_\nu(z) \qquad (7.25)$$

where

$$\kappa_{\ell\nu}^t(z) = \frac{\omega\varepsilon_0}{4} \int_S \Delta\varepsilon_r(x, y, z) \mathbf{e}_{\nu t} \cdot \mathbf{e}_{\ell t}^* \, ds \qquad (7.26)$$

$$\kappa_{\ell\nu}^z(z) = \frac{\omega\varepsilon_0}{4} \int_S \frac{\varepsilon_r(x, y)\Delta\varepsilon_r(x, y, z)}{\varepsilon_r(x, y) + \Delta\varepsilon_r(x, y, z)} e_{\nu z} e_{\ell z}^* \, ds \qquad (7.27)$$

are the *coupling coefficients* of the *grating-assisted interaction* of the transverse and longitudinal field components. We make two observations.

1. There is no interaction between orthogonal field components.
2. Even though integrals in (7.26) and (7.27) are over the whole waveguide cross section, only fields in the perturbed regions where $\Delta\varepsilon_r(x,y,z)$ is finite contribute to the coupling coefficients.

In short, the interaction or the coupling coefficient is negligible unless two interacting modes are present and overlap in regions where the perturbation $\Delta\varepsilon_r(x,y,z)$ is not zero and unless the two mode fields point to the same direction or have the same polarizations.

Recall that a guided-wave grating is a periodic topological structure on, or index variation in, an unperturbed waveguide. Let the *grating periodicity*, also known as the *grating constant*, of the periodic perturbation be Λ. Since $\Delta\varepsilon_r(x,y,z)$ is periodic in z, we can express it as a Fourier series

$$\Delta\varepsilon_r(x, y, z) = \sum_{q=-\infty}^{\infty} \Delta\varepsilon_{rq}(x, y) e^{-jqKz} \qquad (7.28)$$

where $K = 2\pi/\Lambda$ is the *grating wave vector* and $\Delta\varepsilon_{rq}(x,y)$ is a Fourier coefficient of the periodic perturbation. Substituting (7.28) into (7.25), we obtain

$$\pm\frac{da_\ell(z)}{dz} = -j \sum_\nu \sum_{q=-\infty}^{\infty} (\kappa_{\ell\nu q}^t + \kappa_{\ell\nu q}^z) e^{j(\beta_\ell - \beta_\nu - qK)z} a_\nu(z) \qquad (7.29)$$

where

$$\kappa_{\ell\nu q}^t = \frac{\omega\varepsilon_0}{4} \int_S \Delta\varepsilon_{rq}(x, y) \mathbf{e}_{\nu t} \cdot \mathbf{e}_{\ell t}^* \, ds \qquad (7.30)$$

$$\kappa_{\ell\nu q}^z \approx \frac{\omega\varepsilon_0}{4} \int_S \frac{\varepsilon_r(x, y)\Delta\varepsilon_{rq}(x, y)}{\varepsilon_r(x, y) + \Delta\varepsilon_{rq}(x, y)} e_{\nu z} e_{\ell z}^* \, ds \qquad (7.31)$$

Equation (7.29) is the coupled-mode equation for guided-wave gratings. The left-hand side of (7.29) is the rate of change of the amplitude function $a_\ell(z)$. The right-hand side of the equation is a double summation of terms summed over v and q. Although $a_v(z)$ varies slowly, the exponential functions fluctuate periodically and rapidly unless the exponent is small or zero. Because of the rapid swing of the exponential functions, contribution to the double summation from most terms is minuscule. Significant contribution to the double summation comes only from a few terms that have a small or zero exponent. In other words, significant power exchange comes only from modes with a particular combination of β_ℓ, β_v, and qK. For that particular combination, $|\beta_\ell - \beta_v + qK|$ is small or zero. The condition

$$\beta_\ell - \beta_v + qK = 0 \qquad (7.32)$$

is known as the *grating-assisted phase match condition* or the *Bragg condition*. When the Bragg condition is satisfied, power coupled from mode ℓ to mode v accumulates coherently, and this leads to a significant power exchanged between the two modes. Further discussion on the Bragg condition is postponed until Section 7.3.

Usually the lower-order Fourier coefficients, corresponding to the lower-order spatial harmonics, are larger than the higher-order Fourier coefficients. But the Fourier coefficient with $q = 0$ is simply the average value of $\Delta\varepsilon_r(x, y, z)$ and it does not contribute to the grating-assisted interaction. Thus the effect of grating arises mainly from the fundamental or second spatial harmonic term.

7.2.4 Co-directional Coupling

Suppose the Bragg condition (7.32) is met, or nearly met, by two modes propagating in the same direction in the guided-wave grating region. We identify the two modes as modes ℓ' and v'. We also assume that the two modes propagate in the $+z$ direction. We choose the $+$ sign on the left-hand side of (7.29). Let the Fourier component satisfying the Bragg condition be the q'th Fourier component. We set ℓ to ℓ' in the left-hand side of (7.29) and ignore all terms in the right-hand side except the term with $v = v'$ and $q = q'$. Then, (7.29) becomes

$$+\frac{da_{\ell'}(z)}{dz} \approx -j(\kappa^t_{\ell'v'q'} + \kappa^z_{\ell'v'q'})e^{j(\beta_{\ell'} - \beta_{v'} - q'K)z}a_{v'}(z) \qquad (7.33)$$

The equation describes the rate of change of $a_{\ell'}(z)$ due to the interaction with mode v' in the grating region. Obviously, the interaction also affects $a_{v'}(z)$. The equation governing the rate of change of $a_{v'}(z)$ can be obtained directly from (7.33) by interchanging ℓ' with v' and replacing q' by $-q'$:

$$+\frac{da_{v'}(z)}{dz} \approx -j(\kappa^t_{v'\ell'-q'} + \kappa^z_{v'\ell'-q'})e^{j(\beta_{v'} - \beta_{\ell'} + q'K)z}a_{v'}(z) \qquad (7.34)$$

Equations (7.33) and (7.34) describe the effect of the grating-assisted coupling between two modes propagating in the $+z$ direction. If the two modes move in the $-z$ direction, then the two plus signs in the left hand sides of (7.33) and (7.34) become two minus signs. In either case, the coupled-mode equations for two *co-propagating modes* have the same signs in the left hand sides. Except for the presence of $\mp jq'K$ term in the exponents, (7.33) and (7.34) are formally identical to the coupled-mode equations (6.35) and (6.36) studied in Chapter 6. All results based on (6.35) and (6.36) for directional couplers are valid for grating-assisted coupling between two co-propagating modes, provided the phase match condition is replaced by the Bragg condition. No further discussion on the *co-directional coupling* is necessary. However, we note a crucial difference. In the conventional directional couplers discussed in Chapter 6, significant power exchange occurs only if the two interacting modes have identical, or nearly identical, propagation constants. In guided-wave gratings, the phase match condition is replaced by the Bragg condition. The modes with different propagation constants may interact strongly if a suitable grating structure is present in the interaction region and if the Bragg condition is met [33, 34].

7.2.5 Contra-directional Coupling

In considering the two modes propagating in the opposite directions in the grating region, we choose two opposite signs for the $\pm$ sign in the left-hand side of (7.29). Suppose mode ℓ'' propagates in the $+z$ direction and mode v'' moves in the $-z$ direction. That is $\beta_{\ell''} > 0$ and $\beta_{v''} < 0$. Then (7.29) for the two modes are

$$+\frac{da_{\ell''}(z)}{dz} = -j(\kappa^t_{\ell''v''q''} + \kappa^z_{\ell''v''q''})e^{j(\beta_{\ell''}-\beta_{v''}-q''K)z}a_{v''}(z) \tag{7.35}$$

$$-\frac{da_{v''}(z)}{dz} = -j(\kappa^t_{v''\ell''-q''} + \kappa^z_{v''\ell''-q''})e^{j(\beta_{v''}-\beta_{\ell''}+q''K)z}a_{v''}(z) \tag{7.36}$$

Since the two signs on the left-hand sides are different, the *contradirectional coupling* is fundamentally different from the co-directional coupling. Further discussion on the contradirectional coupling is postponed until Sections 7.5 and 7.6.

7.3 COUPLING COEFFICIENTS OF A RECTANGULAR GRATING—AN EXAMPLE

As indicated in (7.30) and (7.31), the strength of the grating-assisted coupling depends on the position and profile of the grating elements and the field distribution of modes involved in the interaction at the grating elements. In this section, we consider the coupling coefficient of gratings with a rectangular grating profile. For gratings with other profiles, the reader is referred to Ref. [43].

Consider a thin grating having rectangular corrugations on a thin-film waveguide. The cover, film, and substrate indices of the unperturbed waveguide are n_c, n_f and n_s and the film thickness is h (Fig. 7.8(a)). The waveguide parameters are the same as that of the thin-film waveguide analyzed in Section 2.2.1. In terms of the relative dielectric constants, we have

$$\varepsilon_r(x, y) = \begin{cases} n_c^2 & x > 0 \\ n_f^2 & -h \leq x \leq 0 \\ n_s^2 & x < -h \end{cases} \qquad (7.37)$$

Suppose the grating is an array of rectangular dielectric bars with index n_f formed on the top of the thin-film waveguide (Fig. 7.8(b)). Each bar has a thickness Δh and a width $\Lambda/2$ (Fig. 7.8(c)). The spacing between bars is also $\Lambda/2$. Thus, the periodicity of the corrugation structure is Λ. The grating is thin in that $\Delta h \ll \lambda$. Clearly, the perturbation $\Delta\varepsilon_r(x, y, z)$ is zero everywhere except for a thin layer

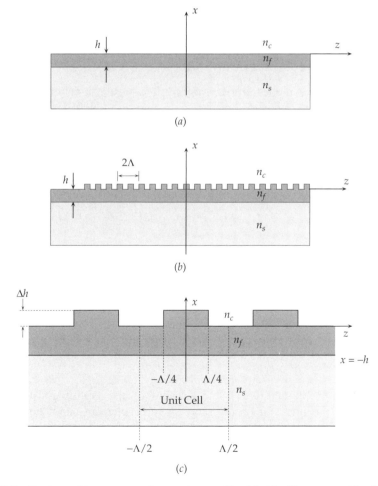

Figure 7.8 Grating with a rectangular grating profile: (a) thin-film waveguide, (b) grating on a thin-film waveguide, and (c) grating geometry in a unit cell.

of $0 < x < \Delta h$. For convenience, we choose the centerline of one of the bars as $z = 0$ (Fig. 7.8(c)). Since the permittivity perturbation is periodic in z, it is only necessary to consider a unit cell of the periodic structure. We choose the region $-\Lambda/2 < z < \Lambda/2$ as a unit cell. This is also shown in Figure 7.8 (c). In the unit cell, the relative permittivity perturbation is

$$\Delta\varepsilon_r(x, y, z) = \begin{cases} n_f^2 - n_c^2, & 0 < x < \Delta h, & |z| < \Lambda/4 \\ 0, & 0 < x < \Delta h, & \Lambda/4 < |z| < \Lambda/2 \end{cases} \quad (7.38)$$

A straightforward Fourier analysis of $\Delta\varepsilon_r(x,y,z)$ with respect to z leads to

$$\Delta\varepsilon_r(x, y, z) = (n_f^2 - n_c^2)\left[\frac{1}{2} - \frac{1}{\pi}\sum_{q=1}^{\infty}\frac{(-1)^q}{2q-1}(e^{j(2q-1)Kz} + e^{-j(2q-1)Kz})\right]$$

$$0 < x < \Delta h \quad (7.39)$$

The Fourier coefficient of the qth spatial harmonic is

$$\Delta\varepsilon_{rq}(x, y) = (-1)^{q+1}\frac{n_f^2 - n_c^2}{\pi(2q-1)} \quad 0 < x < \Delta h \quad (7.40)$$

provided q is greater than 0. The transverse coupling coefficient is, from (7.30),

$$\kappa_{\ell\nu q}^t = (-1)^{q+1}\frac{\omega\varepsilon_0}{4\pi}\frac{n_f^2 - n_c^2}{2q-1}\int_0^{\Delta h}\mathbf{e}_{\nu t}\cdot\mathbf{e}_{\ell t}^*\,ds \quad q = 1, 2, 3\ldots \quad (7.41)$$

Similar expression can be written for $\kappa_{\ell\nu q}^z$.

As a specific example, we consider the contradirectional coupling of two TE$_\nu$ modes propagating in the opposite directions. Since e_z of TE modes is zero, $\kappa_{\nu\nu q}^z$ vanishes for all values of ν and q. It is only necessary to evaluate $\kappa_{\nu\nu q}^t$. The transverse fields $\mathbf{e}_t$ of TE modes in the cover region is given by (2.12). Making use of (2.12), we obtain an expression for the transverse coupling coefficient:

$$\kappa_{\nu\nu q}^t = (-1)^{q+1}\frac{\omega\varepsilon_0}{4\pi}\frac{n_f^2 - n_c^2}{2q-1}\int_0^{\Delta h}E_c^2e^{-2\gamma_c x}\,dx$$

$$= (-1)^{q+1}\frac{\omega\varepsilon_0}{4\pi}\frac{n_f^2 - n_c^2}{2q-1}E_c^2\frac{1 - e^{-2\gamma_c\Delta h}}{2\gamma_c} \quad (7.42)$$

Since the grating is very thin, $|\gamma_c\Delta h| \ll 1$, we approximate the exponential function by the two leading terms of the series expansion and obtain

$$\kappa_{\nu\nu q}^t \approx (-1)^{q+1}\frac{\omega\varepsilon_0}{4\pi}\frac{n_f^2 - n_c^2}{2q-1}E_c^2\Delta h \quad q = 1, 2, 3\ldots \quad (7.43)$$

The amplitude constant E_c of a normalized $\mathbf{e}_v$ can be obtained from (2.16), (2.46) and (2.47) and it is

$$E_c^2 = \frac{4\eta_o}{Nh_{\text{eff}}} \frac{n_f^2 - N^2}{n_f^2 - n_c^2} \qquad (7.44)$$

where h_{eff} is the effective thickness and N is the effective refractive index of the TE_v mode. Substituting (7.44) into (7.42), we have

$$\kappa_{vvq}^t = (-1)^{q+1} k \frac{n_f^2 - N^2}{\pi(2q-1)N} \frac{\Delta h}{h_{\text{eff}}} \qquad (7.45)$$

Obviously, the coupling coefficient increases linearly with the grating thickness Δh provided the grating is thin, that is, $\gamma_c \Delta h \ll 1$. In Figure 7.9, we plot $\kappa_{vvq}^t/(k^2 \Delta h)$ as a function of h for three lowest TE modes. Since N is different for a different mode, the coupling coefficient is also different for a different mode, as shown in Figure 7.9. If the interacting mode is close to the cutoff, h_{eff} is large and the coupling coefficient becomes very small. On the other hand, if the interacting mode is far above the cutoff, N is close to n_f, the coefficient is also small. This is due to the presence of $n_f^2 - N_f^2$ in (7.45). We also note that the peak of $\kappa_{\text{TE0 TE0 1}}$ is larger than that of $\kappa_{\text{TE1 TE1 1}}$ and that in turn is larger than $\kappa_{\text{TE2 TE2 1}}$.

As a numerical example, we consider a guided-wave grating with the following parameters [21]:

$$\begin{aligned} &n_f = 1.54 &&n_s = 1.515 &&n_c = 1.0 \\ &\lambda = 0.5658 \ \mu\text{m} &&h = 0.8479 \ \mu\text{m} &&\Delta h = 0.046 \ \mu\text{m} \end{aligned} \qquad (7.46)$$

For these guided-wave grating parameters, we have $\kappa_{001}^t \sim 3.0 \times 10^3 \text{m}^{-1}$.

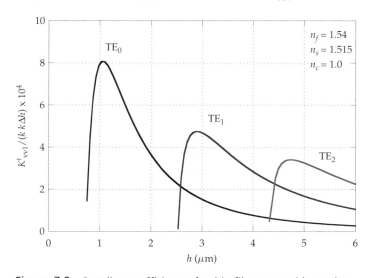

Figure 7.9   Coupling coefficients of a thin-film waveguide grating.

7.4   GRAPHICAL REPRESENTATION OF GRATING EQUATION

As noted in Section 7.1, gratings may be used to couple light into or out of thin-film waveguides [12]. Gratings used for this application are known as *grating couplers*. In this section, we use grating couplers as a vehicle to introduce a graphical representation of the Bragg condition [32, 44]. Suppose a grating structure is built on a thin-film waveguide. The indices of the three regions are n_c, n_f and n_s. Let N be the effective refractive index of a mode guided by the thin-film waveguide without the grating. Thus the propagation constant of the guided mode is $\beta = kN$. Each grating element is a small perturbation of the original waveguide. The incoming wave is scattered by each and every grating element and the scattered field spreads in all directions. Since the perturbation is weak, the field scattered by each grating element is also weak. But if fields scattered by all grating elements superimpose constructively in a certain direction θ, the total field diffracted by the grating structure would have a peak in this direction. Then a significant fraction of incoming power is diverted to the direction θ. Consider two consecutive grating elements, A and B, separated by the grating periodicity Λ, as shown in Figure 7.10. The incident guided mode arrives at element A earlier than that at element B. But the field scattered by element A must travel a longer path length in the cover region before it reaches an observer at the direction θ than that from element B. The extra path length is $\overline{AC}$ as shown in Figure 7.10. Thus a constructive interference occurs if

$$\beta\overline{AB} - kn_c\overline{AC} = q2\pi \tag{7.47}$$

where q is an integer. A simple geometrical consideration (Fig. 7.10) shows that

$$\overline{AC} = \overline{AB}\cos\left(\frac{\pi}{2} - \theta\right) = \overline{AB}\sin\theta = \Lambda\sin\theta \tag{7.48}$$

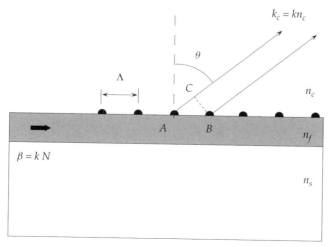

Figure 7.10   Graphic representation of Bragg condition.

Thus the condition for a constructive interference is

$$\beta = qK + kn_c \sin\theta \tag{7.49}$$

Equation (7.49) is often referred to as the *grating equation*. When the grating equation is satisfied, the Bragg condition is met by the incoming guided mode with a propagation constant β and the plane wave propagating in the cover region with an index n_c in the direction θ. Figure 7.11 depicts two graphical constructions of the grating equation with $q = 1$ and $q = 2$. Waves radiated in the two directions are referred to as the *first-order* and the *second-order Bragg diffractions* respectively. If the grating periodicity is longer, the grating wave vector K is shorter, the grating equation may also be met with $q = 3$, $q = 4$ and so on. for the same incoming guided mode. Then additional peaks ap pear and they are to the *third-, fourth-* and *higher-order Bragg diffractions.*

Gratings are also useful to convert a guided mode to a different guided mode or to reflect a guided mode propagating in the opposite direction. In Figure 7.12(*a*), we show schematically the Bragg condition for converting a TE_0 mode to a TE_1 mode, and *vice versa*, propagating in the same direction. The grating wave vector required for the TE_0–TE_1 conversion, $K = \beta_{TE0} - \beta_{TE1}$, is small. Thus a grating with a large grating periodicity Λ is required. In Figure 7.12(*b*), we show the Bragg condition for reflecting the TE_0 mode. In this case, a portion of the incoming TE_0 mode propagating in $+z$ direction is converted to the TE_0 mode traveling in the $-z$ direction. The grating wave vector required for an efficient reflection is $K = 2\beta_{TE0}$

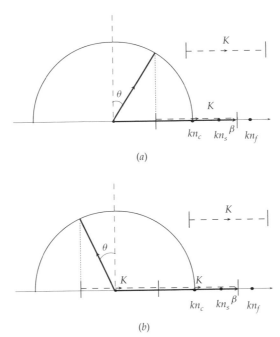

(*a*)

(*b*)

Figure 7.11   (a) First- and (b) second-order Bragg diffractions.

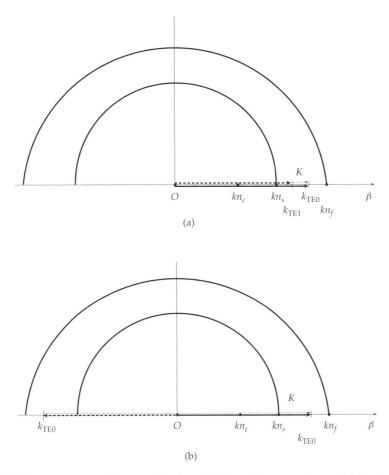

Figure 7.12 Wave vector diagrams showing (a) $TE_0 \leftrightarrow TE_1$ conversion and (b) reflection of TE_0 mode.

and the grating periodicity is $\lambda_{TE0}/2$, which is very small. Physically, a strong reflection is the result of a constructive superposition of fields scattered by each and every grating element in the backward direction. Since the total reflection is the superposition of waves reflected by each grating elements, the total reflection is highly frequency dependent and has a very narrow bandwidth. We will analyze the filter response of grating reflectors in detail in the next section.

7.5 GRATING FILTERS [20, 21, 44, 45, 46]

7.5.1 Coupled-Mode Equations

As an example of grating devices, we consider a guided-wave grating as a narrow-band optical filter. Suppose a mode of unit amplitude impinges upon a grating of

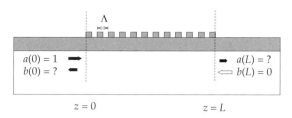

Figure 7.13 Waveguide grating reflector as an optical filter.

length L (Fig. 7.13) from the left-hand side. We study the power reflected by and transmitted through the grating. Clearly, waves propagating in the two opposite directions are involved. The incident and reflected waves may have the same modal field in the transverse plane. We label the modes propagating in the $+z$ and $-z$ direction as mode ℓ'' and mode ν'', respectively. The interaction is governed by (7.35) and (7.36). To simplify the equations, we make the following substitutions:

$$a_{\ell''}(z) \Rightarrow a(z) \qquad\qquad \beta_{\ell''} \Rightarrow \beta$$
$$a_{\nu''}(z) \Rightarrow b(z) \qquad\qquad \beta_{\nu''} \Rightarrow -\beta$$
$$\kappa^{z}_{\nu''\ell'' 1} + \kappa^{t}_{\nu''\ell'' -1} \Rightarrow \kappa$$

We also define the *detuning parameter* or *Bragg parameter*:

$$\delta = \beta - \frac{K}{2} \qquad\qquad (7.50)$$

With these substitutions, (7.35) and (7.36) are greatly simplified:

$$\frac{da(z)}{dz} = -j\kappa\, b(z)e^{j2\delta z} \qquad\qquad (7.51)$$

$$\frac{db(z)}{dz} = +j\kappa\, a(z)e^{-j2\delta z} \qquad\qquad (7.52)$$

Although the mode amplitude functions $a(z)$ and $b(z)$ vary slowly, the two exponential functions may fluctuate rapidly. To minimize the effect of rapid fluctuations, we introduce two auxiliary functions $\mathcal{R}(z)$ and $\mathcal{S}(z)$:

$$\mathcal{R}(z) = a(z)e^{-j\delta z} \qquad\qquad (7.53)$$

$$\mathcal{S}(z) = b(z)e^{+j\delta z} \qquad\qquad (7.54)$$

Functions $\mathcal{R}(z)$ and $\mathcal{S}(z)$ represent the modes propagating in the $+z$ and $-z$ direction, respectively. In terms of $\mathcal{R}(z)$ and $\mathcal{S}(z)$, (7.51) and (7.52) become

$$\frac{d\mathcal{R}(z)}{dz} + j\delta\mathcal{R}(z) = -j\kappa\mathcal{S}(z) \qquad\qquad (7.55)$$

$$\frac{d\mathcal{S}(z)}{dz} - j\delta\mathcal{S}(z) = +j\kappa\mathcal{R}(z) \qquad\qquad (7.56)$$

7.5.2   Filter Response of Grating Reflectors

Combining the two first-order differential equations, we obtain a second-order differential equation:

$$\frac{d^2 S(z)}{dz^2} = \sigma^2 S(z) \tag{7.57}$$

where

$$\sigma^2 = \kappa^2 - \delta^2 \tag{7.58}$$

The solution for $S(z)$ can be written as

$$S(z) = C_1 \sinh[\sigma(L - z)] + C_2 \cosh[\sigma(L - z)] \tag{7.59}$$

The two constants C_1 and C_2 are determined by the boundary conditions. As noted earlier, a mode of unity amplitude impinges on the grating reflector from the left-hand side of the grating, and there is no wave coming from the right-hand side. Therefore, the boundary conditions are $a(0) = 1$ and $b(L) = 0$. In terms of the auxiliary functions, the boundary conditions are $\mathcal{R}(0) = 1$ and $\mathcal{S}(L) = 0$. It follows from (7.59) that $\mathcal{S}(L)$ vanishes if C_2 is 0. Thus

$$S(z) = C_1 \sinh[\sigma(L - z)] \tag{7.60}$$

To determine C_1, we make use of the boundary condition of $\mathcal{R}(0) = 1$. For this purpose, we need to express $\mathcal{R}(z)$ in terms of $\mathcal{S}(z)$. This is done by substituting (7.60) into (7.56):

$$\mathcal{R}(z) = j\frac{C_1}{\kappa}\{\sigma \cosh[\sigma(L - z)] + j\delta \sinh[\sigma(L - z)]\} \tag{7.61}$$

We obtain an explicit expression for C_1 by enforcing the condition that $\mathcal{R}(0) = 1$. Using the expression for C_1 so obtained, we rewrite (7.60) and (7.61) as

$$\mathcal{R}(z) = \frac{\sigma \cosh[\sigma(L - z)] + j\delta \sinh[\sigma(L - z)]}{\sigma \cosh(\sigma L) + j\delta \sinh(\sigma L)} \tag{7.62}$$

$$\mathcal{S}(z) = \frac{-j\kappa \sinh[\sigma(L - z)]}{\sigma \cosh(\sigma L) + j\delta \sinh(\sigma L)} \tag{7.63}$$

The *reflection coefficient* and the *transmission coefficient* of the grating reflector of length L are, respectively,

$$\Gamma = \mathcal{S}(0) = \frac{-j\kappa \sinh(\sigma L)}{\sigma \cosh(\sigma L) + j\delta \sinh(\sigma L)} \tag{7.64}$$

$$T = \mathcal{R}(L) = \frac{\sigma}{\sigma \cosh(\sigma L) + j\delta \sinh(\sigma L)} \tag{7.65}$$

The power reflection and transmission coefficients, also known as *reflectance* and *transmittance*, respectively, are

$$|\Gamma|^2 \equiv |b(0)|^2 = |\mathcal{S}(0)|^2 = \frac{\kappa^2 \sinh^2(\sigma L)}{\sigma^2 + \kappa^2 \sinh^2(\sigma L)} \qquad (7.66)$$

$$|T|^2 \equiv |a(L)|^2 = |\mathcal{R}(L)|^2 = \frac{\sigma^2}{\sigma^2 + \kappa^2 \sinh^2(\sigma L)} \qquad (7.67)$$

In Figure 7.14(*a*), we plot $|a(z)|^2$ and $|b(z)|^2$, which are also, respectively, $|\mathcal{R}(z)|^2$ and $|\mathcal{S}(z)|^2$, for a grating with $\kappa L = 1.5$ and $\delta = 0$. It shows the $|a(z)|^2$ decreases continuously as the mode travels in the $+z$ direction. On the other hand, $|b(z)|^2$ grows as the mode moves in the $-z$ direction. Similar plots can also be obtained for gratings with a different δ. An example with $\delta/\kappa = 1.2$ is shown in Figure 7.14(*b*). In the two plots, note that $|a(0)|^2 = 1$ and $|b(L)|^2 = 0$ as specified by the boundary conditions. Also recall that $|a(L)|^2$ and $|b(0)|^2$ are, respectively, the transmittance and reflectance of the grating reflector. In Figure 7.15, we plot the reflectance (solid curve) and the transmittance (dashed curve) as functions of δ/κ for three values of κL. It is evident from these plots that the total power is

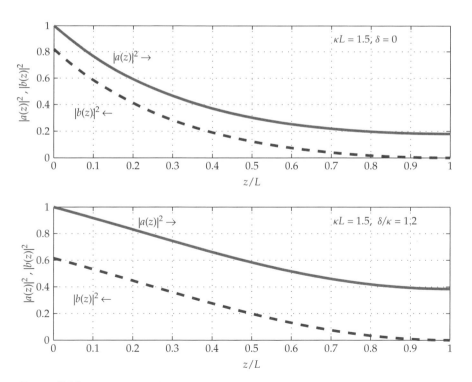

Figure 7.14   Amplitudes of modes propagating in forward and backward directions.

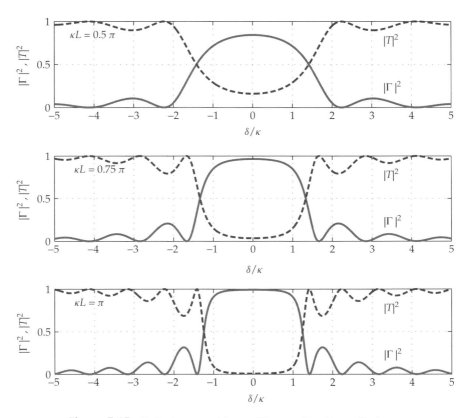

Figure 7.15   Reflectance and transmittance of grating reflectors.

conserved. In fact, it is rather simple to show that

$$|\Gamma|^2 + |T|^2 = 1 \tag{7.68}$$

It follows from Figure 7.15 that the power reflection coefficient reaches its peak when the Bragg condition ($\delta = 0$) is met. The peak reflectance is, from (7.66),

$$|\Gamma|^2_{\text{max}} = \tanh^2(\kappa L) \tag{7.69}$$

Clearly, the maximum reflectance increases monotonically with κL. The peak reflectance is greater than 0.95 if κL is greater than 2.2 as shown in Figure 7.16.

In general, the reflectance curve has a flat plateau, and the transmittance curve has a flat valley near the Bragg condition as long as the grating is long, that is, $\kappa L \gg 1$. To understand this feature, we suppose that the operating wavelength or frequency is slightly off the Bragg condition, $0 < \delta < \kappa$, and that the grating is sufficiently long, $\kappa L \gg 1$. Then the reflectance is approximately

$$|\Gamma|^2 \sim 1 - \frac{4\sigma^2}{\kappa^2}e^{-2\sigma L} \tag{7.70}$$

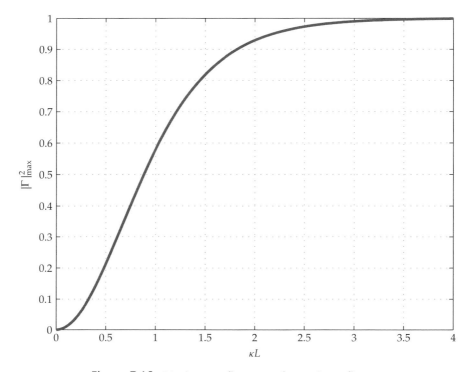

Figure 7.16   Maximum reflectance of a grating reflector.

This explains the existence of a flat plateau in the reflectance curve near the Bragg condition if κL is large. This is shown in the second and third sets of curves of Figure 7.15. It also explains that the transmittance curve has a flat valley near the Bragg condition, since the total power is conserved.

When δ increases further beyond κ, σ becomes imaginary and $|\Gamma|^2$ becomes an oscillatory function of σL:

$$|\Gamma|^2 = \frac{\kappa^2 \sin^2(|\sigma L|)}{\delta^2 - \kappa^2 \cos^2(|\sigma L|)} \tag{7.71}$$

This is evident in Figure 7.15. Note in particular that $|\Gamma|^2$ vanishes at $|\sigma L| = \pi, 2\pi, 3\pi, \ldots$, which can be written as

$$\delta = \pm\kappa\sqrt{1 + \left(\frac{m\pi}{\kappa L}\right)^2} \qquad m = 1, 2, 3, \ldots \tag{7.72}$$

In the next subsection, we use the nulls of $|\Gamma|^2$ given in (7.72) to estimate the bandwidth of grating reflectors.

7.5.3 Bandwidth of Grating Reflectors

Since we use grating reflectors as optical filters, we prefer to express the filter response directly in terms of the wavelength. Let the *center wavelength* and the *full width between nearest nulls* of the filter be λ_0 and $\Delta\lambda$. In terms of λ_0 and $\Delta\lambda$, the nearest nulls of the filter response are at $\lambda_0 \pm (\Delta\lambda/2)$. We choose the waveguide parameters and the grating periodicity Λ such that the Bragg condition is met at λ_0. That is $\beta(\lambda_0) = K/2$, or equivalently $\lambda_0 = 2N(\lambda_0)\Lambda$. At $\lambda_0 \pm (\Delta\lambda/2)$, the Bragg parameters are from (7.50)

$$
\delta\left(\lambda_0 \pm \frac{\Delta\lambda}{2}\right) \equiv \beta\left(\lambda_0 \pm \frac{\Delta\lambda}{2}\right) - \frac{K}{2}
$$

$$
= \beta(\lambda_0) \pm \frac{\Delta\lambda}{2}\frac{d\beta}{d\lambda}\bigg|_{\lambda_0} + \cdots - \frac{K}{2} \approx \pm\frac{\Delta\lambda}{2}\frac{d\beta}{d\lambda}\bigg|_{\lambda_0} \qquad (7.73)
$$

We also deduce from the expression for the group velocity that

$$
\frac{d\beta}{d\omega}\bigg|_{\lambda_0} = \frac{N_{gr}(\lambda_0)}{c} = \frac{d\beta}{d\lambda}\frac{d\lambda}{d\omega}\bigg|_{\lambda_0} \qquad (7.74)
$$

where N_{gr} is the effective group velocity index of the guided mode. Thus, we write

$$
\frac{d\beta}{d\lambda}\bigg|_{\lambda_0} = -\frac{2\pi}{\lambda_0^2}N_{gr}(\lambda_0) \qquad (7.75)
$$

Using this equation in conjunction with (7.73), we obtain

$$
\delta\left(\lambda_0 \pm \frac{\Delta\lambda}{2}\right) \sim \mp\pi N_{gr}\frac{\Delta\lambda}{\lambda_0^2} \qquad (7.76)
$$

Combining (7.76) with (7.72), we obtain the *full bandwidth between nearest nulls*:

$$
\Delta\lambda = \frac{\lambda_0^2}{L N_{gr}(\lambda_0)}\sqrt{1 + \left(\frac{\kappa L}{\pi}\right)^2} \qquad (7.77)
$$

Recall that $\lambda_0 = 2N(\lambda_0)\Lambda$. Thus the *fractional filter bandwidth* is [21]

$$
\frac{\Delta\lambda}{\lambda_0} = 2\frac{N(\lambda_0)\Lambda}{N_{gr}(\lambda_0)L}\sqrt{1 + \left(\frac{\kappa L}{\pi}\right)^2} \qquad (7.78)
$$

For the numerical example considered in Section 7.3, the full bandwidth between nearest nulls, $\Delta\lambda$, is 0.4 nm, assuming $L = 0.57$ mm and that N/N_{gr} is a little greater than 1. This agrees with the result reported by Ref. [21].

7.6   DISTRIBUTED FEEDBACK LASERS

As the second example of grating devices, we consider *distributed feedback* (DFB) *lasers* [22–27, 47–50]. A laser has three essential components, and they are an active medium, a mechanism to cause optical feedback, and a means to supply energy into the laser. In a conventional laser, there are two reflectors that provide the optical feedback. The two mirrors form a Fabry–Perot cavity or resonator. The conventional lasers are commonly referred to as *Fabry–Perot lasers*. In a DFB laser, a grating, in lieu of the two reflectors, is built in or near the active semiconductor heterostructure to provide the optical feedback. Thus a DFB laser has no discrete mirrors. Instead, it has a guided-wave grating. The operation of a DFB laser can be understood as follows. In the presence of the current injection, the semiconductor heterostructure becomes a gain medium. Due to the optical gain and the interaction between two contrapropagating modes in the grating region, fields in a DFB laser grow from noise until a self-sustained oscillation is established. Light radiates from the two sides of the grating structure. In this section, we follow Kogelnik and Shank's work and analyze the operation of DFB lasers [25]. In particular, we study the oscillation frequency and the optical gain required to sustain the oscillations.

Suppose the DFB laser extends from $z = -L/2$ to $z = L/2$, as shown in Figure 7.17. We take the center of the central grating element as the origin of the z coordinate.

7.6.1   Coupled-Mode Equations with Optical Gain

To account for the optical gain in the grating structure, we amend the coupled-mode equations (7.51) and (7.52) by adding $+ga(z)$ and $-gb(z)$ to the equations:

$$\frac{da(z)}{dz} = -j\kappa b(z)e^{j2\delta z} + ga(z) \tag{7.79}$$

$$\frac{db(z)}{dz} = +j\kappa a(z)e^{-j2\delta z} - gb(z) \tag{7.80}$$

where g is the *gain constant* of the active grating region. A positive g signifies the optical gain. For simplicity, we assume κ and g are constants independent of z. To verify if the added terms indeed represent optical amplification, we suppose the

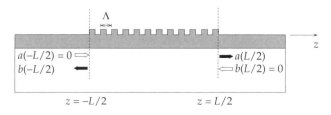

Figure 7.17   Geometry of a DFB laser.

grating is absent and set κ to zero. Then (7.79) and (7.80) become

$$\frac{da(z)}{dz} = +ga(z) \tag{7.81}$$

$$\frac{db(z)}{dz} = -gb(z) \tag{7.82}$$

The solutions of (7.81) and (7.82) are $Ce^{\pm gz}$, and they represent waves growing exponentially as they move in the $\pm z$ direction. In short, g indeed represents the optical gain in the grating region.

7.6.2   Boundary Conditions and Symmetric Condition

As noted earlier, light radiates from the two sides of a DFB structure, and there is no light impinging upon either side of the grating. Since there is no forward-propagating field coming from the left side of the DFB structure, the boundary condition at the left boundary is

$$a(-L/2) = 0 \tag{7.83}$$

Similarly, there is no backward-propagating wave coming from the right-hand side of the grating. The condition at the right boundary is

$$b(L/2) = 0 \tag{7.84}$$

Since the DFB structure is symmetric with respect to the centerline, $E(z)$ must be either *symmetric* or *antisymmetric* with respect to the origin. That is

$$E(z) = \pm E(-z) \tag{7.85}$$

Our task is to solve (7.79) and (7.80) subjected to the boundary conditions (7.83) and (7.84) and the symmetry condition (7.85).

7.6.3   Eigenvalue Equations

To solve the coupled-mode equations, we again make use of the auxiliary functions introduced in (7.53) and (7.54) and the Bragg parameter defined in (7.50) to simplify (7.79) and (7.80). In terms of $\mathcal{R}(z)$, $\mathcal{S}(z)$, and δ, the coupled-mode equations with an optical gain become

$$\frac{d\mathcal{R}(z)}{dz} + (j\delta - g)\mathcal{R}(z) = -j\kappa\mathcal{S}(z) \tag{7.86}$$

$$\frac{d\mathcal{S}(z)}{dz} - (j\delta - g)\mathcal{S}(z) = j\kappa\mathcal{R}(z) \tag{7.87}$$

By combining the two equations, we obtain two second-order differential equations:

$$\frac{d^2 \mathcal{R}(z)}{dz^2} - \gamma^2 \mathcal{R}(z) = 0 \tag{7.88}$$

and

$$\frac{d^2 \mathcal{S}(z)}{dz^2} - \gamma^2 \mathcal{S}(z) = 0 \tag{7.89}$$

where

$$\gamma^2 = \kappa^2 + (j\delta - g)^2 \tag{7.90}$$

Since γ is a constant, the solutions of (7.88) and (7.89) are linear combinations of hyperbolic sine and cosine functions. Accordingly, we cast $\mathcal{R}(z)$ and $\mathcal{S}(z)$ in the form of

$$\mathcal{R}(z) = R_0 \sinh \gamma \left(\frac{L}{2} + z \right) + R_1 \cosh \gamma \left(\frac{L}{2} + z \right) \tag{7.91}$$

$$\mathcal{S}(z) = S_0 \sinh \gamma \left(\frac{L}{2} - z \right) + S_1 \cosh \gamma \left(\frac{L}{2} - z \right) \tag{7.92}$$

where R_0, R_1, S_0, and S_1 are constants to be determined. It follows from the boundary conditions (7.83) and (7.84) that $R_1 = S_1 = 0$.

In the grating region, waves propagate in both directions. Thus

$$E(z) \equiv a(z)e^{-j\beta z} + b(z)e^{+j\beta z} = \mathcal{R}(z)e^{-jKz/2} + \mathcal{S}(z)e^{+jKz/2} \tag{7.93}$$

Upon substituting (7.91) and (7.92) into (7.93), we see that the symmetry condition (7.85) is met if and only if $R_0 = \pm S_0$. By setting R_1 and S_1 to zero and expressing S_0 as $\pm R_0$, (7.91) and (7.92) become

$$\mathcal{R}(z) = R_0 \sinh \gamma \left(\frac{L}{2} + z \right)$$

$$\mathcal{S}(z) = \pm R_0 \sinh \gamma \left(\frac{L}{2} - z \right)$$

So far, we have determined $\mathcal{R}(z)$ and $\mathcal{S}(z)$ independently. But $\mathcal{R}(z)$ and $\mathcal{S}(z)$ must be related. To establish a relation between them, we substitute the two expressions into the two differential equations (7.86) and (7.87) and obtain

$$\gamma \cosh \gamma \left(\frac{L}{2} + z \right) + (j\delta - g) \sinh \gamma \left(\frac{L}{2} + z \right) = \mp j\kappa \sinh \gamma \left(\frac{L}{2} - z \right) \tag{7.94}$$

$$\gamma \cosh \gamma \left(\frac{L}{2} - z \right) + (j\delta - g) \sinh \gamma \left(\frac{L}{2} - z \right) = \mp j\kappa \sinh \gamma \left(\frac{L}{2} + z \right) \tag{7.95}$$

The two equations must hold for all values of z. The two equations can hold for all values of z only if

$$\gamma \cosh \frac{\gamma L}{2} + (j\delta - g) \sinh \frac{\gamma L}{2} = \mp j\kappa \sinh \frac{\gamma L}{2} \tag{7.96}$$

$$\gamma \sinh \frac{\gamma L}{2} + (j\delta - g) \cosh \frac{\gamma L}{2} = \pm j\kappa \cosh \frac{\gamma L}{2} \tag{7.97}$$

are valid simultaneously. We simplify (7.96) and (7.97) further by adding and sub-tracting the two equations. The resulting equations are

$$\gamma + (j\delta - g) = \pm j\kappa e^{-\gamma L} \tag{7.98}$$

$$\gamma - (j\delta - g) = \mp j\kappa e^{\gamma L} \tag{7.99}$$

By adding and subtracting the two equations again, we obtain

$$\gamma L = \mp j\kappa L \sinh \gamma L \tag{7.100}$$

$$j\delta L - gL = \pm j\kappa L \cosh \gamma L \tag{7.101}$$

Equation (7.100) is the *eigenvalue equation* of the DFB laser [25]. For a given DFB structure, κL is known. We solve for γL from (7.100) for a given κL. But no exact analytical solution is known. Thus we solve the eigenvalue equation numer-ically. Once γL is determined, we use (7.101) or (7.90) to evaluate δL and gL. There may be several solutions. Each solution corresponds to an *oscillation mode*. We refer to the solution with the smallest δL as the first and lowest oscillation mode and set q to 1. Typical results for three values of q are shown in Figure 7.18 or equivalently in Figure 7.19. From δL and gL, we determine the *oscillation frequency* of, and the *threshold gain* required to maintain, the self-sustained oscillations in the DFB laser. We will return to these subjects in Section 7.6.5.

To gain further insight of the DFB problem, it would be instructive to derive and approximate solution to the eigenvalue equation. Toward this end, we obtain from (7.100) and (7.101) that

$$\gamma L = (g - j\delta)L \tanh \gamma L \tag{7.102}$$

If the gain constant of the DFB laser is quite high, the real part of γL is much greater than the imaginary part. Thus $\tanh \gamma L$ is approximately 1. Then we can approximate (7.102) as

$$\gamma L \sim gL - j\delta L \tag{7.103}$$

Similarly, the $\cosh \gamma L$ in (7.101) may be approximated by $\frac{1}{2}e^{\gamma L}$. Then we obtain

$$2(gL - j\delta L) \sim \mp j\kappa L e^{(g-j\delta)L} \tag{7.104}$$

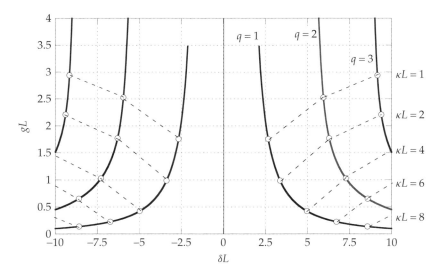

Figure 7.18   Solution of eigenvalue equation (7.100).

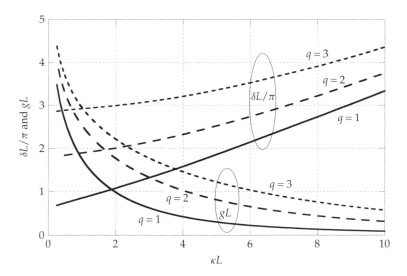

Figure 7.19 $\delta L/\pi$ and gL as function of κL.

Recognizing that $\mp j = e^{j(q+1/2)\pi}$, where q is an integer, we can rewrite the above equation as

$$2(gL - j\delta L) \sim \kappa L e^{gL} e^{j[(q+(1/2))\pi - \delta L]} \qquad (7.105)$$

In arriving at (7.105), we make use of the fact that the real part of γL is much larger than the imaginary part. It means that optical gain is quite high, that is, $gL \gg \delta L$. Therefore the real part of the right-hand side of (7.105) must be greater than the imaginary part too. Thus, the complex exponent in the right-hand side of (7.105) must be negligibly small. Then

$$\delta L \sim \left(q + \tfrac{1}{2}\right) \pi \tag{7.106}$$

Each value of q corresponds to a different δL and a different oscillation mode with a different oscillation frequency.

We also obtain from (7.101) that

$$4(g^2 L^2 + \delta^2 L^2) \approx \kappa^2 L^2 e^{2gL} \tag{7.107}$$

By substituting δL given in (7.106) into (7.107), we can solve for the gain needed to sustain oscillations for a given κL and q.

7.6.4 Mode Patterns

It is instructive to examine the intensity distribution of an oscillation mode. The intensity distribution in the DFB laser is, from (7.93),

$$|E(z)|^2 = |\mathcal{R}(z)|^2 + |\mathcal{S}(z)|^2 + [\mathcal{R}^*(z)\mathcal{S}(z)e^{+jKz} + \mathcal{R}(z)\mathcal{S}^*(z)e^{-jKz}] \tag{7.108}$$

This may be rewritten as

$$|E(z)|^2 = |\mathcal{R}(z)|^2 + |\mathcal{S}(z)|^2 + 2|\mathcal{R}(z)\mathcal{S}(z)| \cos[Kz + \Theta(z)] \tag{7.109}$$

where $\Theta(z)$ is the phase of $\mathcal{R}^*(z)\mathcal{S}(z)$. A typical plot of $|E(z)|^2$ is shown in Figure 7.20. The rapid fluctuation represents the standing-wave pattern in the grating structure. Mathematically, it comes from the cosine term of (7.109). Since the rapid fluctuation is rather confusing, we examine the envelop of $|E(z)|^2$ instead. The envelop of the standing-wave pattern is

$$|E(z)|^2_{\text{max}} = |\mathcal{R}(z)|^2 + |\mathcal{S}(z)|^2 + 2|\mathcal{R}(z)\mathcal{S}(z)| \tag{7.110}$$

In Figure 7.21, we plot the envelop of $|E(z)|^2$ as a function of z/L for the first three oscillation modes for three values of κL. In each plot, $|E(z)|^2_{\text{max}}$ is normalized so that $|E(\pm L/2)|^2_{\text{max}}$ is 1. For oscillation modes with $q = 1, 2$, and 3, the envelopes have one, two, and three peaks or minima, respectively. In all cases, the envelopes are symmetric with respect to the center of the DFB structure.

7.6.5 Oscillation Frequency and Threshold Gain

Now we return to the question of the oscillation frequency and the threshold gain needed to sustain laser oscillations. It is also instructive to compare the oscillation frequencies and threshold gains of DFB lasers with the counterparts of Fabry–Perot lasers.

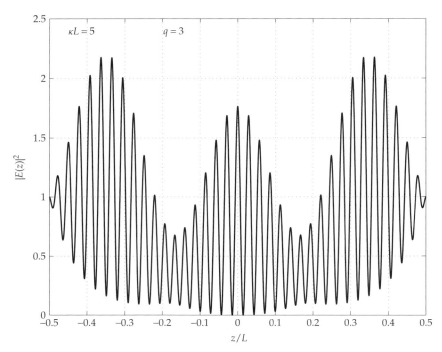

Figure 7.20   Distribution of $|E(z)|^2$ in a DFB laser.

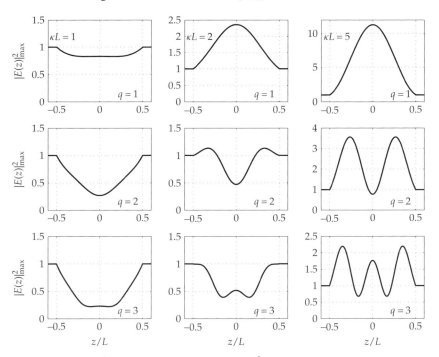

Figure 7.21   Envelope of $|E(z)|^2$ in a DFB laser.

In preparation for the comparison, we briefly summarize the properties of Fabry–Perot lasers having mirror reflectivities Γ_1 and Γ_2 and a cavity length L [51]. We assume that the gain medium is nondispersive and the refractive index is n. Then the oscillation frequency of a longitudinal mode of a Fabry–Perot laser is

$$f_{\mathrm{FP}_q} = q\frac{c}{2nL} \tag{7.111}$$

where q is a large integer even for semiconductor injection lasers. The frequency separation between two neighboring longitudinal modes is

$$\Delta f_{\mathrm{FP}} = \frac{c}{2nL} \tag{7.112}$$

The threshold gain needed to sustain a longitudinal mode of a Fabry–Perot laser is

$$g_{\mathrm{FP}} = -\frac{1}{2L}\ln(\Gamma_1\Gamma_2) \tag{7.113}$$

Two key properties of Fabry–Perot lasers are clear. The frequency separation Δf_{FP} between two neighboring longitudinal modes of a Fabry–Perot (FP) laser is independent of the longitudinal mode numbers. The threshold gain needed to sustain oscillations is the same for all longitudinal modes.

As noted earlier, for a given DFB laser, κL is known and γL is determined numerically from the eigenvalue equation. From γL, we obtain δL and gL. To relate the oscillation frequency to δ, we recall the expression of the Bragg parameter given in (7.50) and that $\beta = kN$ where N is the effective refractive index of guided mode. Therefore the *oscillation frequency* of a DFB laser is

$$f_{\mathrm{DFB}_q} = \frac{cK}{4\pi N} + \frac{c}{2NL}\frac{(\delta L)_q}{\pi} \tag{7.114}$$

The first part of f_{DFB_q} depends on the grating wave vector K and the effective refractive index N of the guided mode. This part is commonly known as the *Bragg frequency*. The second part of f_{DFB_q} is the frequency deviation from the Bragg frequency. As shown in Figures 7.18 and 7.19, δL is never zero. In other words, the second part of (7.114) is always present. Thus the oscillation frequency of the DFB laser cannot be precisely the Bragg frequency. In Figure 7.22, we plot the mode spectra and the threshold gain for DFB lasers with $\kappa L = 2$ and 4. The frequency separation between two neighboring modes is

$$\Delta f_{\mathrm{DFB}_q} = f_{\mathrm{DFB}_{q+1}} - f_{\mathrm{DFB}_q} = \frac{c}{2NL}\frac{(\delta L)_{q+1} - (\delta L)_q}{\pi} \tag{7.115}$$

Clearly, $\Delta f_{\mathrm{DFB}_q}$ depends on q. This is fundamentally different from the longitudinal mode separation (7.112) of FP lasers.

Also shown in Figure 7.22 is the *threshold gain* gL required to sustain the oscillations in DFB lasers. As the oscillation frequency moves away from the Bragg

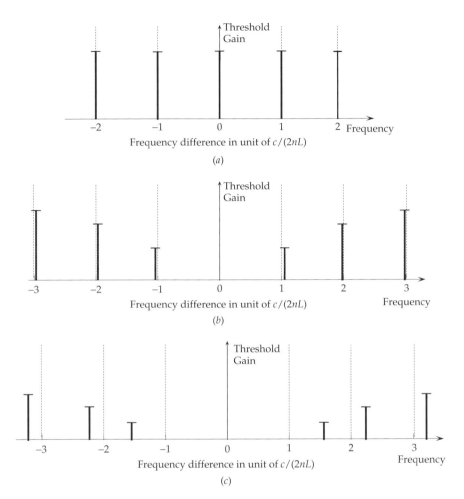

Figure 7.22 Mode spectra and threshold gain of (a) Fabry–Perot lasers, (b) DFB lasers with $\kappa L = 2.0$, and (c) DFB lasers with $\kappa L = 4.0$. (After [25].)

frequency, the gain needed to maintain the oscillations increases. In other words, an amplitude discrimination is built in the DFB laser oscillation condition.

In summary, we note that the frequency separation between two longitudinal modes of DFB lasers depends on the modes involved. In other words, $\Delta f_{\mathrm{DFB}_q}$ varies with q. We also note that the threshold gain is smaller for lower-order modes.

REFERENCES

1. M. Neviere, *Electromagnetic Theory of Gratings*, Springer, Berlin, 1980, Chapter 5.
2. K. O. Hill, Y. Fujii, D. C. Johnson, and B. S. Kawaski, "Photosensitivity in optical fiber waveguides: Application to reflection filter fabrication," *Appl. Phys. Lett.*, Vol. 32, No. 10, pp. 647–649 (1978).

3. G. Meltz, W. W. Morey, and W. H. Glenn, "Formation of Bragg gratings in optical fibers by a transverse holographic method," *Opt. Lett.*, Vol. 14, pp. 823–825 (1989).

4. K. O. Hill, B. Malo, F. Bilodeau, and D. C. Johnson, "Photosensitivity in optical fibers," *Ann. Rev. Mat. Sci.*, Vol. 23, pp. 125–157 (1993).

5. K. O. Hill and G. Meltz, "Fiber Bragg grating technology fundamentals and overview," *IEEE J. Lightwave Technol.*, Vol. 15, pp. 1263–1275 (1997).

6. R. C. Alferness and L. L. Buhl, "Electro-optic waveguide TE ↔ TM mode converter with low drive voltage," *Opt. Lett.*, Vol. 5, pp. 473–475 (Nov. 1980).

7. R. C. Alferness, "Efficient waveguide electro-optic TE ↔ TM mode converter/wavelength filter," *Appl. Phys. Lett.*, Vol. 36, pp. 513–515 (April 1, 1980).

8. R. C. Alferness, "Guided-wave devices for optical communication," *IEEE J. Quantum Electron.*, Vol. QE-17, No. 6, pp. 946–959 (June 1981).

9. P. K. Tien, R. J. Martin, R. Wolfe, R. C. LeCraw, and S. L. Blank, "Switching and modulation of light in magneto-optics waveguide of garnet films," *Appl. Phys. Lett.*, Vol. 21, No. 8, pp. 394–396 (Oct., 1972).

10. P. K. Tien, D. P. Shinke, and S. L. Blank, "Magneto-optics and motion of the magnetization in a film-waveguide optical switch," *J. Appl. Phys.*, Vol. 45, No. 7, pp. 3059–3068 (1974).

11. A. Korpel, "Acousto-optic," in *Applied Solid State Science, Advances in Materials and Devices Research*, Vol. 3, R. Wolfe (ed.), Academic, New York, 1972; and also A. Korpel, "Acousto-optics—a review of fundamentals," *Proc. IEEE*, Vol. 69, pp. 48–53 (1981).

12. L. Kuhn, P. F. Heidrich, and E. G. Lean, "Optical guided wave mode conversion by an acoustic surface wave," *Appl. Phys. Lett.*, Vol. 19, No. 10, pp. 428–430 (Nov. 15, 1971).

13. D. A. Smith, J. E. Baran, K. W. Cheung, and J. J. Johnson, "Polarization-independent acoustically tunable optical filter," *Appl. Phys. Lett.*, Vol. 56, No. 3, pp. 209–211 (Jan. 15, 1990).

14. D. B. Anderson, J. T. Boyd, M. C. Hamilton, and R. R. August, "An integrated-optical approach to the Fourier transform," *IEEE J. Quantum Electron.*, Vol. QE-13, No. 4, pp. 268–275 (April 1977).

15. M. C. Hamilton, D. A. Wille, and W. J. Miceli, "An integrated optical RF spectrum analyzer," *Opt. Eng.*, Vol. 16, No. 52, pp. 475–478 (Sept/Oct. 1977).

16. D. Mergerian, E. C. Malarkey, R. P. Pautlenus, J. C. Bradley, G. E. Max, L. D. Hutcheson, and A. L. Kellner, "Operational integrated optical r.f. spectrum analyzer," *Appl. Opt.*, Vol. 19, No. 19, pp. 3033–3034 (Sept. 1980).

17. C. S. Tsai, "Guided-wave acoustic Bragg modulators for wide-band integrated optic communications and signal processing," *IEEE Trans. Circuits Systems*, Vol. CAS-26, No. 12, pp. 1072–1098 (Dec. 1979).

18. C. J. Lii, C. S. Tsai, and C. C. Lee, "Wide-band guided-wave acoustooptic Bragg cells in GaAs-GaAlAs waveguides," *IEEE J. Quantum Electron.*, Vol. QE-22, pp. 868–872 (June, 1986).

19. M. L. Dakes, L. Kuhn, P. F. Heidrich, and B. A. Scott, "Grating coupler for efficient excitation of optical guided waves in thin films," *Appl. Phys. Lett.*, Vol. 16, pp. 523–525 (June, 15, 1970).

20. D. C. Flanders, H. Kogelnik, R. V. Schmidt, and C. V. Shank, "Grating filters for thin-film optical waveguides," *Appl. Phys. Lett.,* Vol. 24, No. 4, pp. 194–196 (Feb. 15, 1972).

21. R. V. Schmidt, D. C. Flanders, C. V. Shank, and R. D. Standley, "Narrow-band grating filters for thin-film optical waveguides," *Appl. Phys. Lett.*, Vol. 25, No. 11, pp. 651–652 (Dec. 1, 1974).

22. R. Shubert, "Theory of optical-waveguide distributed lasers with nonuniform gain and coupling," *J. Appl. Phys.*, Vol. 45, No. 1, pp. 209–215 (Jan. 1974).

23. M. Okuda and K. Kubo, "Analysis of the distributed Bragg reflector laser of asymmetrical geometry," *Japan. J. Appl. Phys.*, Vol. 14, No. 6, pp. 855–860 (June 1975).

24. H. Kogelnik and C. V. Shank, "Stimulated emission in a periodic structure," *Appl. Phys. Lett.*, Vol. 18, No. 4, pp. 152–154 (Feb. 15, 1971).

25. H. Kogelnik and C. V. Shank, "Coupled-wave theory of distributed feedback lasers," *J. Appl. Phys.*, Vol. 43, No. 5, pp. 2327–2335 (May 1972).

26. J. E. Bjorkholm and C. V. Shank, "Distributed-feedback lasers in thin-film optical waveguides," *IEEE J. Quantum Electron.*, Vol. QE-8, No. 11, pp. 833–838 (Nov. 1972).

27. J. E. Bjorkholm, T. P. Sosnowski, and C. V. Shank, "Distributed-feedback lasers in optical waveguides deposited on anisotropic substrates," *Appl. Phys. Lett.*, Vol. 22, No. 4, pp. 132–134 (Feb. 15, 1973).

28. S. Ura, T. Suhara, H. Nishihara, and J. Koyama, "An integrated-optic disk pickup device," *IEEE J. Lightwave Technol.*, Vol. LT-4, No. 7, pp. 913–918 (July 1986).

29. K. Kobayashi and M. Seki, "Microoptic grating multipliers and optical isolators for fiber-optic communications," *IEEE J. Quantum Electron.*, Vol. QE-16, pp. 11–22 (1980).

30. K. Wagatsuma, H. Sakaki, and S. Saito, "Mode conversion and optical filtering of obliquely incident waves in corrugated waveguide filters," *IEEE J. Quantum Electron.*, Vol. QE-15, pp. 632–637 (1979).

31. T. Fukuzawa and N. Nakamura, "Mode coupling in thin-film chirped gratings," *Opt. Lett.*, Vol. 4 No. 1 pp. 343–345 (Nov. 1979).

32. T. Suhara and H. Nishihara, "Integrated optics components and devices using periodic structures," *IEEE J. Quantum Electron.*, Vol. QE-22, pp. 845–867 (June 1986).

33. R. R. Syms, "Optical directional coupler with a grating overlay," *Appl. Opt.*, Vol. 24, pp. 717–726 (1985).

34. D. Marcuse, "Directional couplers made of nonidentical asymmetric slabs, Part II: Grating-assisted couplers," *IEEE J. Lightwave Technol.*, Vol. LT-5, pp. 268–273 (Feb. 1987).

35. P. Zory, "Laser oscillation in leaky corrugated optical waveguides," *Appl. Phys. Lett.*, Vol. 22, No. 4, pp. 125–128 (Feb. 15, 1973).

36. C. Gill, "Lightwave applications of fiber gratings," *IEEE J. Lightwave Technol.*, Vol. 15, No. 8, pp. 1391–1404 (Aug. 1997).

37. T. Erdogan, "Fiber grating spectra," *IEEE J. Lightwave Technol.*, Vol. 15, No. 8, pp. 1277–1294 (Aug. 1997).

38. F. W. Dabby, A. Kestenbaum, and U. C. Paek, "Periodic dielectric waveguides," *Opt. Comm.*, Vol. 6, pp. 125–130 (1972).

39. S. Wang, "Principles of distributed feedback and distributed Bragg-reflector lasers," *IEEE J. Quantum Electron.*, QE-10, No. 4, pp. 413–427 (1974).

40. A. Yariv, "Coupled-mode theory for guided-wave optics," *IEEE J. Quantum Electron.*, Vol. QE-9, pp. 919–933 (1973).

41. D. Marcuse, *Theory of Dielectric Optical Waveguides*, Academic, New York, 1974, pp. 95–126, 132–145.

42. A. Yariv and A. Gover, "Equivalence of the coupled-mode and Floquet-Bloch formalisms in periodic optical waveguides," *Appl. Phys. Lett.*, Vol. 26, pp. 537–539 (May 1975).

43. W. Streifer, D. R. Sciferes, and R. D. Burnham, "Coupling coefficients for distributed feedback single- and double- heterostructure diode lasers," *IEEE J. Quantum Electron.*, Vol., QE-11, No. 11, pp. 867–873 (Nov. 1975).

44. N. Nishihara, M. Harana, and T. Suhara, *Optical Integrated Circuits*, McGraw-Hill, New York, 1989.

45. H. Kogelnik, "Theory of dielectric waveguides," in *Integrated Optics*, T. Tamir (ed.), Springer, New York, 1982, Chapter 2; and *Guided-Wave Optoelectronics*, 2nd ed., T. Tamir (ed.), Springer, New York, 1990, Chapter 2.

46. H. A. Haus, *Waves and Fields in Optoelectronics*, Prentice Hall, Englewood Cliffs, NJ, 1984.

47. G. H. B. Thompson, *Physics of Semiconductor Laser Devices*, Wiley, New York, 1980, Chapter 8.

48. H. Kressel and J. K. Butler, *Semiconductor Lasers and Heterojunction LEDs*, Academic, New York, 1977, Chapter 15.

49. H. C. Casey, Jr., and M. B. Panish, *Heterostructure Lasers, Part A, Fundamental Principles*, Academic, New York, 1978, Chapter 2.

50. M. Nakamura and A. Yariv, "Analysis of the threshold of double heterojunction GaAs-GaAlAs lasers with a corrugated interface," *Opt. Commun.*, Vol. 11, pp. 18–20 (May 1974).

51. Chin-Lin Chen, *Elements of Optoelectronics & Fiber Optics*, Richard D. Irwin, 1996.

8

ARRAYED-WAVEGUIDE GRATINGS

8.1 INTRODUCTION

Lenses, reflectors, prisms, and gratings are the basic building blocks of optical devices. Lenses and curved reflectors focus and transform light beams. Gratings and prisms disperse light beams spatially according to their spectral contents. The focusing and dispersion properties of these components are combined in concave gratings [1, 2]. In this chapter, we study integrated optic components that also focus and disperse light in a single and passive unit. They are the *arrayed-waveguide gratings*. In this sense, we may view the arrayed-waveguide gratings as the guided-wave optic counterpart of concave gratings.

Many integrated optic lenses have been studied. This includes geodesic, planar Luneberg, Fresnel and grating lenses and curved reflectors [3]. In 1988, Smit proposed a new type of integrated optic lenses and referred to them as the *optical phase arrays*, or *PHASARs* [4–6]. The new integrated optic lenses are arrays of three-dimensional waveguides. He also built a simple optical phase array to demonstrate the focusing properties of waveguide arrays. The dispersion properties of arrayed-waveguide gratings have been demonstrated by Takahashi et al. [7]. They also mentioned the use of the arrayed-waveguide grating (AWG) devices

Foundations for Guided-Wave Optics, by Chin-Lin Chen
Copyright © 2007 John Wiley & Sons, Inc.

in wavelength division multiplexing (WDM) applications. Since then many AWG components and devices have been designed and fabricated for fiber-optic communication applications [8–15].

In the next section, we consider the radiation patterns of arrays of isotropic radiators. We use arrays of radiators as a vehicle to introduce the focusing and dispersion properties of arrayed-waveguide gratings. In Section 8.3, we discuss the two arrayed-waveguide gratings reported by Smit [4] and by Takahashi et al. [7]. Then we describe the configuration and properties of a simple 1×2 AWG demultiplexer. It is followed by a description of $N \times N$ AWG demultiplexers. This chapter concludes with a brief discussion of the applications of AWG devices.

8.2   ARRAYS OF ISOTROPIC RADIATORS

To introduce the basic concepts underlining the operation of AWG devices, we study the radiation by an array of isotropic radiators. An isotropic radiator emits spherical waves uniformly in all directions. Figure 8.1 depicts an array of $\mathcal{N}$ radiators equally spaced along the x axis. We assume that $\mathcal{N}$ is a large number. To simplify the expressions, we also take $\mathcal{N}$ as an odd integer. Although the spacing d between adjacent radiators is on the order of the wavelength λ, the total array width, $(\mathcal{N} - 1)d$, is much longer than λ. Let the excitation for the radiator at $(ld, 0)$ be $E_0 e^{-j\phi_l}$. Then the total field radiated by the array is

$$E(x, z) = E_0 \sum_{l=-(\mathcal{N}-1)/2}^{(\mathcal{N}-1)/2} \frac{1}{r_l} e^{-j\phi_l} e^{-jkr_l} \tag{8.1}$$

where E_0 is a constant and $r_l = [(x - ld)^2 + z^2]^{1/2}$. We study the intensity pattern on an observation plane that is far away from the array. In particular, the distance

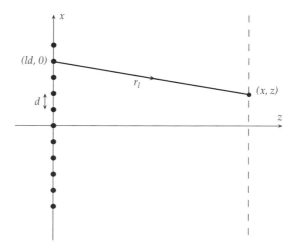

Figure 8.1   Array of isotropic radiators equally spaced along a straight line.

between the array and the observation plane is much larger than $(\mathcal{N} - 1)d$. We are particularly interested in the effect of the phase of excitation ϕ_l on the intensity pattern.

We examine three cases. The case of an equal phase excitation is studied first. Since the phase of excitation is the same for all radiators, we set ϕ_l to 0 for all radiators. In Figure 8.2(a), we plot $|E(x, z)/\mathcal{N}|^2$ due to a 17-element array as a function of x for a fixed z. Next, we suppose the excitation phase changes linearly with ld. Specifically, ϕ_l is $kld \sin \alpha$ where α is a parameter. The normalized intensity pattern is shown in Figure 8.2(b). We observe that the normalized intensity pattern in Figure 8.2(b) shifts with α. In the third case, the phase of excitation is quadratic in ld. We choose $\phi_l = k(ld)^2/(2\rho)$ where ρ is a parameter. We will comment on the meaning of ρ shortly. As shown in Figure 8.2(c), the radiation intensity is sharply concentrated in the center region and the peak intensity is much stronger than the peaks of Figures 8.2(a) and 8.2(b).

We may view the array of $\mathcal{N}$ radiators as a transmission grating having $\mathcal{N}$ grating elements. Then $|E(x, z)|^2$ is the interference fringes on an observation screen at z. The three cases considered correspond to incident waves impinging upon the transmission grating under three different conditions. The first two cases correspond

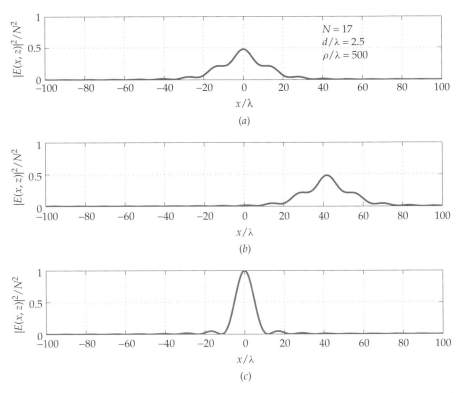

Figure 8.2 Normalized intensity pattern of a 17-element array equally spaced on a straight line: (a) equal phase excitation; (b) ϕ_l varies linearly with l, and (c) ϕ_l varies a l^2.

to normal and oblique incidence. The third case pertains to the excitation by spherical converging waves.

To elaborate on the focusing effect of quadratic phase excitation further, we consider spherical waves converging to $(0, \rho)$. Consider an exponential function $e^{jk\sqrt{x^2+(\rho-z)^2}}$. For points to the right of and near the x axis, the exponential function is approximately

$$e^{jk\sqrt{x^2+(\rho-z)^2}} \approx e^{jk(\rho-z)}e^{jkx^2/[2(\rho-z)]} \tag{8.2}$$

Clearly the exponential function in question represents a spherical wavefront moving in the $+z$ direction. At any plane of constant z, the phase term is quadratic in x. On the x axis, that is, $z = 0$, the field is $e^{jk\rho}e^{jkx^2/(2\rho)}$. The constant phase factor $e^{jk\rho}$ has no effect on the intensity pattern. But we note the presence of the quadratic phase term $e^{jkx^2/(2\rho)}$. In short, (8.2) describes a spherical wave converging to $(0, \rho)$. And ρ is also the radius of curvature of the spherical wave front at $z = 0$.

Placing the radiators along a circular arc of radius ρ produces the same focusing effect. Consider an array shown in Figure 8.3. Let the lth radiator be at (x_l, z_l) where $x_l = ld$ and $z_l = r - \sqrt{\rho^2 - (ld)^2}$. Then the total field radiated by the array

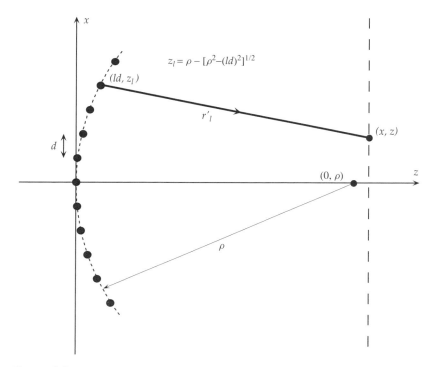

Figure 8.3 Array of isotropic radiators positioned along a circular arc of radius ρ.

of radiators distributed along a circular arc is

$$E(x, z) = E_0 \sum_{l=-(N-1)/2}^{(N-1)/2} \frac{1}{r'_l} e^{-j\phi_l} e^{-jkr'_l} \tag{8.3}$$

where $r'_l = [(x - x_l)^2 + (z - z_l)^2]^{1/2}$.

Two normalized intensity patterns of the array on a circular arc are shown in Figure 8.4. The case of equal phase excitation is shown in Figure 8.4(a) and that of a linear phase excitation is depicted in Figure 8.4(b). Again, we note that the normalized intensity pattern shifts with the linear phase excitation.

By comparing the normalized intensity patterns shown in Figures 8.2 and 8.4, effects of linear and quadratic phase excitations are evident. A linearly varying phase excitation has the effect of shifting the intensity pattern. A quadratic phase excitation concentrates and directs the field to a focal point.

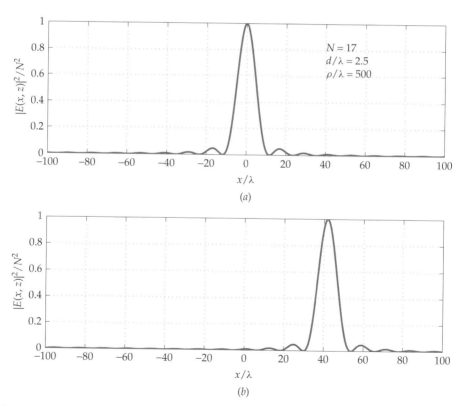

Figure 8.4 Normalized intensity pattern of a 17-element array positioned on a circular arc: (a) equal phase excitation and (b) ϕ_l varies linearly with l.

8.3   TWO EXAMPLES

8.3.1   Arrayed-Waveguide Gratings as Dispersive Components

As noted earlier, Takahashi et al. have built a simple AWG to demonstrate the dispersiveness of the arrayed-waveguide gratings [7]. Figure 8.5 depicts schematically the layout of their AWG and the basic experimental setup. The AWG is an array of $SiO_2/C7059/SiO_2$ channel waveguides on a Si substrate. Each channel waveguide has five straight-line segments and four right-angle bends. All $90°$ bends have the same bending radius. At the input and output planes of the AWG, the adjacent channel waveguides are spaced by d that is several wavelengths long. The path length difference ΔL between any two neighboring channel waveguides is $2(D - d)$. The channel waveguides operate at the fundamental mode and have an effective index of refraction $N_c(\lambda)$ at a wavelength λ. Thus, the phase difference of waves propagating through the two adjacent channel waveguides is $kN_c(\lambda)\,\Delta L$, where ΔL is chosen so that the phase difference at the center wavelength λ_0 is an integer multiple of 2π, that is, $k_0 N_c(\lambda_0)\,\Delta L = 2m\pi$ here k_0 is $2\pi/\lambda_0$. The integer m is a key parameter characterizing the properties of the AWG. We refer to m as the *diffraction order* of the AWG. In their experiment, a collimated light beam shines upon the front face of the AWG. Thus the excitations to the channel waveguides have equal amplitudes and phases. However, due to the path length difference ΔL, waves arriving at and radiated by the distal ends of the channel waveguides have different phase. At the center wavelength λ_0, the phase difference between two neighboring channel waveguides is $2m\pi$. Therefore the intensity pattern has a peak in the forward direction, that is, $\theta = 0$. For λ different from λ_0, the phase difference between adjacent channel waveguides is no longer an integer multiple of 2π. As a result, the intensity pattern shifts away from $\theta = 0°$. Thus, a wavelength change leads to an angular shift of the intensity pattern. This is the basic physics underlining the dispersion of arrayed-waveguide gratings. Many guided-wave devices and components are designed to take advantage of the dispersive property.

To estimate the wavelength resolution of an AWG, we consider waves emitted by two neighboring channel waveguides. At wavelength λ, the phase difference between waves arriving at the far ends of the two adjacent channel waveguides is

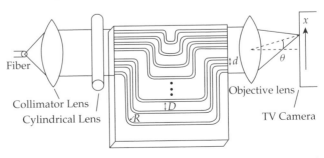

Figure 8.5   Arrayed-waveguide grating multi/demultiplexer and experimental setup. (From [7].)

$kN_c(\lambda)\,\Delta L$. We consider waves emitted from the output face of the AWG. If the wave fronts tilt by an angle θ, then the phase delay due to the wavefront tilting is $kd\sin\theta$. The intensity pattern would have a peak at angle θ if the total phase change is

$$kN_c(\lambda)\,\Delta L + kd\sin\theta = 2m\pi \tag{8.4}$$

If waves emitted by the channel waveguides propagate in a slab waveguide with an effective refractive index N_s, or a medium with an effective refractive index N_s, then the above equation becomes

$$kN_c(\lambda)\,\Delta L + kN_s(\lambda)d\sin\theta = 2m\pi \tag{8.5}$$

Since the observation plane, the TV camera depicted in Figure 8.5, for example, is far from the output plane of the AWG, θ is a small angle. Thus we approximate $\sin\theta$ by θ. A simple differentiation of (8.5) yields

$$\frac{d\theta}{d\lambda} \approx \frac{m - \dfrac{dN_c}{d\lambda}\Delta L - \dfrac{dN_s}{d\lambda}\theta d}{N_s d} \tag{8.6}$$

If the minimum angle resolvable by the experimental setup is $\Delta\theta_{min}$, then the minimum wavelength variation resolvable by the AWG is

$$\Delta\lambda_{min} \approx \frac{N_s d}{m - \dfrac{dN_c}{d\lambda}\Delta L - \dfrac{dN_s}{d\lambda}\theta d}\,\Delta\theta_{min} \tag{8.7}$$

For the arrayed-waveguide gratings with $\mathcal{N}$ grating elements, the total aperture width is $(\mathcal{N}-1)d$. The minimum resolvable angle is

$$\Delta\theta_{min} \approx \frac{\lambda}{N_s(\mathcal{N}-1)d}$$

[16]. Thus the *wavelength resolution* of the AWG is

$$\Delta\lambda_{min} \approx \frac{\lambda}{(\mathcal{N}-1)\left(m - \dfrac{dN_c}{d\lambda}\Delta L - \dfrac{dN_s}{d\lambda}\theta d\right)} \tag{8.8}$$

Most arrayed-waveguide gratings have a large diffraction order m. Thus, the effect due to the waveguide dispersion is negligible in comparison to that arising from the diffraction order. Therefore we may ignore $dN_c/d\lambda$ and $dN_s/d\lambda$ in (8.8). Most arrayed-waveguide gratings also have a large number of grating elements. We use $\mathcal{N}$ in lieu of $\mathcal{N}-1$. Then (8.8) is greatly simplified and we obtain the wavelength resolution of the AWG:

$$\Delta\lambda_{min} \approx \frac{\lambda}{\mathcal{N}m} \tag{8.9}$$

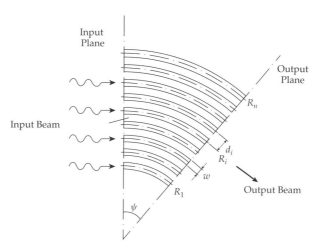

Figure 8.6   Arrayed-waveguide grating lens. (From [4].)

For the AWG constructed by Takahashi et al. [5], there are 150 channel waveguides and ΔL is 17.54 μm. When operating at 1.3 μm, $k \, \Delta L \, N_c(\lambda)$ is about 40π. Thus the diffraction order is about 20. Then $\Delta\lambda_{\min}$ is about 0.43 nm. In their experiment, they were able to resolve wavelength differences as small as 0.63 nm. Clearly, they have established the feasibility of using AWG devices as wavelength demultiplexers.

8.3.2   Arrayed-Waveguide Gratings as Focusing Components

Smit built an AWG lens [4] to demonstrate the focusing property of arrayed-waveguide gratings. Figure 8.6 depicts the geometry of the AWG lens. It consists of 31 concentric Al_2O_3 ridge waveguides on a Si substrate. Each ridge waveguide is 3 μm wide. The radii of curvature of the ridge waveguides vary from 910 to 1090 μm with an average radius difference of 6 μm. His experiment shows that the focal length of the AWG lens is about 1 mm at 633 nm.

8.4   1 × 2 ARRAYED-WAVEGUIDE GRATING MULTIPLEXERS AND DEMULTIPLEXERS

The simplest demultiplexers are 1×2 demultiplexers. Figure 8.7 depicts the basic configuration of 1×2 AWG demultiplexers. It consists of an input waveguide, two output waveguides, two slab waveguides, and an AWG. The slab waveguides, being two-dimensional waveguides, are wide in the transverse direction. Therefore, waves propagating in the slab waveguides are not confined in the transverse direction. In much of the literature, the two slab waveguides are referred to as the *free propagation regions*. The AWG comprises $\mathcal{N}$ three-dimensional waveguides that may be channel or ridge waveguides. Different grating elements have different lengths. Specifically, the grating element lengths increase at a constant increment ΔL. For convenience,

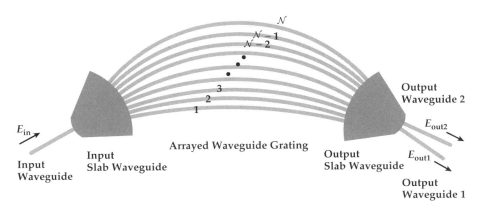

Figure 8.7   1 × 2 demultiplexer with an arrayed-waveguide grating.

we label the grating elements consecutively with the lowest, and also the shortest, grating element as grating element 1. Let the length of the lth grating element be L_l. Then $L_l = L_1 + (l - 1)\,\Delta L$. Entrances to the grating elements are on a circular arc facing the input slab waveguide. The exits from the grating elements are on a circular arc in the output slab waveguide. The output slab waveguide is terminated by two output waveguides. Detail geometry of the slab waveguides and waveguide grating elements are shown in Figures 8.8(a) and 8.8(b) [8].

8.4.1   Waveguide Grating Elements

To appreciate the subtlety of the arrangement, it is necessary to understand the radiation by a truncated thin-film waveguide into a free space region or a truncated channel or ridge waveguide into a slab waveguide. The intensity pattern of a truncated thin-film waveguide radiating into free space has been studied by Boucouvalas [17]. He showed that the radiated field in the far-field zone varies with θ and is confined mainly in a narrow cone in the forward direction. Thus, power coupled from the input waveguide to the grating elements is different for different grating elements. Detailed numerical calculation by Adar et al. shows that fields coupled to the grating elements away from the center region are much weaker than that in the grating elements near the center of the AWG structures [8]. However, for grating elements near the center, the excitation amplitudes are nearly the same. Although the radiated field spreads radially outward, the wavefronts are not perfectly spherical wavefronts. This is understandable since the end face of truncated thin-film waveguide has a finite size. To a good approximation, we may view the radiated field as the field emanating for a point radiator deep inside the truncated waveguide. It is convenient to refer to the point radiator as the *phase center* [10, 11].

As shown in Figure 8.8(a), the input waveguide feeds into the input slab waveguide and is symmetric with respect to the input ports of the grating elements. At the opposite side of the input slab waveguide, waves impinge upon and enter the grating elements. Entrances to the grating elements are on a circular arc. The center

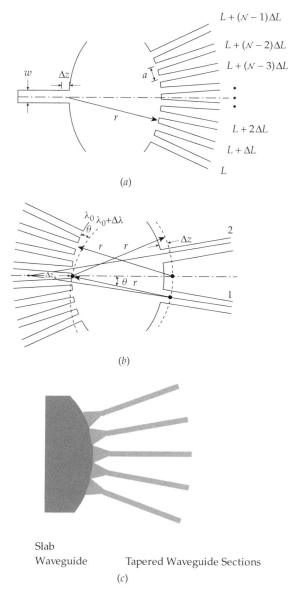

Figure 8.8 (a) Input and (b) output slab waveguides of a 1×2 AWG demultiplexer. (From [8].); (c) waveguide grating elements with tapered input sections.

of the circular arc is the phase center at a distance Δz inside the input waveguide. As noted earlier, the amplitude distribution of waves excited in grating elements is not uniform. In addition, the grating elements are closely spaced, and the interaction between the grating elements is inevitable. The interaction between grating elements also affects the amplitude and phase of excitation [8].

As shown in Figure 8.8(*b*), exits of the grating elements into the output slab waveguide are on a circular arc. The center of the arc is a phase center at a distance Δz further to the right edge of the output slab waveguide. The two output waveguides are on a circular arc centered at a phase center inside one of the grating elements.

To reduce the geometrical mismatch between the slab waveguides and the grating elements and thus improving the coupling efficiency, waveguide horns or fan-out sections are built onto the ends of the grating elements. Figure 8.8(*c*) depicts schematically the tapering of the grating elements.

8.4.2   Output Waveguides

Next we consider the placement of the two output waveguides. Suppose the AWG demultiplexer is designed to sort signals of wavelengths λ_1 and λ_2. Then we choose the average value of λ_1 and λ_2 as the center wavelength λ_0. Based on the center wavelength λ_0, the diffraction order m and the effective refractive index $N_c(\lambda_0)$ of the grating elements, we determine ΔL. When radiated by the grating elements, waves of wavelength λ_0 converge to the center near the right edge of the output slab waveguide. At wavelength λ, wavefronts tilt by an angle θ. This is due to the difference in the phase delay as waves propagate through grating elements of different lengths. From (8.5), we obtain

$$\theta = \frac{N_c(\lambda)}{N_s(\lambda)} \left[\frac{\lambda}{\lambda_0} \frac{N_c(\lambda_0)}{N_c(\lambda)} - 1 \right] \frac{\Delta L}{d} \qquad (8.10)$$

We use (8.10) to determine the optimal position to capture signals of λ_1 and λ_2. As noted earlier, λ_0 is the average of λ_1 and λ_2. Thus the two output waveguides are on the opposite sides of the central line.

8.4.3   Spectral Response

To consider the spectral response of 1×2 AWG demultiplexers, we examine the output of waveguide 1. Let f_l be the coupling efficiency describing the coupling of waves from the input waveguide into the lth grating element, and g_l the coupling efficiency quantifying the coupling between the lth grating element and the output waveguide 1. Thus, the field coupled from the input waveguide to the output waveguide 1 through all grating elements of the AWG is

$$E_1(\lambda) = E_0 e^{-jkN_c(\lambda)L_1} \sum_{l=1}^{N} f_l g_l e^{-jkN_c(\lambda)(l-1)\Delta L} e^{-jkN_s(\lambda)(l-1)d\theta_1} \qquad (8.11)$$

Since the grating elements are not excited uniformly, f_l and g_l vary with l. But if a uniform amplitude distribution and a constant phase excitation were achieved somehow, then f_l and g_l would be constants independent of l. Then (8.11) could be summed analytically and we would obtain

$$\left|\frac{E_1(\lambda)}{E_0 f_1 g_1}\right|^2 = \frac{\sin^2\left\{\pi\mathcal{N}m\dfrac{\lambda_0}{\lambda}\dfrac{N_c(\lambda)}{N_c(\lambda_0)}\left[1+\dfrac{N_s(\lambda)\theta_1 d}{N_c(\lambda)\,\Delta L}\right]\right\}}{\sin^2\left\{\pi m\dfrac{\lambda_0}{\lambda}\dfrac{N_c(\lambda)}{N_c(\lambda_0)}\left[1+\dfrac{N_s(\lambda)\theta_1 d}{N_c(\lambda)\,\Delta L}\right]\right\}} \qquad (8.12)$$

For many arrayed-waveguide gratings, $\theta_1 d$ is much smaller than ΔL. Within a narrow wavelength range, we may also ignore the dispersion of $N_s(\lambda)$ and $N_c(\lambda)$. Then the above equation becomes

$$\left|\frac{E_1(\lambda)}{E_0 f_1 g_1}\right|^2 \approx \frac{\sin^2\left\{\pi\mathcal{N}m\dfrac{\lambda_0}{\lambda}\right\}}{\sin^2\left\{\pi m\dfrac{\lambda_0}{\lambda}\right\}} \qquad (8.13)$$

From (8.12) and (8.13), it is clear that the spectral response depends on the number of grating elements $\mathcal{N}$, the diffraction order m, and the product $\mathcal{N}m$. As noted by Adar et al. [8], the spectral response narrows as the number of grating elements and/or the diffraction order increase. It will be seen shortly that m and $\mathcal{N}$

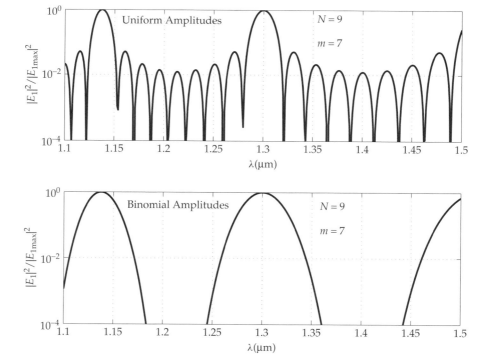

Figure 8.9   Effect of Amplitude distribution on the spectral response.

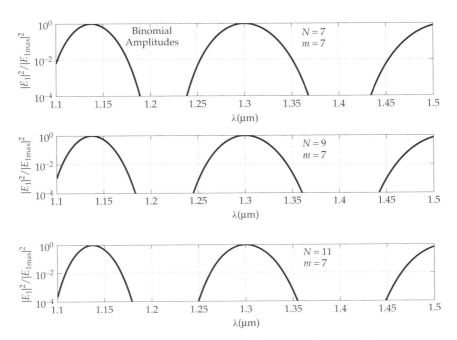

Figure 8.10   Effect of number of grating elements on the spectral response.

affect the spectral response differently. For large diffraction orders, the responses due to neighboring diffraction orders $m - 1$ and/or $m + 1$ may become "visible." This leads to a smaller *free spectral range* of AWG. While these conclusions are established for arrayed-waveguide gratings with constant amplitudes of excitation, they remain valid for nonuniform excitations as well.

As noted earlier, f_l and g_l are not constants. In fact, f_l and g_l decrease drastically for the grating elements near the edges of the AWG. To study the effects of nonuniform excitation on the spectral response, we keep f_l and g_l as l-dependent coefficients and evaluate (8.11) numerically. An example is given in Figure 8.9. In the example, we assume that $f_l g_l$ varies as the coefficients of a binomial series [19]. For a 9-element AWG, $f_l g_l$ for $l = 1, 2, \ldots, 9$ are, respectively, 1, 8, 28, 56, 70, 56, 28, 8, and 1. While the main response peak is broadened, the side lobe disappears completely. For comparison, we also show the response of a AWG with a constant amplitude distribution. The effects of $\mathcal{N}$ and m on the spectral response with a binomial distribution of $f_l g_l$ are shown in Figures 8.10 and 8.11. The effect of m on the free spectral range is clearly evident.

8.5 $N \times N$ ARRAYED-WAVEGUIDE GRATING MULTIPLEXERS AND DEMULTIPLEXERS

Figure 8.12 shows the basic configuration of $N \times N$ AWG multiplexers and demultiplexers. It is essentially the same as that of 1×2 AWG multiplexers and demultiplexers shown in Figure 8.7. It has N input waveguides, N output waveguides,

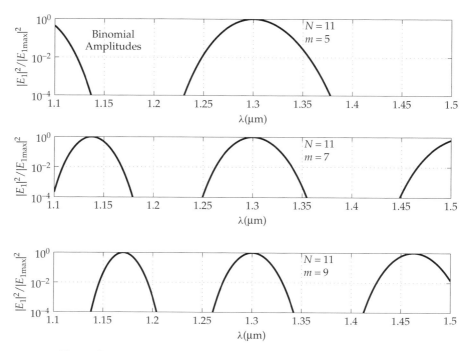

Figure 8.11   Effect of the diffraction order on the spectral response.

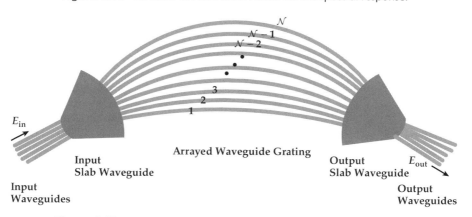

Figure 8.12 $N \times N$ demultiplexer with an arrayed-waveguide grating.

two slab waveguides, and an AWG with $\mathcal{N}$ grating elements. Since there may be a large number of input and output waveguides, they have to be arranged carefully. In particular, the entrances to the $\mathcal{N}$ grating elements are arranged on a circular arc centered at the phase center of the center input waveguide. The output ports of the N input waveguides are also on a circular arc. The two circular arcs have the same radii. The positioning of the $\mathcal{N}$ grating elements and N input waveguides are similar to the arrangement of a concave grating and a photographic plate in a classical grating spectrographs with a *Rowland mounting* [1, 2]. Since the radii of

TABLE 8.1   Key Parameters of a 256-Channel AWG Multi/Demultiplexer

Center wavelength, λ_0	1552 nm
Frequency spacing between channels	25 GHz
Diffraction order, m	26
Path length difference, ΔL	27.7 μm
Input/output slab waveguide focal length, ρ	41.1 mm
Number of waveguide grating elements, $\mathcal{N}$	712
Circuit size	75×55 mm

Source: From Ref. [14].

curvature of the two circular arcs are also the spacing between the two arcs, the arrangement is also a *confocal arrangement*.

Table 8.1 lists the key parameters of a 256-channel AWG multiplexer and demultiplexer built by Hibino et al. on a 4-inch Si wafer [14].

8.6   APPLICATIONS IN WDM COMMUNICATIONS

For most arrayed-waveguide gratings, the diffraction orders are very large. This is an advantage of arrayed-waveguide gratings over conventional gratings that typically operate with low diffraction orders. As discussed in Section 8.2, the wavelength resolution of AWG varies inversely with $m\mathcal{N}$. Since arrayed-waveguide gratings can resolve small wavelength differences, they are used extensively in WDM communications. Figure 8.13 shows schematically the used of 4×4 AWG devices as multiplexers, demultiplexers, drop/add multiplexers, and full interconnections [12]. In Figure 8.13(a), signals having four different wavelengths and impinging upon the four input ports are combined and "multiplexed" in an output port. In a demultiplexer, Figure 8.13(b), an input signal containing four wavelengths λ_1, λ_2, λ_3 and λ_4 is sorted and routed to ports 1, 2, 3, and 4, respectively. In a drop–add multiplexer [Fig. 8.13(c)], information contained in a light beam of wavelength λ_2, for example, is dropped and replaced by new and different data before the beam exiting from the output port. In a full interconnect [Fig. 8.13(d)], a signal arriving at input port 1 with different spectral components is distributed to the output ports according to the signal wavelengths. A signal of wavelength λ_1 goes to output port 1, wavelength λ_2 to output port 2, and so forth. For signals impinging upon input port 2 with wavelengths $\lambda_1, \lambda_2, \lambda_3$, and λ_4 going to output ports 2, 3, 4, and 1, respectively. In short, arrayed-waveguide gratings can perform many functions and are capable of resolving fine wavelength differences. As a result, they find applications in many WDM communications.

Since the path lengths of different grating elements are different, and the difference are defined and determined lithographically. Arrayed-waveguide gratings are also useful in generating and shaping femtosecond pulses [18].

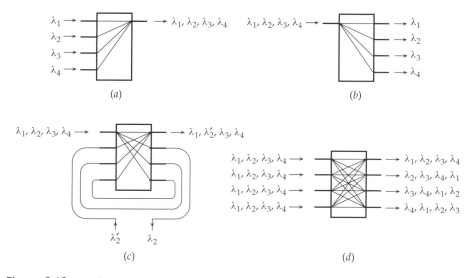

Figure 8.13   Applications of arrayed-waveguide grating devices: (a) multiplexer, (b) demultiplexer, (c) drop/add multiplexer, and (d) full interconnection. (After [12].)

REFERENCES

1. F. A. Jenkins and H. E. White, *Fundamentals of Optics*, 4th ed., McGraw-Hill, New York, 1976.

2. R. W. Ditchiburn, *Light*, 3rd ed., Vol. 1, Academic, London, New York, 1976.

3. D. B. Anderson, R. L. Davis, J. T. Boyd, and R. R. August, "A comparison of waveguide lens technologies," *IEEE J. Quantum Electron.*, Vol. QE-13, pp. 275–282 (1977).

4. M. K. Smit, "New focusing and dispersive planar component based on an optical phased array," *Electron. Lett.*, Vol. 24, No. 7, pp. 385–386 (Mar. 31, 1988).

5. A. H. Vellekoop and M. K. Smit, "Four-channel integrated-optic wavelength demultiplexer with weak polarization dependence," *J. IEEE Lightwave Technol.*, Vol. 9, No. 3, pp. 310–314 (March 1991).

6. M. K. Smit and C. van Dam, "Phasar-based WDM-devices: Principles, design and applications," *IEEE J. Selected Topics Quantum Electron.*, Vol. 2, No. 2 pp. 236–250 (June 1996).

7. H. Takahashi, S. Suzuki, K. Kato, and I. Nishi, "Arrayed-waveguide grating for wavelength division multi/demultiplexer with nanometer resolution," *Electron. Lett.*, Vol. 26, No. 2, pp. 87–88 (Jan. 18, 1990).

8. R. Adar, C. H. Henry, C. Dragone, R. C. Kistler, and M. A. Milbrodt, "Broad-band array multiplers made with silica waveguides on silicon," *J. IEEE Lightwave Technol.*, Vol. 11, No. 2, pp. 212–219 (Feb. 1993).

9. H. Takahashi, I. Nishi, and Y. Hibina, "10GHz spacing optical frequency division multiplexer based on arrayed-waveguide grating," *Electron. Lett.*, Vol. 28, No. 4, pp. 380–382 (Feb. 13, 1992).

10. C. Dragone, "An $N \times N$ optical multiplexer using a planar arrangement of two star couplers," *IEEE Photonics Technol. Lett.*, Vol. 3, No. 9, pp. 812–815 (Sept. 1991).

11. C. Dragone, C. A. Edwards, and R. C. Kistler, "Integrated optics $N \times N$ multiplexer on silicon," *IEEE Photonics Technol. Lett.*, Vol. 3, No. 10, pp. 896–899 (Sept. 1991).

12. H. Takahashi, K. Oda, H. Toba, and Y. Inoue, "Transmission characteristics of arrayed waveguide $N \times N$ wavelength multiplexer," *J. IEEE Lightwave Technol.*, Vol. 13, No. 3, pp. 447–455 (March 1995).

13. K. Okamoto, A. Himeno, Y. Ohmori, and M. Okuno, "16-channel optical add/drop multiplexer consisting of arrayed-waveguide gratings and double-gate switches," *Electron. Lett.*, Vol. 32, No. 16, pp. 1471–1472 (Aug. 1, 1996).

14. Y. Hibino, "Recent advances in high-density and large-scale AWG multi/demultipexers with higher index-contrast silica-based PLCs," *IEEE J. Selected Topics Quantum Electron.*, Vol. 8, No. 6, pp. 236–250 (Nov./Dec. 2002).

15. H. Uetsuka, "AWG technologies for dense WDM applications," *IEEE J. Selected Topics Quantum Electron.*, Vol. 10, No. 2, pp. 393–402 (Mar./Apr. 2004).

16. E. Hecht, *Optics*, 4th ed., Addison-Wesley, New York, 2002.

17. A. C. Boucouvalas, "Use of far-field radiation pattern to characterize single-mode symmetric slab waveguides," *Electron. Lett.*, Vol. 19, No. 3, pp. 120–121 (Feb. 3, 1983).

18. D. E. Leaird, A. M. Weiner, S. Kamei, M. Ishii, A. Sugita, and K. Okamoto, "Generation of flat-topped 500-GHz pulse bursts using loss engineered arrayed waveguide gratings," *IEEE Photonics Lett.*, Vol. 14, No. 6, pp. 816–818 (June 2002).

19. M. Abromowitz and I. A. Stegun, *Handbook of Mathematical Functions with Formulas, Graphs and Mathematical Tables*, Dover, New York, 1964.

9

TRANSMISSION CHARACTERISTICS OF STEP-INDEX OPTICAL FIBERS

9.1 INTRODUCTION

The basic structure of an optical fiber is a core surrounded by a cladding. To protect the core and cladding against physical damage, to seal out moisture and contamination, and to provide mechanical strength and support, one or more coating layers may be placed outside the cladding. Since the cladding region is very thick optically, the coating layers have minimal or no effect on the optical properties of the fiber. In short, the transmission characteristics of an optical fiber depend primarily on the size, shape, and index profile of the core and the cladding index. Although a fiber core can be of any shape, most fibers have a nominally circular core. Specialty fibers may have an elliptical or other noncircular cross section. These fibers will be discussed in Chapter 11. In this chapter, we discuss fibers with a circular core and we take the core radius as a (Fig. 9.1). In all fibers of practical interest, fields are confined mainly in the core and a thin cladding region near the core–cladding boundary. Fields near the outer cladding boundary are extremely weak since the cladding is thick optically. Therefore, we assume that, as an approximation and for simplicity, the cladding is infinitely thick. We also assume that the fiber is infinitely long in the axial direction.

Foundations for Guided-Wave Optics, by Chin-Lin Chen
Copyright © 2007 John Wiley & Sons, Inc.

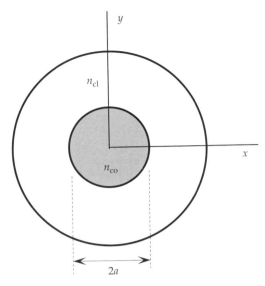

Figure 9.1   Fiber with a circular core and cladding.

Depending on the variation of the core index, a fiber may be classified as a *step-index* or a *graded-index fiber*. In a step-index fiber, the core and cladding indices, n_{co} and n_{cl}, are constants. In terms of these indices, the *index difference* Δ is

$$\Delta = (n_{co} - n_{cl})/n_{co} \qquad (9.1)$$

Typically, Δ is a few percent or a fraction of a percent. For a graded-index fiber, the core index varies with position. In many analysis, it is necessary to represent the core index distribution by a simple function. For example, we approximate the index distribution as

$$[n(r)]^2 = n_{co}^2[1 - 2\Delta(r/a)^\alpha] \qquad r \le a \qquad (9.2)$$

where n_{co} is the peak core index and Δ is the index difference defined in (9.1). The exponent α is the *index profile parameter* or *grading parameter*, and it quantifies the index distribution. In (9.2), $[n(r)]^2$ is an algebraic equation in power of r if α is an integer. Thus fibers with index profiles given by (9.2) are known as power-law index fibers. We may view a step-index profile as a power-law index profile with an infinitely large α. A linear or triangular index profile corresponds to $\alpha = 1$. A *parabolic index profile* is given by $\alpha = 2$. Parabolic and approximately parabolic index profiles are of particular interest in telecommunications since multimode fibers with an index profile parameter close to 2 have a broad transmission bandwidth. In fact, all multimode fibers used in optical communications have a parabolic or approximately parabolic index profile. We will return to the power-law index fibers in Chapter 12.

As noted earlier, the field is confined primarily in the core region and a cladding layer near the inner cladding boundary. Clearly, the index distribution in the cladding region is not as important except for the index in the region near the core–cladding boundary. Thus we approximate the cladding index as a constant throughout the cladding region:

$$n_{cl} = n_{co}(1 - 2\Delta)^{1/2} \quad r > a \tag{9.3}$$

In Section 9.2, we analyze fields of step-index fibers with a circular core. The results, while exact, are too complicated for many applications. Considerable simplification can be achieved if the index difference is very small and the core radius is much larger than the wavelength. Fibers satisfying these conditions are referred to as *weakly guiding fibers*. Most fibers in use are weakly guiding fibers. Fields guided by these fibers are approximately linearly polarized in the core region. We refer to linearly polarized modes as the LP modes. The properties of LP modes are discussed in Section 9.3. In Section 9.4, we discuss the phase and group velocities and the group velocity dispersion of LP modes guided by step-index fibers.

9.2 FIELDS AND PROPAGATION CHARACTERISTIC OF MODES GUIDED BY STEP-INDEX FIBERS

9.2.1 Electromagnetic Fields

To study the wave propagation in fibers, we begin with the time harmonic ($e^{+j\omega t}$) Maxwell's equations (1.7) and (1.8). For a propagating mode, all field components vary like $e^{-j\beta z}$. Thus, we write

$$\mathbf{E}(r, \phi, z) = [\mathbf{e}_t(r, \phi) + \hat{\mathbf{z}} e_z(r, \phi)]e^{-j\beta z} \tag{9.4}$$

and

$$\mathbf{H}(r, \phi, z) = [\mathbf{h}_t(r, \phi) + \hat{\mathbf{z}} h_z(r, \phi)]e^{-j\beta z}. \tag{9.5}$$

Since the fiber cross section and the index profile are independent of z, all transverse field components can be expressed in terms of the longitudinal field components. Explicit expressions for the transverse field components in terms of $e_z(r, \phi)$ and $h_z(r, \phi)$ have been given in (1.76) and (1.77) [1, 2]. The remaining task is to determine $e_z(r, \phi)$ and $h_z(r, \phi)$.

To proceed further, we recall the notion of the *effective guide index N* and the *generalized parameters V* and *b* introduced in Chapter 2 for thin-film waveguides. We use these parameters to characterize fields in step-index fibers as well. Explicitly, we introduce the effective guide index N such that $\beta = kN$. We also define the *generalized frequency V* and *generalized guide index b* as

$$V = ka\sqrt{n_{co}^2 - n_{cl}^2} \tag{9.6}$$

$$b = \frac{N^2 - n_{cl}^2}{n_{co}^2 - n_{cl}^2} \tag{9.7}$$

In terms of V and b, we have

$$k^2 n_{co}^2 - \beta^2 = \frac{V^2}{a^2}(1 - b)$$

for the core region and

$$\beta^2 - k^2 n_{cl}^2 = \frac{V^2}{a^2}b$$

for the cladding region. For the core region, $0 \le r \le a$, the wave equation for e_z becomes

$$\nabla_t^2 e_z(r, \phi) + \frac{V^2(1 - b)}{a^2} e_z(r, \phi) = 0 \qquad (9.8)$$

For the cladding region, $r \ge a$, the wave equation for e_z is

$$\nabla_t^2 e_z(r, \phi) - \frac{V^2 b}{a^2} e_z(r, \phi) = 0 \qquad (9.9)$$

Similar wave equations can be written for h_z as well.

As mentioned earlier, most fibers have a nominally circular core and a ϕ-independent index profile. We express $e_z(r, \phi)$ in terms of a Fourier sine or cosine series. Since $\sin l\phi$ and $\cos l\phi$ with an integer l are orthogonal functions in the range $(0, 2\pi)$, the continuation of tangential field components is tantamount to the continuation of the each individual Fourier component. In short, all boundary conditions are satisfied by each term of the Fourier series. Thus we consider each Fourier series individually. Without loss of generality, we consider a single term of the Fourier cosine series. For the core region, we write

$$e_z(r, \phi) = e_{zco}(r) \cos l\phi \qquad (9.10)$$

To determine $e_{zco}(r)$, we substitute (9.10) into (9.8) and obtain a differential equation for the newly defined function

$$\frac{1}{r}\frac{d}{dr}\left[r\frac{de_{zco}(r)}{dr}\right] + \left[\frac{V^2(1 - b)}{a^2} - \frac{l^2}{r^2}\right] e_{zco}(r) = 0 \qquad (9.11)$$

Equation (9.11) is a Bessel differential equation of order l [3]. It has two linearly independent solutions, and they are the Bessel functions of the first and second kinds of order l, J_l, and Y_l. We have to discard Y_l since it is singular at the origin; J_l is the only acceptable function for $e_{zco}(r)$. Thus we write for $r \le a$,

$$e_{zco}(r) = A_l J_l\left(V\sqrt{1 - b}\frac{r}{a}\right) \qquad (9.12)$$

where A_l is a yet undetermined amplitude constant.

For the cladding region, $r \geq a$, we write

$$e_z(r, \phi) = e_{z\,cl}(r) \cos l\phi \tag{9.13}$$

and obtain a differential equation for $e_{z\,cl}(r)$:

$$\frac{1}{r}\frac{d}{dr}\left[r\frac{de_{z\,cl}(r)}{dr}\right] - \left(\frac{V^2 b}{a^2} + \frac{l^2}{r^2}\right)e_{z\,cl}(r) = 0 \tag{9.14}$$

This is a modified Bessel differential equation of order l. It has two linearly independent solutions, I_l and K_l, the modified Bessel functions of order l [3]. Since we expect $e_{z\,cl}(r)$ to decay exponentially in r, we drop I_l and keep K_l. Thus for $r \geq a$,

$$e_{z\,cl}(r) = C_l K_l\left(V\sqrt{b}\frac{r}{a}\right) \tag{9.15}$$

where C_l is also an unknown amplitude constant. Since K_l decays exponentially for large r, the radiation condition for large r is satisfied automatically.

Combining (9.10), (9.12), (9.13), and (9.15), we write

$$e_z(r, \phi) = \begin{cases} A_l J_l\left(V\sqrt{1-b}\frac{r}{a}\right)\cos l\phi & 0 \leq r \leq a \\ C_l K_l\left(V\sqrt{b}\frac{r}{a}\right)\cos l\phi & r \geq a \end{cases} \tag{9.16}$$

Similarly, we obtain an expression for $h_z(r,\phi)$ with two new unknown constants:

$$h_z(r, \phi) = \begin{cases} B_l J_l\left(V\sqrt{1-b}\frac{r}{a}\right)\sin l\phi & 0 \leq r \leq a \\ D_l K_l\left(V\sqrt{b}\frac{r}{a}\right)\sin l\phi & r \geq a \end{cases} \tag{9.17}$$

By substituting $e_z(r, \phi)$ and $h_z(r, \phi)$ into (1.76) and (1.77), we obtain expressions for $e_\phi(r, \phi)$ and $h_\phi(r, \phi)$:

$$e_\phi(r, \phi) = \begin{cases} \dfrac{jka}{V^2(1-b)}\left[A_l\dfrac{aNl}{r}J_l\left(V\sqrt{1-b}\dfrac{r}{a}\right)\right. \\ \left. +B_l\eta_o V\sqrt{1-b}J_l'\left(V\sqrt{1-b}\dfrac{r}{a}\right)\right]\sin l\phi & 0 \leq r \leq a \\ -\dfrac{jka}{V^2 b}\left[C_l\dfrac{aNl}{r}C_l K_l\left(V\sqrt{b}\dfrac{r}{a}\right)\right. \\ \left. +D_l\eta_o V\sqrt{b}K_l'\left(V\sqrt{b}\dfrac{r}{a}\right)\right]\sin l\phi & r \geq a \end{cases} \tag{9.18}$$

$$
h_\phi(r, \phi) = \begin{cases}
\dfrac{-jka}{V^2(1-b)}\left[A_l \dfrac{n_{co}^2}{\eta_o} V \sqrt{1-b} J_l'\left(V\sqrt{1-b}\dfrac{r}{a}\right) \right. \\
\left. \quad + B_l \dfrac{lNa}{r} J_l\left(V\sqrt{1-b}\dfrac{r}{a}\right) \right]\cos l\phi & 0 \le r \le a \\[2ex]
\dfrac{jka}{V^2 b}\left[C_l \dfrac{n_{cl}^2}{\eta_o} V\sqrt{b} K_l'\left(V\sqrt{b}\dfrac{r}{a}\right) \right. \\
\left. \quad + D_l \dfrac{lNa}{r} K_l\left(V\sqrt{b}\dfrac{r}{a}\right) \right]\cos l\phi & r \ge a
\end{cases}
\tag{9.19}
$$

In the above equations, a prime indicates the differentiation with respect to the argument of the Bessel or modified Bessel functions. The four coefficients A_l, B_l, C_l, and D_l and b are determined by imposing the boundary conditions. Similar expression can also be written for $e_r(r, \phi)$ and $h_r(r, \phi)$. We note shortly that it is only necessary to impose the conditions on $e_\phi(r, \phi)$, $e_z(r, \phi)$, $h_\phi(r, \phi)$, and $h_z(r, \phi)$ at the core–cladding boundary. Therefore, we omit the expressions for $e_r(r, \phi)$ and $h_r(r, \phi)$ here.

In (9.16) and (9.17), we have expressed $e_z(r, \phi)$ in terms of a Fourier cosine series and $h_z(r, \phi)$ a Fourier sine series. We could just as well cast $e_z(r, \phi)$ in a Fourier sine series and $h_z(r, \phi)$ in a Fourier cosine series, and we would obtain similar results. But if we had expressed two z components in terms of two Fourier sine series or two Fourier cosine series, we would not be able to satisfy the boundary conditions.

9.2.2 Characteristic Equation

Since the core–cladding boundary is free of the surface charge density and surface current density, the tangential components $e_z(r, \phi)$, $e_\phi(r, \phi)$, $h_z(r, \phi)$, and $h_\phi(r,\phi)$, and the normal components $d_r(r, \phi)$ and $b_r(r, \phi)$ must be continuous at the core–cladding boundary. Actually, it is only necessary to impose the conditions on the tangential field components. When the tangential components are continuous at the interface, the normal components are continuous at the boundary automatically [4].

By requiring $e_z(a^+, \phi) = e_z(a^-, \phi)$ for all values of ϕ, we obtain

$$
A_l J_l(V\sqrt{1-b}) - C_l K_l(V\sqrt{b}) = 0
\tag{9.20}
$$

From the continuation of $e_\phi(r, \phi)$, $h_z(r, \phi)$, and $h_\phi(r, \phi)$ at $r = a$, we obtain three additional equations. Thus, we have four equations and four unknowns A_l, B_l, C_l, and D_l. The four equations can be cast in a matrix form:

$$
\begin{bmatrix}
J_l(V\sqrt{1-b}) & 0 & -K_l(V\sqrt{b}) & 0 \\[1.5ex]
0 & J_l(V\sqrt{1-b}) & 0 & -K_l(V\sqrt{b}) \\[1.5ex]
Nl\dfrac{J_l(V\sqrt{1-b})}{1-b} & \eta_o V \dfrac{J_l'(V\sqrt{1-b})}{\sqrt{1-b}} & \dfrac{Nl}{b}K_l(V\sqrt{b}) & \dfrac{\eta_o V}{\sqrt{b}}K_l'(V\sqrt{b}) \\[2ex]
\dfrac{n_{co}^2}{\eta_o}\dfrac{V J_l'(V\sqrt{1-b})}{\sqrt{1-b}} & Nl\dfrac{J_l(V\sqrt{1-b})}{1-b} & \dfrac{n_{cl}^2}{\eta_o}\dfrac{V}{\sqrt{b}}K_l'(V\sqrt{b}) & \dfrac{Nl}{b}K_l(V\sqrt{b})
\end{bmatrix}
$$

$$\times \begin{bmatrix} A_l \\ B_l \\ C_l \\ D_l \end{bmatrix} = 0 \tag{9.21}$$

A nontrivial solution for the four unknowns exists only if the 4×4 determinant vanishes. Thus, we obtain the *characteristic equation* of step-index fibers with a circular core:

$$\left[\frac{1}{V\sqrt{1-b}} \frac{J_l'(V\sqrt{1-b})}{J_l(V\sqrt{1-b})} + \frac{(1-\Delta)^2}{V\sqrt{b}} \frac{K_l'(V\sqrt{b})}{K_l(V\sqrt{b})} \right]$$

$$\times \left[\frac{1}{V\sqrt{1-b}} \frac{J_l'(V\sqrt{1-b})}{J_l(V\sqrt{1-b})} + \frac{1}{V\sqrt{b}} \frac{K_l'(V\sqrt{b})}{K_l(V\sqrt{b})} \right]$$

$$= l^2 \frac{b + (1-\Delta)^2(1-b)}{[V^2 b(1-b)]^2} \tag{9.22}$$

The equation is also known as the *dispersion relation*. For a given fiber and operating at a given wavelength, V and Δ are known; l is an integer. With these parameters known, b can be solved numerically. As an example, we plot the two sides of (9.22) as functions of b for a fiber with $V = 5.0$, $\Delta = 0.2$, and $l = 1$. This is presented in Figure 9.2. From the intersections of the two sets of curves, we obtain solutions to the dispersion relation. For the example shown in Figure 9.2, there are three intersections, and they correspond to three solutions for the generalized guide index b. The three b's are approximately 0.18, 0.32, and 0.83.

From the numerical solutions of the characteristic equation, we can construct b versus V curves for the fiber under investigation. Figure 9.3 depicts the bV plots for the seven lowest order modes guided by a step-index fiber with $\Delta = 0.2$. For most fibers, Δ is rather small, and many curves merge into one curve. In Figure 9.3, we purposely choose a large Δ to keep the curves apart.

Once b is known, the four coefficients can be determined, except for a multiplicative constant representing the field amplitude. When coefficients are known, the field of the guided mode is determined completely. In the inserts of Figure 9.3, we show the simplified sketches of the transverse electric field in the core region associated with each bV curve.

9.2.3 Traditional Mode Designation and Fields

In transmission lines and metallic waveguides, guided waves may be classified as *transverse electromagnetic* (TEM), *transverse electric* (TE), and *transverse magnetic* (TM) *modes*. For TEM modes, $e_z(r, \phi)$ and $h_z(r, \phi)$ are both zero. For TE modes, $e_z(r, \phi)$ vanishes everywhere in the waveguide. For TM modes, $h_z(r, \phi)$ vanishes completely. As noted in Chapter 2, waves guided by thin-film waveguides can be categorized as TE and TM modes. For optical fibers, the field classification

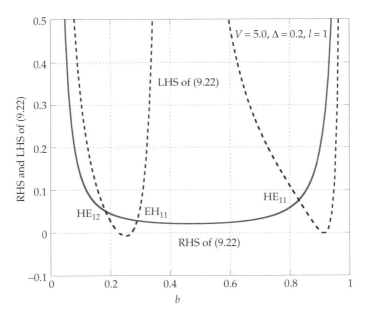

Figure 9.2 Numerical solution of the dispersion relation (9.22).

is more complicated. There are rotationally symmetric TE and TM modes. But, there is no ϕ-dependent TE nor TM mode. All ϕ-dependent fields are *hybrid modes* since neither $e_z(r, \phi)$ nor $h_z(r, \phi)$ vanishes. The designation for hybrid modes is somewhat arbitrary, and we follow the practice commonly used in microwave technology. Consider the field on the fiber axis and at a wavelength far above cutoff. If $e_z(r, \phi)$ and $h_z(r, \phi)$ have the same polarity, the hybrid mode is identified as a *HE mode*. If $e_z(r, \phi)$ and $h_z(r, \phi)$ have opposite polarity, the mode is an *EH mode* [5–7].

To describe the field distributions, we use two integer subscripts, l and m. The first subscript l is the *azimuthal mode number*. It relates to the angular variation of the field. If l is zero, $e_z(r, \phi)$ and $h_z(r, \phi)$ are independent of ϕ. For $l \neq 0$, $e_z(r, \phi)$ and $h_z(r, \phi)$ vary as $\cos l\phi$ and $\sin l\phi$, respectively. Or, they vary as $\sin l\phi$ and $\cos l\phi$, respectively. The second subscript m is the *radial mode number*. It describes the field variation in the radial direction. In particular, $|e_z(r, \phi)|$ and $|h_z(r, \phi)|$ have m maxima each in the radial direction, including the peak, if it exists, at the origin. As noted earlier, there is no TE_{lm}, or TM_{lm}, mode if l is not zero. If l is not zero, neither $e_z(r, \phi)$ nor $h_z(r, \phi)$ vanishes. Therefore the modes are hybrid modes. In short, waves guided by a step-index fiber are TE_{0m}, TM_{0m}, HE_{lm}, and EH_{lm} modes. TE_{0m} and TM_{0m} modes are nondegenerate modes independent of ϕ. All HE_{lm} and EH_{lm} modes are doubly degenerate. Electric and magnetic field lines of HE_{12}, EH_{11}, and TE_{02} modes in the core region are shown in Figures 9.4, 9.5, and 9.6 [5]. The fields of HE_{11} fields are the same as that depicted in the central portion of Figure 9.4.

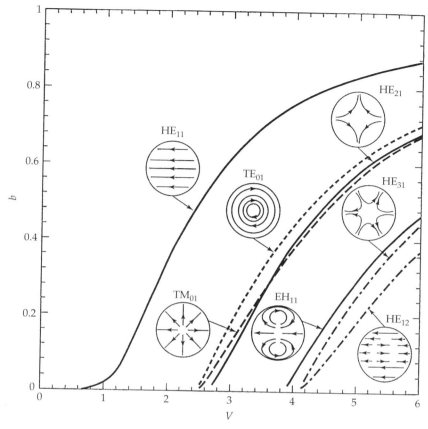

Figure 9.3 Dispersion of a fiber with $\Delta = 0.2$ and electric field lines of various modes.

An examination of Figure 9.3 reveals that all modes, except the HE_{11} mode, are cut off if V is smaller than 2.4048. Thus, a step-index fiber supports one and only one mode if V is less than 2.4048. The HE_{11} mode is the *dominant* or *fundamental mode* of the circular-core step-index fiber. In the core region, the transverse electric and magnetic fields of HE_{11} mode are approximately linearly polarized and mutually orthogonal. This is depicted in Figure 9.4. In addition, there are two degenerate, mutually orthogonal and independent HE_{11} modes. That is, fields of the two degenerate HE_{11} modes are nearly linearly polarized in the x or y direction. For a fiber with a small Δ, the transverse field is approximately linearly polarized in the cladding as well. For typical telecommunication-grade fibers, one of the transverse field components, say the x component, is much stronger than the other transverse field component and the longitudinal field component. The difference may be as much as 60–70 dB [8]. For a fiber with a large Δ, the field is linearly polarized in much of the core region. But near the core–cladding boundary, the electric field lines turn in the direction of the boundary. Outside the core region, the field is not linearly polarized as shown in Figure 9.7 [9].

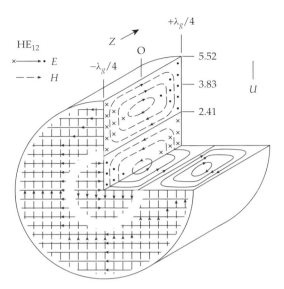

Figure 9.4 Field plot in the core for the HE$_{12}$ mode far from cutoff and for a small index difference. (From [5].)

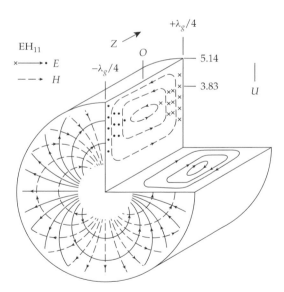

Figure 9.5 Field plot in the core for the EH$_{11}$ mode far from cutoff and for a small index difference. (From [5].)

For a circular-core step-index fiber, the next three modes are the TE$_{01}$, TM$_{01}$, and HE$_{21}$ modes. The cutoffs for TE$_{01}$ and TM$_{01}$ are exactly $V = 2.4048$. But the cutoff of HE$_{21}$, while near 2.4048, depends on Δ. As shown in Figure 9.3, the bV curves of the TE$_{01}$, TM$_{01}$, and HE$_{21}$ modes of fibers with $\Delta = 0.2$ are very close. They become indistinguishable if Δ is very small. The three modes

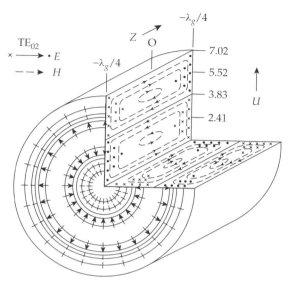

Figure 9.6 Field plot in the core for the TE_{02} mode far from cutoff and for a small index difference. (From [5].)

individually are not linearly polarized. But they can be combined to form nearly linearly polarized fields in the core region. Figure 9.8 depicts schematically linearly polarized fields formed by superimposing TM_{01} and HE_{21} modes, or TE_{01} and HE_{21} modes. For $3.8317 < V < 5.1356$, HE_{12}, EH_{11}, and HE_{31} appear. The cutoffs of HE_{12} and HE_{31} modes are functions of Δ while that of EH_{11} mode is independent of Δ. If Δ is very small, these modes become nearly degenerate. When properly combined, the fields of EH_{11} and HE_{31} modes can be superimposed to form linearly polarized fields. In short, modes of step-index fibers with a very small Δ are either approximately linearly polarized or two nearly degenerate modes may be combined to form approximately linearly polarized fields. Gloge labeled the linearly polarized modes as LP modes [10, 11]. Many expressions of fields and key quantities of LP modes can be expressed in terms of elementary functions and in closed forms. We will discuss LP modes exclusively in the next section.

9.3 LINEARLY POLARIZED MODES GUIDED BY WEAKLY GUIDING STEP-INDEX FIBERS

As mentioned earlier, most fibers in use are weakly guiding fibers. In the core region of a weakly guiding fiber, the field is approximately linearly polarized [8]. In this section, we study the characteristic equation and the field of *linearly polarized (LP) modes* guided by weakly guiding fibers. We follow closely the analysis presented originally by Gloge [10]. To begin, we note that for a linearly polarized mode, a transverse Cartesian electric field component is much stronger than other electric field components. In addition, the longitudinal electric field component and

Figure 9.7 Lines of the transverse electric fields (*top*) and magnetic fields (*bottom*) of a step-index fiber with $V = 2.4$ and $\Delta = 0.4$. (From [9].)

all magnetic field components can be expressed in terms of the strong transverse Cartesian electric field component. Therefore, all field components of LP modes are determined when the dominant transverse electric field component is known.

9.3.1 Basic Properties of Fields of Weakly Guiding Fibers

Fibers are considered as weakly guiding if $\Delta \rightarrow 0$ while the generalized frequency V remains finite. The conditions are met if $n_{co} - n_{cl}$ is much smaller than n_{co} and the core radius is large compared with λ. Under these conditions, we can show that one of the transverse Cartesian field components is much stronger than the other

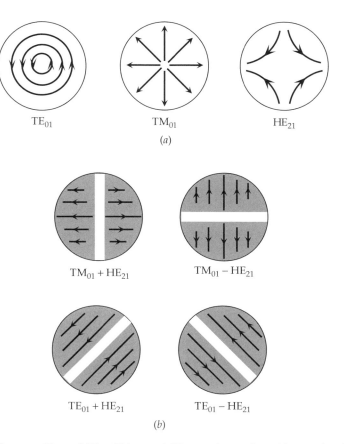

Figure 9.8 Superposition of TE_{01}, TM_{01}, and HE_{21} modes to form LP_{11} modes: (a) Electric fields of TE_{01}, TM_{01}, and HE_{21} modes and (b) four degenerate LP_{11} modes. (After [5].)

transverse field component and the longitudinal component. Furthermore, the field variation in the transverse direction is much smoother or gentler than the change in the longitudinal direction. In other words, $|\nabla_t e_z| \ll |\beta \mathbf{e}_t|$ and $|\nabla_t h_z| \ll |\beta \mathbf{h}_t|$. Then we deduce from (1.72) and (1.74):

$$\mathbf{h}_t \approx \frac{\beta}{\omega \mu_o} \hat{\mathbf{z}} \times \mathbf{e}_t \qquad (9.23)$$

$$\mathbf{h}_t \approx \frac{\omega \varepsilon_o n^2}{\beta} \hat{\mathbf{z}} \times \mathbf{e}_t \qquad (9.24)$$

The two equations hold simultaneously if and only if

$$\frac{\beta}{\omega \mu_o} \approx \frac{\omega \varepsilon_o n^2}{\beta}$$

Thus, $\beta \approx kn$ and (9.23) and (9.24) become

$$\mathbf{h}_t \approx \frac{n}{\eta_o} \hat{\mathbf{z}} \times \mathbf{e}_t \tag{9.25}$$

where $\eta_o = \sqrt{\mu_o/\varepsilon_o}$ is the intrinsic impedance of free space. Equation (9.25) is accurate to the order of Δ. In view of (9.25), we see that the transverse field components $\mathbf{e}_t$ and $\mathbf{h}_t$ of a weakly guiding fiber are related in the same manner as the transverse field components of a uniform plane wave in isotropic, nonmagnetic media with an index n [1, 2, 10].

9.3.2 Fields and Boundary Conditions

From (1.10) and (1.73), we have

$$e_z = -\frac{j}{\beta} \nabla_t \cdot \mathbf{e}_t \tag{9.26}$$

$$h_z = \frac{j}{\omega \mu_o} \hat{\mathbf{z}} \cdot \nabla_t \times \mathbf{e}_t \tag{9.27}$$

The two equations are exact. From (9.25), (9.26), and (9.27), we conclude that e_z, $\mathbf{h}_t$, and h_z can be expressed in terms of $\mathbf{e}_t$. The remaining task is to determine $\mathbf{e}_t$. In general, there are two linearly independent and mutually orthogonal solutions, corresponding to two orthogonal polarizations. In the following discussions, we consider the x-polarized electric field only. The y-polarized electric field can be studied in the same manner.

Again, we consider a single term of the Fourier cosine series. For the core and cladding regions respectively, we write

$$\mathbf{e}_t(r, \phi) = e_{\text{co}}(r) \cos l\phi \, \hat{\mathbf{x}} \qquad \text{for } r \leq a \tag{9.28}$$

$$\mathbf{e}_t(r, \phi) = e_{\text{cl}}(r) \cos l\phi \, \hat{\mathbf{x}} \qquad \text{for } r \geq a \tag{9.29}$$

where $e_{\text{co}}(r)$ and $e_{\text{cl}}(r)$ are yet to be determined.

In terms of $e_{\text{co}}(r)$, the other field components in the core region are

$$e_z(r, \phi) = -\frac{j}{\beta} \left[\frac{de_{\text{co}}(r)}{dr} \cos l\phi \cos \phi + \frac{l}{r} e_{\text{co}}(r) \sin l\phi \sin \phi \right] \tag{9.30}$$

$$h_z(r, \phi) = -\frac{j}{\omega \mu_o} \left[\frac{de_{\text{co}}(r)}{dr} \cos l\phi \sin \phi - \frac{l}{r} e_{\text{co}}(r) \sin l\phi \cos \phi \right] \tag{9.31}$$

$$\mathbf{h}_t(r, \phi) \approx \frac{\beta}{\omega \mu_o} e_{\text{co}}(r) \cos l\phi \, \hat{\mathbf{y}} \tag{9.32}$$

Corresponding expressions can be written for the field components in the cladding region.

The continuation of e_ϕ and e_z at $r = a$ requires

$$e_{co}(a^-) = e_{cl}(a^+) \tag{9.33}$$

and

$$\left.\frac{de_{co}}{dr}\right|_{r=a-} = \left.\frac{de_{cl}}{dr}\right|_{r=a+} \tag{9.34}$$

When the two conditions are satisfied, h_ϕ and h_z are also continuous at $r = a$. Thus all boundary conditions are met when (9.33) and (9.34) are satisfied.

9.3.3 Characteristic Equation and Mode Designation

To determine $e_{co}(r)$ and $e_{cl}(r)$, we substitute (9.28) and (9.29) into the wave equation and obtain differential equations for the two functions. The differential equation for $e_{co}(r)$ is

$$\frac{1}{r}\frac{d}{dr}\left[r\frac{de_{co}(r)}{dr}\right] + \left[\frac{V^2(1-b)}{a^2} - \frac{l^2}{r^2}\right]e_{co}(r) = 0 \tag{9.35}$$

Again, this is a Bessel differential equation of order l. Since Y_l is singular at the origin, J_l is the only acceptable function. Thus we write for $r \leq a$,

$$e_{co}(r) = E_l \frac{J_l\left(V\sqrt{1-b}\,\dfrac{r}{a}\right)}{J_l(V\sqrt{1-b})} \tag{9.36}$$

where E_l is a yet undetermined amplitude constant.

The differential equation for $e_{cl}(r)$ is a modified Bessel differential equation of order l. We expect $e_{cl}(r)$ to decay exponentially as r increases and express $e_{cl}(r)$ in terms of the modified Bessel function K_l:

$$e_{cl}(r) = E_l' \frac{K_l\left(V\sqrt{b}\,\dfrac{r}{a}\right)}{K_l(V\sqrt{b})} \tag{9.37}$$

where E_l' is also an unknown amplitude constant. From the boundary conditions (9.33) and (9.34), we obtain

$$E_l = E_l' \tag{9.38}$$

$$E_l \frac{V\sqrt{1-b}}{a}\frac{J_l'(V\sqrt{1-b})}{J_l(V\sqrt{1-b})} = E_l' \frac{V\sqrt{b}}{a}\frac{K_l'(V\sqrt{b})}{K_l(V\sqrt{b})} \tag{9.39}$$

We combine the two equations and obtain

$$V\sqrt{1-b}\frac{J_{l-1}(V\sqrt{1-b})}{J_l(V\sqrt{1-b})} + V\sqrt{b}\frac{K_{l-1}(V\sqrt{b})}{K_l(V\sqrt{b})} = 0 \tag{9.40}$$

or equivalently

$$V\sqrt{1-b}\frac{J_{l+1}(V\sqrt{1-b})}{J_l(V\sqrt{1-b})} - V\sqrt{b}\frac{K_{l+1}(V\sqrt{b})}{K_l(V\sqrt{b})} = 0 \qquad (9.41)$$

Equation (9.40) or (9.41) is the characteristic equation for a weakly guiding step-index fiber. For a given V and l, we solve for b. No analytic, closed-form solution is known. The equation has to be solved numerically. Depending on V and l, there may be one or more solutions. Or there is no solution at all. Each solution, if it exists, corresponds to a linearly polarized mode. Following Gloge, we label these modes as the LP_{lm} modes [10]. The azimuthal mode number l pertains to the angular variation of the transverse electric field. Specifically, the transverse electric field of a LP_{lm} mode varies like $\cos l\phi$ or $\sin l\phi$. If l is zero, the transverse electric field is independent of ϕ. The radial mode number m corresponds to the mth root of (9.40) and (9.41) for a given V and l. We take the solution with the largest b as the first solution and set m to 1. Once b is determined, we obtain, from (9.28), (9.29), (9.36), and (9.37), an expression for $\mathbf{e}_t$. When $\mathbf{e}_t$ is known, other field components are readily deduced from (9.30) to (9.32). In Figures 9.9, 9.10, 9.11, and 9.12 we depict the intensity distribution ($|\mathbf{e}_t(r, \phi)|^2$) of the four lowest LP modes with $V = 4.5$. In general, the field distribution, $\mathbf{e}_t(r, \phi)$, of the LP_{lm} mode has l maxima and l minima as a function of ϕ. The intensity distribution, $|\mathbf{e}_t(r, \phi)|^2$, has $2l$ peaks or bright spots on a circle of constant radius, as shown in Figure 9.13. The radial mode number m gives the number of intensity maxima in the radial direction, including the maximum at the origin for modes with $l = 0$.

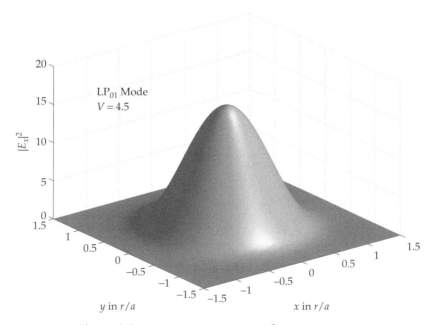

Figure 9.9 Intensity distribution ($|E_x|^2$) of LP_{01} mode.

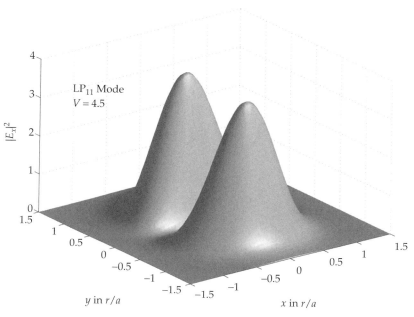

Figure 9.10   Intensity distribution ($|E_x|^2$) of LP$_{11}$ mode.

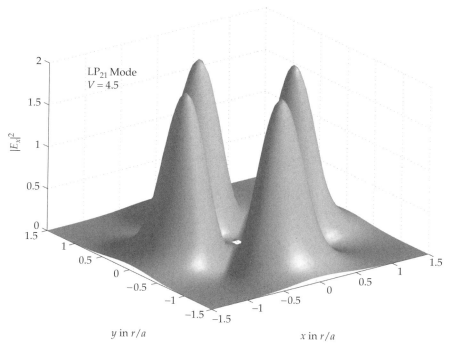

Figure 9.11   Intensity distribution ($|E_x|^2$) of LP$_{21}$ mode.

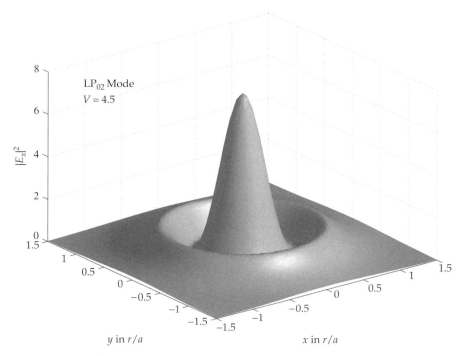

Figure 9.12   Intensity distribution ($|E_x|^2$) of LP$_{02}$ mode.

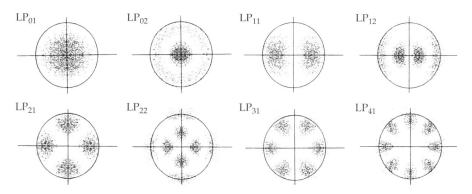

Figure 9.13   Intensity distributions of LP modes guided by a step-index fiber with $V = 7.1$. The density of dots is proportional to $|\mathbf{e}_t|^2$. (From [12].)

To summarize, we note that there are two possible ways to designate modes in weakly guiding fibers. They are the *traditional mode designation* discussed in the last section and the *LP mode designation* introduced in this section. The two designations are equivalent. The equivalence has been established by Gloge [10] and it is

$$LP_{lm} \leftrightarrow HE_{l+1m}, EH_{l-1m} \quad \text{for } l \neq 1$$
$$LP_{1m} \leftrightarrow HE_{2m}, TE_{0m}, TM_{0m} \quad \text{for } l = 1 \qquad (9.42)$$

The dominant or fundamental mode is the HE_{11} mode in the traditional mode designation and the LP_{01} mode in the LP mode designation. The first 20 modes are listed in Table 9.1. The bV curves of the first 13 LP modes of a weakly guiding step-index fiber are shown in Figure 9.14 [10], and they are qualitatively similar to the dispersion curves shown in Figure 9.3. However, there is a crucial difference. Under the weakly guiding approximation, the bV curves are independent of Δ. But the bV curves depicted in Figure 9.3 are for a specific value of Δ, namely $\Delta = 0.2$.

TABLE 9.1   Traditional versus LP Mode Designation

Normalized Frequency V	Traditional Mode Designation	LP Mode Designation	Additional Number of Modes	Total Number of Modes
0–2.4048	HE_{11}	LP_{01}	2	2
2.4048–3.8317	$TE_{01}, TM_{01}, HE_{21}$	LP_{11}	4	6
3.8317–5.1356	$EH_{11}, HE_{31}, HE_{12}$	LP_{21}	4	10
		LP_{02}	2	12
5.1356–5.5201	EH_{21}, HE_{41}	LP_{31}	4	16
5.5201–6.3802	$TE_{02}, TM_{02}, HE_{22}$	LP_{12}	4	20

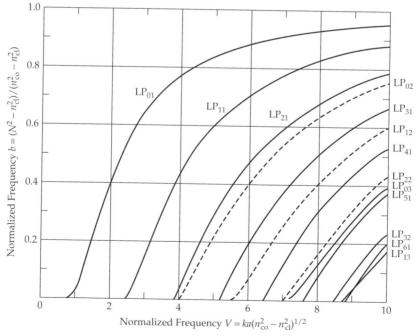

Figure 9.14 bV characteristics of weakly guiding fibers. (From [10].)

9.3.4 Fields of x-Polarized LP$_{0m}$ Modes

For a given V and l, we find b by solving the characteristic equation (9.40) or (9.41) numerically. Using these values of V, l, and b, we calculate $\mathbf{e}_t$ from (9.28), (9.29), (9.36), and (9.37). When $\mathbf{e}_t$ is known, other field components are readily deduced from (9.30) to (9.32). For example, the field components of x-polarized LP$_{0m}$ modes are

$$e_x(r,\phi) = \begin{cases} E_0 \dfrac{J_0(V\sqrt{1-b}\,r/a)}{J_0(V\sqrt{1-b})} & r \le a \\[4mm] E_0 \dfrac{K_0(V\sqrt{b}\,r/a)}{K_0(V\sqrt{b})} & r \ge a \end{cases} \tag{9.43}$$

$$e_z(r,\phi) = \begin{cases} \dfrac{jE_0}{kaN} \dfrac{V\sqrt{1-b}\,J_1(V\sqrt{1-b}\,r/a)}{J_0(V\sqrt{1-b})}\cos\phi & r \le a \\[4mm] \dfrac{jE_0}{kaN} \dfrac{V\sqrt{b}\,K_1(V\sqrt{b}\,r/a)}{K_0(V\sqrt{b})}\cos\phi & r \ge a \end{cases} \tag{9.44}$$

$$h_z(r,\phi) = \begin{cases} \dfrac{jE_0}{ka\eta_o} \dfrac{V\sqrt{1-b}\,J_1(V\sqrt{1-b}\,r/a)}{J_0(V\sqrt{1-b})}\sin\phi & r \le a \\[4mm] \dfrac{jE_0}{ka\eta_o} \dfrac{V\sqrt{b}\,K_1(V\sqrt{b}\,r/a)}{K_0(V\sqrt{b})}\sin\phi & r \ge a \end{cases} \tag{9.45}$$

$$h_y(r,\phi) = \begin{cases} \dfrac{E_0 n_{co}}{\eta_o} \dfrac{J_0(V\sqrt{1-b}\,r/a)}{J_0(V\sqrt{1-b})} & r \le a \\[4mm] \dfrac{E_0 n_{cl}}{\eta_o} \dfrac{K_0(V\sqrt{b}\,r/a)}{K_0(V\sqrt{b})} & r \ge a \end{cases} \tag{9.46}$$

In these expressions, E_0 is an amplitude constant. Following the same procedure, we can derive expressions for fields of LP$_{lm}$ modes with $\ell \ge 1$ and that of y-polarized LP$_{lm}$ modes. This is left as an exercise for the readers.

9.3.5 Time-Average Power

The total time-average power transported by a fiber is

$$P_{\text{total}} = \frac{1}{2} \int_0^{2\pi} \int_0^{\infty} \operatorname{Re}\left[E(r,\phi,z) \times \mathbf{H}^*(r,\phi,z) \cdot \hat{\mathbf{z}} \right] r\,dr\,d\phi \tag{9.47}$$

By making use (9.25), we simplify the above equation to

$$P_{\text{total}} = \frac{1}{2} \frac{\beta}{\omega\mu_o} \int_0^{2\pi} \int_0^{\infty} |\mathbf{e}_t(r,\phi)|^2 \, r\,dr\,d\phi \tag{9.48}$$

Naturally, we partition the integral into two parts. The two parts are

$$I_{co} = \int_0^{2\pi} \int_0^a |\mathbf{e}_t|^2 r\,dr\,d\phi = E_l^2 \int_0^{2\pi} \int_0^a \left[\frac{J_l(V\sqrt{1-b}\,r/a)}{J_l(V\sqrt{1-b})} \right]^2 \cos^2 l\phi\, r\,dr\,d\phi \quad (9.49)$$

$$I_{cl} = \int_0^{2\pi} \int_a^\infty |\mathbf{e}_t|^2 r\,dr\,d\phi = E_l^2 \int_0^{2\pi} \int_a^\infty \left[\frac{K_l(V\sqrt{b}\,r/a)}{K_l(V\sqrt{b})} \right]^2 \cos^2 l\phi\, r\,dr\,d\phi \quad (9.50)$$

where I_{co} pertains to power carried in the core region and I_{cl} power in the cladding region. The integral with respect to ϕ is quite simple since

$$\int_0^{2\pi} \cos^2 l\phi\; d\phi = \begin{cases} 2\pi & \text{for } l = 0 \\ \pi & \text{for } l \geq 1 \end{cases} \quad (9.51)$$

The integral with respect to r can be evaluated in a closed form with the help of two Bessel function identities [3]:

$$\int x\, J_l^2(ax)\, dx = \frac{x^2}{2} [J_l^2(ax) - J_{l-1}(ax) J_{l+1}(ax)] \quad (9.52)$$

$$\int x\, K_l^2(ax)\, dx = \frac{x^2}{2} [K_l^2(ax) - K_{l-1}(ax) K_{l+1}(ax)] \quad (9.53)$$

In short, the two integrals in (9.49) and (9.50) can be evaluated analytically and expressed in closed forms. Explicitly, the time-average power carried by an LP$_{lm}$ mode in a weakly guiding step-index fiber is

$$P_{total} = P_{co} + P_{cl}$$

where

$$P_{co} = \frac{N}{\eta_o} \frac{\pi a^2 E_l^2}{4} \left[1 - \frac{J_{l-1}(V\sqrt{1-b}) J_{l+1}(V\sqrt{1-b})}{J_l^2(V\sqrt{1-b})} \right] \quad (9.54)$$

$$P_{cl} = \frac{N}{\eta_o} \frac{\pi a^2 E_l^2}{4} \left[\frac{K_{l-1}(V\sqrt{b}) K_{l+1}(V\sqrt{b})}{K_l^2(V\sqrt{b})} - 1 \right] \quad (9.55)$$

$$P_{total} = \frac{N}{\eta_o} \frac{\pi a^2 E_l^2}{4} \frac{1}{1-b} \frac{K_{l-1}(V\sqrt{b}) K_{l+1}(V\sqrt{b})}{K_l^2(V\sqrt{b})} \quad (9.56)$$

and P_{co} and P_{cl} are power transported in the core and cladding regions, respectively. The ratio P_{co}/P_{total} is the fraction of power carried in the core region, and it is also

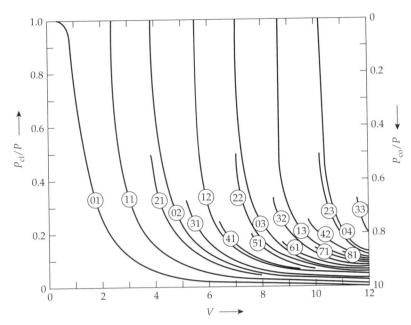

Figure 9.15 Fractional power carried in the core and cladding regions as functions of V. (From [10].)

known as the *confinement factor*. In Figure 9.15, we plot P_{co}/P_{total} and P_{cl}/P_{total} as functions of V [10]. From these plots, we observe:

1. As V approaches the cutoff values of LP_{0m} and LP_{1m} modes, $P_{co}/P_{total} \rightarrow 0$. In other words, for the two lowest LP modes, no power is carried in the core region at cutoff. Power is entirely in the cladding region at cutoff.
2. For LP_{lm} modes with $l \geq 2$, P_{cl}/P_{total} tends $1/l$ as V approaches the cutoff value. Thus for higher-order LP modes, a small and finite fraction of power remains in the core region even at the cutoff.

9.3.6 Single-Mode Operation

Depending on the core radius and index difference, fibers may be classified as single-mode, few-mode, or multimode fibers. The information capacity of step-index few-mode and multimode fibers is rather limited [12]. However, the information capacity is very large if one and only one mode is guided by the fiber. Fibers supporting one and only one mode are known as *single-mode* or *monomode fibers*. It is constructive to estimate the core radius and the index difference required for the single-mode operation. As seen in Figure 9.14, the lowest LP mode has no cutoff. The second lowest LP mode is the LP_{11} mode. If the LP_{11} mode is cut off, all higher-order LP modes are also cut off. To find the cutoff of an LP mode, we let b approach 0. Making use of the power series expansion of $K_l(z)$ [3], we can show that

$$\lim_{b \to 0} \left[V \sqrt{b} \frac{K_{l-1}(V\sqrt{b})}{K_l(V\sqrt{b})} \right] \longrightarrow 0 \tag{9.57}$$

for all values of l. Then, the characteristic equation (9.40) becomes

$$J_{l-1}(V) = 0 \text{ for } l \geq 1 \tag{9.58}$$

when b approaches 0. Thus the cutoff V for the LP_{11} mode is the first root of J_0 (V), which is 2.4048 [3]. It follows that a fiber is single moded if V is less than 2.4048. Then, we find the core radius required for the single-mode operation:

$$\frac{a}{\lambda} < \frac{2.4048}{2\pi n_{co} \sqrt{2\Delta}} = \frac{0.2706}{n_{co}\sqrt{\Delta}} \tag{9.59}$$

As a simple example, we consider a fiber with $n_{co} = 1.48$ operating at $\lambda = 1.55$ μm. We conclude from (9.59) that the fiber is single moded if the core radius is smaller than 2.83 μm if $\Delta = 0.010$, 4.00 μm if $\Delta = 0.005$, and 5.17 μm if $\Delta = 0.003$. It is easy to appreciate that most single-mode fibers have core diameters between 6 and 10 μm if the operating wavelength is about 1.55 μm.

As mentioned earlier, a fiber is single moded if V is less than 2.4048. But if V is too small, a significant portion of power resides in the cladding region, as shown in Figure 9.15. Thus fibers having a small V are vulnerable to external disturbance, bending for example, and contamination. Most single-mode fibers of practical interest are designed to operate with a V between 1.5 and 2.4. Numerical calculations show that for V between 1.5 and 2.5, b of the LP_{01} mode may be approximated accurately by [13]

$$b \approx \left(1.1428 - \frac{0.996}{V} \right)^2 \tag{9.60}$$

and the error is less than 0.1%.

9.4 PHASE VELOCITY, GROUP VELOCITY, AND DISPERSION OF LINEARLY POLARIZED MODES

Signals transmitted in fibers are usually in a digital format. Therefore, we are interested in pulses propagate in fibers. Optical pulses, or packets, travel at the speed of the group velocity. Pulses also become distorted and broadened as they travel in fibers. In single-mode fibers, the distortion and broadening is due to the group velocity dispersion. In this section, we study the phase and group velocities of LP modes in fibers. We also consider the group velocity dispersion of step-index fibers.

9.4.1 Phase Velocity and Group Velocity

First, we examine the phase velocity of LP modes. In a weakly guiding fiber, the effective index is approximately, from (9.7),

$$N \approx n_{co}(1 - \Delta + b\Delta) \tag{9.61}$$

Thus, the *phase velocity* is

$$v_{ph} = \frac{\omega}{\beta} = \frac{c}{N} \approx \frac{c}{n_{co}}(1 + \Delta - b\Delta) \tag{9.62}$$

Since b is between 0 and 1, the phase velocity of an LP mode is between c/n_{co} and c/n_{cl}.

As noted in Chapter 1, the *group velocity* v_{gr} is $d\omega/d\beta$. For a fiber of length L, the *transit time* of a pulse packet, or the *group delay* τ_{gr}, is L/v_{gr}. Instead of discussing the group velocity, it is convenient to discuss the *group delay per unit length*:

$$\frac{\tau_{gr}}{L} = \frac{1}{v_{gr}} = \frac{d\beta}{d\omega} = \frac{1}{c}\frac{d}{dk}(kN) \tag{9.63}$$

Before proceeding further, we make two preliminary observations. First, we note from the definition of Δ that

$$\frac{d\Delta}{dk} = \frac{n_{cl}}{n_{co}}\left(\frac{1}{n_{co}}\frac{dn_{co}}{dk} - \frac{1}{n_{cl}}\frac{dn_{cl}}{dk}\right) \tag{9.64}$$

Since the core and cladding are made of the same basic materials, the dispersion of the core and cladding materials are similar too. Thus the two terms in the brackets are approximately the same and they cancel each other almost completely. In other words, $d\Delta/dk$ is very small. In the following discussions, we take Δ as a wavelength-independent term and ignore $d\Delta/dk$ completely. If the simplifying assumption were not made, the analysis would be very tedious. We will return to this problem again in Chapter 12. In Chapter 12, we consider the contribution to the group velocity dispersion arising from the dispersion of Δ.

For a weakly guiding fiber, V is approximately $ka\sqrt{2\Delta}n_{co}$. It follows that

$$\frac{dV}{dk} \approx \frac{V}{k}\left(1 - \frac{\lambda}{n_{cl}}\frac{dn_{cl}}{d\lambda}\right) \tag{9.65}$$

In the visible and near-infrared regions, $|(\lambda/n)(dn/d\lambda)|$ of most silica glass materials is on the order of 0.01 or smaller [12, 14]. By ignoring the second term of (9.65), we obtain

$$\frac{dV}{dk} \approx \frac{V}{k} \tag{9.66}$$

This is the second observation. Now we consider the group delay. From (9.62) and (9.63), we have

$$\frac{\tau_{gr}}{L} = \frac{1}{c}\left[\frac{d}{dk}(kn_{cl}) + \Delta\frac{d}{dk}(kn_{co}b)\right] \tag{9.67}$$

A little manipulation will show that

$$\frac{d}{dk}(kn_{co}b) = \frac{1}{a\sqrt{2\Delta}}\frac{d}{dk}(Vb) = \frac{1}{a\sqrt{2\Delta}}\frac{d}{dV}(Vb)\frac{dV}{dk}$$

$$\approx \frac{V}{ka\sqrt{2\Delta}}\frac{d}{dV}(Vb) = n_{co}\frac{d}{dV}(Vb) \tag{9.68}$$

Substituting (9.68) into (9.67), we obtain

$$\frac{\tau_{gr}}{L} \approx \frac{1}{c}\frac{d}{dk}(kn_{co}) - \Delta\frac{n_{co}}{c}\left[1 - \frac{d}{dV}(Vb)\right] \tag{9.69}$$

The first term of (9.69) is simply the group delay per unit length of a uniform plane wave in a medium with an index n_{co}. The second term is the contribution to the group delay by wave guiding in the fiber. Chang [15] has derived an explicit expression for $d(Vb)/(dV)$ for the LP_{01} mode. He showed that

$$\frac{d(Vb)}{dV} = 1 - (1-b)\left[1 - 2\frac{K_0^2(V\sqrt{b})}{K_1^2(V\sqrt{b})}\right] \tag{9.70}$$

As shown in Figure 9.16, $d(Vb)/(dV)$ of LP_{01} mode is about 1 for $1.5 < V < 2.5$ and $1 - (d/dV)(Vb)$ is much smaller than 1. We also note the presence of the multiplicative factor Δ in the second term of (9.69). Thus, the effect of wave guiding on the group delay is rather minor. In short, the material dispersion of the core, as represented by the first term of (9.69), is the dominant factor affecting the group delay in a fiber.

9.4.2 Dispersion

Since the group velocity and group delay are wavelength and mode dependent, optical pulses are distorted and broadened as they propagate. In short, optical fibers are *dispersive*. In multimode fibers, a large number of modes may be excited. Each mode propagates with a different group velocity. As a result, pulses evolve as they propagate. This is known as the *intermodal* or *multimode dispersion*, and it is the dominant pulse broadening mechanism in multimode fibers. A single-mode fiber supports one and only one mode. Thus the intermodal broadening does not exist in single-mode fibers. For most single-mode fibers, there may be two mutually orthogonal and independent polarizations. If the fiber geometry and the index profile were perfectly rotationally symmetric, the two polarizations would be degenerate and

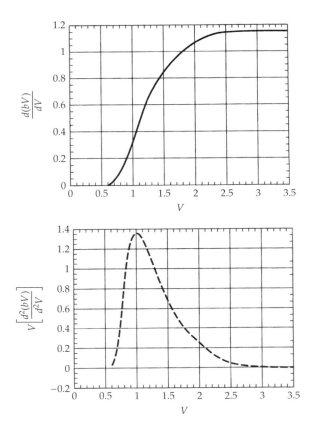

Figure 9.16 $d(bV)/dV$ and $V[d^2(bV)/d^2V]$ vs. V. (After [15].)

would not contribute to the pulse broadening. In real single-mode fibers, however, the fiber geometry and the index profile are not perfect in all respects; the two polarizations are nearly, but not perfectly, degenerate. This leads to the pulse broadening. This is known as the *polarization mode dispersion*. But pulses are still distorted and broadened in *single-mode single-polarization fibers*. This is because all pulses of a finite duration have a finite spectral width, and different spectral components travel with a different group velocity. This is the *intramodal* or *chromatic dispersion*. Strictly speaking, the effects due to the existence of two nondegenerate polarization modes and a finite spectral width also contribute to pulse broadening in multimode fibers. But the effects due to polarization dispersion and intramodal dispersion are negligibly small in comparison with the effects of the intermodal dispersion in multimode fibers. Thus, we ignore the intramodal and the polarization mode dispersions in multimode fibers.

9.4.2.1 *Intermodal Dispersion*

To illustrate the effect of intermodal dispersion qualitatively, we briefly consider a fictitious fiber that supports four modes. Suppose three narrow pulses enter one end

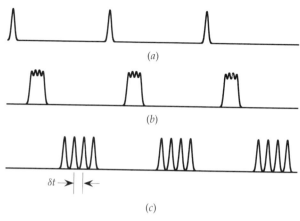

Figure 9.17 Pulse distortion and broadening due to intermodal dispersion: (a) Input pulses, (b) output pulses from a short fiber section, and (c) output pulses from a long fiber section.

of a four-mode fiber, and we examine pulses emerging from the distal end. This is depicted schematically in Figure 9.17. For each input pulse [Fig. 9.17(a)], a pulse packet containing four closely spaced pulses appears at the output [Fig. 9.17(c)] if the fiber is long enough. We refer the closely spaced pulses within a pulse packet as subpulses. The separation δt between the subpulses within a pulse packet varies linearly with the fiber length. If the four-mode fiber is short, the subpulses would be so close that it would be impossible to distinguish the four individual subpulses in the same pulse packet. Instead, a broadened pulse comprising the four subpulses appears at the output. This is shown schematically in Figure 9.17(b). In a typical multimode fiber, a few hundred modes are excited. The separation δt between neighboring subpulses of the same pulse packet at the output end is extremely small even for a very long fiber. Thus, for each input pulse, a broadened pulse packet appears at the output. This is the *intermodal dispersion* and it is the dominant pulse broadening mechanism in multimode fibers. In this book, we are mainly interested single-mode fibers and will not discuss the intermodal broadening further. Details can be found in [12].

9.4.2.2 Intramodal Dispersion

Pulses transmitted by single-mode fibers are broadened and distorted since all pulses of a finite duration have a finite spectral width. As seen from (9.69), different spectral components travel with a different group delay. This is the intramodal dispersion or chromatic dispersion in single-mode fibers. Mathematically speaking, the intramodal dispersion in single-mode fibers originates from two sources. First, the normalized guide index b is a function of V that in turn is a function of λ, n_{co}, and n_{cl}. Second, the indices n_{co} and n_{cl} themselves are also wavelength dependent.

To estimate the intramodal dispersion, we consider a pulse with a finite spectral width $\Delta\lambda$ centering at λ_0. In particular, we consider the group delays of two spectral components $\lambda_0 \pm (\Delta\lambda/2)$. The difference of group delays of the two spectral components is

$$\delta\tau_{gr} \equiv \tau_{gr}\left(\lambda_0 + \frac{\Delta\lambda}{2}\right) - \tau_{gr}\left(\lambda_0 - \frac{\Delta\lambda}{2}\right) \approx \frac{d\tau_{gr}}{d\lambda}\Delta\lambda \qquad (9.71)$$

Clearly, $\delta\tau_{gr}$ increases linearly with $\Delta\lambda$ and length L. We define the *intramodal dispersion* $\mathcal{D}$ as the broadening per unit spectral width per unit length:

$$\mathcal{D} \equiv \frac{1}{L}\frac{\delta\tau_{gr}}{\Delta\lambda} = \frac{1}{L}\frac{d\tau_{gr}}{d\lambda} = -\frac{k^2}{2\pi c^2}\frac{d^2\beta}{dk^2} \qquad (9.72)$$

To evaluate the derivatives of β, we recall that Δ is insensitive to the wavelength change. By making use of (9.64)–(9.66) and (9.68), we obtain from (9.61)

$$\frac{d\beta}{dk} \approx \frac{d}{dk}(kn_{co}) + n_{co}\Delta\left[\frac{d(Vb)}{dV} - 1\right] \qquad (9.73)$$

Further differentiation leads to

$$\mathcal{D} = -\frac{k^2}{2\pi c}\left(\frac{d^2}{dk^2}(kn_{co}) + \left\{\Delta\frac{dn_{co}}{dk}\left[\frac{d(Vb)}{dV} - 1\right]\right\} + \left[\Delta\frac{n_{co}}{k}V\frac{d^2(Vb)}{dV^2}\right]\right) \qquad (9.74)$$

where

$$\frac{d^2}{dk^2}(kn_{co}) = \frac{\lambda^3}{2\pi}\frac{d^2 n_{co}}{d\lambda^2} \qquad (9.75)$$

An analytic expression for $d^2(Vb)/(dV^2)$ has also been obtained by Chang [15] and he showed that

$$\frac{d^2(Vb)}{dV^2} = \frac{2}{V}\frac{K_0^2(V\sqrt{b})}{K_1^2(V\sqrt{b})}\left[1 - \frac{d(Vb)}{dV}\right]$$

$$+ 2(1-b)\frac{d}{dW}\left[\frac{K_0^2(W)}{K_1^2(W)}\right]\frac{dW}{dV}\bigg|_{W=V\sqrt{1-b}} \qquad (9.76)$$

where

$$\frac{d}{dW}\left[\frac{K_0^2(W)}{K_1^2(W)}\right] = -2\frac{K_0(W)}{K_1(W)} + \frac{2}{W}\frac{K_0^2(W)}{K_1^2(W)} + 2\frac{K_0^3(W)}{K_1^3(W)} \qquad (9.77)$$

Plots of $d(Vb)/(dV)$ and $V[d^2(Vb)/(dV^2)]$ are given in Figure 9.16 [11, 15]. As noted earlier, $d(Vb)/(dV)$ is on the order of 1, the second term of (9.74) is negligible in comparison with the other terms in the equation. By ignoring the second term, we obtain an approximate expression for $\mathcal{D}$:

$$\mathcal{D} \approx -\frac{k^2}{2\pi c}\left[\frac{\lambda^3}{2\pi}\frac{d^2 n_{co}}{d\lambda^2} + \Delta\frac{n_{co}}{k}V\frac{d^2(Vb)}{dV^2}\right] \qquad (9.78)$$

Equation (9.78) is the *intramodal dispersion* of weakly guiding, step-index single-mode fibers. To appreciate the physical meanings of the each term of (9.78), we consider single-mode fibers with $V > 2$. For $V > 2$, $V[d^2(Vb)/(dV^2)]$ is less than 0.25 as shown in Figure 9.16. Also, note the presence of the multiplicative factor Δ in the second term. Thus, the dominant term of $\mathcal{D}$ is the first term of (9.78):

$$\mathcal{D}_{\text{mt}} \approx -\frac{k^2}{2\pi c}\frac{\lambda^3}{2\pi}\frac{d^2 n_{\text{co}}}{d\lambda^2} = -\frac{\lambda}{c}\frac{d^2 n_{\text{co}}}{d\lambda^2} \qquad (9.79)$$

As given by (9.79) $\mathcal{D}_{\text{mt}}$ is precisely the dispersion of uniform plane waves propagating in an extended medium having an index n_{co}. Therefore, we identify $\mathcal{D}_{\text{mt}}$ as the *material dispersion*. For a single-mode fiber operating with a V greater than 2.0, the confinement factor is greater than 0.75 (Fig. 9.15). In other words, the time-average power of LP_{01} mode is confined mainly in the core region. Therefore, we expect the intramodal dispersion of LP_{01} mode guided by a single-mode fiber with $V > 2$ as essentially due to the dispersion of the core material. Then we can use the material dispersion of the core material to estimate the dispersion of step-index fibers. In Figure 9.18, we depict the material dispersion of four fiber materials [14]. For fibers with a pure fused silica core, the material dispersion vanishes at 1.28 μm. If the core is doped with GeO_2, the zero dispersion wavelength is lengthened to 1.38 μm. On the other hand, if the core is doped with B_2O_3, the zero dispersion wavelength is shortened to 1.23 μm [14]. Clearly, the material dispersion depends on the material composition.

To discuss the physical meaning of the second term of (9.78), we suppose that the core and cladding materials are nondispersive. Then the pulse broadening is

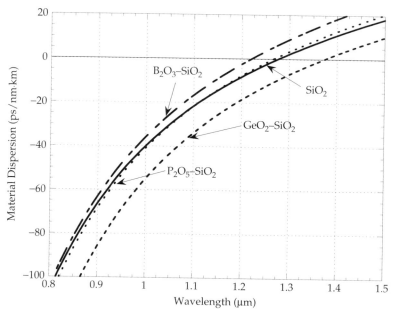

Figure 9.18 Material dispersion of glass fibers [14].

due to the wave guiding only. When all derivatives of n_{co} with respect to k are ignored, (9.78) reduces to

$$\mathcal{D}_{wg} = -\Delta \frac{n_{co}}{\lambda c} V \frac{d^2(Vb)}{dV^2} \tag{9.80}$$

This is the *waveguide dispersion* of the fiber, and it represents the dispersion of a fictitious fiber made of wavelength-independent materials. Note that $\mathcal{D}_{wg}$ depends on the index difference and the core radius. Figure 9.19 shows the waveguide dispersion of fibers having two different core radii [16]. Also shown in Figure 9.19 are the material dispersion and the total dispersion as functions of wavelength.

In summary, we note that when the dispersion of Δ is ignored, the intramodal dispersion may be viewed as the sum of material and waveguide dispersions. We should caution, however, that such a simple interpretation, while intuitively pleasing, is not valid for fibers with a small core or operating with a small V. For fibers operating with a small V, there are other terms in addition to the material and waveguide dispersions, as observed by Marcuse [17].

9.4.2.3 *Zero Dispersion Wavelength*

Loss, dispersion, and nonlinear effects are the key factors limiting the link length and the transmission bandwidth of communication systems. To operate the communication systems over a longer span and at a higher speed, it is necessary to operate the fibers in a spectral region where the fiber attenuation and the group velocity dispersion are minimized. As discussed in the last section, the intramodal dispersion of a weakly guiding single-mode fiber is the sum of the material and the waveguide dispersions. The material dispersion is essentially the dispersion of the core material. As shown in Figure 9.18, $\mathcal{D}_{mt}$ of pure silicate core fibers is positive for wavelength longer than 1.28 μm and negative for shorter wavelengths [14]. On

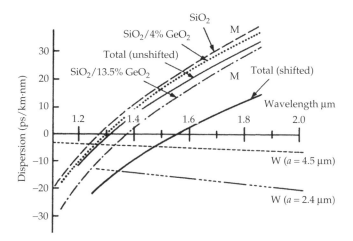

Figure 9.19 Dispersion of silicate-based standard step-index single-mode fibers and dispersion-shifted fibers [16].

the other hand, $\mathcal{D}_{\text{wg}}$ of silicate fibers is always negative and depends on the index difference, core radius, and V. When the material and waveguide dispersions are combined, the two dispersion effects may cancel at a certain wavelength. This is the *zero dispersion wavelength* of the fiber [15, 16, 18, 19]. Two examples are shown in Figure 9.19 [16]. It is possible to design a fiber so that the total fiber dispersion vanishes at a desired wavelength between 1.3 to 1.75 μm [19]. At the zero dispersion wavelength the system span is limited mainly by the fiber loss and nonlinear effects of the system.

Fiber attenuation is highly wavelength dependent. There are two spectral ranges where the loss is particularly small. One spectral range is near 1.3 μm and the other is between 1.5 to 1.6 μm. While the loss of silicate fibers around 1.3 μm is quite low already (about 0.35 dB/km), the fiber loss is even lower near 1.55 μm (about 0.21 dB/km). By choosing the core radius and the core materials, it is possible to force the total dispersion to vanish at 1.55 μm where the fiber attenuation is the smallest. These fibers are known as the *dispersion-shifted fibers.*

PROBLEMS

1. Derive the field expressions of x-polarized LP_{lm} modes having $\ell \geq 1$.
2. A circular fiber with an index profile shown in Figure 9.20 is referred to as the W-type fibers. Since n_1, n_2, and n_3 are much greater than $n_1 - n_3$, the conditions for weakly guided fibers are satisfied. Derive a dispersion relation for y-polarized LP modes.

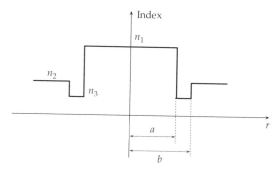

Figure 9.20 Index profile of a W-type fiber.

3. Anisotropic media may be characterized by a relative dielectric tensor $[\varepsilon_r]$. Refer to Section 13.8 of Ref. [2] for details. If the principal axes are the x, y, and z axes, the relative dielectric tensor is

$$[\varepsilon_r] = \begin{bmatrix} n_1^2 & 0 & 0 \\ 0 & n_2^2 & 0 \\ 0 & 0 & n_3^2 \end{bmatrix}$$

Although $\nabla \cdot \mathbf{D}$ is still zero for charge-free anisotropic media, $\nabla \cdot \mathbf{E}$ is not necessarily zero. Thus wave propagation in anisotropic media is much more complicated than that in isotropic media. But time-harmonic wave ($e^{j\omega t}$) propagating in anisotropic media is still governed by Maxwell's equations, provided a relative dielectric tensor $[\varepsilon_r]$ is used in lieu of a scalar dielectric constant:

$$\nabla \times \mathbf{E} = -j\omega\mu_o\mathbf{H}$$
$$\nabla \times \mathbf{H} = j\omega\varepsilon_o[\varepsilon_r]\mathbf{E}$$

In uniaxial materials, $n_1 = n_2 \neq n_3$, the dielectric tensor $[\varepsilon_r]$ becomes

$$[\varepsilon_r] = \begin{bmatrix} n_1^2 & 0 & 0 \\ 0 & n_1^2 & 0 \\ 0 & 0 & n_3^2 \end{bmatrix}$$

Beginning with the Maxwell equations, show that for waves propagating in uniaxial media and along the z direction, the wave equations for E_z and H_z are

$$\left[\nabla_t^2 + \left(k^2 n_3^2 - \frac{n_3^2}{n_1^2}\beta^2\right)\right]E_z = 0$$
$$[\nabla_t^2 + (k^2 n_1^2 - \beta^2)]H_z = 0$$

4. Consider a fiber with a uniaxial core of radius a and a thick and isotropic cladding. The cladding index is n_{cl}. The fiber axis is along the z direction. Derive the dispersion equation for the waves guided by the fiber with a uniaxial core. Assume that the fiber is a weakly guided fiber, that is, n_1, n_3, and n_{cl} are close but not equal.

5. Refer to Gloge's original study on weakly guiding fibers [10]. Derive (31), (32) and the two unnumbered expressions for $\overline{p}(r)$ between Figure 6 and Figure 7 of Gloge's paper.
   Remarks: For convenience, we list the symbols used in this book and the corresponding symbols in Gloge's paper

This book	Gloge's study
n_{co}	n_c
n_{cl}	n
V	v
$V\sqrt{1-b}$	u
$V\sqrt{b}$	w

REFERENCES

1. R. E. Collin, *Field Theory of Guided Waves*, 2nd ed., IEEE Press, New York, 1991.
2. S. Ramo, J. R. Whinnery, and T. Van Duzer, *Fields and Waves in Communication Electronics*, 3rd ed., Wiley, New York, 1994.

3. M. Abramowitz and I. A. Stegun, *Handbook of Mathematical Functions with Formulas, Graphs and Mathematical Tables*, Dover, New York, 1965.

4. C. Yeh, "Boundary conditions in electromagnetics," *Phys. Rev.*, Vol. 48, No. 2, pp. 1426–1427 (Aug. 1993).

5. E. Snitzer, "Cylindrical dielectric waveguide modes," *J. Opt. Soc. Am.*, Vol. 51, pp. 491–498 (1961); E. Snitzer and H. Osterberg, "Observed dielectric waveguide modes in the visible spectrum," *J. Opt. Soc. Am.*, Vol. 51, pp. 499–505 (1961).

6. C. Yeh, "Guided waves modes in cylindrical optical fibers," *IEEE Trans. Education*, Vol. E-30, pp. 43–51 (1987).

7. A. Kapoor and G. S. Singh, "Mode classification in cylindrical dielectric waveguides," *J. Lightwave Technol.*, Vol. 18, pp. 849–852 (May 2000).

8. M. P. Varnham, D. N. Payne, and J. D. Love, "Fundamental limits to the transmission of linearly polarized light by birefringent optical fibres," *Electron. Lett.*, Vol. 20, No. 1, pp. 55–56 (Jan. 5 1984).

9. H. Zheng, W. M. Henry, and A. W. Snyder, "Polarization characteristics of the fundamental modes of optical fibers," *J. Lightwave Technol.*, Vol. 6, pp. 1300–1305 (Aug. 1988).

10. D. Gloge, "Weakly guided fibers," *Appl. Opt.*, Vol. 10, pp. 2252–2258 (1971).

11. D. Gloge, "Dispersion in weakly guiding fibers," *Appl. Opt.*, Vol. 10, pp. 2442–2445 (1971).

12. Chin-Lin Chen, *Elements of Optoelectronics and Fiber Optics*, McGraw-Hill, New York, 1996.

13. H. D. Rudolph and E. G. Neumann, "Approximations for the eigenvalues of the fundamental mode of a step index glass fiber waveguide," *Nachrichtentechn. Z.*, Vol. 29, pp. 328–329 (1976).

14. J. W. Fleming, "Material dispersion in lightguide glasses," *Electron. Lett.*, Vol. 14, pp. 326–327 (May 25, 1978).

15. C. T. Chang, "Minimum dispersion in a single-mode step-index optical fiber," *Appl. Opt.*, Vol. 18, pp. 2516–2522 (1979).

16. B. J. Ainslie and C. R. Day, "A review of single-mode fibers with modified dispersion characteristics," *J. Lightwave Technol.*, Vol. LT-4, No. 8, pp. 967–979 (Aug. 1986).

17. D. Marcuse, "Interdependence of waveguide and material dispersion," *Appl. Opt.*, Vol. 18, pp. 2930–2932 (1979).

18. J. Jeunhomme, "Dispersion minimization in single-mode fibres between 1.3 μm and 1.7 μm," *Electron. Lett.*, Vol. 15, pp. 478–479 (1979).

19. K. I. White and B. P. Nelson, "Zero total dispersion in step-index monomode fibres at 1.30 and 1.55 μm," *Electron. Lett.*, Vol. 15, pp. 396–397 (1979).

10

INPUT AND OUTPUT CHARACTERISTICS OF WEAKLY GUIDING STEP-INDEX FIBERS

10.1 INTRODUCTION

In considering the transmission properties of a fiber in the last chapter, we have assumed that light somehow exists in the fiber. In fact and for most applications, light must be fed into and coupled out of the fiber. Therefore the input and output characteristics of fibers are also of interest. To consider the radiation and excitation problems of fibers, we suppose that a fiber is truncated at $z = 0$ and examine the field radiated by, or fed to, the truncated fiber (Fig. 10.1). The region outside the truncated fiber is taken as air. In considering the field radiated by the fiber [Fig. 10.1(a)], we are interested in the field in the Fraunhofer zone that is also referred to as the far-field zone. In studying the excitation problems, we assume that the wave impinging upon the truncated fiber tip is a Gaussian beam [Fig. 10.1(b)]. Throughout the discussions, we assume the fiber is a weakly guiding step-index fiber with a circular core. The field expressions and the dispersion relation of LP modes given in (9.28), (9.29), (9.36), (9.37), (9.40), and (9.41) are used in this chapter.

Foundations for Guided-Wave Optics, by Chin-Lin Chen
Copyright © 2007 John Wiley & Sons, Inc.

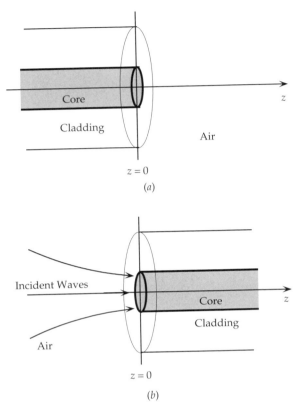

Figure 10.1   (a) Radiation by and (b) excitation of a truncated fiber.

10.2   RADIATION OF LP MODES

10.2.1   Radiated Fields in the Fraunhofer Zone

We use a cylindrical coordinate system $(r, \phi, 0)$ to denote points on the truncated fiber tip and a spherical coordinate system (R, Θ, Φ) for points in air outside the truncated fiber. This is shown in Figure 10.2. In terms of these variables, the *Fresnel–Kirchhoff diffraction formula* [1–3] for the radiation field in the *Fraunhofer* or *far-field zone* is

$$\mathbf{E}_{\mathrm{FF}}(R, \Theta, \Phi) \approx jk \frac{e^{-jkR}}{2\pi R} \frac{1 + \cos \Theta}{2} \int_0^{2\pi} \int_0^\infty \mathbf{E}(r, \phi, 0) e^{jkr \sin \Theta \cos (\Phi - \phi)} r \, dr \, d\phi \tag{10.1}$$

We use subscripts FF to denote the field in the Fraunhofer zone. We will see shortly that the field radiated by a truncated fiber is confined mainly in a small cone in the forward direction. Within this small cone, the *inclination factor* $(1 + \cos \Theta)/2$

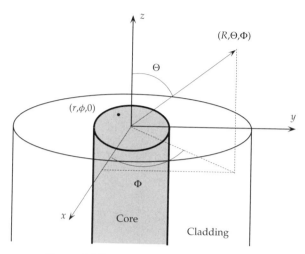

Figure 10.2   Truncated circular core fiber.

is approximately 1. Thus, we can simplify (10.1) further:

$$\mathbf{E}_{FF}(R, \Theta, \Phi) \approx jk \frac{e^{-jkR}}{2\pi R} \int_0^{2\pi} \int_0^{\infty} \mathbf{E}(r, \phi, 0) e^{jkr \sin \Theta \cos (\Phi - \phi)} r \, dr \, d\phi \qquad (10.2)$$

Suppose the field guided by the truncated fiber is an x-polarized LP_{lm} mode. As discussed in Chapter 9, the transverse electric field of an x-polarized LP_{lm} mode is

$$\mathbf{E}(r, \phi, 0) = \hat{\mathbf{x}} e_l(r) \cos l\phi \qquad (10.3)$$

Although $e_l(r)$ has been given in (9.36) and (9.37), it is repeated here for convenience:

$$e_l(r) = \begin{cases} E_l \dfrac{J_l \left(V \sqrt{1 - b} \dfrac{r}{a} \right)}{J_l(V \sqrt{1 - b})} & 0 \le r \le a \\[4mm] E_l \dfrac{K_l \left(V \sqrt{b} \dfrac{r}{a} \right)}{K_l(V \sqrt{b})} & r \ge a \end{cases} \qquad (10.4)$$

The field radiated by an x-polarized field is also polarized in the x direction. Thus, it is only necessary to consider the x component of the field. To find E_{FFx}, we substitute (10.3) and (10.4) into (10.2) and make use of two identities [4, 5]:

$$e^{jx \cos \phi} = J_0(x) + 2 \sum_{p=1}^{\infty} J_p(x) \cos p\phi \qquad (10.5)$$

$$\int_0^{2\pi} \cos l\phi \cos p(\Phi - \phi) \, d\phi = \begin{cases} 0 & l \ne p \\ 2\pi & l = p = 0 \\ \pi \cos l\Phi & l = p \ne 0 \end{cases} \qquad (10.6)$$

and obtain

$$E_{\text{FF}x}(R, \Theta, \Phi) = jk \frac{e^{-jkR}}{R} \cos l\Phi \int_0^\infty e_l(r) J_l(kr \sin \Theta) r \, dr \qquad (10.7)$$

To evaluate the integral analytically, we partition the integral into two parts:

$$\frac{E_l}{J_l(V\sqrt{1-b})} \int_0^a J_l\left(V\sqrt{1-b}\,\frac{r}{a}\right) J_l(kr \sin \Theta) r \, dr \qquad (10.8)$$

and

$$\frac{E_l}{K_l(V\sqrt{b})} \int_a^\infty K_l\left(V\sqrt{b}\,\frac{r}{a}\right) J_l(kr \sin \Theta) r \, dr \qquad (10.9)$$

Each integral can be evaluated in a closed form. This can be done by making use of the dispersion relation (9.40), the recurrence relations of Bessel functions, and the two integral identities of Bessel functions [4, 5]:

$$\int t J_l(pt) J_l(qt) \, dt = \frac{t}{p^2 - q^2} [p J_{l+1}(pt) J_l(qt) - q J_l(pt) J_{l+1}(qt)] \qquad (10.10)$$

$$\int t J_l(pt) K_l(qt) \, dt = \frac{t}{p^2 + q^2} [p J_{l+1}(pt) K_l(qt) - q J_l(pt) K_{l+1}(qt)] \qquad (10.11)$$

When the integrals in (10.8) and (10.9) are evaluated and combined, we obtain [6–8]:

$$E_{\text{FF}x}(R, \Theta, \Phi) = -j E_l \frac{e^{-jkR}}{kR} (kaV)^2 \cos l\Phi F_l(\Theta) \qquad (10.12)$$

where

$$F_l(\Theta) = \frac{ka \sin \Theta J_{l+1}(ka \sin \Theta) - V\sqrt{1-b}\,\dfrac{J_{l+1}(V\sqrt{1-b})}{J_l(V\sqrt{1-b})} J_l(ka \sin \Theta)}{[V^2(1-b) - k^2a^2 \sin^2 \Theta][V^2b + k^2a^2 \sin^2 \Theta]} \qquad (10.13)$$

for $ka \sin \Theta \neq V\sqrt{1-b}$. The limiting value of $F_l(\Theta)$ at $ka \sin \Theta = V\sqrt{1-b}$ is

$$F_l(\Theta)\big|_{ka \sin \Theta = V\sqrt{1-b}}$$

$$= \frac{2l J_l(V\sqrt{1-b}) J_{l+1}(V\sqrt{1-b}) - V\sqrt{1-b}\,[J_l^2(V\sqrt{1-b}) + J_{l+1}^2(V\sqrt{1-b})]}{2V^3\sqrt{1-b} J_l(V\sqrt{1-b})}$$

$$(10.14)$$

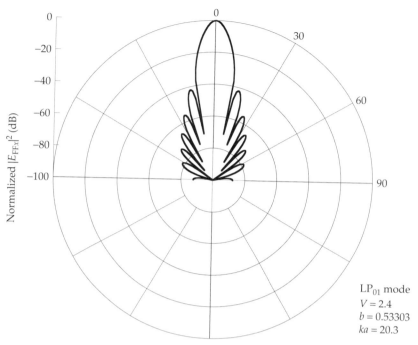

Figure 10.3   Radiation pattern of field radiated by LP_{01} mode.

Except for a multiplicative constant, $\cos l\Phi \, F_l(\Theta)$ is the *field pattern* and $|\cos l\Phi \, F_l(\Theta)|^2$ is the *power* or *irradiance pattern* of the radiated field in the far-field zone. The azimuthal variation of the irradiance pattern is simply $\cos^2 l\Phi$. The irradiance pattern varies with Θ in a complicated manner. In Figure 10.3, we present a polar plot of the normalized irradiance pattern radiated by an LP_{01} mode with $V = 2.4$ and $ka = 20.3$. As noted earlier, the radiated field is mainly in a small cone in the forward direction. In addition, the side lobes are very low. In fact, the largest side lobe is at least −40 dB below the main peak. It is more informative to plot the irradiance pattern as a function of $ka \sin \Theta$ instead of Θ. The normalized irradiance patterns radiated by LP_{01}, LP_{11}, and LP_{21} modes are presented in Figures 10.4–10.6 as functions of $ka \sin \Theta$ for several values of V. Each curve is normalized with respect to its maximum. From these figures, we note:

1. The main peak of the irradiance pattern of LP_{0m} mode is at $\Theta = 0$.
2. The irradiance pattern of LP_{lm} mode with $l \geq 1$ has a null, instead of a peak, at $\Theta = 0$. This can be understood as follows. In a transverse plane, the field of the LP_{lm} mode varies like $\cos l\phi$. Thus, contribution to the radiated field for all points on the z axis from points of various parts of the truncated fiber tip cancel exactly.
3. The main peak of the irradiance pattern of the LP_{lm} mode with $l \geq 1$ is slightly off the z axis. The angular position of the main peak depends on V and the mode.

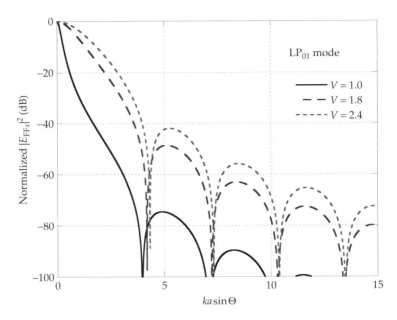

Figure 10.4   Normalized irradiance pattern radiated by LP$_{01}$ mode as a function of $ka \sin \Theta$.

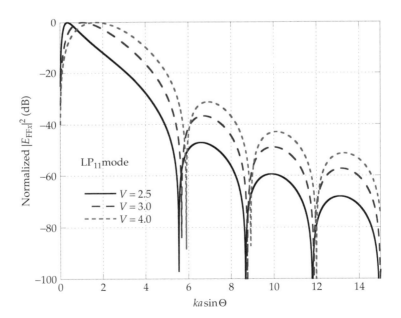

Figure 10.5   Normalized irradiance pattern radiated by LP$_{11}$ mode as a function of $ka \sin \Theta$.

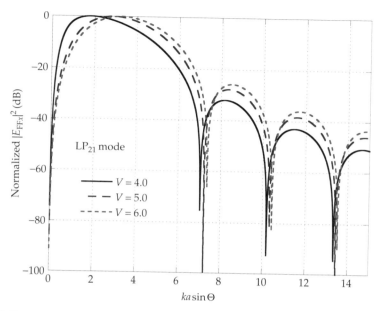

Figure 10.6 Normalized irradiance pattern radiated by LP$_{21}$ mode as a function of $ka \sin \Theta$.

4. After reaching the main peak, $|E_{FFx}|^2$ drops off precipitously for all LP modes.

5. The side lobes depend on the V and the mode. For the fundamental LP mode, the side lobe level is about -50 to -40 dB below the main peak. For higher-order LP modes, the side lobes are larger. For all LP modes, the side-lobe level rises as V increases.

10.2.2 Radiation by a Gaussian Aperture Field

As noted in Chapter 9, the transverse field of the LP$_{01}$ mode has a peak on the fiber axis and the field decays monotonically as r increases. This description fits the Gaussian function in r as well. In fact, for V greater than 2.0, the transverse electric field of the LP$_{01}$ mode may be accurately approximated as a Gaussian function [9]. Therefore it is instructive to compare the field radiated by the LP$_{01}$ mode with that by a Gaussina aperture field. For this purpose, we suppose the aperture field at $z = 0$ plane is the Gaussian function

$$\mathbf{E}_{GB}(r, \phi, 0) = \hat{\mathbf{x}} E_0 e^{-r^2/w^2} \tag{10.15}$$

where w is the beam radius of the Gaussian field. The field radiated by such a Gaussian aperture field in the Fraunhofer zone is obtained by substituting (10.15) into (10.2) and performing the requisite integration. The integral can be evaluated in a closed form with the aid of an integral identity [5]:

$$\int_0^\infty J_0(qt) e^{-p^2 t^2} t \, dt = \frac{1}{2p^2} e^{-q^2/4p^2} \qquad |\arg(p)| < \frac{\pi}{4} \tag{10.16}$$

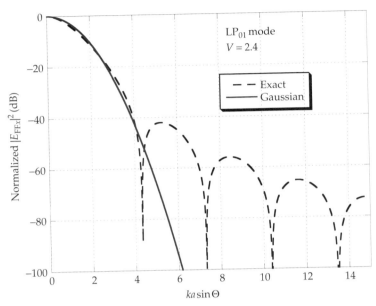

Figure 10.7 Normalized irradiance pattern radiated by LP$_{01}$ mode as compared with that of a Gaussian field.

The result is surprisingly simple:

$$E_{\text{GB FF}x}(R, \Theta, \Phi) = jk\, E_0 \frac{e^{-jkR}}{R} \frac{w^2}{2} e^{-(kw \sin \Theta)^2/4} \qquad (10.17)$$

Note that both the aperture field (10.15) at the $z = 0$ plane and the radiated field (10.17) in the Fraunhofer zone are Gaussian functions. In Figure 10.7, we plot $|E_{\text{GB FF}x}|^2$ as a function of $kw \sin \Theta$ and compare it with the irradiance pattern of the field produced by an LP$_{01}$ mode. We note that

1. $|E_{\text{GB FF}x}|^2$ has a peak at $\Theta = 0$.
2. The radiated field decreases monotonically and it has no side lobe.
3. $|E_{\text{GB FF}x}|^2$ drops to $1/e^2$ of the peak value at $kw \sin \Theta = 2$. It is convenient to define $\Theta_{\text{H}} = \sin^{-1}(2/kw)$ as the *half-power angular width* of the radiated field. Thus, by measuring the half-power angular beam width Θ_{H} in the Fraunhofer zone, we can infer the beam radius w of the Gaussian aperture field.
4. For $2.0 < V < 2.8$, the main lobes of $|E_{\text{GB FF}x}|^2$ and $|E_{\text{FF}x}|^2$ agree very well, as shown in Figure 10.7.

10.2.3 Experimental Determination of *ka* and *V*

As indicated in the last section, it is possible to estimate the effective radius of a Gaussian aperture field by measuring the half-power angular width of the radiated field in the Fraunhofer zone. It would be interesting to inquire if a similar procedure also applies to the field radiated by the LP mode. In this subsection, we will show that it is indeed possible to estimate V, ka, and Δ of a step-index fiber from the far-field measurements [7, 8].

To establish the basis of the experimental procedure, we plot a typical normalized irradiance pattern of the LP_{01} mode in Figure 10.8. Two angles Θ_H and Θ_N are identified. Θ_N is the angle of the null nearest to the main peak. Mathematically, it given by the first root of $F_0(\Theta)$. It follows from (10.13) that $F_0(\Theta)$ vanishes if

$$ka \sin \Theta \frac{J_1(ka \sin \Theta)}{J_0(ka \sin \Theta)} = V\sqrt{1-b} \frac{J_1(V\sqrt{1-b})}{J_0(V\sqrt{1-b})} \tag{10.18}$$

provided $ka \sin \Theta \neq V\sqrt{1-b}$. The right-hand side of (10.18) depends on V and b. But b itself is also a function of V. In other words, the right-hand side of (10.18) is a function of V. By solving $kw \sin \Theta$ from (10.18) numerically for a given V and label the angle Θ so obtained as Θ_N, we obtain $ka \sin \Theta_N$ as a function of V. For convenience, we write

$$ka \sin \Theta_N = G_N(V) \tag{10.19}$$

Next, we consider the *half-power angle* Θ_H. As noted earlier, the peak of the radiated power of an LP_{01} mode is at $\Theta = 0$. Thus, the peak irradiance is proportional to $[F_0(0)]^2$. At the half-power angle Θ_H, the irradiance reduces to half of the peak value

$$[F_0(\Theta_H)]^2 = \tfrac{1}{2}[F_0(0)]^2 \tag{10.20}$$

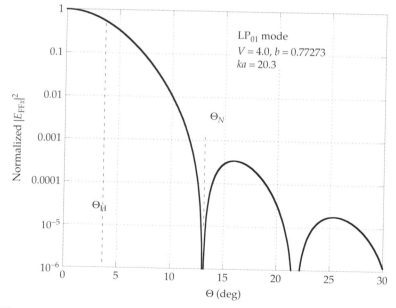

Figure 10.8 Normalized irradiance pattern of LP_{01} mode as a function of Θ.

For a given fiber, we can solve for $ka \sin \Theta_H$ numerically. Clearly $ka \sin \Theta_H$ depends only on V and we write

$$ka \sin \Theta_H = G_H(V) \tag{10.21}$$

Although Θ_H and Θ_N depend on V and ka, the ratio of $\sin \Theta_H / \sin \Theta_N$ is independent of ka. Namely

$$\frac{\sin \Theta_H}{\sin \Theta_N} = \frac{ka \sin \Theta_H}{ka \sin \Theta_N} = \frac{G_H(V)}{G_N(V)} \tag{10.22}$$

In Figure 10.9 we plot $ka \sin \Theta_H$ and $ka \sin \Theta_N$ and the ratio $\sin \Theta_N / \sin \Theta_H$ as functions of V. Equation (10.22), or equivalently Figure 10.9, is the basis for an experimental procedure for determining V and ka. By measuring the *nearest null angle* Θ_N and the half-power angle Θ_H experimentally, we determine V from (10.22) or Figure 10.9. Once V is known, we deduce ka from (10.19) or (10.21). From the values of ka and V so determined, we infer Δ. If the wavelength λ is known by some means, we can estimate the core radius as well. The procedure was first introduced by Gambling et al. [7] for the LP$_{01}$ mode. The same procedure can also be modified for the LP$_{11}$ mode guided by using a different set of curves for G_H, G_N, and G_H / G_N [8].

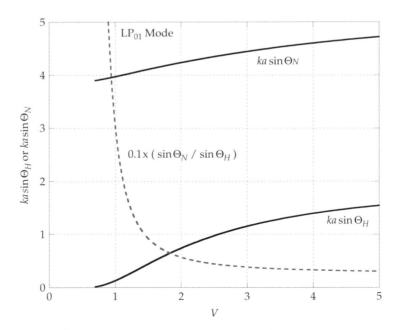

Figure 10.9 $ka \sin \Theta_N$ and $ka \sin \Theta_H$ as functions of V.

10.3 EXCITATION OF LP MODES

10.3.1 Power Coupled to LP Mode

Now we consider the excitation problem. We consider waves impinging upon a truncated fiber tip normally as shown in Figure 10.1(b). The medium to the left of the truncated fiber is taken as air. For $z \leq 0$, the incident wave is

$$\mathbf{E}_{\text{in}}(r, \phi, z) = (\mathbf{e}_{\text{in}\,t} + \hat{z}e_{\text{in}\,z})e^{-jkz}$$
$$\mathbf{H}_{\text{in}}(r, \phi, z) = (\mathbf{h}_{\text{in}\,t} + \hat{z}h_{\text{in}\,z})e^{-jkz} \tag{10.23}$$

The total incident power is

$$
\begin{aligned}
P_{\text{in}} &= \frac{1}{2} \int_0^{2\pi} \int_0^\infty \text{Re}(\mathbf{E}_{\text{in}} \times \mathbf{H}_{\text{in}}^*) \cdot \hat{z} \, r \, dr \, d\phi \\
&= \frac{1}{2} \int_0^{2\pi} \int_0^\infty \text{Re}(\mathbf{e}_{\text{in}\,t} \times \mathbf{h}_{\text{in}\,t}^*) \cdot \hat{z} \, r \, dr \, d\phi \tag{10.24}
\end{aligned}
$$

Due to the presence of the truncated fiber, a portion of the incident wave is reflected. The reflected wave propagates in the $-z$ direction and is given by

$$\mathbf{E}_{\text{rf}}(r, \phi, z) = (\mathbf{e}_{\text{rf}\,t} + \hat{z}e_{\text{rf}\,z})e^{+jkz}$$
$$\mathbf{H}_{\text{rf}}(r, \phi, z) = (\mathbf{h}_{\text{rf}\,t} + \hat{z}h_{\text{rf}\,z})e^{+jkz} \tag{10.25}$$

We are interested in the LP modes excited by the incident field. For simplicity, we use a single subscribe v, instead of two subscribes l and m, to designate an LP mode. We assume that the modal fields $\mathbf{e}_v$ and $\mathbf{h}_v$ and the propagation constants β_v of mode v are known. Our task is to determine the amplitude of the LP mode excited in the fiber. Let the field in the step-index fiber launched by the incident wave be

$$\mathbf{E}_{\text{fb}}(r, \phi, z) = \sum_v C_v(\mathbf{e}_{vt} + \hat{z}e_{vz})e^{-j\beta_v z}$$
$$\mathbf{H}_{\text{fb}}(r, \phi, z) = \sum_v C_v(\mathbf{h}_{vt} + \hat{z}h_{vz})e^{-j\beta_v z} \tag{10.26}$$

where C_v is the mode amplitude. The summation covers all possible LP modes and the radiation mode, if excited. Power coupled into mode v is

$$P_v = \tfrac{1}{2}|C_v|^2 \int_0^{2\pi} \int_0^\infty \text{Re}(\mathbf{e}_{vt} \times \mathbf{h}_{vt}^*) \cdot \hat{z} \, r \, dr \, d\phi \tag{10.27}$$

To determine C_ν, we make use of the continuity of the tangential components of electric and magnetic fields at the truncated fiber surface $z = 0$ and obtain

$$\mathbf{e}_{\text{in}\,t} + \mathbf{e}_{\text{rf}\,t} = \sum_\nu C_\nu \mathbf{e}_{\nu\,t} \tag{10.28}$$

$$\mathbf{h}_{\text{in}\,t} + \mathbf{h}_{\text{rf}\,t} = \sum_\nu C_\nu \mathbf{h}_{\nu\,t} \tag{10.29}$$

In the above equations, the incident field $\mathbf{e}_{\text{in}}$ and $\mathbf{h}_{\text{in}}$ and modal field $\mathbf{e}_\nu$ and $\mathbf{h}_\nu$ of the LP mode are known. But the reflected fields $\mathbf{e}_{\text{rf}}$ and $\mathbf{h}_{\text{rf}}$ and mode amplitudes C_ν are yet unknown. To determine C_ν, we make a vector product of both sides of (10.28) with $\mathbf{h}_{\nu t}^*$, integrate the resulting expression over the entire fiber cross section, apply the orthogonality relation, and obtain

$$C_\nu = \frac{\int_0^{2\pi} \int_0^\infty (\mathbf{e}_{\text{in}\,t} + \mathbf{e}_{\text{rf}\,t}) \times \mathbf{h}_{\nu\,t}^* \cdot \hat{\mathbf{z}}\, r\, dr\, d\phi}{\int_0^{2\pi} \int_0^\infty \mathbf{e}_{\nu\,t} \times \mathbf{h}_{\nu\,t}^* \cdot \hat{\mathbf{z}}\, r\, dr\, d\phi} \tag{10.30}$$

But $\mathbf{e}_{\text{rf}}$ is yet unknown. Thus, expression (10.30), while exact, is not useful as it stands. For weakly guiding fibers, n_{co} and n_{cl} are nearly the same. Besides, the core region is much smaller than the cladding region. Thus we expect the reflected field is essentially the same as the waves reflected by a large planar boundary separating air and a dielectric medium with an index n_{cl}. In other words, we may approximate $\mathbf{e}_{\text{rf}\,t}$ as

$$\mathbf{e}_{\text{rf}\,t} \approx \frac{1 - n_{\text{cl}}}{1 + n_{\text{cl}}} \mathbf{e}_{\text{in}\,t} \tag{10.31}$$

Making use of this approximation, we obtain

$$C_\nu \approx \frac{2}{1 + n_{\text{cl}}} \frac{\int_0^{2\pi} \int_0^\infty \mathbf{e}_{\text{in}\,t} \times \mathbf{h}_{\nu\,t}^* \cdot \hat{\mathbf{z}}\, r\, dr\, d\phi}{\int_0^{2\pi} \int_0^\infty \mathbf{e}_{\nu\,t} \times \mathbf{h}_{\nu\,t}^* \cdot \hat{\mathbf{z}}\, r\, dr\, d\phi} \tag{10.32}$$

Since $\mathbf{e}_{\text{in}}$ and $\mathbf{h}_\nu$ are known, the integrals in (10.32) can be evaluated, and C_ν can be determined. Once C_ν is known, we have the percentage of power coupled into mode ν. The result is

$$\eta_\nu = \frac{P_\nu}{P_{\text{in}}}$$

$$\approx \frac{4}{(1 + n_{\text{cl}})^2} \frac{\left(\int_0^{2\pi} \int_0^\infty \mathbf{e}_{\text{in}\,t} \times \mathbf{h}_{\nu\,t}^* \cdot \hat{\mathbf{z}}\, r\, dr\, d\phi \right)^2}{\left(\int_0^{2\pi} \int_0^\infty \mathbf{e}_{\nu\,t} \times \mathbf{h}_{\nu\,t}^* \cdot \hat{\mathbf{z}}\, r\, dr\, d\phi \right) \left(\int_0^{2\pi} \int_0^\infty \mathbf{e}_{\text{in}\,t} \times \mathbf{h}_{\text{in}\,t}^* \cdot \hat{\mathbf{z}}\, r\, dr\, d\phi \right)}$$

$$\tag{10.33}$$

10.3.2 Gaussian Beam Excitation

Suppose the input to the truncated fiber is an x-polarized Gaussian beam. We assume that the Gaussian beam aligns perfectly with the fiber axis, the beam waist is at $z = 0$, and the waist radius is w_0. Thus the input field at $z = 0$ is

$$\mathbf{e}_{\text{in}} = \hat{\mathbf{x}} E_{\text{in}} e^{-r^2/w_0^2} \tag{10.34}$$

$$\mathbf{h}_{\text{in}} = \hat{\mathbf{y}} \frac{E_{\text{in}}}{\eta_o} e^{-r^2/w_0^2} \tag{10.35}$$

Since the input field and the fiber geometry are rotationally symmetric, the mode excited in the fiber must also be rotational by symmetric. Therefore, we consider LP_{0m} modes only. The field of LP_{0m} mode in a weakly guiding step-index fiber is

$$\mathbf{e}_{0t} = \hat{\mathbf{x}} e_0(r) \tag{10.36}$$

$$\mathbf{h}_{0t} \approx \hat{\mathbf{y}} \frac{N}{\eta_o} e_0(r) \tag{10.37}$$

where $e_0(r)$ is given in (10.4).

Substituting (10.34)–(10.37) into (10.33), evaluating the integral with respect to ϕ, we obtain an expression for the percentage power fed into the LP_{0m} mode:

$$\eta_{\text{LP}0m} = \frac{P_{\text{LP}0m}}{P_{\text{in}}} \approx \frac{4N}{(1+n_{\text{cl}})^2} \frac{\left(\int_0^\infty e^{-r^2/w_0^2} e_0(r)\, r\, dr\right)^2}{\left(\int_0^\infty e^{-2r^2/w_0^2}\, r\, dr\right)\left(\int_0^\infty e_0^2(r)\, r\, dr\right)} \tag{10.38}$$

The first factor of (10.38) is essentially the *power transmission coefficient*, or *transmittance*, of plane waves incident upon a planar boundary between two dielectric media. The second factor is the *launching efficiency*:

$$T_{\text{LP}0m} = \frac{\left(\int_0^\infty e^{-r^2/w_0^2} e_0(r)\, r\, dr\right)^2}{\left(\int_0^\infty e^{-2r^2/w_0^2}\, r\, dr\right)\left(\int_0^\infty e_0^2(r)\, r\, dr\right)} \tag{10.39}$$

The numerator is an integral of the product of the input field and that of the LP_{0m} mode. Often, it is referred to as the *overlap integral*.

As a simple exercise of (10.39), we approximate the modal field $e_0(r)$ by a Gaussian function with a beam waist radius w_{eff},

$$e_{\text{GB}}(r) = E_0 e^{-r^2/w_{\text{eff}}^2} \qquad 0 \le r \le \infty \tag{10.40}$$

Substituting (10.40) into (10.39) and evaluating the integrals, we obtain

$$T_{\text{GB}} = \frac{4w_0^2 w_{\text{eff}}^2}{(w_0^2 + w_{\text{eff}}^2)^2} \tag{10.41}$$

Clearly, the launching efficiency depends on the two waist radii, w_0 and w_{eff}. The maximum launching efficiency is 1.0 and it is reached when $w_{\text{eff}} = w_0$.

When the exact expression for $e_0(r)$ as given in (10.4) is used in the expression (10.39), the integral has to be evaluated numerically. Marcuse showed that the launching efficiency depends on V and w_0 [9, 10]. For the LP_{01} mode with a given V, the launching efficiency may be maximized by varying w_0. Based on the numerical calculation, Marcuse obtained an empirical equation for the optimum Gaussian beam waist radius w_0 for the maximum launching efficiency. The empirical equation is [9, 10]

$$\frac{w_0}{a} \approx 0.65 + \frac{1.619}{V^{3/2}} + \frac{2.879}{V^6} \tag{10.42}$$

The maximum launching efficiency so obtained is plotted in Figure 10.10 as function of V. It is worthy noting:

1. For $V > 1.2$, the maximum launching efficiency is greater than 0.944.
2. For $V = 2.4$, very close to the cutoff of the LP_{11} mode, the maximum launching efficiency is 0.9965.
3. The maximum launching efficiency reaches its peak value when $V = 2.8$.

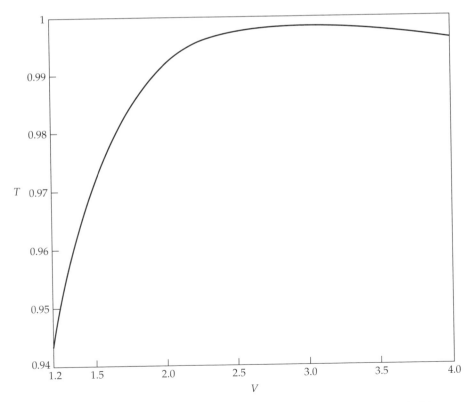

Figure 10.10   Maximum launching efficiency as a function of V. (From [10].)

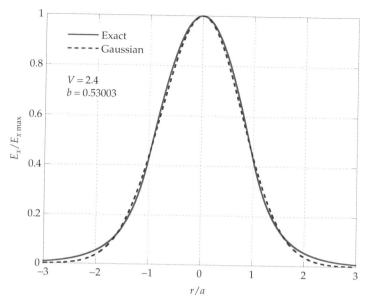

Figure 10.11 Field distributions of LP_{01} mode and Gaussian beams.

Note that the maximum launching efficiency is extremely close to unity for V between 2.0 and 2.8. In other words, the transverse field of the LP_{01} mode matches very well with that of a Gaussian beam. In Figure 10.11, we plot and compare the field of $e_0(r)$ of the LP_{01} mode with $V = 2.4$ and that of a Gaussian beam with a beam waist given by (10.42). It clearly shows that the modal field of the LP_{01} mode can be accurately represented by a Gaussian function with a waist radius given by (10.42).

PROBLEMS

1. Starting from Fresnel–Kirchhoff diffraction equation (10.1), derive an expression of the field radiated by LP_{11} mode guided by a two-mode or few-mode fiber and cast the equation in terms of V, b, and $ka \sin \Theta$. Compare your result to (6) of Ref. [8].

2. Two angles Θ_0 and Θ_M have been mentioned in Ref. [8]. Rename the two angles as Θ_0 and Θ_M and plot $\sin \Theta_0 / \sin \Theta_M$ as a function of V. Compare your curve to Figure 3 of Ref. [8].

REFERENCES

1. M. Born and E. Wolf, *Principles of Optics*, 6th ed., Pergamon Press, Oxford, UK, 1980, Section 9.3.

2. R. E. Collin, *Field Theory of Guided Waves*, 2nd ed., IEEE Press, New York, 1991.

3. S. Ramo, J. R. Whinnery, and T. Van Duzer, *Fields and Waves in Communication Electronics*, 3rd ed., Wiley, New York, 1994.

4. L. Luke, *Integrals of Bessel Functions*, McGraw-Hill Book, New York, 1962.

5. M. Abramowitz and I. A. Stegun, *Handbook of Mathematical Functions with Formulas, Graphs and Mathematical Tables*, Dover, New York, 1965.

6. W. A. Gambling, D. N. Payne, H. Matsumura, and R. B. Dyott, "Determination of core diameter and refractive-index difference of single-mode fibres by observation of the fair-field pattern," *IEE J. Microwaves, Optics Acoustics*, Vol. 1, No. 1, pp. 13–17 (Sept. 1976).

7. W. A. Gambling, D. N. Payne, and H. Matsumura, "Routine characterization of single mode fibres," *Electron. Lett.*, Vol. 12, No. 21, pp. 546–547 (1976).

8. P. Pocholle, "Single mode optical fiber characterization by the LP_{11} mode radiation pattern," *Opt. Comm.*, Vol. 31, No. 2, pp. 143–147 (Nov. 1979).

9. D. Marcuse, "Gaussian approximation of the fundamental modes of graded-index fibers," *J. Opt. Soc. Am.*, Vol. 68, No. 1, pp. 103–109 (Jan. 1978).

10. D. Marcuse, "Loss analysis of single-mode fiber splices," *Bell Syst. Tech. J.*, Vol. 56, No. 5, pp. 703–718 (May–June 1977).

11

BIREFRINGENCE IN SINGLE-MODE FIBERS

11.1 INTRODUCTION

An ideal fiber has a perfectly circular cross section, a truly rotationally symmetric index profile, and is free of mechanical, electric, and magnetic disturbances. But such an ideal fiber does not exist. Inevitably, imperfections are introduced in the manufacturing processes. As a result, the fiber cross section is not perfectly circular. Nor is the index profile truly rotationally symmetric. Even if an ideal fiber were made somehow, perturbation may be introduced in the coating, cabling, and other postfabrication processes. In addition, fibers may be pressed, bent, or twisted during the installation or in experiments. In the presence of imperfections and disturbances, the fibers cease to be ideal.

An ideal single-mode fiber would support two *degenerate* and *orthogonal polarization modes*. All real single-mode fibers support two *nearly degenerate polarization modes* and should be labeled as *single-mode double-polarization fibers*. Although single-mode single-polarization fibers have been proposed [1], most single-mode fibers currently in use, or to be used in the near future, are single-mode double-polarization fibers. It is customary to refer single-mode double-polarization fibers as *single-mode fibers*. In these single-mode fibers, the two propagation constants of the two polarization modes are slightly different. Therefore, single-mode

Foundations for Guided-Wave Optics, by Chin-Lin Chen
Copyright © 2007 John Wiley & Sons, Inc.

fibers are birefringent. The physical origins of the fiber birefringence may be traced
to the *noncircular cross section, built-in stress, clamping, bending, twisting, axis
rotation*, or other *externally applied stress* or *fields* [2–6]. These topics are discussed in Sections 11.2–11.5. In Section 11.6, we introduce the Jones matrices to
describe the effects of birefringent fibers on the state of polarization of propagating
waves under various conditions.

Suppose the propagation constants of two linearly polarized modes are β_x and
β_y. If $\beta_x \neq \beta_y$, the fiber is *linearly birefringent*. The difference $\beta_x - \beta_y$ is known as
the *linear birefringence*, or simply *birefringence*. If $\beta_y > \beta_x$, the phase velocity of
the x-polarized mode is faster than that of the y-polarized mode. Then the x axis is
the *fast axis* of the birefringent fiber. On the other hand, if $\beta_y < \beta_x$, the y axis is the
fast axis. To compare the birefringence of various fibers at different wavelengths,
we introduce the *modal birefringence* or the *normalized birefringence*:

$$B = \frac{\beta_x - \beta_y}{k} \tag{11.1}$$

Or we use the *polarization beat length* or simply the *beat length*, to quantify
the fiber birefringence:

$$L_B = \frac{2\pi}{|\beta_x - \beta_y|} = \frac{\lambda}{|B|} \tag{11.2}$$

A single-mode fiber also supports two circularly polarized modes that propa-
gate with two propagation constants β_R and β_L. If $\beta_R \neq \beta_L$, the fiber is *circularly
birefringent*. We label $\beta_R - \beta_L$ as the *circular birefringence*.

Depending on the linear birefringence, single-mode fibers may be classified as
the conventional single-mode fibers, high-birefringence fibers, and low-birefringence
fibers [6]. For the conventional single-mode fibers, the modal birefringence is between 10^{-6} and 10^{-5}. Fibers with a modal birefringence greater than 10^{-5} are
referred to as *high-birefringence fibers*. They are also known as *polarization main-
taining fibers*. Fibers having a modal birefringence less than 10^{-6} are *low-birefringence fibers*. In the visible and near-infrared spectra, the polarization beat length
of the conventional single-mode fibers ranges from a few centimeters to a few
meters [2, 5, 7]. For *high-birefringence fibers*, the beat length may be as short as
a few millimeters. In all cases, $|\beta_x - \beta_y|$ is much smaller than β_x and β_y. Take a
fiber with a polarization beat length of 10 cm operating at 1.55 µm as an example.
$|\beta_x - \beta_y|$ is 62.8 m^{-1}. The propagation constants β_x and β_y are on the order of
10^7 m^{-1}. Although the difference $|\beta_x - \beta_y|$ is small in comparison with β_x or β_y,
the fiber length L is many wavelengths long. Thus $|\beta_x - \beta_y|L$ is a large number.
Therefore, the birefringent effects are considerable. The phase or polarization fluctu-
ation due to fiber birefringence must be compensated for if information contained in
the pulses is to be preserved. In short, the fiber birefringence is of practical interest.

The birefringence due to applied stress and electric and magnetic fields may
be understood or modeled in terms of the perturbation on the core index. Consider
a weakly guiding fiber with core and cladding indices n_{co} and n_{cl}. For a weakly

guiding fiber, the effective index of refraction N may be approximated as

$$N \approx n_{cl} + b(n_{co} - n_{cl}) \qquad (11.3)$$

where b is the generalized guide index defined in (9.7). If for some reason, the core index changes by Δn_{co}, the effect on the effective index is

$$\Delta N \approx \left(b + \frac{V}{2} \frac{db}{dV} \right) \Delta n_{co} \qquad (11.4)$$

If a fiber is perturbed such that the effective refractive indices "seen" by x- and y-polarized modes are different, then, ΔN of the x- and y-polarized modes are also different. The resulting modal birefringence and beat length are

$$B \approx \left(b + \frac{V}{2} \frac{db}{dV} \right) |\Delta n_{cox} - \Delta n_{coy}| \qquad (11.5)$$

$$L_B \approx \frac{\lambda}{\left(b + \frac{V}{2} \frac{db}{dV} \right) |\Delta n_{cox} - \Delta n_{coy}|} \qquad (11.6)$$

In the following sections, we use these equations to estimate fiber birefringence caused by external stress or perturbation. For $1.8 < V < 2.4$, $b + (V/2)(db/dV)$ is between 0.67 and 0.82 as shown in Figure 11.1. We may simply take $b + (V/2)(db/dV)$ as 1 if a simple estimate of the modal birefringence and beat length is desired. Then the modal birefringence and beat length are simply $\Delta n_{cox} - \Delta n_{coy}$ and $\lambda/|\Delta n_{cox} - \Delta n_{coy}|$.

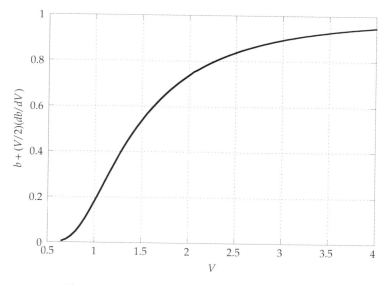

Figure 11.1 $b + (V/2)(db/dV)$ as a function of V.

11.2 GEOMETRICAL BIREFRINGENCE OF SINGLE-MODE FIBERS

Due to variations in the manufacturing processes, the fiber cross section may not
be perfectly circular. For example, the core–cladding concentricity of typical communication grade fibers is within ± 1.0 μm [8], but not zero. In short, a real fiber
has a nominally, not perfectly, circular cross section. A nominally circular cross
section can be viewed as an elliptical cross section. In some specialty fibers, the
core and/or cladding are made elliptical on purpose. Birefringence due to the noncircular cross section is referred to as the *geometrical, form*, or *shape birefringence*.
It is possible to analyze the propagation and dispersion characteristics of elliptical core fibers rigorously. But the results, while exact, are very complicated [4, 6,
9, 10]. In this section, we follow Ramaswamy and French's work and look for
an order-of-magnitude estimation of the form birefringence of nominally circular
fibers [11].

Suppose we are interested in an elliptical core fiber having a major axis $2a_{mj}$
and a minor axis $2a_{mn}$ as shown in Figure 11.2(a). As an approximation, we treat
the elliptical core as a rectangular core of the dimension $2a_{mj} \times 2a_{mn}$ shown in
Figure 11.2(b). For a nominally circular core, a_{mj} is only slightly greater than a_{mn}.
In other words, the aspect ratio of the rectangular core is nearly 1. We apply the
Marcatili method discussed in Chapter 5 to determine propagation constants β_x and
β_y of E_{11}^x and E_{11}^y modes guided by the rectangular waveguide, and use β_x and

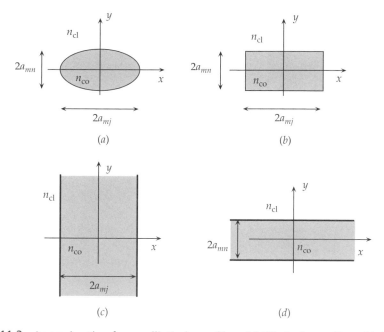

Figure 11.2 Approximation for an elliptical core fiber: (*a*) Elliptical core fiber, (*b*) dielectric
waveguide with rectangular core, (*c*) thin-film waveguide of width $2a_{mj}$, and (*d*) thin-film
waveguide of thickness $2a_{mm}$.

β_y so determined to estimate the geometrical birefringence of the elliptical core fiber. Admittedly, a rectangular core is a poor approximation for an elliptical core. Besides, the Marcatili method itself is more accurate for waveguides with a large core aspect ratio. Despite the shortcoming, we prefer Ramaswamy and French's [11] approach because it is intuitive and simple. It also gives the correct dependence of ellipticity and index difference.

To determine the propagation constant of the E_{11}^y mode guided by the rectangular core waveguide shown in Figure 11.2(b), we begin with the dispersion relation for the TE modes guided by a symmetric thin-film waveguide of width $2a_{mj}$ shown in Figure 11.2(c):

$$\kappa_x 2a_{mj} = \pi - 2\tan^{-1} \frac{\kappa_x}{\sqrt{k^2(n_{co}^2 - n_{cl}^2) - \kappa_x^2}} \tag{11.7}$$

An exact and analytic solution of κ_x from (11.7) is not known. We look for an approximate expression for κ_x. To evaluate κ_x approximately, we assume that the E_{11}^y mode in question is far from the cutoff. Then $k^2(n_{co}^2 - n_{cl}^2)$ is much greater than κ_x^2, and (11.7) may be approximated as

$$\kappa_x 2a_{mj} \approx \pi - \frac{2\kappa_x}{k\sqrt{n_{co}^2 - n_{cl}^2}} \tag{11.8}$$

A simple approximate expression for κ_x is now obvious:

$$\kappa_x \approx \frac{\pi k \sqrt{n_{co}^2 - n_{cl}^2}}{2(V_{mj} + 1)} \tag{11.9}$$

where $V_{mj} = ka_{mj}\sqrt{n_{co}^2 - n_{cl}^2}$.

Similarly, we determine κ_y from the dispersion relation of the TM mode guided by the waveguide of thickness $2a_{mn}$ depicted in Figure 11.2(d):

$$\kappa_y 2a_{mn} = \pi - 2\tan^{-1} \frac{n_{cl}^2 \kappa_y}{n_{co}^2 \sqrt{k^2(n_{co}^2 - n_{cl}^2) - \kappa_y^2}} \tag{11.10}$$

An approximate expression for κ_y is

$$\kappa_y \approx \frac{\pi k \sqrt{n_{co}^2 - n_{cl}^2}}{2[V_{mn} + (n_{cl}^2/n_{co}^2)]} \tag{11.11}$$

where $V_{mn} = ka_{mn}\sqrt{n_{co}^2 - n_{cl}^2}$. Following Marcatili's prescription, the propagation constant of E_{11}^y is

$$\beta_y = [k^2 n_{co}^2 - (\kappa_x^2 + \kappa_y^2)]^{1/2} \tag{11.12}$$

From (11.9), (11.11), and (11.12), we obtain

$$\beta_y \approx kn_{co} - \frac{\pi^2}{8} \frac{k(n_{co}^2 - n_{cl}^2)}{n_{co}} \left\{ \frac{1}{(V_{mj} + 1)^2} + \frac{1}{[V_{mn} + (n_{cl}^2/n_{co}^2)]^2} \right\} \qquad (11.13)$$

Following the same procedure, we obtain an approximate expression for the propagation constant of E_{11}^x modes:

$$\beta_x \approx kn_{co} - \frac{\pi^2}{8} \frac{k(n_{co}^2 - n_{cl}^2)}{n_{co}} \left\{ \frac{1}{[V_{mj} + (n_{cl}^2/n_{co}^2)^2]} + \frac{1}{(V_{mn} + 1)^2} \right\} \qquad (11.14)$$

The modal birefringence of a rectangular core waveguide is, from (11.5),

$$B_G \approx \frac{\pi^2}{8} \frac{n_{co}^2 - n_{cl}^2}{n_{co}} \left\{ \frac{1}{(V_{mj} + 1)^2} - \frac{1}{(V_{mn} + 1)^2} - \frac{1}{[V_{mj} + (n_{cl}^2/n_{co}^2)]^2} \right.$$
$$\left. + \frac{1}{[V_{mn} + (n_{cl}^2/n_{co}^2)]^2} \right\} \qquad (11.15)$$

As noted earlier, we use it to estimate the *geometrical birefringence* B_G of fibers with noncircular core. Since $a_{mj} > a_{mn}$, B_G given by (11.15) is always positive. Therefore, the fast axis of the waveguide shown in Figure 11.2(b) is along the y axis. In general, the fast axis of the form birefringence is along the direction of the minor axis [5].

For fibers with a nominally circular core, a_{mj} and V_{mj} are only slightly larger than a_{mn} and V_{mn}, we can simplify (11.15) as

$$B_G \approx \frac{3\pi^2}{2} \frac{V_{mj}}{(V_{mj} + 1)^4} n_{co} \Delta^2 e^2 \qquad (11.16)$$

where

$$e^2 = 1 - \frac{a_{mn}^2}{a_{mj}^2}$$

is the *core eccentricity*. The linear birefringence can also be written as

$$(\beta_x - \beta_y) a_{mj} \approx 3\pi^2 \frac{V_{mj}^2}{(V_{mj} + 1)^4} \left(\frac{\Delta}{2} \right)^{3/2} e^2 \qquad (11.17)$$

As expected, the geometrical birefringence of a nominally circular core fiber depends on the core eccentricity and the index difference. Since we treat a nominally circular core fiber as a rectangular core waveguide and apply the Marcatili method that is best suited for waveguides with a large aspect ratio, we cannot expect (11.15)–(11.17) to be very accurate. Nevertheless, these equations have the correct dependence of Δ and e, even though the numerical values are not as accurate as we would wish.

As noted previously, it is possible to study elliptical core fibers rigorously as a boundary value problem and obtain an exact dispersion relation involving Mathieu functions [9, 10]. Based on the exact dispersion relations, it is possible to determine the linear birefringence of weakly guiding elliptical fibers rigorously. The result is [12, 13]

$$(\beta_x - \beta_y)a \approx F(V)\left(\frac{\Delta}{2}\right)^{3/2} e^2 \tag{11.18}$$

where

$$F(V) = Vb\left\{1 - b + (1 - 2b)\left[\frac{J_0(V\sqrt{1-b})}{J_1(V\sqrt{1-b})}\right]^2 \right.$$
$$\left. + Vb\sqrt{1-b}\left[\frac{J_0(V\sqrt{1-b})}{J_1(V\sqrt{1-b})}\right]^3\right\} \tag{11.19}$$

Note that (11.17) and (11.18) have the same functional form and the same dependence on e and Δ. In Figure 11.3, we plot $(\beta_x - \beta_y)a_{mj}/[(\Delta/2)^{3/2}e^2]$ as a function of V_{mj} and $(\beta_x - \beta_y)a/[(\Delta/2)^{3/2}e^2]$ as a function of V. The two curves

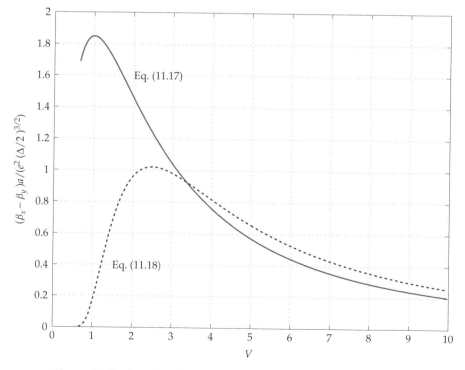

Figure 11.3 Form birefringence of fibers with slightly elliptical cores.

have the similar shape. Although the numerical values are different, the results are on the same order of magnitude for $V > 1.5$. In other words, the basic physics of the geometrical birefringence as modeled by Ramaswamy and French [11] is correct.

Take a fiber with $V_{mj} = 2.0$, $a_{mn}/a_{mj} = 0.99$, $\Delta = 0.005$, and $n_{co} \approx 1.48$ operating at 1.55 μm as an example. We obtain from (11.16) a modal birefringence of 2.7×10^{-7} and a beat length of 5.7 m. Clearly, the geometrical birefringence due to a nominally circular core is rather small. As noted in Section 11.1, the modal birefringence of the conventional single-mode fibers is between 10^{-6} and 10^{-5}. Thus the linear birefringence of the conventional single-mode fibers must originate from effects other than a noncircular core. It is also clear from (11.16) to (11.18) that the birefringence increases with the index difference. When fibers with a large modal birefringence and a short beat length are desired, it is necessary to increase the core eccentricity and/or index difference. Unfortunately, fibers having a large index difference and for core eccentricity are usually lossy. The alternative is to introduce a built-in stress in the fibers. This is the subject of the next section.

11.3   BIREFRINGENCE DUE TO BUILT-IN STRESS

In optical fibers, the core index has to be greater than the cladding index. To realize the index difference, the core and cladding are doped with different dopants and different concentrations. In Figure 11.4, we show the index variations of several doped silica glass as functions of the dopant concentration [8]. When dopants are added to vary the refractive index, other physical parameters also change. In particular, the thermal expansion coefficient changes with dopants and dopant concentration. The linear thermal expansion coefficient of undoped silica glass, that is, the pure SiO_2, is 4.9×10^{-7} $(°C)^{-1}$. When silica glass is doped with P_2O_5, GeO_2, B_2O_3, or Al_2O_3, the thermal expansion coefficient increases with the dopant concentration. If silica glass is doped with TiO_2, the thermal expansion coefficient decreases as TiO_2 concentration increases as shown in Figure 11.4 [8]. In short, the thermal expansion coefficients of the fiber core and cladding may be, and usually are, different because of different doping. While there is no internal stress in the fiber when the fiber is hot and softened, a built-in stress appears when the fiber cools to room temperature. If the fiber cross section or the index profile is asymmetric, the elastic stress arising from different thermal expansion coefficients would be asymmetric and anisotropic. Due to the *anisotropic stress* and through *photoelastic effects*, the fiber becomes *optically anisotropic*. This is the basic physics behind many polarization-maintaining fibers. Figure 11.5 shows the cross sections of four high-birefringence fibers [5, 6].

To estimate the stress-induced birefringence, we consider a thin slab model shown in Figure 11.6 [3, 14]. Let the thermal expansion coefficients of the core and cladding be α_{co} and α_{cl}, respectively. We assume α_{co} is greater than α_{cl}. To simplify matters, we also assume that the core and cladding widths are the same at the softening temperature T_s:

$$W_{co}(T_s) = W_{cl}(T_s) = W \qquad (11.20)$$

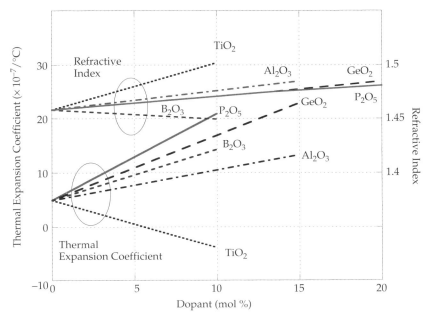

Figure 11.4   Refractive index and thermal expansion coefficient of doped silica glasses [8].

When the composite structure cools to room temperature T_r, the core and cladding widths would be

$$W_{co}(T_r) = W[1 - \alpha_{co}(T_s - T_r)] \tag{11.21}$$

and

$$W_{cl}(T_r) = W[1 - \alpha_{cl}(T_s - T_r)] \tag{11.22}$$

if the core and cladding were free to contract independently. But the core and cladding are fused together and they are not free to contract. Since the cladding is much more bulky and massive than the core, the core is stretched by the cladding when the structure cools down. The tensile strain component in the core region in the x direction developed is

$$S_{xx} = \frac{W_{cl}(T_r) - W_{co}(T_r)}{W_{co}(T_s)} \approx (\alpha_{co} - \alpha_{cl})(T_s - T_r) \tag{11.23}$$

As shown in Figure 11.6, there is no material or structure outside the slabs to prevent the core and cladding to contract in the y direction. Thus S_{yy} is zero when the fiber cools down. There is no need to consider S_{zz} since it produces no fiber birefringence effect.

Because of the tensile strain component S_{xx} and through the photoelastic effect discussed in Appendix D, we obtain, from (D.47) and (D.48),

$$\Delta n_x \approx -\tfrac{1}{2}n^3 p_{11} S_{xx} \tag{11.24}$$

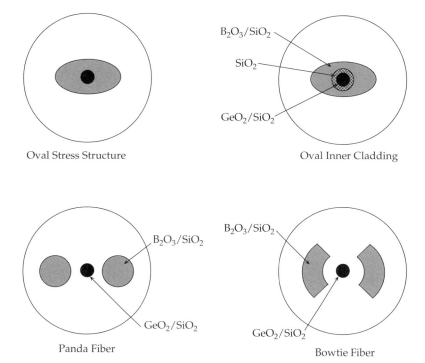

Figure 11.5 Cross sections of four high-birefringence fibers.

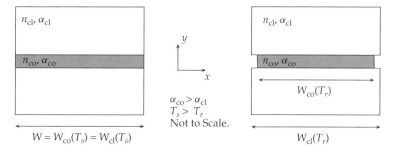

Figure 11.6 Thin-film model for single-mode fibers with built-in stress.

$$\Delta n_y \approx -\tfrac{1}{2}\, n^3 p_{12} S_{xx} \qquad (11.25)$$

where p_{11} and p_{12} are the photoelastic constants of the core material. Numerical values of p_{11}, p_{12}, and other physical parameters of silica fibers are listed in Table 11.1.

Using (11.5), (11.6), (11.24), and (11.25), we have the *modal birefringence* due to *built-in stress*:

$$B_T = \frac{1}{2}\left(b + \frac{V}{2}\frac{db}{dV}\right) n_{\text{co}}^3 (p_{11} - p_{12})(\alpha_{\text{co}} - \alpha_{\text{cl}})(T_s - T_r) \qquad (11.26)$$

TABLE 11.1   Selected Material Constants of Silica Fibers

	Symbol	Value	Unit	Remarks
Young's modulus	Y	$7.6 \times 10^{+10}$	N/m^2	
Poisson ratio	σ	0.17		
Photoelastic constant	p_{11}	0.113		@ 0.633 μm [40]
Photoelastic constant	p_{12}	0.252		@ 0.633 μm [40]
Kerr constant	K	5.3×10^{-16}	M/V^2	@ 0.633 μm [23]
Verdet constant	V	4.53×10^{-6}	rad/A	@ 0.633 μm [30]
Verdet constant	V	0.756×10^{-6}	rad/A	@ 1.55 μm [30]

The beat length due to the built-in stress is

$$L_B = \frac{2\lambda}{n_{co}^3 \left(b + \dfrac{V}{2} \dfrac{db}{dV} \right) |p_{11} - p_{12}| (\alpha_{co} - \alpha_{cl})(T_s - T_r)} \tag{11.27}$$

As an estimate, we consider a fiber with a pure SiO$_2$ core and a B$_2$O$_3$–SiO$_2$ cladding. Based on (11.26) and (11.27), we deduce that the built-in stress is $S_{xx} \approx 9.4 \times 10^{-4}$, the model birefringence is 2.0×10^{-4}, and the beat length is 7.8 mm at 1.55 μm. Such a beat length is much shorter than the beat length originated from the shape birefringence discussed in Section 11.2.

Clearly, the model used to derive (11.26) and (11.27) is a simple one. For an analysis based on realistic and accurate models, readers are referred to Eickhoff's work [15]. In particular, he studied the stress-induced birefringence of fibers with a circular core and an elliptical inner cladding and that of fibers with an elliptical core and a circular cladding.

11.4   BIREFRINGENCE DUE TO EXTERNALLY APPLIED MECHANICAL STRESS

When an ideal single-mode fiber is subjected to a lateral compression, bending, or strong electric field, it becomes *linearly birefringent* [2–6]. When a circular fiber is twisted mechanically or subjected to a strong axial magnetic field, it becomes *circularly birefringent* [5, 6]. In this section, we estimate the birefringence induced by the mechanical disturbances.

11.4.1   Lateral Stress

Clamping a circular fiber between two flat plates, as shown schematically in Figure 11.7(a), produces two effects. First, the lateral force compresses and deforms the fiber. As a result, a circular core becomes an elliptical core. As discussed in Section 11.2, the geometrical birefringence of a fiber having a slightly deformed core is rather small. We will not consider the geometrical birefringence due to the

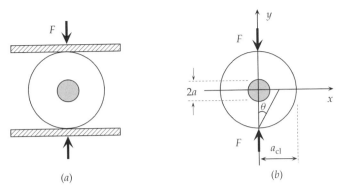

Figure 11.7   Single-mode fiber clamped by a lateral force.

core deformation any further. Second, the lateral force also produces strain, which in turn leads to an index change through the *photoelastic effects*. This is the dominant birefringence effect produced by laterally clamping. We consider this effect in this section.

We examine the birefringence caused by a lateral compression in two steps. First, we study the elastic problem of the clamping of a circular cylinder. Then, we use the stress and strain in a clamped cylinder to estimate the refractive index change. In our study of optical fibers so far, we have taken the cladding as infinitely thick. To consider the elastic problem of clamping, however, we have to take the cladding radius as finite. Let the cladding radius be a_{cl}. As the fiber core and cladding are made of the same basic materials, it is reasonable to treat the core and cladding as a single and homogeneous elastic cylinder of radius a_{cl} in the elastic consideration. Suppose a force F in the y direction is applied uniformly to a cylinder of length L [Fig. 11.7(b)]. The lateral force per length is F/L. Since the force is uniformly distributed in the z direction, stress and strain components are also independent of z. Therefore we may treat the cylinder as a two-dimensional elastic disk. The clamping of a circular elastic disk is a well-known elasticity problem and has been examined thoroughly in many textbooks on elasticity (e.g., [16–18]). Here, we summarize the key results. Refer to Figure 11.7(b). When a lateral force per unit length F/L acts on a circular disk of radius a_{cl}, the tensile stress components at points along the x axis are

$$T_{xx} = \frac{\cos^2 2\theta}{\pi a_{cl}} \frac{F}{L} \tag{11.28}$$

$$T_{yy} = \frac{1 - 4\cos^4 \theta}{\pi a_{cl}} \frac{F}{L} \tag{11.29}$$

$$T_{zz} = 0 \tag{11.30}$$

where θ is an angle defined in Figure 11.7(b) [4, 19]. Plots of T_{xx} and T_{yy} as functions of x are shown in Figure 11.8. For all points along the x axis, T_{xx} is positive and T_{yy} is negative. Thus the cylinder is compressed in the y direction and elongated

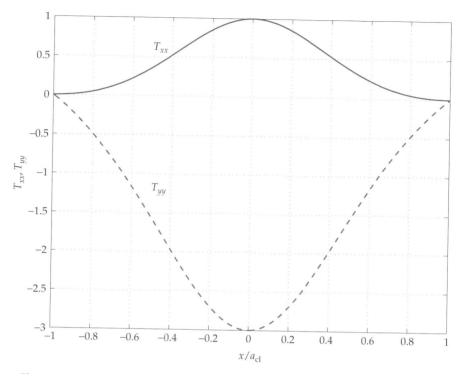

Figure 11.8 Stress components T_{xx} and T_{yy} along the axis of lateral stressed fibers.

or stretched in the x direction. Also note that T_{xx} and T_{yy} are essentially constant in the central region. In fact, T_{xx} and $|T_{yy}|$ are within 97% of the respective peak values for $|x| < a_{cl}/10$. For communication-grade single-mode fibers, the cladding diameter is typically 125 μm and the core diameter is 12 μm or less. In other words, the core is at the center of the cladding and well within 10% of the cladding diameter. Thus we may approximate the stress components in the core region by the peak values

$$T_{xx} \approx \frac{F}{\pi a_{cl} L} \tag{11.31}$$

$$T_{yy} \approx -\frac{3F}{\pi a_{cl} L} \tag{11.32}$$

Substituting these results in (D.50), we obtain

$$\Delta n_x - \Delta n_y = -2n_{co}^3 (p_{11} - p_{12}) \frac{1+\sigma}{\pi Y} \frac{1}{a_{cl}} \frac{F}{L} \tag{11.33}$$

where Y and σ are the Young's modulus and the Poisson ratio of the elastic disk.

The *linear birefringence* due to the *lateral compressive force* is from (11.5) and (11.6) [3, 6, 19]:

$$\Delta\beta_{cp} = \beta_x - \beta_y = -2\left(b + \frac{V}{2}\frac{db}{dV}\right)kn_{co}^3\frac{(p_{11} - p_{12})(1 + \sigma)}{\pi Y}\frac{1}{a_{cl}}\frac{F}{L} \qquad (11.34)$$

Since p_{11} is smaller than p_{12}, $\Delta\beta_{cp}$ is positive. In other words, the *fast axis* is along the y axis, that is, the direction of the compressive force.

When the numerical values of Y, σ, p_{11} and p_{12} of silica glass are used, we have the total linear retardation of a fiber of length L at $\lambda = 1.55$ μm:

$$\Delta\beta_{cp}L \approx 1.71 \times 10^{-5}\left(b + \frac{V}{2}\frac{db}{dV}\right)\frac{F}{a_{cl}} \quad \text{rad} \qquad (11.35)$$

In the expression, a_{cl} and L are in meters and F is in newtons. Note that the total linear retardation $\Delta\beta_{cp}L$ depends mainly on F/a_{cl}. Take a single-mode fiber with a cladding diameter of 125 μm and operating with a V of 2.3 as an example. A linear retardation of π rad is realized when a lateral force of 14.3 N is applied to the fiber section. Such a stressed fiber section can be used as a half-wave plate. If the applied force is 7.2 N, the fiber section acts as a quarter-wave plate.

In many applications, a lateral force is often applied to hold the fiber in place. The lateral force used to hold the fiber also introduces linear birefringence. To minimize the linear birefringence induced by a lateral force, a fiber may be held in a V groove instead of between two flat plates (Fig. 11.9). The problem of a single-mode fiber pressed against a V-groove has been studied by Kumar and Ulrich [20]. They showed that the linear retardation of a single-mode fiber held in a V groove of angle 2δ by a lateral force F is

$$\Delta\beta_V L = -kn_{co}^3\frac{(1 + \sigma)(p_{11} - p_{12})}{Y\pi}(1 - \cos 2\delta \sin \delta)\frac{F}{a_{cl}} \qquad (11.36)$$

A plot of $\Delta\beta_V L$ as a function of δ is shown in Figure 11.10. Clearly, it is possible to reduce the compression-induced linear birefringence by keeping the fiber

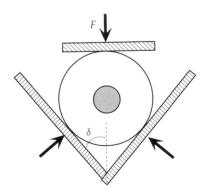

Figure 11.9   Single-mode fiber pressed against a V groove.

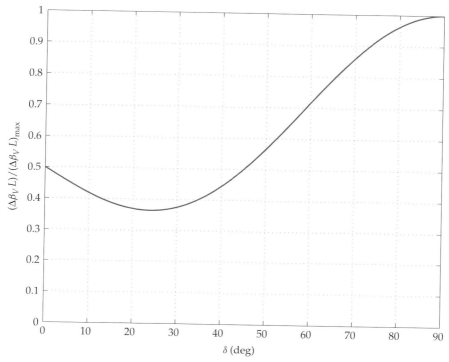

Figure 11.10   Linear retardation of a fiber pressed against a V groove.

in a V groove with an angle δ between $20°$ and $30°$. In fact, the linear retardation
is minimized if δ is $24.1°$.

11.4.2   Bending

From the mechanics viewpoint and to the first-order approximation, a circular fiber
is an isotropic elastic cylinder. An axial stretching without bending produces no
anisotropic effect and would not lead to the linear birefringence. But a lateral bending, with or without stretching, breaks the rotational symmetry and would induce the
linear birefringence. A bending without stretching is referred to as a *pure bending*,
and a bending by pulling and pressing the fiber against a solid cylinder as a *bending
under tension*.

   First, we consider a pure bending. Suppose a fiber is bent into a circular arc
of radius R that is much larger than the cladding radius a_{cl} [Fig. 11.11(a)]. To
simplify the elasticity consideration, we study an elastic tube of thickness $2a_{cl}$ and
radius R in lieu of a fiber coil [Fig. 11.11(b)]. To simplify the elastic problem, we
ignore the tube width W completely. We focus our attention on a differential arc
element of angle $\delta\theta$ identified by the shaded area in Figure 11.11(c). We label the
local coordinates as the x and z axes. For the local coordinates, we choose the
tube centerline, shown as a dashed line in Figure 11.11(c), as $x = 0$. The inner
and outer boundaries of the differential arc element are x and $x + \Delta x$. In a pure

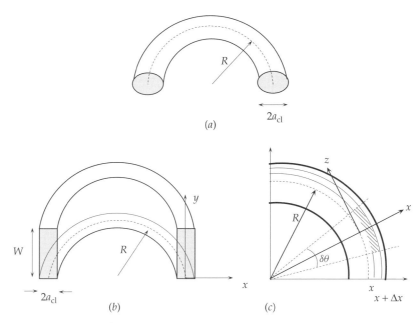

Figure 11.11 Slab model for a bent fiber.

bending, the fiber is not stretched axially and the tube centerline is not elongated. The arc length along the centerline is $R\delta\theta$. Away from the centerline, the arc length is $(R + x)\,\delta\theta$. Depending on x, the arc length at x is either elongated or compressed by bending. The fractional change of the arc length, relative to the centerline, is $(x\,\delta\theta)/(R\,\delta\theta) = x/R$. Thus, the tensile strain component at x due to a pure bending is

$$S_{zz}(x) = \frac{x}{R} \qquad (11.37)$$

When a fiber is bent under tension, much of the tube, including the centerline, is stretched. Let $\overline{S}_{zz}$ be the axial strain of the centerline due to stretching. Then the tensile strain component at x produced by the bending under tension is

$$S_{zz}(x) = \overline{S}_{zz} + \frac{x}{R} \qquad (11.38)$$

The tensile stress accompanying the tensile strain is, from Hooke's law (D.18),

$$T_{zz}(x) = \frac{Y(1 - \sigma)}{(1 + \sigma)(1 - 2\sigma)} \left[\overline{S}_{zz} + \frac{x}{R}\right] \qquad (11.39)$$

For a circular fiber with a rotationally symmetric index profile, the axial strain or stress produces no linear birefringence directly. However, the lateral stress and strain arising from the axial stress would lead to the linear birefringence. To consider the lateral strain component accompanying S_{zz}, we return to the differential arc element shown in Figure 11.11(c). In static equilibrium, the net force acting on the differential arc element must vanish. In particular, the net force in the x direction is zero:

$$2T_{zz}(x)(W\Delta x)\sin\frac{\delta\theta}{2} + T_{xx}(x)[(R+x)\ \delta\theta\ W]$$
$$- T_{xx}(x+\Delta x)[(R+x+\Delta x)\ \delta\theta W] = 0 \qquad (11.40)$$

Since $\delta\theta$ is small, we may approximate the sine function by its argument. After factoring out the common factor $W\delta\theta$, we obtain

$$T_{xx}(x+\Delta x)\left(1+\frac{x+\Delta x}{R}\right) - T_{xx}(x)\left(1+\frac{x}{R}\right) = \frac{1}{R}T_{zz}(x)\Delta x \qquad (11.41)$$

Since R is much larger than a_{cl}, we ignore terms on order of, or smaller than, a_{cl}/R and obtain

$$T_{xx}(x+\Delta x) - T_{xx}(x) = \frac{1}{R}T_{zz}(x)\Delta x \qquad (11.42)$$

In the limit of infinitesimal Δx, we obtain a differential equation for T_{xx} in terms of T_{zz}:

$$\frac{dT_{xx}}{dx} = \frac{1}{R}T_{zz}(x) = \frac{1}{R}\frac{(1-\sigma)Y}{(1+\sigma)(1-2\sigma)}\left(\overline{S}_{zz} + \frac{x}{R}\right) \qquad (11.43)$$

A simple integration leads to

$$T_{xx}(x) = \frac{(1-\sigma)Y}{(1+\sigma)(1-2\sigma)}\left(\overline{S}_{zz}\frac{x}{R} + \frac{x^2}{2R^2} + C'\right) \qquad (11.44)$$

where C' is a constant of integration. To determine C', we make use of the fact that there is no lateral stress acting on the outer surface ($x = a_{cl}$) of the tube. Thus, we choose C' such that $T_{xx}(a_{cl}) = 0$ and obtain

$$T_{xx}(x) = \frac{(1-\sigma)Y}{(1+\sigma)(1-2\sigma)}\left(\overline{S}_{zz}\frac{x-a_{cl}}{R} + \frac{x^2-a_{cl}^2}{2R^2}\right) \qquad (11.45)$$

For a pure bending, the inner surface ($x = -a_{cl}$) is free of lateral stress, that is, $T_{xx}(-a_{cl}) = 0$. But the lateral stress $T_{xx}(-a_{cl})$ is not zero when the fiber is bent under tension.

In regions near the tube centerline, the lateral stress component is approximately

$$T_{xx} \approx T_{xx}(0) = -\frac{(1-\sigma)Y}{(1+\sigma)(1-2\sigma)}\frac{a_{cl}}{R}\left(\overline{S}_{zz} + \frac{a_{cl}}{2R}\right) \qquad (11.46)$$

For the simple model depicted in Figure 11.11, T_{yy} and S_{yy} vanish completely.

11.4.2.1 Pure Bending [21]

For a bending without stretching, $\overline{S}_{zz} = 0$. We obtain from (11.46)

$$T_{xx} \approx -\frac{1}{2}\frac{(1-\sigma)Y}{(1+\sigma)(1-2\sigma)}\frac{a_{cl}^2}{R^2} \tag{11.47}$$

From (11.47) and (D.50), we obtain

$$\Delta n_x - \Delta n_y = \frac{1}{4}n_{co}^3(p_{11}-p_{12})\frac{1-\sigma}{1-2\sigma}\frac{a_{cl}^2}{R^2} \tag{11.48}$$

Thus, the *linear birefringence* induced by a *pure bending* is, from (11.5),

$$\Delta\beta_B = \beta_x - \beta_y = kn_{co}^3\left(b + \frac{V}{2}\frac{db}{dV}\right)\frac{(p_{11}-p_{12})(1-\sigma)}{4(1-2\sigma)}\frac{a_{cl}^2}{R^2} \tag{11.49}$$

Numerically, the linear birefringence induced by a pure bending at 1.55 µm is

$$\Delta\beta_B = \beta_x - \beta_y = -5.49 \times 10^5\left(b + \frac{V}{2}\frac{db}{dV}\right)\frac{a_{cl}^2}{R^2} \quad \text{rad/m} \tag{11.50}$$

The fast axis is in the plane of the fiber coil and normal to the fiber axis.

11.4.2.2 Bending under Tension [21, 22]

When a fiber is bent and stretched against a solid cylinder of radius R, $\overline{S}_{zz}$ in (11.46) is not zero. Suppose a fiber of length L is stretched by δL in the bending and stretching processes, then $\overline{S}_{zz} \approx \delta L/L$. The *linear birefringence* due to *bending under tension* is, from (11.5), (11.46), and (D.50),

$$\Delta\beta_{BT} = \beta_x - \beta_y = kn_{co}^3\left(b + \frac{V}{2}\frac{db}{dV}\right)\frac{(p_{11}-p_{12})(1-\sigma)}{4(1-2\sigma)}\frac{a_{cl}}{R}\left(\frac{a_{cl}}{R}+2\frac{\delta L}{L}\right) \tag{11.51}$$

Numerically, the linear birefringence due to bending under tension at 1.55 µm is

$$\Delta\beta_{BT} = \beta_x - \beta_y = -5.49 \times 10^5\left(b + \frac{V}{2}\frac{db}{dV}\right)\frac{a_{cl}}{R}\left(\frac{a_{cl}}{R}+2\frac{\delta L}{L}\right) \quad \text{rad/m} \tag{11.52}$$

In deriving (11.46), (11.50), (11.51), and (11.52) we have approximated a fiber coil by an elastic tube. Ulrich, Rashleigh, and Eickhoff [21] used an accurate fiber model in their study of the bending problem. They treated a fiber as a circular rod and showed [21, 22]

$$\Delta\beta_{BT} = \beta_x - \beta_y = -k\frac{n_{co}^3(p_{11}-p_{12})(1+\sigma)}{4}\frac{a_{cl}}{R}\left[\frac{a_{cl}}{R}+\frac{2(2-3\sigma)}{1-\sigma}\frac{\delta L}{L}\right] \tag{11.53}$$

Numerically, $\Delta\beta_{BT}$ at 1.55 μm is

$$\Delta\beta_{BT} = \beta_x - \beta_y = -5.11 \times 10^5 \frac{a_{cl}}{R}\left(\frac{a_{cl}}{R} + 3.59\frac{\delta L}{L}\right) \quad \text{rad/m} \qquad (11.54)$$

As given by (11.52) and (11.54), the linear birefringence due to pure bending and that arising from stretching are additive. For moderate stretching ($\delta L/L \leq$ 0.002), the birefringence effect due to pure bending is small unless the bending radius R is less than $140a_{cl}$. For fibers with a cladding diameter of 125 μm, the bending radius has to be smaller than 9 mm. In other words, two birefringence components are comparable only if the bending radius is on order of 1 cm or smaller.

As a numerical example, we consider a single-mode fiber with $V = 2.3$ at 1.55 μm. We also take the cladding diameter as 125 μm. Suppose the fiber is wound into a coil of radius of 1 cm. The linear birefringence due to pure bending is -19.96 rad/m, which corresponds to a beat length of 31.5 cm. If the coil is formed under tension and the axial strain is 0.002, the linear birefringence due to bending and stretching is -42.9 rad/m, which leads to a beat length of 14.6 cm. If the coil radius is reduced to 0.5 cm, then $\Delta\beta_B$ and $\Delta\beta_{BT}$ are, respectively, -79.8 and -125.7 rad/m and the beat lengths are 7.9 and 5.0 cm.

11.4.3 Mechanical Twisting

Long fiber sections are often twisted unintentionally. In some applications, fibers are twisted purposely. Mechanical twisting leads to two effects: a *circular birefringence* and an *axis rotation*. In this section, we briefly discuss the circular birefringence induced by a mechanical twisting. A detailed study of the effect of twisting is presented in Appendix E.

When a single-mode fiber of length L is twisted by an angle Θ, a shearing strain is induced. Due to the shearing strain and through photoelastic effects, the fiber becomes circularly birefringent. As shown in Appendix E, the circular birefringence induced by twisting is proportional to the rate of twisting Θ/L:

$$\Delta\beta_{TW} = \beta_R - \beta_L = \frac{1}{2}n_{co}^2(p_{11} - p_{12})\frac{\Theta}{L}F_{TW}(V) \qquad (11.55)$$

where $F_{TW}(V)$ is a function of the generalized frequency V, and Θ is taken as positive for a counterclockwise twist. When the numerical values of n_{co}, p_{11}, and p_{12} of silica glass fibers are used, we have the circular birefringence

$$\Delta\beta_{TW} = \beta_R - \beta_L = -0.148\frac{\Theta}{L}F_{TW}(V) \quad \text{rad/m} \qquad (11.56)$$

A twisted fiber works as a circular retarder with a *circular retardation* of ($\beta_R - \beta_L)L = -0.148\Theta F_{TW}(V)$. When a linearly polarized field propagates through a circular retarder with a circular retardation ($\beta_R - \beta_L)L$, the field vector rotates by an angle $-(\beta_R - \beta_L)L/2$. Thus, a mechanical twisting of angle Θ will cause a linearly polarized field to rotate by an angle of $0.074\Theta F_{TW}(V)$. In other words, a mechanical twist of $360°$ will cause a linearly polarized field to rotate by $26.6° F_{TW}(V)$. This is in qualitative agreement with Smith's observation [2].

11.5 BIREFRINGENCE DUE TO APPLIED ELECTRIC AND MAGNETIC FIELDS

11.5.1 Strong Transverse Electric Fields

When a solid is subjected to a strong electric field, the refractive index changes. In amorphous materials, glass, for example, the index change is proportional to the square of the electric field intensity. This is the *quadratic electrooptical effect*, or *Kerr effect*. A brief discussion on the Kerr effect can be found in Appendix D. Let $\mathbf{E}_m$ be the applied electric field intensity. Then, the linear birefringence, from (D.59), is

$$\Delta\beta_K = 2\pi K |\mathbf{E}_m|^2 \qquad (11.57)$$

where K is the *Kerr constant* of the material. The fast axis is along the direction of the applied electric field. As listed in Table 11.1, the Kerr constant of silica fibers is very small. In addition, the Kerr constant is also temperature dependent [23, 24]. Other than an experiment mentioned below, the Kerr-effect-induced birefringence is rarely used in practical applications. The experiment is to use the Kerr effect to measure the fiber beat length [25].

11.5.2 Strong Axial Magnetic Fields

In this section, we consider the circular birefringence induced by a strong axial magnetic field [26–29]. We begin by considering waves propagating in an isotropic medium under the influence of a strong magnetic field. Then we apply the results to fibers subjected to a strong axial magnetic field intensity H_m.

Due to the interaction of magnetic fields with orbital electrons, an isotropic material becomes circularly birefringent. If the strong magnetic field is in the z direction, the relative permittivity tensor of the medium is [26–29]

$$\varepsilon_r = \begin{bmatrix} n^2 & j\gamma & 0 \\ -j\gamma & n^2 & 0 \\ 0 & 0 & n^2 \end{bmatrix} \qquad (11.58)$$

where γ is proportional to the axial magnetic field and n is the index of refraction of the medium.

To consider the wave propagation in a medium under the influence of an axial magnetic field, we begin with the time-harmonic $(e^{+j\omega t})$ Maxwell equations with a complex relative permittivity tensor given in (11.58)

$$\nabla \times \mathbf{E} = -j\omega\mu_0\mathbf{H} \tag{11.59}$$

$$\nabla \times \mathbf{H} = j\omega\varepsilon_0\boldsymbol{\varepsilon}_r \cdot \mathbf{E} \tag{11.60}$$

and determine the propagation constants and field vectors of the eigenpolarization modes. We consider plane waves propagating along the direction of the strong magnetic field. Let the electric field intensity be

$$\mathbf{E} = (\hat{\mathbf{x}}E_{x0} + \hat{\mathbf{y}}E_{y0})e^{-j\beta z} \tag{11.61}$$

To determine β, E_{x0}, and E_{y0}, we substitute (11.61) into (11.59) and (11.60) and obtain a dispersion relation:

$$(\beta^2 - k^2n^2)(\beta^2 - k^2n^2) - k^4\gamma^2 = 0 \tag{11.62}$$

There are two independent and distinct solutions for β and we label them as β_R and β_L:

$$\beta_{R,L} = k\sqrt{n^2 \pm \gamma} \tag{11.63}$$

The eigenpolarization modes associated with β_R and β_L are

$$\mathbf{E}_R(z) = E_R\frac{\hat{\mathbf{x}} - j\hat{\mathbf{y}}}{\sqrt{2}}e^{-j\beta_R z} \tag{11.64}$$

$$\mathbf{E}_L(z) = E_L\frac{\hat{\mathbf{x}} + j\hat{\mathbf{y}}}{\sqrt{2}}e^{-j\beta_L z} \tag{11.65}$$

where E_R and E_L are the amplitude constants. By examining the evolution of the $\text{Re}[\mathbf{E}_R(z)e^{j\omega t}]$ and $\text{Re}[\mathbf{E}_L(z)e^{j\omega t}]$ as functions of time, we establish that $\mathbf{E}_R(z)$ and $\mathbf{E}_L(z)$ represent, respectively, right- and left-hand circularly polarized fields.

Since $\beta_R \neq \beta_L$, the isotropic medium under the influence of a strong axial magnetic field is circularly birefringent, and $\beta_R - \beta_L$ is the *circular birefringence*. An interesting property of circularly birefringent media is as follows. Upon entering a circular birefringent medium, the field vector of a linearly polarized wave rotates continuously as the field propagates. This is the *Faraday rotation* or *Faraday effect*. To demonstrate the field rotation, we suppose the input electric field is polarized in the x direction and write

$$\mathbf{E}(0) = E_0\hat{\mathbf{x}} = \frac{E_0}{\sqrt{2}}\left(\frac{\hat{\mathbf{x}} - j\hat{\mathbf{y}}}{\sqrt{2}} + \frac{\hat{\mathbf{x}} + j\hat{\mathbf{y}}}{\sqrt{2}}\right) \tag{11.66}$$

At point z, the electric field is

$$\mathbf{E}(z) = \frac{E_0}{\sqrt{2}} \left(\frac{\hat{\mathbf{x}} - j\hat{\mathbf{y}}}{\sqrt{2}} e^{-j\beta_R z} + \frac{\hat{\mathbf{x}} + j\hat{\mathbf{y}}}{\sqrt{2}} e^{-j\beta_L z} \right) \tag{11.67}$$

A little manipulation will convert the expression to

$$\mathbf{E}(z) = E_0 \left(\hat{\mathbf{x}} \cos \frac{\beta_R - \beta_L}{2} z - \hat{\mathbf{y}} \sin \frac{\beta_R - \beta_L}{2} z \right) e^{-j(\beta_R + \beta_L)z/2} \tag{11.68}$$

Clearly, $\mathbf{E}(z)$ is at an angle $-(\beta_R - \beta_L)z/2$ with respect to the x axis. In other words, the linearly polarized field rotates continuously as it propagates in an isotropic medium under the influence of a strong magnetic field.

Since γ given in (11.58) is much smaller than n, we approximate β_R and β_L as $k[n \pm (\gamma/2n)]$. Thus the circular birefringence is

$$\beta_R - \beta_L \approx k \frac{\gamma}{n} \tag{11.69}$$

The angle of rotation per unit length is $-k\gamma/(2n)$. In dielectrics, the rotation per unit length is proportional to the axial magnetic field intensity H_m. The proportionality constant is the *Verdet constant* V_H. In other words, $k\gamma/(2n) = V_H H_m$.

The Faraday effect exists in all dielectrics when the materials are subjected to a strong axial magnetic field. This includes optical fibers. Let the strong axial magnetic field intensity be H_m. Then the circular birefringence due to Faraday effect in the single mode fibers is

$$\beta_R - \beta_L = -V_H H_m \tag{11.70}$$

For silica glass fibers, $V_H = 2.021 \times 10^{-35} f^2$ rad/A where f is the optical frequency in hertz [30, 31]. At 0.633 and 1.55 μm, V_H is 4.53×10^{-6} rad/A and 0.756×10^{-6} rad/A, respectively.

Linear birefringence induced by lateral compression, bending, and electric fields and circular birefringence induced by twisting are the same for waves propagating in either direction. These effects are the *reciprocal effects*. In contrast, circular birefringence due to axial magnetic field is *nonreciprocal* in that the effects are different for waves propagating in opposite directions. Thus the circular birefringence induced by twisting discussed in Section 11.4.3 and that caused by axial magnetic fields discussed in this section are fundamentally different.

11.6   JONES MATRICES FOR BIREFRINGENT FIBERS

Having discussed the origins of fiber birefringence, we study the effects of the fiber birefringence on the state of polarization of propagating waves. The state of polarization can be quantified by the *Jones vectors* or the *Stokes parameters* [32, 33].

The effect of a fiber section or a polarization component on the state of polarization is described by a *Jones matrix* or a *Muller matrix* [32, 33]. In the following sections, we discuss the Jones matrices of single-mode fibers under various conditions [34].

11.6.1   Linearly Birefringent Fibers with Stationary Birefringent Axes

A linearly birefringent fiber can be viewed as a *linear retarder*. Suppose the fiber linear birefringence is $\Delta\beta_{LB}$ and the fast axis is in the x direction. Then the Jones matrix for the linearly birefringent fiber of length z is

$$[J_{LB}(0)] = \begin{bmatrix} e^{j\Delta\beta_{LB}z/2} & 0 \\ 0 & e^{-j\Delta\beta_{LB}z/2} \end{bmatrix} \tag{11.71}$$

As discussed in the preceding sections, the fast axis depends on the fiber geometry and the applied stress or field. If the fast axis is at an angle θ relative to the x axis, we use a *rotation matrix*

$$[S(\theta)] = \begin{bmatrix} \cos\theta & \sin\theta \\ -\sin\theta & \cos\theta \end{bmatrix} \tag{11.72}$$

and similarity transformation [33] to find $J_{LB}(\theta)$. Specifically, the Jones matrix of a linearly birefringent fiber with a fast axis in θ is

$$[J_{LB}(\theta)] = [S(-\theta)][J_{LB}(0)][S(\theta)] \tag{11.73}$$

By carrying out matrix multiplications, we obtain an explicit expression for $J_{LB}(\theta)$:

$$[J_{LB}(\theta)] = \begin{bmatrix} \cos\dfrac{\Delta\beta_{LB}z}{2} + j\cos 2\theta \sin\dfrac{\Delta\beta_{LB}z}{2} & j\sin 2\theta \sin\dfrac{\Delta\beta_{LB}z}{2} \\ j\sin 2\theta \sin\dfrac{\Delta\beta_{LB}z}{2} & \cos\dfrac{\Delta\beta_{LB}z}{2} - j\cos 2\theta \sin\dfrac{\Delta\beta_{LB}z}{2} \end{bmatrix} \tag{11.74}$$

11.6.2   Linearly Birefringent Fibers with a Continuous Rotating Axis

The similarity transformation (11.73) can be applied repeatedly to fibers with a finite number of discrete axis rotations. But it is not practical to apply the similarity transformation to fibers with a continuously turning axis. To study a linearly bire-fringent medium with a continuous changing fast axis, McIntyre and Snyder [35] have set up two coupled-mode equations for the medium in question. By solving the coupled-mode equations, they have obtained the desired Jones matrix [35, 36].

   If the fast axis is fixed, the two orthogonal field components propagate independently with the effective indices n_x and n_y, respectively. If the axis rotates continuously, the two field components are coupled. To study the coupling, we consider the fields at two neighboring points, z and $z + \Delta z$. Let the transverse field components at the two points be $[E_x(z), E_y(z)]$ and $[E_x(z + \Delta z), E_y(z + \Delta z)]$. The field components differ on two accounts: wave propagation and axis rotation. Since Δz is infinitesimally small, the two mutually independent effects may be treated separately and then combined.

In the absence of axis rotation, the changes in the field components are due to the propagation alone. Since the x-component propagates with an index n_x, E_x varies as $e^{-jkn_x z}$. The change of E_x due to the propagation effect is

$$E_x(z + \Delta z) - E_x(z) \approx \frac{dE_x(z)}{dz}\Delta z = -jkn_x E_x(z)\Delta z \tag{11.75}$$

Similarly, the change of the y component due to the propagation effect is

$$E_y(z + \Delta z) - E_y(z) \approx \frac{dE_y(z)}{dz}\Delta z = -jkn_y E_y(z)\Delta z \tag{11.76}$$

Next, we consider the effect of axis rotation. We label the axes at z as the x and y axes, and at $z + \Delta z$ as the x' and y' axes. As shown in Figure 11.12, the x' and y' axes are rotated with respect to the x and y axes by an infinitesimal angle $\Delta\theta$:

$$\hat{\mathbf{x}} = \hat{\mathbf{x}}' \cos \Delta\theta - \hat{\mathbf{y}}' \sin \Delta\theta \tag{11.77}$$

$$\hat{\mathbf{y}} = \hat{\mathbf{x}}' \sin \Delta\theta + \hat{\mathbf{y}}' \cos \Delta\theta \tag{11.78}$$

Since $\Delta\theta$ is infinitesimally small, the two equations can be approximated by

$$\hat{\mathbf{x}} \approx \hat{\mathbf{x}}' - \hat{\mathbf{y}}'\Delta\theta \tag{11.79}$$

$$\hat{\mathbf{y}} \approx \hat{\mathbf{x}}'\Delta\theta + \hat{\mathbf{y}}' \tag{11.80}$$

The transverse field components must be continuous:

$$E_x(z)\hat{\mathbf{x}} + E_y(z)\hat{\mathbf{y}} = E_x(z + \Delta z)\hat{\mathbf{x}}' + E_y(z + \Delta z)\hat{\mathbf{y}}' \tag{11.81}$$

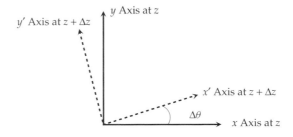

Figure 11.12   Axis rotation.

By substituting (11.79) and (11.80) into (11.81), we obtain

$$E_x(z) + \Delta\theta E_y(z) \approx E_x(z + \Delta z) \tag{11.82}$$

$$E_y(z) - \Delta\theta E_x(z) \approx E_y(z + \Delta z) \tag{11.83}$$

Thus, the coupling between the two field components due to the axis rotation is

$$E_x(z + \Delta z) - E_x(z) \approx \Delta\theta E_y(z) \tag{11.84}$$

$$E_y(z + \Delta z) - E_y(z) \approx -\Delta\theta E_x(z) \tag{11.85}$$

The two effects exist simultaneously. We combine (11.75), (11.76), (11.84), and (11.85) and obtain

$$E_x(z + \Delta z) - E_x(z) \approx -jkn_x\Delta z E_x(z) + \Delta\theta E_y(z) \tag{11.86}$$

$$E_y(z + \Delta z) - E_y(z) \approx -jkn_y\Delta z E_y(z) - \Delta\theta E_x(z) \tag{11.87}$$

By letting Δz approach 0, we obtain the coupled mode equations [35]

$$\frac{dE_x(z)}{dz} = -jkn_x E_x(z) + \frac{d\theta}{dz} E_y(z) \tag{11.88}$$

$$\frac{dE_y(z)}{dz} = -jkn_y E_y(z) - \frac{d\theta}{dz} E_x(z) \tag{11.89}$$

If the fiber axis rotates at a constant rate, $d\theta/dz$ is a constant ξ. For constant n_x, n_y, and ξ, we combine (11.88) and (11.89) and obtain a second-order differential equation for $E_x(z)$:

$$\frac{d^2 E_x}{dz^2} + jk(n_x + n_y)\frac{dE_x}{dz} + (\xi^2 - k^2 n_x n_y)E_x = 0 \tag{11.90}$$

A similar differential equation is also obtained for E_y. It is simple to solve for E_x and E_y subject to the boundary conditions $E_{x\text{in}} = E_x(0)$ and $E_{y\text{in}} = E_y(0)$. Finally we obtain expressions for the field components at z:

$$E_x(z) = e^{-jk\bar{n}z}\left[\left(\cos\frac{\delta_t}{2} + j\frac{\Delta\beta_{\text{LB}}z}{\delta_t}\cos\frac{\delta_t}{2}\right)E_{x\text{in}} + \frac{2\theta}{\delta_t}\sin\frac{\delta_t}{2}E_{y\text{in}}\right] \tag{11.91}$$

$$E_y(z) = e^{-jk\bar{n}z}\left[-\frac{2\theta}{\delta_t}\sin\frac{\delta_t}{2}E_{x\text{in}} + \left(\cos\frac{\delta_t}{2} - j\frac{\Delta\beta_{\text{LB}}z}{\delta_t}\cos\frac{\delta_t}{2}\right)E_{y\text{in}}\right] \tag{11.92}$$

where $\bar{n} = (n_x + n_y)/2$, $\Delta\beta_{\text{LB}} = \beta_y - \beta_x = k(n_y - n_x)$, $\theta = \xi z$, and $\delta_t = \sqrt{(\Delta\beta_{\text{LB}}z)^2 + 4\theta^2}$. Thus the Jones matrix for a linear birefringent medium with a continuously rotating axis [35, 36] is

$$[J_{LB\xi}] = \begin{bmatrix} \cos\dfrac{\delta_t}{2} + j\dfrac{\Delta\beta_{LB}z}{\delta_t}\sin\dfrac{\delta_t}{2} & \dfrac{2\theta}{\delta_t}\sin\dfrac{\delta_t}{2} \\ -\dfrac{2\theta}{\delta_t}\sin\dfrac{\delta_t}{2} & \cos\dfrac{\delta_t}{2} - j\dfrac{\Delta\beta_{LB}z}{\delta_t}\sin\dfrac{\delta_t}{2} \end{bmatrix} \tag{11.93}$$

To verify the results, we consider two special cases. If the birefringent axes are fixed, then $\xi = 0$, $\delta_t = \Delta\beta_{LB}z$, and the 2×2 matrix in (11.93) reduces to $[J_{LB}(0)]$ given in (11.71) for a linearly birefringent fiber with the fast axis fixed in the x direction. If the fiber is nonbirefringent but the axes turn at a constant rate ξ, then $\Delta\beta_{LB} = 0$, $\delta_t = 2\theta$. Thus the 2×2 matrix in (11.93) becomes the rotation matrix given in (11.72).

11.6.3 Circularly Birefringent Fibers

A circularly birefringent fiber can be treated as a *circular retarder*. Let the fiber circular birefringence be $\Delta\beta_{CB} = \beta_R - \beta_L$ and the fiber length be z, then the Jones matrix is

$$[J_{CB}] = \begin{bmatrix} \cos\dfrac{\Delta\beta_{CB}z}{2} & \sin\dfrac{\Delta\beta_{CB}z}{2} \\ -\sin\dfrac{\Delta\beta_{CB}z}{2} & \cos\dfrac{\Delta\beta_{CB}z}{2} \end{bmatrix} \tag{11.94}$$

11.6.4 Linearly and Circularly Birefringent Fibers

In this subsection, we consider a medium that is linearly and circularly birefringent. We model the medium as an anisotropic medium under the influence of a strong magnetic field along the z axis. Let the principal indices of the anisotropic medium be n_x, n_y and n_z. Then, in the absence of a strong magnetic field, the relative dielectric constant tensor is

$$\boldsymbol{\varepsilon}_r = \begin{bmatrix} n_x^2 & 0 & 0 \\ 0 & n_y^2 & 0 \\ 0 & 0 & n_z^2 \end{bmatrix}$$

In the presence of a strong magnetic field, the relative dielectric constant tensor becomes [26–28]

$$\boldsymbol{\varepsilon}_r = \begin{bmatrix} n_x^2 & j\gamma & 0 \\ -j\gamma & n_y^2 & 0 \\ 0 & 0 & n_z^2 \end{bmatrix} \tag{11.95}$$

and γ is proportional to the axial magnetic field, as noted in connection with (11.58). We are interested in plane waves propagating along the direction of the strong magnetic field. We begin with the time-harmonic Maxwell equations (11.59) and

(11.60) and determine the propagation constants and field vectors of the eigenpolarization modes, as we did in Section 11.5.2. To simplify the expressions, we cast
the propagation constant β in terms of an effective index N and write

$$\mathbf{E}(z) = (\hat{\mathbf{x}}E_{x0} + \hat{\mathbf{y}}E_{y0})e^{-jkNz} \tag{11.96}$$

By substituting (11.96) into (11.59), (11.60), and (11.95), we obtain two algebraic equations:

$$(n_x^2 - N^2)E_{x0} + j\gamma E_{y0} = 0 \tag{11.97}$$

$$-j\gamma E_{x0} + (n_y^2 - N^2)E_{y0} = 0 \tag{11.98}$$

There are two independent solutions for N and we label them as N_p and N_n:

$$N_{p,n}^2 = \left\{ \tfrac{1}{2}[(n_x^2 + n_y^2) \pm \sqrt{(n_x^2 - n_y^2)^2 + 4\gamma^2}] \right\}^{1/2} \tag{11.99}$$

The field vectors propagating with kN_p and kN_n are

$$\mathbf{E}_p(z) = E_{p0}\frac{\hat{\mathbf{x}} - j\alpha\hat{\mathbf{y}}}{\sqrt{1 + \alpha^2}}e^{-jkN_pz} \tag{11.100}$$

$$\mathbf{E}_n(z) = E_{n0}\frac{\alpha\hat{\mathbf{x}} + j\hat{\mathbf{y}}}{\sqrt{1 + \alpha^2}}e^{-jkN_nz} \tag{11.101}$$

where

$$\alpha = \frac{2\gamma}{n_x^2 - n_y^2 + \sqrt{(n_x^2 - n_y^2)^2 + 4\gamma^2}} \tag{11.102}$$

and E_{p0} and E_{n0} are the two amplitude constants.
It is convenient to introduce two basis vectors

$$\hat{\mathbf{p}} = \frac{\hat{\mathbf{x}} - j\alpha\hat{\mathbf{y}}}{\sqrt{1 + \alpha^2}} \tag{11.103}$$

$$\hat{\mathbf{n}} = \frac{\alpha\hat{\mathbf{x}} + j\hat{\mathbf{y}}}{\sqrt{1 + \alpha^2}} \tag{11.104}$$

In terms of the two basis vectors $\hat{\mathbf{p}}$ and $\hat{\mathbf{n}}$, unit vectors $\hat{\mathbf{x}}$ and $\hat{\mathbf{y}}$ can be written as

$$\hat{\mathbf{x}} = \frac{\hat{\mathbf{p}} + \alpha\hat{\mathbf{n}}}{\sqrt{1 + \alpha^2}} \tag{11.105}$$

$$\hat{\mathbf{y}} = j\frac{\alpha\hat{\mathbf{p}} - \hat{\mathbf{n}}}{\sqrt{1 + \alpha^2}} \tag{11.106}$$

Suppose the input electric field intensity vector at $z = 0$ is

$$\mathbf{E}(0) = \hat{\mathbf{x}} E_{x\text{in}} + \hat{\mathbf{y}} E_{y\text{in}} = \hat{\mathbf{p}} \frac{E_{x\text{in}} + j\alpha E_{y\text{in}}}{\sqrt{1 + \alpha^2}} + \hat{\mathbf{n}} \frac{\alpha E_{x\text{in}} - j E_{y\text{in}}}{\sqrt{1 + \alpha^2}} \qquad (11.107)$$

Then, the field at z is

$$\mathbf{E}(z) = \hat{\mathbf{p}} \frac{E_{x\text{in}} + j\alpha E_{y\text{in}}}{\sqrt{1 + \alpha^2}} e^{-jkN_p z} + \hat{\mathbf{n}} \frac{\alpha E_{x\text{in}} - j E_{y\text{in}}}{\sqrt{1 + \alpha^2}} e^{-jkN_n z} \qquad (11.108)$$

The Cartesian components of $\mathbf{E}(z)$ are

$$E_x(z) = \left[\left(\cos \frac{k\Delta N z}{2} + j \frac{1 - \alpha^2}{1 + \alpha^2} \sin \frac{k\Delta N z}{2} \right) E_{x\text{in}} - \frac{2\alpha}{1 + \alpha^2} \sin \frac{k\Delta N z}{2} E_{y\text{in}} \right] e^{-jk\overline{N} z}$$

$$\qquad (11.109)$$

$$E_y(z) = \left[\frac{2\alpha}{1 + \alpha^2} \sin \frac{k\Delta N z}{2} E_{z\text{in}} + \left(\cos \frac{k\Delta N z}{2} - j \frac{1 - \alpha^2}{1 + \alpha^2} \sin \frac{k\Delta N z}{2} \right) E_{y\text{in}} \right] e^{-jk\overline{N} z}$$

$$\qquad (11.110)$$

where $\overline{N} = (N_n + N_p)/2$ and $\Delta N = N_n - N_p$.

By rewriting (11.109) and (11.110) in a matrix form, we obtain the Jones matrix for a linearly and circularly birefringent medium [29]:

$$[J_{\text{LCB}}] = \begin{bmatrix} \cos \frac{k\Delta N z}{2} + j \frac{1 - \alpha^2}{1 + \alpha^2} \sin \frac{k\Delta N z}{2} & -\frac{2\alpha}{1 + \alpha^2} \sin \frac{k\Delta N z}{2} \\ \frac{2\alpha}{1 + \alpha^2} \sin \frac{k\Delta N z}{2} & \cos \frac{k\Delta N z}{2} - j \frac{1 - \alpha^2}{1 + \alpha^2} \sin \frac{k\Delta N z}{2} \end{bmatrix}$$

$$\qquad (11.111)$$

To check the validity of (11.111), again we consider two special cases. If the strong magnetic field is absent, γ and α are zero. N_p and N_n become n_x and n_y and the medium is linearly birefringent with a linear birefringence $\Delta\beta_{\text{LB}} = k\Delta N$. The eigenpolarization modes $\mathbf{E}_p$ and $\mathbf{E}_n$ given in (11.100) and (11.101) are linearly polarized fields in the x and y directions. In addition, (11.111) reduces to (11.71) that is, the Jones matrix for a linearly birefringent medium with the fast axis in the x direction.

If the medium is nonbirefringent and subjected to a strong axial magnetic field, then $n_x = n_y$, $\gamma \neq 0$, and $\alpha = 1$. The fields $\mathbf{E}_p$ and $\mathbf{E}_n$ given in (11.100) and (11.101) are, respectively, the right- and left-hand circularly polarized waves studied in Section 11.5.2. In other words, a nonbirefringent medium becomes circularly birefringent when it is subjected to a strong magnetic field. The circular birefringence is $\beta_R - \beta_L = k(N_p - N_n) = -k\Delta N$. Under this circumstance, (11.111) becomes (11.94) that is, the Jones matrix for circularly birefringent medium. In short, we have demonstrated that (11.71) and (11.94) are the two special cases of (11.111).

11.6.5 Fibers with Linear and Circular Birefringence and Axis Rotation

In general, single-mode fibers are linearly as well as circularly birefringent. In addition, the birefringence axes turn continuously. The Jones matrix for these fibers has been studied by Kapron, Borrelli, and Keck [37], Song and Choi [38], and Tsao [39]. In this subsection, we follow Tsao's work and derive the Jones matrix for such a fiber.

Conceptually, we divide a single-mode fiber of length L into a large number of sections of infinitesimal fiber sections. Suppose the ith section has a length Δz_i and is at z_i [Fig. 11.13(a)]. Let the local linear and circular birefringence of the infinitesimal fiber section be $\Delta \beta_{LB}(z_i)$ and $\Delta \beta_{CB}(z_i)$. The local fast axis is at an angle $\theta(z_i)$ relative to the x axis. To model the infinitesimal fiber section, we make use of a result established by Jones in his original work on Jones calculus [34]. Jones has shown that a polarization component can always be represented by the concatenation of a linear retarder and a circular retarder. Therefore, we expect that the infinitesimal fiber section can be represented by the series combination of a linear retarder with a linear retardation $\Delta \beta_{LB}(z_i) \Delta z_i$ and a circular retarder with a circular retardation $\Delta \beta_{CB}(z_i) \Delta z_i$. There are two possible arrangements as shown in Figures 11.13(b) and 11.13(c). It is not possible to determine a priori which arrangement would be more accurate. Therefore, we will consider both combinations before making the final selection. First we consider the arrangement shown in Figure 11.13(b). A rotation matrix $[S(\theta_i - \theta_{i-1})]$ is added to account for the axis rotation between neighboring sections. The Jones matrix is

$$[J_{bi}(z_i)] = \begin{bmatrix} \cos \dfrac{\Delta \beta_{CB} \Delta z_i}{2} & \sin \dfrac{\Delta \beta_{CB} \Delta z_i}{2} \\ -\sin \dfrac{\Delta \beta_{CB} \Delta z_i}{2} & \cos \dfrac{\Delta \beta_{CB} \Delta z_i}{2} \end{bmatrix} \begin{bmatrix} e^{j \Delta \beta_{LB} \Delta z_i / 2} & 0 \\ 0 & e^{-j \Delta \beta_{LB} \Delta z_i / 2} \end{bmatrix}$$

$$\times \begin{bmatrix} \cos \Delta \theta_i & \sin \Delta \theta_i \\ -\sin \Delta \theta_i & \cos \Delta \theta_i \end{bmatrix} \tag{11.112}$$

where $\Delta \theta_i = \theta_i - \theta_{i-1}$.

In the arrangement depicted in Figure 11.13(c), the order of the two retarders is reversed. The Jones matrix is

$$[J_{ci}(z_i)] = \begin{bmatrix} e^{j \Delta \beta_{LB} \Delta z_i / 2} & 0 \\ 0 & e^{-j \Delta \beta_{LB} z_i / 2} \end{bmatrix} \begin{bmatrix} \cos \dfrac{\Delta \beta_{CB} \Delta z_i}{2} & \sin \dfrac{\Delta \beta_{CB} \Delta z_i}{2} \\ -\sin \dfrac{\Delta \beta_{CB} \Delta z_i}{2} & \cos \dfrac{\Delta \beta_{CB} \Delta z_i}{2} \end{bmatrix}$$

$$\times \begin{bmatrix} \cos \Delta \theta_i & \sin \Delta \theta_i \\ -\sin \Delta \theta_i & \cos \Delta \theta_i \end{bmatrix} \tag{11.113}$$

If Δz_i and $\Delta \theta_i$ are finite, the polarization characteristics of the two arrangements are different as evidenced by the fact that $[J_{bi}]$ and $[J_{ci}]$ are different. However, if

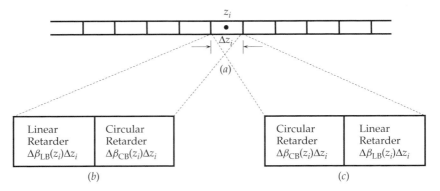

Figure 11.13 Two models for a fiber with linear and circular birefringence and axis rotation: (a) Short fiber section, (b) polarization model for short fiber section, and (c) alternate polarization model for short fiber section.

Δz_i and $\Delta \theta_i$ are infinitesimally small, (11.112) and (11.113) become

$$[J_{bi}(z_i)] = \begin{bmatrix} 1 & 0 \\ 0 & 1 \end{bmatrix} + \begin{bmatrix} j \dfrac{\Delta \beta_{LB}}{2} & \dfrac{\Delta \beta_{CB}}{2} + \dfrac{\Delta \theta_i}{\Delta z_i} \\ -\dfrac{\Delta \beta_{CB}}{2} - \dfrac{\Delta \theta_i}{\Delta z_i} & -j \dfrac{\Delta \beta_{LB}}{2} \end{bmatrix} \Delta z_i$$

$$+ O[(\Delta z_i)^2] = [J_{ci}(z_i)] \qquad (11.114)$$

When terms on the order of $(\Delta z_i)^2$ or smaller are ignored, $[J_{bi}]$ and $[J_{ci}]$ become identical. In other words, the polarization characteristic of a fiber section is independent of the order of arrangement if the fiber section is infinitesimally short. Thus, we can use either arrangement to represent the polarization characteristic of an infinitesimal fiber section. Accurate to the first order of Δz_i, $[J_{bi}(z_i)]$ and $[J_{ci}(z_i)]$ are

$$[J_i(z_i)] = [I] + [j(z_i)]\Delta z_i \qquad (11.115)$$

where $[I]$ is the identity matrix and

$$[j(z_i)] = \begin{bmatrix} j \dfrac{\Delta \beta_{LB}(z_i)}{2} & \dfrac{\Delta \beta_{CB}(z_i)}{2} + \dfrac{\Delta \theta(z_i)}{\Delta z} \\ -\dfrac{\Delta \beta_{CB}(z_i)}{2} - \dfrac{\Delta \theta(z_i)}{\Delta z_i} & -j \dfrac{\Delta \beta_{LB}(z_i)}{2} \end{bmatrix} \qquad (11.116)$$

The polarization characteristic of a fiber section of a finite length is then the ordered product of $[J_i(z_i)]$:

$$[J_{fb}] = \prod_i [J_i(z_i)] = \prod_i ([I] + [j(z_i)]\Delta z_i) = \exp\left(\sum_i [j(z_i)]\Delta z_i\right) \qquad (11.117)$$

As Δz_i approaches 0, the summation of the 2×2 matrices in the exponent becomes an integral of matrices

$$[M] = \sum_i [j(z_i)]\Delta z_i \longrightarrow \int_0^L [j(z)]\,dz \tag{11.118}$$

We evaluate the matrix integral element by element and obtain

$$[M] = \frac{1}{2}\begin{bmatrix} j\Delta_L & \Delta_C + 2\Delta_\Theta \\ -(\Delta_C + 2\Delta_\Theta) & -j\Delta_L \end{bmatrix} \tag{11.119}$$

where

$$\Delta_L = \int_0^L \Delta\beta_{\mathrm{LB}}(z)\,dz \tag{11.120}$$

$$\Delta_C = \int_0^L \Delta\beta_{\mathrm{CB}}(z)\,dz \tag{11.121}$$

are the total linear and circular retardation of the fiber of length L and

$$\Delta_\Theta = \int_0^L \frac{d\theta(z)}{dz}\,dz = \theta(L) - \theta(0) \tag{11.122}$$

is the total angle of rotation.

An explicit expression for $[J_{\mathrm{fb}}]$, in lieu of $[M]$, is obtained by applying the following theorem [40]. Let m_1 and m_2 be the eigenvalues, and $[P]$ be the diagonalizing transformation matrix, of $[M]$:

$$[P]^{-1}[M][P] = \begin{bmatrix} m_1 & 0 \\ 0 & m_2 \end{bmatrix} \tag{11.123}$$

Then, the exponential function of $[M]$ is

$$e^{[M]} = [P]\begin{bmatrix} e^{m_1} & \\ & e^{m_2} \end{bmatrix}[P]^{-1} \tag{11.124}$$

For the problem at hand, the eigenvalues of $[M]$ are $\pm j\Gamma$ and

$$\Gamma = \tfrac{1}{2}[\Delta_L^2 + (\Delta_C + 2\Delta_\Theta)^2]^{1/2} \tag{11.125}$$

In terms of Δ_L, Δ_C, Δ_Θ, and Γ, the diagonalizing transformation matrix and its inverse matrix are

$$[P] = \frac{1}{2}\begin{bmatrix} -j(\Delta_L + 2\Gamma) & -j(\Delta_L - 2\Gamma) \\ \Delta_C + 2\Delta_\Theta & \Delta_C + 2\Delta_\Theta \end{bmatrix} \tag{11.126}$$

$$[P]^{-1} = \frac{j}{2\Gamma(\Delta_C + 2\Delta_\Theta)}\begin{bmatrix} \Delta_C + 2\Delta_\Theta & j(\Delta_L - 2\Gamma) \\ -(\Delta_C + 2\Delta_\Theta) & -j(\Delta_L + 2\Gamma) \end{bmatrix} \tag{11.127}$$

A straightforward and tedious matrix manipulation leads to

$$[J_{fb}] = e^{[M]} = \begin{bmatrix} \cos\Gamma + j\dfrac{\Delta_L}{2}\dfrac{\sin\Gamma}{\Gamma} & \left(\dfrac{\Delta_C}{2} + \Delta_\Theta\right)\dfrac{\sin\Gamma}{\Gamma} \\ -\left(\dfrac{\Delta_C}{2} + \Delta_\Theta\right)\dfrac{\sin\Gamma}{\Gamma} & \cos\Gamma - j\dfrac{\Delta_L}{2}\dfrac{\sin\Gamma}{\Gamma} \end{bmatrix} \qquad (11.128)$$

This is the Jones matrix of a single-mode fiber that is linearly and circularly birefringent and has continuously rotating birefringence axes [39]. No assumption is made on the nature of the birefringence or the dependence of $\Delta\beta_{LB}$, $\Delta\beta_{CB}$, and θ on z.

To verify the veracity of (11.128), we consider three special cases. For a linear birefringent fiber with the fast axis along the x axis, Δ_Θ and Δ_C vanish and (11.128) reduces to (11.71). For a linearly birefringent fiber with constant turning axes, $\Delta_C = 0$, $\Delta_\Theta \neq 0$, (11.128) reduces to (11.93). Finally, for a circularly birefringent fiber without the axis rotation, Δ_L and Δ_Θ are zero, (11.128) reduces to (11.94).

REFERENCES

1. T. Okoshi, "Single-polarization single-mode optical fibers," *IEEE J. Quantum Electron.*, Vol. QE-17, No. 6, pp. 879–884 (1981).

2. A. M. Smith, "Birefringence induced by bends and twists in single-mode optical fiber," *Appl. Opt.*, Vol. 19, pp. 2606–2611 (1980).

3. I. P. Kaminow, "Polarization in optical fibers," *IEEE J. Quantum Electron.*, Vol. QE-17, pp. 15–22 (1981).

4. J. I. Sakai and T. Kimura, "Birefringence and polarization characteristics of single-mode optical fibers under elastic deformations," *IEEE J. Quantum Electron.*, Vol. QE-17, No. 6, pp. 1041–1051 (June 1981).

5. S. C. Rashleigh, "Origins and control of polarization effects in single-mode fiber," *IEEE J. Lightwave Technol.*, Vol. LT-1, pp. 312–331 (1983).

6. J. Noda, K. Okamoto, and Y. Sasaki, "Polarization-maintaining fibers and their applications," *IEEE J. Lightwave Technol.*, Vol. LT-4, No. 8, pp. 1071–1089 (Aug. 1986).

7. F. I. Akers and R. E. Thompson, "Polarization-maintaining single-mode fibers," *Appl. Opt.*, Vol. 21, pp. 1720–1721 (1982).

8. A. J. Morrow, "Introduction for optical fibers," 1986 Conference on Optical Fiber Communication, Minitutorial Session MF1 (Feb. 24, 1986).

9. C. Yeh, "Elliptical dielectric waveguides," *J. Appl. Phys.*, Vol. 33, pp. 3235–3243 (1962).

10. C. Yeh, "Modes in weakly guided elliptical optical fibers," *Opt. Quantum Electron.*, Vol. 8, pp. 43–47 (1976).

11. V. Ramaswamy and W. G. French, "Influence of noncircular core on the polarization performance of single mode fibers," *Electron. Lett.*, Vol. 14, pp. 143–144 (1979).

12. M. J. Adams, D. N. Payne, and C. M. Ragdale, "Birefringence in optical fibers with elliptical cross-section," *Electron. Lett.*, Vol. 15, pp. 298–299 (1979).

13. J. D. Love, R. A. Sammut, and A. W. Snyder, "Birefringence in elliptically deformed optical fibers," *Electron. Lett.*, Vol. 15, No. 20, pp. 615–616 (1979).

14. I. P. Kaminow and V. Ramaswamy, "Single-polarization optical fibers: Slab model," *Appl. Phys. Lett.*, Vol. 34, No. 4, pp. 268–270 (1979).

15. W. Eickhoff, "Stress-induced single-polarization single-mode fiber," *Opt. Lett.*, Vol. 7, No. 12, pp. 629–631 (1982).

16. S. Timoshenko and J. N. Goodier, *Theory of Elasticity*, 3rd ed., Sec. 37, McGraw-Hill, New York, (1970).

17. A. P. Boresi and K. P. Chong, *Elasticity in Engineering Mechanics*, 2nd ed., Chapter 6, Wiley, New York, (2000), Chapter 6.

18. G. Nadeau, *Introduction to Elasticity*, Sec. 10.7, Holt, Rinehart and Winston, New York, (1964).

19. Y. Namihira, M. Kudo, and Y. Mushiako, "Effect of mechanical stress on the transmission characteristics of optical fiber," *Trans. Inst. Electronics Commun. Japan,* Vol. 60C, No. 7, pp. 107–115 (1977).

20. A. Kumar and R. Ulrich, "Birefringence of optical fiber pressed into a V groove," *Opt. Lett.*, Vol. 6, No. 12, pp. 644–646 (1981).

21. R. Ulrich, S. C. Rashleigh, and W. Eickhoff, "Bending-induced birefringence in single-mode fibers," *Opt. Lett.*, Vol. 5, No. 6, pp. 273–275 (1980).

22. S. C. Rashleigh and R. Ulrich, "High birefringence in tension-coiled single-mode fibers," *Opt. Lett.*, Vol. 5, No. 8, pp. 354–356 (1980).

23. M. C. Farris and A. J. Rogers, "Temperature dependence of Kerr effect in silica optical fiber," *Electron. Lett.*, Vol. 19, No. 21, pp. 890–891 (1983).

24. A. J. Rogers, "Polarization-optical time domain reflectometry: A technique for the measurement of field distributions," *Appl. Opt.*, Vol. 20, No. 6, pp. 1060–1074 (March 15, 1981).

25. A. Simon and R. Ulrich, "Evolution of polarization along a single-mode fiber," *Appl. Phys., Lett.*, Vol. 31, No. 8, pp. 517–520 (1977).

26. P. S. Pershan, "Magneto-optical effects," *J. Appl. Phys.*, Vol. 38, No. 3, pp. 1482–1490 (1967).

27. M. J. Freiser, "A survey of magnetooptic effects," *IEEE Trans. Magnetics*, Vol. MAG-4, No. 2, pp. 152–161 (1968).

28. S. Donati, V. Annovazzi-Lodi, and T. Tambosso, "Magneto-optical fibre sensors for electrical industry: Analysis of performance," *IEE Proc.*, Vol. 135, Pt. J, No. 5 (Oct. 1988).

29. W. J. Tabor and F. S. Chen, "Electromagnetic propagation through materials possessing both Faraday rotation and birefringence: Experiments with Ytterbium Orthoferrite," *J. Appl. Phys.*, Vol. 40, No. 7, pp. 2760–2765 (1969).

30. J. Noda, T. Hosaka, Y. Sasaki, and R. Ulrich, "Dispersion of Verdet constant in stress-birefringent silica fibre," *Electron. Lett.*, Vol. 20, No. 22, pp. 906–908, (Oct. 25, 1984).

31. A. H. Rose, S. M. Etzel, and C. M. Wang, "Verdet constant dispersion in annealed optical fiber current sensors," *IEEE J. Lightwave Technol.*, Vol. 15, No. 5, pp. 803–807 (1997).

32. E. Hecht, *Optics*, 4th ed., Addison Wesley, Reading, MA, (2002).

33. W. A. Shurcliff, *Polarized Light, Production and Use*, Harvard University Press, Cambridge, MA, (1962).

34. R. C. Jones, "A new calculus for the treatment of optical systems, Part I, Description and discussion of the calculus," *J. Opt. Soc. Am.*, Vol. 31, pp. 488–493 (1941); and

H. Hurwits, Jr., and R. C. Jones, "Part II Proof of three general equivalent theorems," *J. Opt. Soc. Am.*, Vol. 31, pp. 493–499 (1941).

35. P. McIntyre and A. W. Snyder, "Light propagation in twisted anisotropic media: Application to photoreceptors," *J. Opt. Soc. Am.*, Vol. 68, No. 2, pp. 149–157, (Feb. 1978).

36. A. J. Barlow, J. J. Ramskov-Hansen, and D. N. Payne, "Birefringence and polarization mode dispersion in spun single mode fibers," *Appl. Opt.*, Vol. 20, pp. 2962–2968 (1981).

37. F. P. Kapron, N. F. Borrelli, and D. B. Keck, "Birefringence in dielectric optical waveguides," *IEEE J. Quantum Electron.*, Vol. QE-8, pp. 222–225 (1972).

38. G. H. Song and S. S. Choi, "Analysis of birefringence in single-mode fibers and theory for the backscattering measurement," *J. Opt. Soc. Am., A.*, Vol. 2, No. 2, pp. 167–170 (Feb. 1985).

39. C. Y. Tsao, "Polarization parameters of plane waves in hybrid birefringent optical fibers," *J. Opt. Soc. Am. A*, Vol. 4, pp. 1407–14127 (1987).

40. K. Ogata, *State Space Analysis for Control Systems*, Prentice-Hall, Englewood Cliffs, NJ, (1967).

12

MANUFACTURED FIBERS

12.1 INTRODUCTION

In Chapters 9 and 10, we studied the properties of ideal step-index fibers. For ideal step-index fibers, the index profiles are very simple that we are able to obtain closed-form expressions for many quantities of interest. But manufactured fibers are not ideal fibers. Most, if not all, manufactured fibers have complicated and position-dependent index profiles. For many manufactured fibers, the index profiles are so complicated, by design or due to the manufacturing processes, that they cannot be expressed in terms of "simple" or "nice" functions. For some fibers, the index profiles are not known at all. Naturally, we are interested in the propagation and dispersion properties of manufactured fibers. Of particular interest are the characteristics of the fundamental modes guided by these fibers.

Because of the manufacturing processes, the index distributions in real fibers are independent of the azimuthal variable ϕ. Thus we consider fibers having radially inhomogeneous and angularly independent index profiles. We take the peak index in the core region as n_{co}. For simplicity, we also assume the cladding index is a constant n_{cl}. To facilitate the analysis and to compare results, we define the *normalized frequency V* and the *generalized guide index b* in terms of the peak core

Foundations for Guided-Wave Optics, by Chin-Lin Chen
Copyright © 2007 John Wiley & Sons, Inc.

index n_{co}, the cladding index n_{cl}, the effective refractive index N, and the core radius a:

$$V = ka\sqrt{n_{co}^2 - n_{cl}^2} \qquad (12.1)$$

and

$$b = \frac{N^2 - n_{cl}^2}{n_{co}^2 - n_{cl}^2} \qquad (12.2)$$

The generalized parameters so defined are identical to those introduced in Chapter 9 for ideal step-index fibers. Once b is known, we can evaluate the propagation constant of the fiber,

$$\beta \approx \frac{2\pi}{\lambda} n_{cl}(1 + b\Delta) \qquad (12.3)$$

where the index difference Δ is

$$\Delta = \frac{n_{co}^2 - n_{cl}^2}{2n_{cl}^2} \qquad (12.4)$$

The transit time of an optical pulse traversing a fiber of length L is

$$\tau_{gr} = L\frac{d\beta}{d\omega}$$

The group delay per unit length is $\tau_{gr}/L = d\beta/d\omega$. A straightforward differentiation of (12.3) leads to

$$\tau_{gr} = -L\frac{\lambda^2}{2\pi c}\frac{d\beta}{d\lambda} = \frac{L}{c}\left\{\left(n_{cl} - \lambda\frac{dn_{cl}}{d\lambda}\right)\left[1 + \Delta\frac{d(Vb)}{dV}\right]\right.$$
$$\left. -\frac{1}{2}\lambda n_{cl}\frac{d\Delta}{d\lambda}\left[b + \frac{d(Vb)}{dV}\right]\right\} \qquad (12.5)$$

For most fibers of interest, n_{co} and n_{cl} are slightly dispersive, that is, $|dn/d\lambda| \ll n/\lambda$. If we ignore terms involving $d\Delta/d\lambda$ and $\Delta[dn_{cl}/d\lambda]$, then (12.5) reduces to

$$\tau_{gr} \approx \frac{L}{c}\left[\left(n_{cl} - \lambda\frac{dn_{cl}}{d\lambda}\right) + n_{cl}\Delta\frac{d(Vb)}{dV}\right] \qquad (12.6)$$

The first term of (12.6) is the elapse time of a pulse of plane waves traveling in a medium with an index n_{cl} and thickness L. The second term is the correction to the group delay due to the wave guiding in the fiber. Note the presence of $d(Vb)/dV$ that depends on the fiber geometry and wavelength of operation. We refer to $d(Vb)/dV$ as the *normalized group delay*.

Clearly, the transit time given in (12.5) or (12.6) is wavelength dependent. All pulses of a finite temporal duration have a finite spectral width. Different spectral components move with different group velocities and have different transit times. As a result, pulses are distorted and broadened as they propagate. As discussed in Section 9.4, we define the change of the group delay per unit fiber length per unit spectral width,

$$\frac{1}{L}\frac{\Delta\tau_{gr}}{\Delta\lambda} \sim \frac{1}{L}\frac{d\tau_{gr}}{d\lambda}$$

as the *dispersion* $\mathcal{D}$. Gambling et al. [1] showed that the dispersion $\mathcal{D}$ can be partitioned into three parts: the *material dispersion* $\mathcal{D}_{mt}$, the *waveguide dispersion* $\mathcal{D}_{wg}$, and the *profile dispersion* $\mathcal{D}_{pf}$. The three dispersion components are approximately ([1] and Appendix F)

$$\mathcal{D}_{mt} \approx -\frac{\lambda}{c}\frac{d^2 n_{cl}}{d\lambda^2}\left\{1 - \frac{1}{2}\left[b + \frac{d(Vb)}{dV}\right]\right\} - \frac{\lambda}{2c}\frac{d^2 n_{co}}{d\lambda^2}\left[b + \frac{d(Vb)}{dV}\right] \quad (12.7)$$

$$\mathcal{D}_{wg} \approx -\frac{n_{cl}\Delta}{c\lambda}\left(1 - \frac{\lambda}{n_{cl}}\frac{dn_{cl}}{d\lambda}\right)^2 V\frac{d^2(Vb)}{dV^2} \quad (12.8)$$

$$\mathcal{D}_{pf} \approx \frac{n_{cl}}{c}\left\{\left(1 - \frac{\lambda}{n_{cl}}\frac{dn_{cl}}{d\lambda} + \frac{\lambda}{4\Delta}\frac{d\Delta}{d\lambda}\right)\left[V\frac{d^2(Vb)}{dV^2} + \frac{d(Vb)}{dV} - b\right]\right.$$
$$\left. + \frac{\lambda}{n_{cl}}\frac{dn_{cl}}{d\lambda}\left[\frac{d(Vb)}{dV} + b\right]\right\}\frac{d\Delta}{d\lambda} \quad (12.9)$$

Clearly, $d^2 n_{co}/d\lambda^2$, $d^2 n_{cl}/d\lambda^2$, and $d\Delta/d\lambda$ are properties of the core and cladding materials. On the other hand, b, $d(Vb)/dV$, and $V[d^2(Vb)/dV^2]$ are the propagation and dispersion properties of the fiber, and they depend on the fiber geometry and index profile. We note the presence of $V[d^2(Vb)/dV^2]$ in the material dispersion $\mathcal{D}_{wg}$ and the profile dispersion $\mathcal{D}_{pf}$. Clearly $V[d^2(Vb)/dV^2]$ is an important property of the waveguide. We refer to $V[d^2(Vb)/dV^2]$ as the *normalized waveguide dispersion*. In summary, b, $d(Vb)/dV$, and $V[d^2(Vb)/dV^2]$ are the three key propagation and dispersion parameters. They are the focus of our study in the next two sections.

Since most manufactured fibers have angularly independent index profiles, the modal fields of, and the fields radiated by, the fundamental modes of the manufactured fibers are also angularly independent. In other words, fields in the transverse planes depend on r only. It is reasonable to expect the key fiber properties may be characterized or represented by a single parameter. Such a parameter is the *mode field radius*, *effective mode radius*, or *spot size*. We will study the mode field radius in the last two sections of this chapter.

12.2 POWER-LAW INDEX FIBERS

12.2.1 Kurtz and Striefer's Theory of Waves Guided by Inhomogeneous Media

The usual methods for analyzing waves in homogeneous media are not applicable to graded-index fibers since graded-index fibers are inhomogeneous media. Kurtz and Streifer [2] have showed that waves guided by inhomogeneous media can be expressed in terms of two potential functions. In the special case of a weak inhomogeneity, the two potential functions can be combined into a single potential function. It follows that transverse fields guided by weakly inhomogeneous media can be expressed as one, instead of two, potential function. This special case applies exactly to the weakly guiding graded-index fibers. This is the starting point of Gambling and Matsumura's analysis of weakly guiding fibers with power-law index profiles [3]. We follow their analysis and discuss key results they obtained.

12.2.2   Fields and Dispersion of LP Modes

Since the fiber geometry and index profile are independent of ϕ, we write the potential function as $\mathcal{R}_\ell(r)e^{\pm j\ell\phi}e^{-j\beta z}$, where $\ell = 0, 1, 2, \ldots$. We express the dominant transverse field component E_y and H_x of LP modes as

$$E_y(r, \phi, z) = \mathcal{R}_\ell(r)e^{\pm j\ell\phi}e^{-j\beta z} \tag{12.10}$$

$$H_x(r, \phi, z) = -\frac{n(r)}{\eta_0}\mathcal{R}_\ell(r)e^{\pm j\ell\phi}e^{-j\beta z} \tag{12.11}$$

where $\eta_0 = \sqrt{\mu_0/\varepsilon_0}$ is the intrinsic impedance of free space. $\mathcal{R}_\ell(r)$ satisfies a scalar wave equation:

$$\left[\frac{1}{r}\frac{d}{dr}\left(r\frac{d}{dr}\right) + k^2 n^2(r) - \beta^2 - \frac{\ell^2}{r^2}\right]\mathcal{R}_\ell(r) = 0 \tag{12.12}$$

For convenience, we write the *power-law index profile* of a weakly guiding fiber as

$$n^2(r) = \begin{cases} n_{co}^2(1 - 2\Delta'(r/a)^\alpha) & r \leq a \\ n_{cl}^2 = n_{co}^2(1 - 2\Delta') & r \geq a \end{cases} \tag{12.13}$$

where $\Delta' = (n_{co}^2 - n_{cl}^2)/(2n_{co}^2)$. Strictly speaking, Δ' defined here and Δ defined in (12.4) are different. For weakly guiding fibers, Δ and Δ' are the same to the first order of $(n_{co} - n_{cl})/n_{co}$. As noted in Chapter 9, the exponent α is the *index profile parameter* or the *grading parameter* describing the index variation in the core region. The cases of $\alpha = 1, 2$, and ∞ correspond to the triangular, parabolic, and step-index index profiles, as shown in Figure 12.1. In general, a large α corresponds to an index profile that turns sharply near the core–cladding boundary.

To proceed, we substitute (12.13) into (12.12) and cast terms in terms of V and b. We also introduce a normalized variable $\rho = r/a$. Then we obtain a differential equation for the core region

$$\left[\frac{1}{\rho}\frac{d}{d\rho}\left(\rho\frac{d}{d\rho}\right) + V^2(1 - b) - V^2\rho^\alpha - \frac{\ell^2}{\rho^2}\right]\mathcal{R}_\ell(\rho) = 0 \tag{12.14}$$

In the limit of an infinite α, corresponding to step-index fibers, $V^2\rho^\alpha$ approaches to zero for $\rho < 1$. Then the differential equation reduces to the well-known Bessel

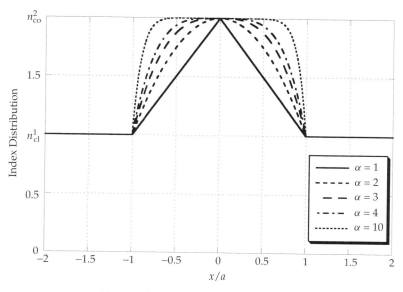

Figure 12.1 Power-law index profiles.

differential equation of order ℓ that has $J_\ell(V\sqrt{1-b}\rho)$ and $Y_\ell(V\sqrt{1-b}\rho)$ as the two linearly independent solutions. This is exactly the differential equation and solutions we discussed in Chapter 9 for step-index fibers. For power-law index fibers with a finite α, we have to keep $V^2\rho^\alpha$ in the differential equation. Solutions to the differential equation with the $V^2\rho^\alpha$ term cannot be expressed in terms of simple or known functions. Thus, we search for a power series representation. Since the origin, $\rho = 0$, is a point in the region of interest and the field must be finite at the origin, we are interested in a power series that converges at $\rho = 0$. For simplicity, we take α as an integer and write $\mathcal{R}_\ell(\rho)$ as a power series

$$R_\ell(\rho) = E_0\rho^\ell \sum_{i=0}^{\infty} a_i\rho^i \tag{12.15}$$

where the coefficients a_i and amplitude constant E_0 are yet undetermined. To determine a_i, we substitute (12.15) into the differential equation (12.14), arrange terms in ascending power of ρ and obtain

$$\rho^\ell \sum_{i=0}^{\infty} \left[i(i + 2\ell) + V^2(1 - b) - V^2\rho^\alpha \right] a_i\rho^i = 0 \tag{12.16}$$

If and only if the coefficient of each power of ρ in (12.16) vanishes individually, $\mathcal{R}_\ell(\rho)$ is a valid solution of (12.14). By setting the coefficient of each power of ρ in (12.16) to zero, we obtain a *recurrence relation* connecting the coefficients a_i. We leave the amplitude constant E_0 unspecified. Thus we are free to specify one of the coefficients. For convenience, we set a_0 to 1, and obtain a recurrence relation [3, 4]:

$$a_0 = 1$$

$$a_1 = 0$$

$$a_2 = -\frac{1}{2(2 + 2\ell)} V^2 (1 - b) a_0$$

$$\cdots \qquad (12.17)$$

$$a_i = -\frac{1}{i(i + 2\ell)} [V^2 (1 - b) a_{i-2}] \qquad 2 \leq i < \alpha + 2$$

$$a_i = -\frac{1}{i(i + 2\ell)} [V^2 (1 - b) a_{i-2} - V^2 a_{i-\alpha-2}] \qquad \alpha + 2 \leq i$$

Since the cladding region has a constant index $n_{cl.}$, the differential equation (12.12) for the cladding region is simply the modified Bessel differential equation of order ℓ as discussed in Chapter 9:

$$\left[\frac{1}{\rho} \frac{d}{d\rho} \left(\rho \frac{d}{d\rho} \right) - V^2 b - \frac{\ell^2}{\rho^2} \right] \mathcal{R}_\ell(\rho) = 0 \qquad (12.18)$$

It has two independent solutions, $I_\ell(V \sqrt{b} \rho)$ and $K_\ell(V \sqrt{b} \rho)$. Since the field must decay exponentially in the cladding region far from the core–cladding boundary, we write for the cladding region, $\rho > 1$,

$$\mathcal{R}_\ell(\rho) = E_0' K_\ell(V \sqrt{b} \rho) \qquad (12.19)$$

In (12.15) and (12.19), the two amplitude constants E_0 and E_0' and b are yet unknown. To determine b, we make use of the conditions that $\mathcal{R}_\ell(\rho)$ and $d\mathcal{R}_\ell(\rho)/d\rho$ are continuous at the core–cladding boundary, that is, at $\rho = 1$,

$$E_0 \sum_{i=0}^{\infty} a_i = E_0' K_\ell(V \sqrt{b}) \qquad (12.20)$$

$$E_0 \left(\ell \sum_{i=0}^{\infty} a_i + \sum_{i=0}^{\infty} i a_i \right) = E_0' V \sqrt{b} K_\ell'(V \sqrt{b}) \qquad (12.21)$$

Upon eliminating E_0 and E_0' from the two equations, we obtain a *dispersion relation*:

$$\ell + \frac{\sum_{i=0}^{\infty} i a_i}{\sum_{i=0}^{\infty} a_i} = V \sqrt{b} \frac{K_\ell'(V \sqrt{b})}{K_\ell(V \sqrt{b})} \qquad (12.22)$$

As indicated by the recurrence relation (12.17), a_i depends on V, α, ℓ, and b. For a given power-law index fiber having a specific α and V, all terms, except b, in the dispersion relation are known. We solve (12.22) numerically for b. Once

b is known, we obtain the bV curves and modal fields of LP modes guided by weakly guiding power-law index fibers. In Figures 12.2(a) and 12.2(b), we display the intensity distributions ($|E_y|^2$) of the fundamental LP modes for five values of α. As observed by Gambling and Matsumura [3], distributions of transverse electric fields of LP_{10} modes of various values of α are very similar. In particular, all fields of the fundamental LP modes have a peak on the fiber axis and fields decay monotonically as r increases.

In (12.15) and (12.17), the potential function $\mathcal{R}_\ell(r)$ and the recurrence relation are derived for integer α only. Love [5] and Krumbholz et al., [6] have studied modes guided by fibers with a rational number α. Gambling et al. have also studied the LP modes guided by fibers with an index dip on the fiber axis [7].

12.2.3 Cutoff of Higher-Order LP Modes

For communication applications, it is preferable to operate fibers in the fundamental modes. To ensure operation at the LP_{01} mode, we choose fibers and wavelengths such that all higher-order LP modes are cutoff. Accordingly, we are particularly interested in the cutoff condition of the second lowest mode, LP_{11} mode. At the cutoff, b is 0. As b tends to zero,

$$V\sqrt{b}\,\frac{K'_\ell(V\sqrt{b})}{K_\ell(V\sqrt{b})} \to -\ell \qquad (12.23)$$

Then, we obtain from (12.22) the cutoff V_c for the $LP_{\ell m}$ mode [8]:

$$2\ell + \frac{\sum_{i=0}^{\infty} i a_i}{\sum_{i=0}^{\infty} a_i} = 0 \qquad (12.24)$$

By solving (12.24) numerically, we have the cutoff V_c of the $LP_{\ell m}$ mode. Gambling et al. [8] have evaluated V_c of the LP_{11} mode guided by power-law index fibers with integer α. They found that for fibers with a *triangular* and *parabolic index profiles*, corresponding to $\alpha = 1$ and 2, V_c is 4.38 and 3.518, respectively. For comparison, we recall that the cutoff of the LP_{11} mode of step-index fibers is 2.4048. In general, V_c decreases as α increases. Gambling et al. also obtained an empirical expression for V_c of LP_{11} modes as a function of α [8]:

$$V_c \approx 2.405\,\sqrt{1 + \frac{2}{\alpha}} \qquad (12.25)$$

In Figure 12.3, we plot V_c as a function of α as given by (12.25). Also shown in the figure are the exact values of V_c for $\alpha = 1$, 2, and ∞. Note that fibers having a smaller α have a wider single-mode bandwidth.

As mentioned earlier, Gambling and Matsumura [3] have determined b of fibers with power-law index profiles (see Fig. 12.2). They also obtained $d(Vb)/dV$ and $V[d^2(Vb)/dV^2]$ by the numerical differentiation. Their results for the LP_{01} mode for four values of α are shown in Figure 12.4 [1]. Clearly, within each group, the curves have the same shape.

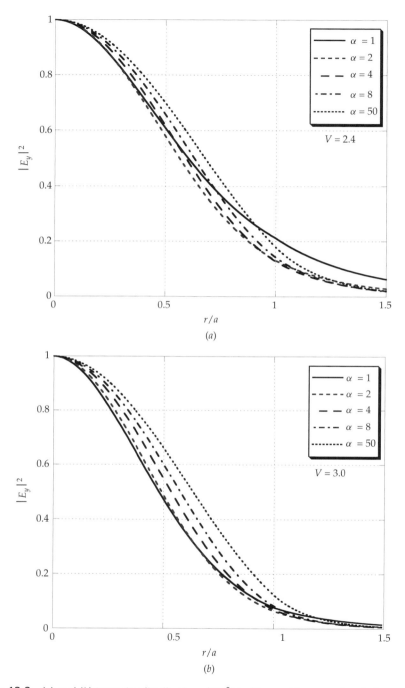

Figure 12.2 (a) and (b) Intensity distributions ($|E_y|^2$) in fibers with power-law index profiles. (After [3].)

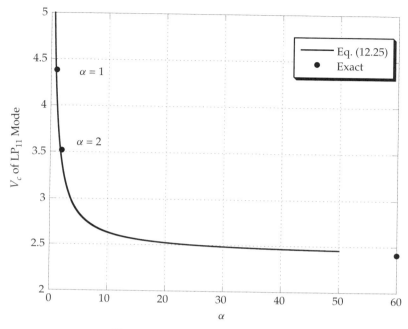

Figure 12.3   Cutoff V of LP$_{11}$ mode.

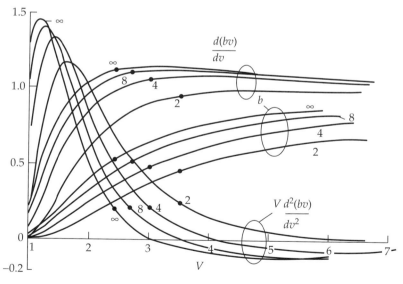

Figure 12.4   Propagation and dispersion characteristics of fibers with power-law index profiles. ●●● indicates cutoff of LP$_{11}$ modes. (From [1].)

12.3   KEY PROPAGATION AND DISPERSION PARAMETERS OF GRADED-INDEX FIBERS

As noted earlier, an accurate analytic representation of the index profile is often unknown or does not exist. Without an explicit expression for the index profile, it would be impossible to analyze the fiber from the differential equation approach. On the other hand, the modal field of a manufactured fiber can always be measured experimentally or inferred somehow. It is natural to inquire if the key propagation and dispersion parameters may be estimated if the modal field is known. In this section, we establish relations between the modal field and the key parameters.

Again, we consider a fiber having a radially dependent and angularly independent index profile $n(r)$. In addition, we are mainly interested in the fundamental mode for which the modal field is also independent of ϕ. Thus we write the model field as $\mathcal{R}_0(r)e^{-j\beta z}$.

12.3.1   Generalized Guide Index *b*

We set ℓ in the differential equation (12.12) to zero, multiply the differential equation by $\mathcal{R}_0(r)$, and integrate both sides of the resulting equation over the entire fiber cross section and obtain

$$\int_{co+cl} \left[\frac{d\mathcal{R}_0(r)}{dr} \right]^2 ds = \int_{co+cl} [k^2 n^2(r) - \beta^2] \mathcal{R}_0^2(r)\, ds \qquad (12.26)$$

where $ds = r\, dr\, d\phi$ is the differential surface element. In the limits of the integrals, co and cl stand for the core and cladding regions, respectively. In the manipulation, we have made use of the fact that (1) $\mathcal{R}_0(r)$ and $d\mathcal{R}_0(r)/dr$ are finite at $r = 0$; and (2) the modal field decays exponentially as r increases. We also assume that $\mathcal{R}_0(r)$ is normalized in that

$$\int_{co+cl} \mathcal{R}_0^2(r)\, ds = 1 \qquad (12.27)$$

From the definitions of b and V defined in (12.1) and (12.2), we cast (12.26) as

$$b = \int_{co} \frac{n^2(r) - n_{cl}^2}{n_{co}^2 - n_{cl}^2} \mathcal{R}_0^2(r)\, ds - \frac{a^2}{V^2} \int_{co+cl} \left[\frac{d\mathcal{R}_0(r)}{dr} \right]^2 ds \qquad (12.28)$$

Thus, the generalized guide index b can be expressed in terms of an integral of $\mathcal{R}_0^2(r)$ of the core region and that of $(d\mathcal{R}_0/dr)^2$ of the core and cladding regions.

12.3.2   Normalized Group Delay *d(Vb)/dV*

To find an integral expression for the normalized group delay, we make use the Brown identity established in Appendix A. We also assume that the indices n_{co} and

n_{cl} are not dispersive. Thus we ignore all derivatives of the index with respect to λ. To proceed, we begin with (12.2) and write

$$\beta^2 = \frac{\omega^2}{c^2}[n_{cl}^2 + b(n_{co}^2 - n_{cl}^2)] \tag{12.29}$$

Thus

$$\frac{d(\beta^2)}{d\omega} = \frac{2\omega}{c^2}[n_{cl}^2 + b(n_{co}^2 - n_{cl}^2)] + \frac{\omega}{c^2}(n_{co}^2 - n_{cl}^2)V\frac{db}{dV} \tag{12.30}$$

The equation can be rewritten as

$$c^2\frac{d(\beta^2)}{d(\omega^2)} = n_{cl}^2 + \frac{1}{2}(n_{co}^2 - n_{cl}^2)\left[b + \frac{d(Vb)}{dV}\right] \tag{12.31}$$

On the other hand, we substitute (12.10) and (12.11) into the Brown identity (A.10) and obtain

$$c^2\frac{d(\beta^2)}{d(\omega^2)} = \int_{co+cl} n^2(r)\mathcal{R}_0^2(r)\,ds = \int_{co}[n^2(r) - n_{cl}^2]\mathcal{R}_0^2(r)\,ds + n_{cl}^2 \tag{12.32}$$

By comparing (12.31) with (12.32), we obtain immediately

$$b + \frac{d(Vb)}{dV} = 2\int_{co}\frac{n^2(r) - n_{cl}^2}{n_{co}^2 - n_{cl}^2}\mathcal{R}_0^2(r)\,ds \tag{12.33}$$

Upon combining (12.28) and (12.33), we have the desired integral expression for the normalized group delay:

$$\frac{d(Vb)}{dV} = \int_{co}\frac{n^2(r) - n_{cl}^2}{n_{co}^2 - n_{cl}^2}\mathcal{R}_0^2(r)\,ds + \frac{a^2}{V^2}\int_{co+cl}\left[\frac{d\mathcal{R}_0(r)}{dr}\right]^2 ds \tag{12.34}$$

It connects the normalized group delay to the integrals of the square of the mode field and its derivative.

For power-law index fibers, (12.33) becomes

$$b + \frac{d(Vb)}{dV} = \frac{2}{a^\alpha}\int_{co}r^\alpha\mathcal{R}_0^2(r)\,ds \tag{12.35}$$

Krumbholz et al. [6] have studied the integral on the right-hand side and they showed that

$$\frac{1}{a^\alpha}\int_{co}r^\alpha\mathcal{R}_0^2(r)\,ds = \frac{\alpha}{\alpha + 2}\frac{P_{co}}{P_{total}} + \frac{2b}{\alpha + 2} \tag{12.36}$$

where

$$\frac{P_{co}}{P_{total}} = \int_{co}\mathcal{R}_0^2(r)\,ds \tag{12.37}$$

is the *confinement factor*, or the fractional power carried in the core region, discussed in Chapter 9.

Making use of (12.36), we rewrite (12.35) in terms of the confinement factor

$$b + \frac{d(Vb)}{dV} = \frac{2\alpha}{\alpha + 2} \frac{P_{co}}{P_{total}} + \frac{4}{\alpha + 2} b \qquad (12.38)$$

Conversely, we can express the confinement factor in terms of b and $d(Vb)/dV$:

$$\frac{P_{co}}{P_{total}} = \frac{\alpha + 2}{2\alpha} \frac{d(Vb)}{dV} + \frac{\alpha - 2}{2\alpha} b \qquad (12.39)$$

This is a simple relation connecting $b + [d(Vb)/dV]$ to the confinement factor of power-law index fibers with an arbitrary α.

The relation (12.38) is particularly simple for fibers with a step-index or parabolic-index profile. For step-index fibers ($\alpha \to \infty$), (12.38) reduces to

$$b + \frac{d(Vb)}{dV} = 2 \frac{P_{co}}{P_{total}} \qquad (12.40)$$

For fibers with a parabolic index profile ($\alpha = 2$), (12.38) becomes

$$\frac{d(Vb)}{dV} = \frac{P_{co}}{P_{total}} \qquad (12.41)$$

12.3.3 Group Delay and the Confinement Factor

It is also possible to relate the group delay directly to the confinement factor. Recall that the transit time for a pulse to traverse a fiber of length L is

$$\tau_{gr} = \frac{L}{v_{gr}} = L \frac{d\beta}{d\omega} = L \frac{\omega}{\beta} \frac{1}{2\omega} \frac{d(\beta^2)}{d\omega} \qquad (12.42)$$

A simple differentiation of (12.29) leads to

$$\tau_{gr} = \frac{L}{c} \frac{k}{\beta} \left\{ n_{cl}^2 + \frac{1}{2} (n_{co}^2 - n_{cl}^2) \left[\frac{d(Vb)}{dV} + b \right] \right\} \qquad (12.43)$$

provided that the indices are nondispersive.

For power-law index fibers, $[d(Vb)/dV] + b$ can be expressed in terms of b and the confinement factor P_{co}/P_{total} as given by (12.38). Thus, we obtain

$$\tau_{gr} = \frac{L}{c} \frac{k}{\beta} \left[n_{cl}^2 + (n_{co}^2 - n_{cl}^2) \left(\frac{\alpha}{\alpha + 2} \frac{P_{co}}{P_{total}} + \frac{2b}{\alpha + 2} \right) \right] \qquad (12.44)$$

Alternatively, τ_{gr} can be expressed as

$$\tau_{gr} = \frac{L}{c}\frac{k}{\beta}\left\{ n_{co}^2 - (n_{co}^2 - n_{cl}^2)\left[\frac{\alpha}{\alpha+2}\frac{P_{cl}}{P_{total}} + \frac{2}{\alpha+2}(1-b)\right]\right\} \qquad (12.45)$$

where P_{cl}/P_{total} is the *fractional power carried in the cladding region*. In other words, the group delay relates to the fractional power transported in the core region, or that in the cladding region.

12.3.4 Normalized Waveguide Dispersion $V[d^2(Vb)/dV^2]$

Before deriving an integral expression for the normalized waveguide dispersion, we need to establish two identities. As noted in (12.27), the potential function $\mathcal{R}_0(r)$ is normalized. Therefore,

$$\frac{1}{2}\frac{\partial}{\partial V}\int_{co+cl}\mathcal{R}_0^2(r)\,ds = \int_{co+cl}\mathcal{R}_0(r)\frac{\partial\mathcal{R}_0(r)}{\partial V}\,ds = 0 \qquad (12.46)$$

By a straightforward and tedious differentiation, we can show

$$\frac{\partial\mathcal{R}_0(r)}{\partial V}\frac{d^2\mathcal{R}_0(r)}{dr^2} + \frac{1}{r}\frac{\partial\mathcal{R}_0(r)}{\partial V}\frac{d\mathcal{R}_0(r)}{dr}$$

$$= \frac{1}{r}\frac{d}{dr}\left[r\frac{\partial\mathcal{R}_0(r)}{\partial V}\frac{d\mathcal{R}_0(r)}{dr}\right] - \frac{d\mathcal{R}_0(r)}{dr}\frac{d}{dr}\left[\frac{\partial\mathcal{R}_0(r)}{\partial V}\right]$$

$$= \frac{1}{r}\frac{d}{dr}\left[r\frac{\partial\mathcal{R}_0(r)}{\partial V}\frac{d\mathcal{R}_0(r)}{dr}\right] - \frac{1}{2}\frac{\partial}{\partial V}\left[\frac{d\mathcal{R}_0(r)}{dr}\right]^2 \qquad (12.47)$$

We multiply (12.12) with $\partial\mathcal{R}_0/\partial V$, use the above equation and obtain

$$\frac{1}{r}\frac{d}{dr}\left[r\frac{\partial\mathcal{R}_0(r)}{\partial V}\frac{d\mathcal{R}_0(r)}{dr}\right] - \frac{1}{2}\frac{\partial}{\partial V}\left[\frac{d\mathcal{R}_0(r)}{dr}\right]^2$$

$$+ [k^2n^2(r) - \beta^2]\mathcal{R}_0(r)\frac{\partial\mathcal{R}_0(r)}{\partial V} = 0 \qquad (12.48)$$

An integration of (12.48) over the entire fiber cross section would lead to

$$\int_{co+cl}\frac{1}{r}\frac{d}{dr}\left[r\frac{\partial\mathcal{R}_0(r)}{\partial V}\frac{d\mathcal{R}_0(r)}{dr}\right]ds - \frac{1}{2}\frac{\partial}{\partial V}\int_{co+cl}\left[\frac{d\mathcal{R}_0(r)}{dr}\right]^2 ds$$

$$+ k^2\int_{co+cl}n^2(r)\mathcal{R}_0(r)\frac{\partial\mathcal{R}_0(r)}{\partial V}\,ds - \beta^2\int_{co+cl}\mathcal{R}_0(r)\frac{\partial\mathcal{R}_0(r)}{\partial V}\,ds = 0 \qquad (12.49)$$

Obviously, the first integral of (12.49) is zero. The last integral vanishes as indicated in (12.46). It then follows that

$$k^2 \int_{co+cl} n^2(r) \mathcal{R}_0(r) \frac{\partial \mathcal{R}_0(r)}{\partial V} \, ds = \int_{co+cl} \frac{d\mathcal{R}_0(r)}{dr} \frac{\partial}{\partial V} \left[\frac{d\mathcal{R}_0(r)}{dr} \right] ds \qquad (12.50)$$

Further manipulation leads to

$$\int_{co} \frac{n^2(r) - n_{cl}^2}{n_{co}^2 - n_{cl}^2} \mathcal{R}_0(r) \frac{d\mathcal{R}_0(r)}{dV} \, ds = \frac{a^2}{V^2} \int_{co+cl} \frac{d\mathcal{R}_0(r)}{dr} \frac{\partial}{\partial V} \left[\frac{d\mathcal{R}_0(r)}{dr} \right] ds \qquad (12.51)$$

On the other hand, by differentiating both sides of (12.34) with respect to V, we obtain

$$\frac{d^2(Vb)}{dV^2} = 2 \int_{co} \frac{n^2(r) - n_{cl}^2}{n_{co}^2 - n_{cl}^2} \mathcal{R}_0(r) \frac{\partial \mathcal{R}_0(r)}{\partial V} \, ds + \frac{2a^2}{V^2} \int_{co+cl} \frac{d\mathcal{R}_0(r)}{dr} \frac{\partial}{\partial V} \left[\frac{d\mathcal{R}_0(r)}{dr} \right] ds$$
$$- \frac{2a^2}{V^3} \int_{co+cl} \left[\frac{d\mathcal{R}_0(r)}{dr} \right]^2 ds \qquad (12.52)$$

Combining the two equations, we obtain

$$V \frac{d^2(Vb)}{dV^2} = 2a^2 \frac{\partial}{\partial V} \left\{ \frac{1}{V} \int_{co+cl} \left[\frac{d\mathcal{R}_0(r)}{dr} \right]^2 ds \right\} \qquad (12.53)$$

This is the desired integral expression for $V[d^2(Vb)/dV^2]$. It relates the normalized waveguide dispersion to the integral of the derivative of $\mathcal{R}_0$.

12.3.5 An Example

Clearly, if the modal field is known somehow, we can use (12.28), (12.34), and (12.53) to estimate the key propagation and dispersion parameters of the fibers. As an example, we again consider a power-law index fiber with an arbitrary α. We suppose the modal field is approximated by a Gaussian function with a beam radius w_0 and label it as $\tilde{\mathcal{R}}_0(r)$:

$$\tilde{\mathcal{R}}_0(r) = \frac{1}{w_0} \sqrt{\frac{2}{\pi}} e^{-r^2/w_0^2}, \qquad 0 \leq r < \infty \qquad (12.54)$$

A simple calculation will show that $\tilde{R}_0(r)$ is indeed normalized. To estimate b, we substitute (12.54) into (12.28) and obtain

$$b = \frac{4}{w_0^2} \int_0^a \left[1 - \left(\frac{r}{a} \right)^{\alpha} \right] e^{-2r^2/w_0^2} r \, dr - \frac{16a^2}{V^2 w_0^6} \int_0^{\infty} e^{-2r^2/w_0^2} r^3 \, dr \qquad (12.55)$$

Similarly, we obtain the normalized group delay and waveguide dispersion from (12.34) and (12.53):

$$\frac{d(Vb)}{dV} = \frac{4}{w_0^2} \int_0^a \left[1 - \left(\frac{r}{a}\right)^\alpha\right] e^{-2r^2/w_0^2} r\, dr + \frac{16a^2}{V^2 w_0^6} \int_0^\infty e^{-2r^2/w_0^2} r^3\, dr \qquad (12.56)$$

and

$$V\frac{d^2(Vb)}{dV^2} = 32a^2 \frac{d}{dV}\left(\frac{1}{Vw_0^2}\int_0^\infty e^{-2r^2/w_0^2} r^3\, dr\right) \qquad (12.57)$$

For integer α, the integrals in the above equations can be evaluated in closed forms, as reported by Sansonetti [9, 10]. In particular for $\alpha = 4$, the results are

$$b = \left(1 - \frac{w_0^4}{2a^4} - \frac{4a^2}{V^2 w_0^2}\right) + \left(\frac{w_0^2}{a^2} + \frac{w_0^4}{2a^4}\right) e^{-2a^2/w_0^2} \qquad (12.58)$$

$$\frac{d(Vb)}{dV} = \left(1 - \frac{w_0^4}{2a^4} + \frac{4a^2}{V^2 w_0^2}\right) + \left(\frac{w_0^2}{a^2} + \frac{w_0^4}{2a^4}\right) e^{-2a^2/w_0^2} \qquad (12.59)$$

$$V\frac{d^2(Vb)}{dV^2} = 2a^2 \frac{d}{dV}\left(\frac{w_0^2}{V}\right) \qquad (12.60)$$

The equations given in (12.58)–(12.60) are approximate expressions for the normalized parameters, since $\tilde{\mathcal{R}}_0(r)$ is an approximate expression of the modal field. In Figure 12.5, we compare the estimated values (dashed curves) based on (12.58), (12.59), and (12.60) with numerically exact results (solid curves) [9, 10]. Clearly, the estimated values are reasonable accurate. It is worthy to stress that we are able to estimate b, $d(Vb)/dV$, and $V[d^2(Vb)/dV^2]$ without solving the wave equation for the modal field $\mathcal{R}_0(r)$.

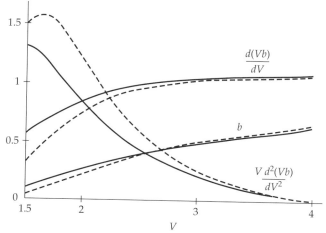

Figure 12.5 Propagation and dispersion characteristics of fibers with $\alpha = 4$. —theoretical results from [1], --- Gaussian approximation based on (12.58), (12.59), and (12.60)]. (From [10].)

12.4 RADIATION AND EXCITATION CHARACTERISTIC OF GRADED-INDEX FIBERS

Having studied the propagation and dispersion characteristics of fundamental modes of power-law index fibers, we turn our attention to the input/output characteristics of these fibers. We have discussed the radiation and excitation problems of step-index fibers in Chapter 10. The basic integral expressions developed there remain valid for any fiber having an angularly independent index profile. It is only necessary to substitute the pertinent expressions for the modal and incident fields in the equations and evaluate the integrals. This is done in this section. The results show that the radiation and excitation characteristics of fundamental modes of all weakly guiding fibers are essentially similar irrespective of the index profiles, so long as the index distributions are independent of ϕ. This is not surprising since the modal fields of all fundamental modes of weakly guiding fibers are basically similar, as observed by Gambling and Matsumura [3].

12.4.1 Radiation of Fundamental Modes of Graded-Index Fibers

Suppose a graded-index fiber is truncated at $z = 0$ and we study the field radiated by the truncated fiber [Fig. 10.1(a)]. Since the modal field $\mathcal{R}_0(r)e^{-j\beta z}$ is independent of ϕ, so is the field radiated by the fundamental mode. We consider a Cartesian component of the radiation field in the Fraunhofer zone and write, from (10.7),

$$E_{\mathrm{FF}x}(R, \Theta, \Phi) = jk\frac{e^{-jkR}}{R}e_{\mathrm{FF}}(\Theta) \tag{12.61}$$

where

$$e_{\mathrm{FF}}(\Theta) = \int_0^\infty \mathcal{R}_0(r)J_0(kr\sin\Theta)r\,dr \tag{12.62}$$

Since the radiated field is confined mainly in the forward direction along the fiber axis, we are justified to approximate the obliquity factor in (12.62) by 1 [11, 12].

By writing $q = k\sin\Theta$, (12.62) becomes

$$e_{\mathrm{FF}}(q) = \int_0^\infty \mathcal{R}_0(r)J_0(qr)r\,dr \tag{12.63}$$

Conversely, we make use of the Hankel inversion theorem [13], and express $\mathcal{R}_0(r)$ in terms of $e_{\mathrm{FF}}(q)$:

$$\mathcal{R}_0(r) = \int_0^\infty e_{\mathrm{FF}}(q)J_0(qr)q\,dq \tag{12.64}$$

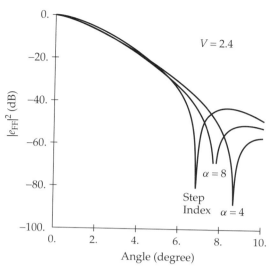

Figure 12.6 Intensity patterns of fields radiated by power-law index fibers with $a = 4$, 8, and ∞ and $V = 2.4$. (From [15].)

In other words, $\mathcal{R}_0(r)$ and $e_{\text{FF}x}(q)$ form a Hankel transform pair of the zeroth order [13, 14]. Explicitly $e_{\text{FF}}(q) = \mathcal{H}_0[\mathcal{R}_0(r)]$ and $\mathcal{R}_0(r) = \mathcal{H}_0^{-1}[e_{\text{FF}}(q)]$.

For power-law index fibers, expressions for the modal field $\mathcal{R}_0(r)$ in the core and that in the cladding are known as given in (12.15) and (12.19). Thus (12.62) can be integrated analytically. But the resulting expressions are very complicated. This approach will not work if $n(r)$ and $\mathcal{R}_0(r)$ are not known. Danielsen has reported an alternate method [15]. He partitioned a fiber with an arbitrary index profile into a series of dielectric shells of constant indices and calculated the fields radiated by the multiple shell fiber. He applied this method to study fields radiated by power-law index fibers. His results for fibers with $\alpha = 4$, 8, and ∞ are shown in Figure 12.6 [15]. As noted earlier, the radiated fields of power-law index fibers are confined mainly in a small cone in the forward direction. As depicted in Figure 12.6, the radiation intensity patterns of power-law index fibers are essentially the same as that of step-index fibers. This is particularly true for the main lobe. As α gets smaller, the null nearest the major peak moves away from the fiber axis and the side lobes also become weaker. For all fibers studied, the strongest side lobe is at least -40 dB below the major peak.

12.4.2 Excitation by a Linearly Polarized Gaussian Beam

For the excitation problem, we suppose that a linearly polarized Gaussian beam impinges upon a truncated fiber tip normally [Fig. 10.1(b)]. We assume that the incoming Gaussian beam aligns perfectly with the fiber axis and the beam waist is at the fiber tip ($z = 0$). Let the beam waist radius be w. Then the percentage power

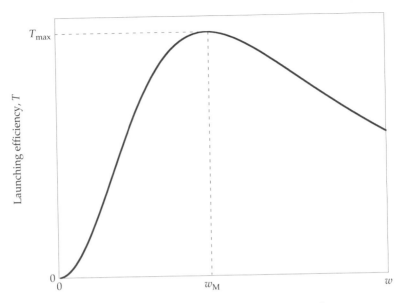

Figure 12.7   Launching efficiency as a function of w.

coupled into the fundamental mode is, from (10.38),

$$\eta_{\text{LP01}} = \frac{P_{\text{LP01}}}{P_{\text{in}}} = \frac{4N}{(1 + n_{\text{cl}})^2} T \tag{12.65}$$

where N is the effective index of the fundamental mode and T is the *launching efficiency*:

$$T = \left[\frac{2\sqrt{2\pi}}{w} \int_0^\infty e^{-r^2/w^2} \mathcal{R}_0(r) r \, dr \right]^2 \tag{12.66}$$

The launching efficiency depends on the index profile, the normalized frequency of the fiber and the beam waist radius of the incident Gaussian beam. For an arbitrary fiber, it is always possible to maximize T by varying w. We denote the maximum value of T as $T_{\max}$. An example is shown schematically in Figure 12.7. Marcuse has studied the excitation problem in great details [16, 17]. Figures 12.8 and 12.9 are two examples reported by him. In Figure 12.8, $T_{\max}$ is plotted as a function of V for five values of α [17]. He noted that for a given α, $T_{\max}$ reaches to its maximum value at V slightly greater than the cutoff V_c of the next higher-order mode. After reaching the peak value, $T_{\max}$ decreases slightly and slowly. For all fibers studied, the peaks are rather broad. Fibers with a parabolic index profile, $\alpha = 2$, is slightly different. For $\alpha = 2$, $T_{\max}$ increases monotonically with V. More importantly, $T_{\max}$ is greater than 0.99 for $V > 2$ for all values of α. In other words, the modal fields of the fundamental modes of power-law fibers resemble the Gaussian field for all values of α so long as V is greater than 2.

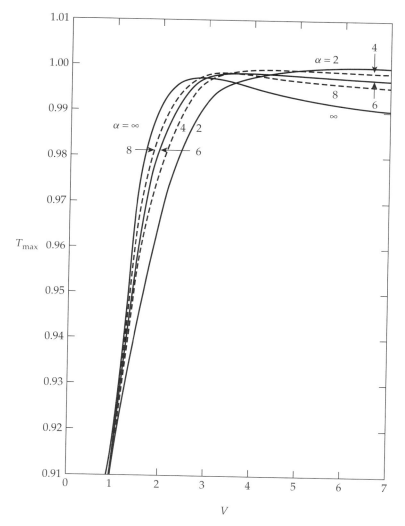

Figure 12.8   Maximum launching efficiency for power-law index fibers. (From [17].)

12.5   MODE FIELD RADIUS

As noted earlier, the modal fields of, and fields radiated by, the fundamental modes of manufactured fibers are independent of ϕ. Thus, fields in the transverse planes depend on r only. We wonder if a single parameter may be found to characterize the transverse fields of these fibers. If the transverse fields were truly a Gaussian function, we could use the beam waist as the characterizing parameter. But the modal fields are only approximately, not exactly, Gaussian, the parameter, if it exists, must be different. This is the mode field radius, effective mode radius, or spot size to be discussed in this section.

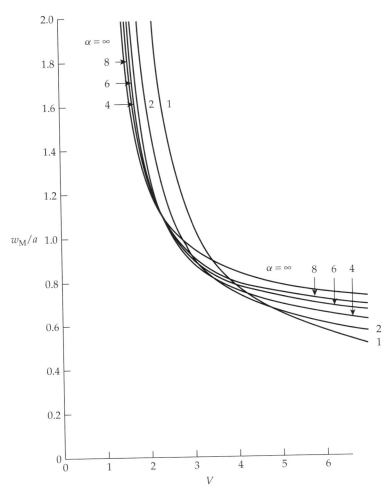

Figure 12.9   Marcuse's mode field radius of power-law index fibers. (From [17].)

Several mode field radii have been proposed. In fact, eight different mode field radii have been examined in Neumann [18]. Practicing engineers commonly use three of the eight mode field radii to describe the modal fields. One mode field radius was introduced by Marcuse and two by Petermann. In this section, we will discuss the three mode field radii and label them as w_M, w_{PI} and w_{PII}, respectively.

Experimental procedures and commercial instruments have also been developed to measure the mode field radius [19]. The procedures involve either the near-field or the far-field measurements. It is easier, and more accurate and reliable, to measure the far-field quantities. Thus, the mode field radius based on the far-field quantities is preferred. Indeed, the Telecommunications Industry Association recommends the second Petermann mode radius, w_{PII}, as the parameter to quantify the fiber properties and w_{PII} is based on the radiated fields in the far-field zone [20].

12.5.1   Marcuse Mode Field Radius

As discussed in Section 12.4.2, the percentage power coupled to the fundamental mode from a Gaussian beam depends on the beam waist radius w of the incoming beam. Marcuse defined the mode field radius as the Gaussian beam waist radius that maximizes the launching coefficient [16, 17]. We identify such a mode field radius as w_M. The launching efficiency T has been discussed in the last section and is given by (12.66). Thus, Marcuse's mode field radius can be deduced mathematically:

$$\frac{d}{dw}\left[\frac{2\sqrt{2\pi}}{w}\int_0^\infty e^{-r^2/w^2}\mathcal{R}_0(r)r\,dr\right]=0 \qquad (12.67)$$

A simple differentiation leads to a transcendental equation in w_M [21]:

$$w_M^2 = 2\frac{\int_0^\infty e^{-r^2/w_M^2}\mathcal{R}_0(r)r^3\,dr}{\int_0^\infty e^{-r^2/w_M^2}\mathcal{R}_0(r)r\,dr} \qquad (12.68)$$

Marcuse used (12.68) to evaluate w_M numerically for several power-law index fibers. His results are shown in Figure 12.9. Based on the numerical results, he also obtained an empirical expression for the mode field radii of power-law index fibers [17]:

$$\frac{w_M}{a}=\frac{A}{V^{2/(\alpha+2)}}+\frac{B}{V^{3/2}}+\frac{C}{V^6} \qquad (12.69)$$

where A, B, and C are parameters related to the index profile parameter α:

$$A=\left\{\frac{2}{5}\left[1+4\left(\frac{2}{\alpha}\right)^{5/6}\right]\right\}^{1/2} \qquad (12.70)$$

$$B=e^{0.298/\alpha}-1+1.478(1-e^{-0.077\alpha}) \qquad (12.71)$$

$$C=3.76+\exp(4.19/\alpha^{0.418}) \qquad (12.72)$$

Equation (12.68) is in terms of the modal field $\mathcal{R}_0(r)$ of the fundamental mode and the field of the incoming Gaussian beam at the truncated fiber tip. In other words, (12.68) involves the *near-field* quantities. As noted earlier, it is easier and more accurate to measure the far-field quantities. Therefore, we like to express the mode field radius in terms of the *far-field* quantities. Toward this end, we use (12.64) to rewrite the launching efficiency integral in (12.66) as

$$T=\left\{\frac{2\sqrt{2\pi}}{w}\int_0^\infty e^{-r^2/w^2}\left[\int_0^\infty e_{FF}(q)J_0(qr)q\,dq\right]r\,dr\right\}^2 \qquad (12.73)$$

By exchanging the order of integration and applying the Parseval theorem of Hankel transforms, we obtain a new expression for the launching efficiency:

$$T = \left[\sqrt{2\pi}\, w \int_0^\infty e_{\mathrm{FF}}(q) e^{-q^2 w^2/4} q\, dq \right]^2 \tag{12.74}$$

In the derivation, we also make use of (12.63) and that the Hankel transform of a Gaussian function is also a Gaussian function:

$$\mathcal{H}_0(e^{-r^2/w^2}) = \frac{w^2}{2} e^{-q^2 w^2/4} \tag{12.75}$$

Now we maximize the launching efficiency T given in (12.74):

$$\frac{d}{dw} \left[\sqrt{2\pi}\, w \int_0^\infty e_{\mathrm{FF}}(q) e^{-q^2 w^2/4} q\, dq \right] = 0 \tag{12.76}$$

and obtain a new definition for the mode field radius:

$$w_{\mathrm{M}}^2 = 2 \frac{\int_0^\infty e^{-q^2 w_{\mathrm{M}}^2/4} e_{\mathrm{FF}}(q) q\, dq}{\int_0^\infty e^{-q^2 w_{\mathrm{M}}^2/4} e_{\mathrm{FF}}(q) q^3\, dq} \tag{12.77}$$

Equation (12.77) involves the far-field quantities $e_{\mathrm{FF}}(q)$ and $e^{-q^2 w^2/4}$. Although (12.68) and (12.77) look different, the two equations are equivalent and lead to the same mode field radius if $\mathcal{R}_0(r)$ and $e_{\mathrm{FF}}(q)$ are exact.

12.5.2   First Petermann Mode Field Radius

In the last subsection, the Marcuse mode field radius w_{M} is defined in terms of the Gaussian beam excitation. Although w_{M} so defined is physically intuitive, the definition for w_{M} is rather indirect and arbitrary. We might inquire if a mode field radius could be defined directly in terms of the modal field $\mathcal{R}_0(r)$ itself, or the radiated field $e_{\mathrm{FF}}(q)$ itself, without referring to the input Gaussian beam. This leads to the two Petermann mode field radii w_{PI} and w_{PII}.

The integrands in (12.68) contain the product of the modal field and the input Gaussian field. Petermann replaced the Gaussian function $e^{-r^2/w_{\mathrm{M}}^2}$ by the modal field $\mathcal{R}_0(r)$ and defined the mode field radius as

$$w_{\mathrm{PI}}^2 = 2 \frac{\int_0^\infty [\mathcal{R}_0(r)]^2 r^3\, dr}{\int_0^\infty [\mathcal{R}_0(r)]^2 r\, dr} \tag{12.78}$$

This is the *first Peterman mode field radius* or *Petermann I mode field radius* [22, 23]. From the expression (12.78), we see that $w_{\mathrm{PI}}/\sqrt{2}$ is physically the *root mean square radius* weighted by the power density ($[\mathcal{R}_0(r)]^2$) of the modal field. Once the modal field $\mathcal{R}_0(r)$ is known, the mode field radius can be calculated.

For step-index fibers, the modal field $\mathcal{R}_0(r)$ of the LP_{01} mode is given by (9.43). When (9.43) is substituted in (12.78), a closed-form expression for w_{PI} for the LP_{01} mode of step-index fibers can be deduced [24]:

$$\frac{w_{PI}^2}{a^2} = \frac{4}{3}\left[\frac{J_0(V\sqrt{1-b})}{V\sqrt{1-b}J_1(V\sqrt{1-b})} + \frac{1}{2} + \frac{1}{V^2b} - \frac{1}{V^2(1-b)}\right] \quad (12.79)$$

12.5.3 Second Petermann Mode Field Radius

The mode field radius defined in (12.78) involves the near-field quantities only. As noted earlier, we prefer a definition involving the far-field quantities only. Such a mode field radius has also been introduced by Petermann [25]. Starting from (12.77), Petermann replaced the Gaussian function in the integrand by $e_{FF}(q)$ and obtained

$$w_{PII}^2 = 2\frac{\int_0^\infty [e_{FF}(q)]^2 q\, dq}{\int_0^\infty [e_{FF}(q)]^2 q^3\, dq} \quad (12.80)$$

This is the *second Petermann mode field radius* or *Petermann II mode field radius* [25]. Physically, $\sqrt{2}/w_{PII}$ is the *root mean square angular width* of the radiated field weighted by the power density of the radiated field in the Fraunhofer zone.

Pask has shown that w_{PII} can also be expressed in terms of the modal field $\mathcal{R}_0(r)$ as well [26]. To demonstrate that, we recall that $\mathcal{R}_0(r)$ and $e_{FF}(q)$ are a Hankel transform pair as given in (12.75). By applying the Parseval theorem of Hankel transforms [13], we deduce two identities:

$$\int_0^\infty [\mathcal{R}_0(r)]^2 r\, dr = \int_0^\infty [e_{FF}(q)]^2 q\, dq \quad (12.81)$$

$$\int_0^\infty \left[\frac{d\mathcal{R}_0(r)}{dr}\right]^2 r\, dr = \int_0^\infty \left\{\mathcal{H}_1\left[\frac{d\mathcal{R}_0(r)}{dr}\right]\right\}^2 q\, dq$$

$$= \int_0^\infty \{-q\mathcal{H}_0[\mathcal{R}_0(r)]\}^2 q\, dq = \int_0^\infty [e_{FF}(q)]^2 q^3\, dq \quad (12.82)$$

Substituting (12.81) and (12.82) into (12.80), we cast w_{PII} in terms of the modal field $\mathcal{R}_0(r)$:

$$w_{PII}^2 = 2\frac{\int_0^\infty [\mathcal{R}_0(r)]^2 r\, dr}{\int_0^\infty \left[\frac{d\mathcal{R}_0(r)}{dr}\right]^2 r\, dr} \quad (12.82)$$

This is the alternate definition for w_{PII}, and it is expressed in terms of the modal field $\mathcal{R}_0(r)$. For the LP_{01} mode of step-index fibers, the integrals can be evaluated

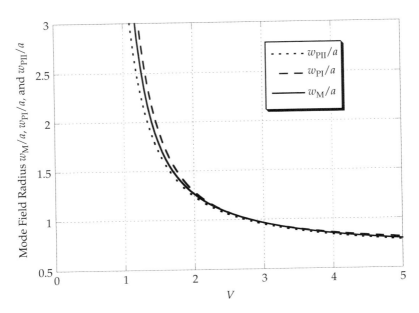

Figure 12.10   Mode field radius w_M/a, w_{PI}/a, and w_{PII}/a of step-index fibers.

in closed form and the result is [27]

$$\frac{w_{PII}}{a} = \frac{\sqrt{2}}{V\sqrt{b}}\frac{J_1(V\sqrt{1-b})}{J_0(V\sqrt{1-b})} \tag{12.83}$$

12.5.4   Comparison of Three Mode Field Radii

In the above subsections, we have introduced three mode field radii w_M, w_{PI}, and w_{PII}. If the modal field were truly a Gaussian function, then the three mode field radii would be identical. But $\mathcal{R}_0(r)$ is not exactly Gaussian, the three mode field radii are different. In Figure 12.10, we plot the mode field radii of step-index fibers as functions of V. It shows that the three mode field radii, while close, are not identical. In fact, it has been noted that for fibers with an arbitrary index profile [18, 19]

$$w_{PII} < w_M < w_{PI} \tag{12.84}$$

12.6   MODE FIELD RADIUS AND KEY PROPAGATION AND DISPERSION PARAMETERS

As mentioned in Section 12.1, b, $d(Vb)/dV$, and $V[d^2(Vb)/dV^2]$ are the three key propagation and dispersion parameters. In this section, we relate w_{PII} to these parameters [28, 29]. We begin with (12.10), (12.11), and the Brown identity (A.10) and obtain

$$\frac{\beta}{\omega}\frac{d\beta}{d\omega} = \frac{\int_{\text{co+cl}} n^2(r)\mathcal{R}_0^2(r)\,ds}{c^2 \int_{\text{co+cl}} \mathcal{R}_0^2(r)\,ds} = \frac{\int_0^\infty k^2 n^2(r)\mathcal{R}_0^2(r)r\,dr}{k^2 c^2 \int_0^\infty \mathcal{R}_0^2(r)r\,dr} \tag{12.85}$$

From the differential equation (12.12), we rewrite the numerator of (12.85) as

$$\int_0^\infty k^2 n^2(r)\mathcal{R}_0^2(r)r\,dr = \int_0^\infty [\beta^2 \mathcal{R}_0^2(r)]r\,dr$$

$$- \int_0^\infty \left\{ \mathcal{R}_0(r)\frac{1}{r}\frac{d}{dr}\left[r\frac{d\mathcal{R}_0(r)}{dr}\right]\right\} r\,dr \tag{12.86}$$

By integration by parts, the second integral in the right-hand side can be rewritten as

$$\int_0^\infty \left\{ \mathcal{R}_0(r)\frac{1}{r}\frac{d}{dr}\left[r\frac{d\mathcal{R}_0(r)}{dr}\right]\right\} r\,dr = \int_0^\infty \frac{d}{dr}\left[r\mathcal{R}_0(r)\frac{d\mathcal{R}_0(r)}{dr}\right] dr$$

$$- \int_0^\infty \left[\frac{d\mathcal{R}_0(r)}{dr}\right]^2 r\,dr \tag{12.87}$$

The first integral in the right-hand side is zero. Thus

$$\int_0^\infty \left\{ \mathcal{R}_0(r)\frac{1}{r}\frac{d}{dr}\left[r\frac{d\mathcal{R}_0(r)}{dr}\right]\right\} r\,dr = - \int_0^\infty \left[\frac{d\mathcal{R}_0(r)}{dr}\right]^2 r\,dr \tag{12.88}$$

Substituting (12.86) and (12.88) into (12.85) and making use of (12.82), we obtain

$$\frac{\beta}{\omega}\frac{d\beta}{d\omega} = \frac{1}{k^2 c^2}\left\{ \beta^2 + \frac{\int_0^\infty \left[\frac{d\mathcal{R}_0(r)}{dr}\right]^2 r\,dr}{\int_0^\infty \mathcal{R}_0^2(r)r\,dr}\right\} = \frac{1}{k^2 c^2}\left(\beta^2 + \frac{2}{w_{\text{PII}}^2}\right) \tag{12.89}$$

By combining (12.89) with (12.31), we obtain

$$\frac{1}{c^2}[n_{\text{cl}}^2 + b(n_{\text{co}}^2 - n_{\text{cl}}^2)] + \frac{1}{2c^2}(n_{\text{co}}^2 - n_{\text{cl}}^2)V\frac{db}{dV} = \frac{1}{k^2 c^2}\left(\beta^2 + \frac{2}{w_{\text{PII}}^2}\right) \tag{12.90}$$

Therefore,

$$V\frac{db}{dV} = \frac{4a^2}{V^2 w_{\text{PII}}^2} \tag{12.91}$$

By adding b to both sides, the above expression becomes

$$\frac{d(Vb)}{dV} = \frac{4a^2}{V^2 w_{\text{PII}}^2} + b \qquad (12.92)$$

This is the first relation and it connects w_{PII} to the normalized group delay.

By differentiating both sides of (12.92) with respect to V and making use of (12.91), we have the second relation:

$$V\frac{d^2(Vb)}{dV^2} = \frac{d}{dV}\left(\frac{4a^2}{Vw_{\text{PII}}^2}\right) \qquad (12.93)$$

Equation (12.93) relates the change of Vw_{PII}^2 to the normalized waveguide dispersion [28, 29]. It should be stressed that the two relations, (12.92) and (12.93), are exact and valid for all fibers with ϕ-independent index profiles and wavelength-independent n_{co} and n_{cl}.

PROBLEMS

1. Show that if $\mathcal{R}_0(r)$ is a Gaussian function with a beam radius w_1,

$$\mathcal{R}_0(r) = E_0 e^{-r^2/w_1^2}$$

Then $w_{\text{M}} = w_{\text{PI}} = w_{\text{PII}} = w_1$.

2. Calculate w_{M}, w_{PI}, and w_{PII} if the modal field is

$$\mathcal{R}_0(r) = E_0\left(1 - \frac{r^2}{w_1^2}\right)e^{-r^2/w_1^2}$$

3. Show that [26]

$$\frac{1}{v_{\text{gr}}} = \frac{1}{v_{\text{ph}}} + \frac{2}{\omega\beta w_{\text{PII}}^2}$$

4. Show that [29]

$$1 - b = \int_0^\infty n(r)\mathcal{R}_0^2(r)r\,dr + \frac{2a^2}{V^2 w_{\text{PII}}^2}$$

REFERENCES

1. W. A. Gambling, H. Matsumura, and C. M. Ragdale, "Mode dispersion, material dispersion and profile dispersion in graded-index single-mode fibers," *IEE J. Microwaves, Optics Acoustics*, Vol. 3, No. 6, pp. 239–246 (1979).

2. C. N. Kurtz and W. Streifer, "Guided waves in inhomogeneous focusing media, part I: Formulation, solution for quadratic in homogeneity," *IEEE Trans. Microwave Theory Tech.*, Vol. MTT-17, No. 1, pp. 11–15 (1969).

3. W. A. Gambling and H. Matsumura, "Propagation in radially-inhomogeneous single-mode fibre," *Optical Quantum Electronics*, Vol. 10, No. 1, pp. 31–40 (1978).

4. K. Oyamada and T. Okoshi, "High-accuracy numerical data on propagation characteristics of α-power graded-core fibers," *IEEE Trans. Microwave Theory Tech.*, Vol. MTT-28, No. 10, pp. 1113–1980 (1980).

5. J. D. Love, "Power series solutions of the scalar wave equations for cladded, power-law profiles of arbitrary exponent," *Opt. Quantum Electron*, Vol. 11m, pp. 464–466 (1979).

6. D. Krumbholz, E. Brinkmeyer, and E.-G. Neumann, "Core/cladding power distribution, propagation constant, and group delay: Simple relation for power-law graded-index fibers," *J. Opt. Soc. Am.*, Vol. 70, No. 2, pp. 179–183 (Feb. 1980).

7. W. A. Gambling, H. Matsumura, and C. M. Ragdale, "Wave propagation in a single-mode fibre with dip in the refractive index," *Optical Quantum Electronics*, Vol. 10, No. 10, pp. 301–309 (1978).

8. W. A. Gambling, D. N. Payne, and H. Matsumura, "Cutoff frequency in radially inhomogeneous single-mode fibre," *Electron. Lett.*, Vol. 13, No. 5, pp. 139–140 (1977).

9. P. Sansonetti, "Modal dispersion in single-mode fibres: Simple approximation issued from mode spot size spectral behavior," *Electron. Lett.*, Vol. 18, pp. 647–648 (July 22, 1982).

10. P. Sansonetti, "Prediction of modal dispersion in single-mode fibres from spectral behavior of mode spot size," *Electron. Lett.*, Vol. 18, pp. 136–138 (Feb. 4, 1982).

11. E. Hecht, *Optics*, 4th ed., Addison Wesley, San Francisco, 2002, Chapter 10.

12. M. Born and E. Wolf, *Principles of Optics*, 6th ed., Pergamon, Oxford, 1980, Chapter 8.

13. I. N. Sneddon, *Fourier Transforms*, McGraw-Hill, New York, 1951.

14. M. Abramowitz and I. A. Stegun, *Handbook of Mathematical Functions with Formulas, Graphs and Mathematical Tables*, Dover, New York, 1965.

15. P. L. Danielsen, "Analytical expressions for group delay in the far field from an optical fiber having an arbitrary index profile," *IEEE J. Quantum Electron.*, Vol. 17, No. 6, pp. 850–853 (1981).

16. D. Marcuse, "Loss analysis of single-mode fiber splices," *BSTJ*, Vol. 56, No. 5, pp. 703–718 (May–June 1977).

17. D. Marcuse, "Gaussian approximation of the fundamental modes of graded-index fibers," *J. Opt. Soc. Am.*, Vol. 68, pp. 103–109 (Jan. 1978).

18. E. G. Neumann, "Spot size and width of the radiation pattern," in *Single-mode Fibers, Fundamental*, Springer, Berlin, 1988, Chapter 10.

19. B. J. Ainslie and C. R. Day, "A review of single-mode fibers with modified dispersion characteristics," *IEEE J. Lightwave Tech.*, Vol. LT-4, pp. 967–979 (1986).

20. Telecommunications Industry Association, "Measurement of mode-field diameter of single-mode optical fibers," Fiberoptic Test Procedure FOTP-191, Telecommunications Industry Association, Standards and Technology Department, Arlington, VA, 1998.

21. H. Matsumura and T. Suganuma, "Normalization of single-mode fibers having an arbitrary index profile," *Appl. Opt.*, Vol. 19, No. 18, pp. 3151–3158 (1980).

22. K. Petermann, "Microbending loss in monomode fibres," *Electron. Lett.*, Vol. 12, No. 4, pp. 107–109 (1976).

23. K. Petermann, "Fundamental mode microbending loss in graded-index and W fibres," *Opt. Quantum Electronics*, Vol. 9, pp. 167–175 (1977).

24. W. A. Gambling and H. Matsumura, "Simple characterization factor for practical single-mode fibres," *Electron. Lett.*, Vol. 13, No. 23, pp. 691–693 (1977).

25. K. Petermann, "Constraints for fundamental-mode spot size for broadband dispersion-compensated single-mode fibres," *Electron. Lett.*, Vol. 19, No. 18, pp. 712–714 (1983).

26. C. Pask, "Physical interpretation of Petermann's strange spot size for single-mode fibres," *Electron. Lett.*, Vol. 20, No. 3, pp. 144–145 (1984).

27. C. D. Hussey and F. Martinez, "Approximate analytic forms for the propagation characteristics of single-mode optical fibers," *Electron. Lettr.*, Vol. 21, No. 23, pp. 1103–1104 (Nov. 7, 1985).

28. C. D. Hussey, "Field to dispersion relationships in single-mode fibers," *Electron. Lettr.*, Vol. 20, No. 26, pp. 1051–1052 (Nov. 7, 1985).

29. F. Wilczewski, "Relation between new 'field radius' w_∞ and Petermann II field radius w_d in single-mode fibres with arbitrary refractive index profile," *Electron. Lettr.*, Vol. 24, No. 7, pp. 411–412 (Mar. 31, 1988).

13

PROPAGATION OF PULSES IN SINGLE-MODE FIBERS

13.1 INTRODUCTION

Loss, dispersion, and nonlinearity are the three physical effects limiting the overall length and speed of optical communication systems. All real fibers are *lossy*. Therefore, signals attenuate as they propagate in fibers. If signals arriving at the receiving end are so weak that they are comparable to or buried in noise, no useful information can be extracted from the received signals. All fibers, ideal or real, are also *dispersive* in that different modes and spectral components of the optical pulses move with different speed. As a result, pulses evolve as they travel in fibers. All single-mode fibers are birefringent. Thus the two polarization modes move with different group velocities. This is the *polarization mode dispersion*. It also leads to signal degradation. If pulses are distorted beyond recognition, or broadened so much that two or more pulses merge into one, information extracted from the signals would be erroneous. Due to fiber nonlinearity, different spectral components interact and interfere as they propagate. The nonlinear interaction and interference leads to the cross modulation and pulse degradation. We will not elaborate on the nonlinear effects further until the next chapter. With the advent of erbium-doped fiber amplifiers and semiconductor optical amplifiers, the fiber attenuation can be

Foundations for Guided-Wave Optics, by Chin-Lin Chen
Copyright © 2007 John Wiley & Sons, Inc.

compensated by the amplifier gain. Thus the fiber dispersion remains the dominant factor limiting the overall length and the bit rate of communication systems. In this chapter, we concentrate mainly on the dispersion effects arising from the finite spectral width in linear fibers.

To discuss the dispersion effects in fibers or waveguides, we begin with monochromatic waves. *Monochromatic waves* are waves having a well-defined angular frequency and a constant amplitude. We can write a monochromatic electric field at the input as

$$\mathcal{E}(0, t) = \mathbf{e} \mathcal{A}_0 \cos \omega t \tag{13.1}$$

where $\mathbf{e}$, $\mathcal{A}_0$, and ω are the electric field vector, amplitude, and angular frequency, respectively. Let β be the propagation constant at ω; then the electric field at an arbitrary point z in the fiber is

$$\mathcal{E}(z, t) = \mathbf{e} \mathcal{A}_0 \cos(\omega t - \beta z) \tag{13.2}$$

As noted in Chapter 1, the constant-phase surfaces of waves move with the *phase velocity*

$$v_{\mathrm{ph}} = \frac{\omega}{\beta} \tag{13.3}$$

The optical power carried by the waves travels with the *group velocity*

$$v_{\mathrm{gr}} = \frac{d\omega}{d\beta} \tag{13.4}$$

In studying transmission lines and waveguides, it is customary to plot ω as a function of β. Figure 13.1 depicts the $\omega-\beta$ diagram of a metallic waveguide

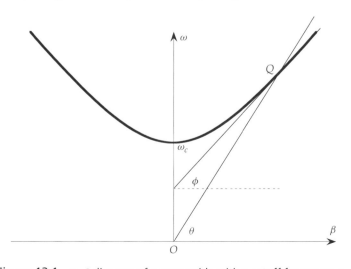

Figure 13.1 $\omega-\beta$ diagram of a waveguide with a cutoff frequency ω_c.

having a cutoff frequency ω_c. In the ω–β diagram, the slope of a straight-line OQ connecting the origin O and a point Q on the ω–β curve is the phase velocity. The slope of the ω–β curve at point Q is the group velocity.

A medium is dispersive if the phase velocity is frequency dependent. In vacuum, plane waves propagate with a propagation constant ω/c. The phase and group velocities are the same and they are c. Thus, vacuum is a nondispersive medium. The TEM mode guided by an evaporated coaxial transmission line made of perfect conductors also has a propagation constant ω/c. Thus an evaporated coaxial line made of perfect conductors is also nondispersive if it operates in the dominant mode. In an infinitely large material medium with an index n, the propagation constant is $\omega n/c$. The TE, TM, EH, and HE, modes guided by metallic waveguides, dielectric waveguides, or optical fibers propagate with a propagation constant $\omega N/c$ where N is an effective index of refraction. In general, n and N are frequency dependent. So are the phase and group velocities of waves or modes propagating in the media or waveguides. Thus the material media, waveguides, or optical fibers are dispersive. We are interested in the dispersion effect on the propagation and distortion of pulses traveling in the material media and waveguides.

A *modulated input signal* can be written as

$$\mathcal{E}(0, t) = \mathbf{e}\mathcal{A}(0, t) \cos \omega_0 t \tag{13.5}$$

where $\mathcal{A}(0, t)$ is the *pulse envelope* or *pulse shape function*. The spectral content of $\mathcal{E}(0, t)$ spreads over a finite bandwidth $\Delta\omega$ centering at the center angular frequency or around the average angular frequency ω_0. For most modulated signals of interest, the spectral bandwidth $\Delta\omega$ is much smaller than ω_0. These signals are referred to as *quasi-monochromatic* signals. They can be viewed as the superposition of signals with continuous spectrum. In the transmission media and waveguides, the pulse envelope varies in position and time. It is appropriate to describe the propagation and evolution of quasi-monochromatic signals in terms of five velocities depending on the temporal width, or equivalently the spectral width of the pulses. The five velocities are the *wavefront, phase, group, energy,* and *signal velocities* [1]. Sommerfeld and Brillouin noted that the wavefront velocity is always c even in media with indices different from 1 [1]. The group, energy, and signal velocities are identical if the pulse energy is confined in a narrow spectral width. If the pulse spectral width is wide, either the energy velocity or the signal velocity should be used in lieu of the group velocity. For details, refer to Ref. [1]. In communication applications, the spectral width involved is very narrow and the group velocity is a convenient and meaningful term describing the transmission of signals.

In a dispersive medium, each spectral component propagates with a different propagation constant. As a result, pulse envelopes evolve as they propagate. A detailed knowledge of β as a function of ω is required if $\mathcal{A}(z, t)$ is to be determined. For metallic waveguides with perfectly conducting walls, an analytic expression for the propagation constant exists and it is

$$\beta(\omega) = \sqrt{\omega^2 \mu\varepsilon - k_c^2} \tag{13.6}$$

where ε and μ are the permittivity and permeability of the medium. The cutoff wave vector k_c depends on the waveguide size, shape, and material medium. For thin-film waveguides (Chapter 2) and optical fibers (Chapter 9), the dispersion relations are known. But no explicit and analytic expression for $\beta(\omega)$ is known. This is due to the complexity of the waveguide geometry and material dispersion. Even if an analytic expression for $\beta(\omega)$ were known explicitly, it would be difficult, if indeed possible, to evaluate many integrals involved in the transient analysis. Although we can use software packages, MATLAB, MATHCAD, or MATHEMATICA, for example, to evaluate integrals numerically, little insight can be gleaned from the numerical results. In addition, the physical meaning and significance of each term are often obscured by the numerical computation. Fortunately, an accurate approximation is sufficient for many applications. In the following sections, we study the pulse propagation in dispersive media analytically, albeit approximately.

Three basic approaches to the pulse propagation problems are presented in this chapter. The first approach is a straightforward application of the *Fourier* and *inverse Fourier transforms* (Section 13.3). We use this direct approach, with suitable approximations to simplify the mathematical manipulations, to study the evolution of Gaussian pulses in waveguides and optical fibers (Section 13.4). In Section 13.5, we introduce the concept of the *impulse response* of a transmission medium and show that the electric field at an arbitrary point in the medium or waveguide is the convolution of the input electric field and the impulse response of the transmission medium. As an example, we use the impulse response concept to demonstrate the degradation of rectangular pulses in waveguides (Section 13.6). The third approach is completely different from the first two approaches. We recognize that $\mathcal{E}(z, t)$ is the product of a carrier term and a pulse shape function. The propagation of carrier waves is simple and well understood. We concentrate mainly on the slow evolution of the pulse envelope. An *envelope equation* is developed in Section 13.7 to describe the evolution of the pulse envelope. A general discussion on the envelope distortion and frequency chirping is then presented. The chapter concludes in a study of a simple dispersion compensation technique.

13.2 DISPERSION AND GROUP VELOCITY DISPERSION

By *phase velocity dispersion* and *group velocity dispersion*, we mean that the phase and group velocities vary with frequency. The phase velocity dispersion is also referred to as the dispersion. All materials are dispersive. Take materials that are transparent in visible and near-infrared (IR) regions as examples. In much of the visible and near-infrared regions, the refractive indices decrease as λ increases. In other words, long-wavelength (low-frequency) plane waves "see" a smaller index and propagate with a faster phase velocity than the short-wavelength (high-frequency) plane waves. Thus, long-wavelength plane waves, red light, for example, are refracted less by a prism than the short-wavelength counterparts, blue or green light, for example [Fig. 13.2(a)]. This is the *normal dispersion effect*. However, in the narrow spectral bands where the material is in resonance, the index increases with

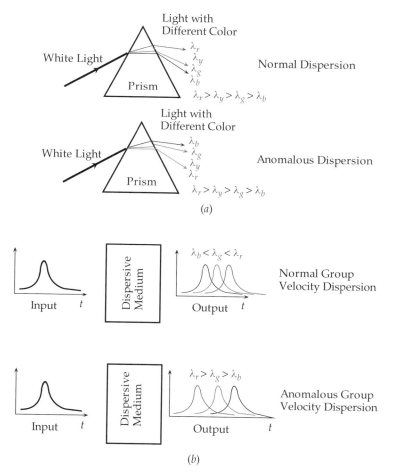

Figure 13.2 (a) Dispersion and (b) group velocity dispersion.

λ in the narrow spectral regions. In the narrow resonance regions, the dispersion effect is just the opposite. Namely, the long-wavelength plane waves "experience" a larger index and travel with a slower phase velocity than the short-wavelength plane waves. This is the *anomalous dispersion effect*.

To examine the pulse propagation and evolution in dispersive media, it is necessary to know β as a function of ω. As noted earlier, a detailed knowledge of $\beta(\omega)$ is often unavailable. Instead, a limited information on $\beta(\omega)$ may be known. Therefore, we look for ways to extract useful information from the limited knowledge on $\beta(\omega)$. For quasi-monochromatic pulses containing several cycles of oscillations, the pulse energy is confined within a narrow spectral width $\Delta\omega$ near ω_0. Within the narrow spectral width, we approximate $\beta(\omega)$ by the four leading terms of the Taylor series expansion and write

$$\beta_E(\omega) = \beta_0 + \beta'(\omega - \omega_0) + \tfrac{1}{2}\beta''(\omega - \omega_0)^2 + \tfrac{1}{6}\beta'''(\omega - \omega_0)^3 \qquad (13.7)$$

where $\beta_0 = \beta(\omega_0)$ is the propagation constant at the center angular frequency ω_0. The derivatives

$$\beta' = \frac{d\beta}{d\omega}\bigg|_{\omega=\omega_0} \qquad \beta'' = \frac{d^2\beta}{d\omega^2}\bigg|_{\omega=\omega_0} \qquad \beta''' = \frac{d^3\beta}{d\omega^3}\bigg|_{\omega=\omega_0}$$

are also evaluated at ω_0. We use β_0, β', β'', and β''' to describe the pulse envelop evolution.

The term β' is the inverse of the group velocity of the waves or modes. It is also the time required by a narrow pulse to travel a medium of unit thickness or a waveguide of unit length. Therefore, β' is also known as the *group delay*. If β' varies with λ, then different spectral components of a pulse travel with different group velocities and arrive at the output at different times. As a result, pulse shape changes continuously as the pulse propagates. This is the *group velocity dispersion* (GVD) *effect*.

It is customary to use the rate of change of the arrival time per unit path length per unit spectral width

$$\mathcal{D}_1 = \frac{d}{d\lambda}\left(\frac{1}{v_{\mathrm{gr}}}\right) = \frac{d\beta'}{d\lambda} \tag{13.8}$$

to quantify the GVD effect. $\mathcal{D}_1$ is the *first-order group velocity dispersion parameter* or simply the *group velocity dispersion parameter*. If $\mathcal{D}_1$ is negative, the long-wavelength (low-frequency) components of a pulse travel with a faster group velocity and arrive sooner than the short-wavelength (high-frequency) parts of the pulse. This is the *normal GVD effect* in materials [Fig. 13.1(*b*)]. On the other hand, if $\mathcal{D}_1$ is positive, the long-wavelength (low-frequency) components move slower than the short-wavelength (high-frequency) parts. This is the *anomalous GVD effect*.

For uniform plane waves propagating in a large material medium with a refractive index n,

$$\beta' = \frac{n}{c} - \frac{\lambda}{c}\frac{dn}{d\lambda} \quad \text{and} \quad \mathcal{D}_1 = -\frac{\lambda}{c}\frac{d^2n}{d\lambda^2}$$

An examination of the n versus λ curves of various materials shows that $\mathcal{D}_1$ can be positive or negative. There may exist discrete wavelengths where $\mathcal{D}_1$ vanishes. These discrete wavelengths are the *zero group velocity dispersion wavelengths*, λ_{ZGVD}, of the material.

Thin-film waveguides and optical fibers are dispersive due to the finite waveguide geometry and the material dispersion. Just like a large material medium, a waveguide may have a positive or a negative $\mathcal{D}_1$, depending on the waveguide geometry, material, and the operating wavelength. It is possible to select waveguide geometry, index profile, and materials so that the various components of GVD have opposite polarities and cancel exactly at a certain wavelength. This is the *zero group velocity dispersion wavelength* of the optical waveguide or fiber.

The rate of change of $\mathcal{D}_1$ per wavelength is the *second-order group velocity dispersion parameter*:

$$D_2 = \frac{d\mathcal{D}_1}{d\lambda} = \frac{d^2}{d\lambda^2}\left(\frac{1}{v_{\text{gr}}}\right) = \frac{d^2\beta'}{d\lambda^2} \tag{13.9}$$

It is also referred to as the *group velocity dispersion slope* or simply *dispersion slope*.

It is simple to relate β'' and β''' introduced in (13.7) to the two GVD parameters:

$$\beta'' = -\frac{\lambda^2 \mathcal{D}_1}{2\pi c} \tag{13.10}$$

$$\beta''' = \frac{\lambda^4}{4\pi^2 c^2}\left(\mathcal{D}_2 - \frac{2}{\lambda}\mathcal{D}_1\right) \tag{13.11}$$

Studies show that pulse distortion and broadening is determined mainly by β'' and therefore $\mathcal{D}_1$. On the other hand, β''', and therefore $\mathcal{D}_2$, is important only in regions near the zero GVD wavelength where β'' is small or zero.

Many existing communication grade single-mode fibers have zero GVD near 1.3 μm. However, the loss of silica fibers is much lower in the 1.5–1.6 μm range. In addition, erbium-doped fiber amplifiers commonly used to compensate for the fiber loss are more efficient at 1.55 μm. Thus, there is ample motivation to shift the operating wavelength from 1.3 to 1.55 μm. For the communication-grade fibers, $\mathcal{D}_1$ is between +15 and +20 ps/(km·nm) at 1.55 μm. Techniques exist to offset the GVD effects. One scheme is to insert a section of specialty fiber having a negative $\mathcal{D}_1$ near 1.55 μm. The specialty fibers are known as the *dispersion compensation fibers* (DCF). $\mathcal{D}_1$ depends on the index difference and the index profile. It also depends on the thickness and number of cladding layers. It is possible to choose the index profile and index difference to realize a specific value of $\mathcal{D}_1$. When properly engineered, a single-mode single-cladding DCF may have a first-order GVD parameter as large as -150 ps/(km·nm) [2, 3]. For single-mode DCF with two cladding layers or for two-mode or few-mode fibers operating near the cutoff of LP_{11} modes, $\mathcal{D}_1$ can be as large as -770 ps/(km·nm) near 1.55 μm [4, 5].

For all fibers of interest, $\mathcal{D}_2$ is positive and it is in the 0.07- to 0.08-ps/[km·(nm)^2] range.

13.3 FOURIER TRANSFORM METHOD

Now, we are ready to study the pulse propagation in waveguides. Let the input to a waveguide be $\mathcal{E}(0, t)$, and we are interested in $\mathcal{E}(z, t)$ at an arbitrary point z in the waveguide. Instead of dealing with the real vector $\mathcal{E}(z, t)$ directly, it is convenient to treat it as the real part of a complex vector $\mathbf{E}(z, t)$

$$\mathcal{E}(z, t) = \text{Re}\{\mathbf{E}(z, t)\} \tag{13.12}$$

A crucial step in analyzing transient problems is to introduce the Fourier and inverse Fourier transform pair:

$$\mathbf{E}(z, \omega) = \frac{1}{2\pi} \int_{-\infty}^{\infty} \mathbf{E}(z, t) e^{-j\omega t} \, dt \tag{13.13}$$

$$\mathbf{E}(z, t) = \int_{-\infty}^{\infty} \mathbf{E}(z, \omega) e^{j\omega t} \, d\omega \tag{13.14}$$

Suppose the input is a quasi-monochromatic wave with a frequency ω_0 and an input pulse shape function $\mathcal{A}(0, t)$:

$$\mathcal{E}(0, t) = \mathbf{e}\mathcal{A}(0, t) \cos \omega_0 t$$

The corresponding complex vector is

$$\mathbf{E}(0, t) = \mathbf{e}\mathcal{A}(0, t) e^{j\omega_0 t}$$

The Fourier transform of $\mathbf{E}(0, t)$ is $\mathbf{E}(0, \omega)$ and $\mathbf{E}(0, \omega)$ is

$$\mathbf{E}(0, \omega) = \mathbf{e}A(0, \omega - \omega_0)$$

where $A(0, \omega)$ is the Fourier transform of the input pulse shape function:

$$A(0, \omega) = \frac{1}{2\pi} \int_{-\infty}^{\infty} \mathcal{A}(0, t) e^{-j\omega t} \, dt \tag{13.15}$$

We assume that the temporal pulse width of $\mathcal{A}(0, t)$ is sufficiently wide that the spectral width $\Delta\omega$ of $A(0, \omega)$ is narrow compared to ω_0. In other words, $|A(0, \omega)|$ is negligibly small except for $|\omega| < \Delta\omega$. We also assume that within the narrow spectral range $\Delta\omega$, the field vector $\mathbf{e}$ remains unchanged.

Let $\beta(\omega)$ be the propagation constant, then we have immediately an expression for the field at z,

$$\mathbf{E}(z, \omega) = \mathbf{E}(0, \omega) e^{-j\beta(\omega)z} \tag{13.16}$$

By taking the inverse Fourier transform, we have

$$\mathbf{E}(z, t) = \int_{-\infty}^{\infty} \mathbf{E}(0, \omega) e^{j[\omega t - \beta(\omega)z]} \, d\omega \tag{13.17}$$

The integral may be rearranged as

$$\mathbf{E}(z, t) = e^{j(\omega_0 t - \beta_0 z)} \int_{-\infty}^{\infty} \mathbf{E}(0, \omega) e^{j(\omega - \omega_0)t - j[\beta(\omega) - \beta_0]z} \, d\omega \tag{13.18}$$

We can rewrite (13.18) as

$$\mathbf{E}(z, t) = \mathbf{e}\, e^{j(\omega_0 t - \beta_0 z)} \mathcal{A}(z, t) \tag{13.19}$$

where

$$\mathcal{A}(z, t) = \int_{-\infty}^{\infty} A(0, \omega - \omega_0) e^{j(\omega - \omega_0)t - j[\beta(\omega) - \beta_0]z} \, d\omega$$

$$= \int_{-\infty}^{\infty} A(0, \omega) e^{j\omega t - j[\beta(\omega + \omega_0) - \beta_0]z} \, d\omega \tag{13.20}$$

Clearly, $\mathbf{E}(z, t)$ is the product of two terms: a *carrier* or *sinusoid* term and a pulse shape function. Since $|A(0, \omega)|$ is appreciable only in a narrow range $\Delta\omega$, the contribution to the integral of (13.20) comes mainly from a narrow band where ω is small. This completes the formal analysis of the pulse propagation problem. No approximation is made other than the fact the field vector $\mathbf{e}$ remains unchanged within a narrow spectral range of $\Delta\omega$. An exact analysis of $\mathcal{A}(z, t)$ and $\mathcal{E}(z, t)$ would require a detailed and explicit knowledge of $\beta(\omega)$. This is possible for a few special cases. The propagation of rectangular pulses in empty rectangular microwave waveguides with perfectly conducting walls is an example [6]. For other waveguides, an exact analysis is rather difficult if at all possible. The carrier phase term is usually known. We also assume that the field vector $\mathbf{e}$ remain unchanged. Thus the only term in (13.19) left to be studied is the pulse shape function.

13.4 PROPAGATION OF GAUSSIAN PULSES IN WAVEGUIDES

As an application of the Fourier transform method described in Section 13.3, we consider the propagation of Gaussian pulses in waveguides [7–10]. Let the input be a cosine wave of frequency ω_0 with a Gaussian pulse shape function

$$\mathcal{A}(0, t) = e^{-(2 \ln 2)t^2/T_0^2} \tag{13.21}$$

The pulse peaks at $t = 0$ and the peak value is 1. At $t = \pm T_0/2$, the envelope decreases to $1/\sqrt{2}$ as shown in Figure 13.3. Thus the envelope has a *full-width between half-power points* (FWHP) of T_0. For brevity, we write $b = 2 \ln 2$. The Fourier transform of $\mathcal{A}(0, t)$ is

$$A(0, \omega) = \frac{T_0}{2\sqrt{\pi b}} e^{-\omega^2 T_0^2/(4b)} \tag{13.22}$$

and it can be established by making use of an identity

$$\int_0^{\infty} e^{-\alpha^2 t^2} \, dt = \frac{\sqrt{\pi}}{2\alpha} \tag{13.23}$$

which holds even for a complex α, provided Re(α^2) is positive [11].

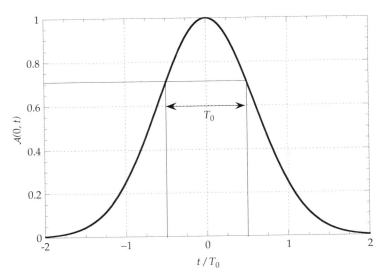

Figure 13.3   Gaussian pulse shape function.

From (13.21) and (13.22), we note that the temporal and spectral widths of the Gaussian envelop are T_0 and $2b/T_0$, respectively. The product of the temporal and spectral widths is independent of T_0. Numerically, it is $2b = 4 \ln 2$ radian or 0.441 s-Hz.

At an arbitrary point z in the waveguide, the pulse envelope function is

$$A(z, t) = \frac{T_0}{2\sqrt{\pi b}} \int_{-\infty}^{\infty} e^{-\omega^2 T_0^2/(4b)} e^{j[\omega t - \beta(\omega + \omega_0)z + \beta_0 z]} \, d\omega \qquad (13.24)$$

As mentioned earlier, it would not be possible to evaluate the integral unless $\beta(\omega)$ is known explicitly. We assume that the pulse width T_0 is not too narrow that the spectral width $\Delta\omega$ is very small in comparison to ω_0. Then, we are justified to use $\beta_E(\omega)$ in lieu of $\beta(\omega)$. In the derivation, we see that β'' and T_0^2 often appear together and T_0^2/β'' has the dimension of length. It is a characteristic distance or length in the study of pulse broadening. Thus we define T_0^2/β'' as the *dispersion length*. To simplify the expressions further, we introduce four dimensionless *normalized variables*: $t_n = (t - \beta'z)/T_0$, $z_n = \beta''z/T_0^2$, $\eta = \beta'''z/(6T_0^3)$, and $\varsigma = \omega T_0$. Physically, t_n is the *normalized time* in a moving time frame, and z_n is the *normalized distance* relative to the dispersion length. In terms of the dimensionless variables, $A(z, t)$ given in (13.24) becomes

$$A(z_n, t_n) \approx \frac{1}{2\sqrt{\pi b}} \int_{-\infty}^{\infty} e^{jt_n \varsigma} e^{-(1 + j2bz_n)\varsigma^2/(4b)} e^{-j\eta\varsigma^3} \, d\varsigma \qquad (13.25)$$

In the following sections, we study the dispersion effects on $A(z, t)$ under various conditions.

13.4.1 Effects of the First-Order Group Velocity Dispersion

As noted earlier, the pulse distortion depends mainly on the first-order GVD para-
meter unless $\mathcal{D}_1$ is vanishingly small. To focus our attention on the effect of $\mathcal{D}_1$, or
equivalently β'', we set β''' and η to zero and obtain from (13.25)

$$A(z_n, t_n) \approx \frac{1}{2\sqrt{\pi b}} \int_{-\infty}^{\infty} e^{jt_n\varsigma} e^{-(1+j2bz_n)\varsigma^2/(4b)} \, d\varsigma$$

The integral can be evaluated by making use of the identity (13.23). To evaluate
the integral, we rearrange the exponent of the integrand as

$$jt_n\varsigma - \frac{1+j2bz_n}{4b}\varsigma^2 = -\frac{1+j2bz_n}{4b}\left[\left(\varsigma - j\frac{2bt_n}{1+j2bz_n}\right)^2 + \left(\frac{2bt_n}{1+j2bz_n}\right)^2\right]$$

and obtain

$$A(z_n, t_n) \approx \frac{1}{\sqrt{1+j2bz_n}} e^{-bt_n^2/(1+j2bz_n)} \tag{13.26}$$

Substituting the above equation into (13.19) and taking the real part of the
resulting expression, we obtain an analytic expression for the instantaneous electric
field at an arbitrary point z:

$$\mathcal{E}(z, t) = \mathbf{e} \frac{1}{(1+4b^2z_n^2)^{1/4}} e^{-bt_n^2/(1+4b^2z_n^2)}$$

$$\cos\left(\omega_o t - \beta_0 z - \frac{1}{2}\tan^{-1} 2bz_n + \frac{2b^2z_n t_n^2}{1+4b^2z_n^2}\right) \tag{13.27}$$

If the group velocity of the transmission medium is not dispersive, β'' and z_n
are zero. Then $\mathcal{E}(z, t)$ reduces to a particularly simple form. In terms of the physical
parameters, $\mathcal{E}(z, t)$ is

$$\mathcal{E}(z, t) = \mathbf{e} \, e^{-b(t-\beta'z)^2/T_o^2} \cos(\omega_o t - \beta_0 z) \tag{13.28}$$

Thus, in a transmission system with a zero GVD, the pulse envelope retains
the Gaussian shape everywhere. The envelope travels with the group velocity $1/\beta'$
and the pulse width remains the same. In short, the pulses are not broadened nor
distorted in a waveguide with a vanishing GVD. It is also reassuring to note that
the carrier sinusoid moves with a phase velocity ω_o/β_0.

If the group velocity is dispersive, two effects become apparent. The pulse
envelope evolves and the instantaneous frequency varies with time and position. The
frequency change is often referred to as *frequency chirping*. An example is shown in
Figure 13.4. For the particular example, we choose $\mathcal{D}_1 = +17.5$ ps/(km·nm), which
is typical for communication-grade optical fibers. However, we purposely choose
an extremely narrow pulse ($T_o = 10$ fs) for the example to accentuate the effect of

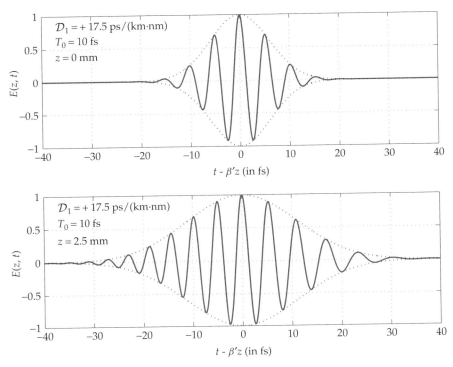

Figure 13.4 Propagation of a narrow Gaussian pulse in a single-mode fiber.

frequency chirping. In Figure 13.4, the pulse envelopes are shown as the dashed curves. The input pulse $\mathcal{E}(0, t)$ is symmetric with respect to the main peak at $t = 0$. Seven to nine cycles of oscillation are discernible in the top plot of Figure 13.4. There are about two cycles of oscillation within the FWHP of 10 fs. Such a narrow pulse broadens quickly as it moves in the fiber. At $z = 2.5$ mm, the FWHP pulse width increases to 18.4 fs as shown in the bottom plot of Figure 13.4. While the pulse envelope at $z = 2.5$ mm still has a Gaussian shape, but the pulse width is much broader. Also note that $\mathcal{E}(0, t)$ is no longer symmetric with respect to the main peak at $t - \beta'z = 0$. The oscillation periods in the trailing edge $(t - \beta'z > 0)$ is longer than those in the leading edge $(t - \beta'z < 0)$. The frequency chirping is clearly evident. The explanation for frequency chirping is as follows. In a fiber with a positive $\mathcal{D}_1$, the short-wavelength (high-frequency) components move faster than the long-wavelength (low-frequency) counterparts. Thus, the instantaneous frequency decreases as a function of $t - \beta'z$. If the fiber has a negative $\mathcal{D}_1$, a DCF, for example, the instantaneous frequency chirps the opposite direction, that is, the oscillation period in the trailing edge is shorter than those in the leading edge.

To quantify the pulse broadening in general, we return to (13.27). At an arbitrary point z in the waveguide, the pulse envelope peaks at $t_n = 0$. At $t_n = \pm\sqrt{1 + 4b^2z_n^2}/2$, the pulse amplitude decreases to $1/\sqrt{2}$ of the peak value. We label this time as the *half-power time* at point z. The *FWHP pulse width* at z is

$$T(z) = T_0(1 + 4b^2 z_n^2)^{1/2} = [T_0^2 + (2b\beta'' z/T_0)^2]^{1/2} \qquad (13.29)$$

The term $T(z)$ increases monotonically with z irrespective of the polarity of β'' and $\mathcal{D}_1$. Initially, the FWHP pulse width increases very slowly. In fact, the FWHP pulse width is essentially the same as T_0 for points near the input. Deep into the waveguide, the FWHP pulse width increases linearly with z and the rate of increase is inversely proportional to T_0.

The instantaneous frequency can be obtained by differentiating the phase term of (13.27) with respect to t. The result is

$$\omega(z,t) = \omega_0 + \frac{4b^2 z_n t_n}{(1 + 4b^2 z_n^2)T_0} = \omega_0 + \frac{4b^2 \beta'' z(t - \beta' z)}{T_0^2[T_0^2 + 4b^2(\beta'' z/T_0)^2]} \qquad (13.30)$$

Note the instantaneous frequency changes linearly with $t - \beta' z$. The rate of frequency change is $\partial \omega / \partial t$, which is given by

$$\frac{\partial \omega(z,t)}{\partial t} = \frac{4b^2 z_n}{(1 + 4b^2 z_n^2)T_0^2} = \frac{4b^2 \beta'' z}{T_0^2[T_0^2 + 4b^2(\beta'' z/T_0)^2]} \qquad (13.31)$$

The frequency change can be positive or negative depending on the sign of $\mathcal{D}_1$ and β''. At the *half-power time* (HPT), the instantaneous frequency change is

$$\omega(z,t)|_{\mathrm{HPT}} = \omega_0 \pm \frac{2b^2 z_n}{T_0[1 + 4b^2 z_n^2]^{1/2}} = \omega_0 \pm \frac{2b^2 \beta'' z}{T_0^2[T_0^2 + 4b^2(\beta'' z/T_0)^2]^{1/2}} \qquad (13.32)$$

As expected, the pulse broadening and frequency chirping in a short fiber are very small even if the fiber is highly dispersive. This can be seen from the fact that as $z \to 0$, $t_n \to t/T_0$, and $z_n \to 0$ and (13.27) and (13.28) reduces to (13.21) for all values of β''. As z increases, $T(z)$ increases. But the frequency may increase or decrease depending on the sign of $\mathcal{D}_1$ and β''. To depict the general behavior of $T(z)$ and $\partial \omega / \partial t$, we plot $T(z)/T_0$ and $[\partial \omega / \partial t]T_0^2$ as functions of the normalized distance in Figure 13.5. As a specific example, we again consider a single-mode fiber with a $\mathcal{D}_1$ of $+17.5$ ps/(km·nm). We take the input as an unchirped Gaussian pulse with a FWHP pulse width of 5 ps. At $z = 1$, 10, and 100 km, the FWHP pulse widths are 13.3 ps, 124 ps, and 1.24 ns, respectively. At the half-power time at 1, 10, and 100 km, the instantaneous frequency change is ± 40.9, ± 44.1, and ± 44.1 GHz, respectively.

13.4.2 Effects of the Second-Order Group Velocity Dispersion

To study the effects of the third-order term of (13.25) on the pulse envelope, Miyagi and Nishida defined a new dimensionless variable [9, 10]

$$\chi = \eta^{1/3} \left(\varsigma - j \frac{1 + j2bz_n}{12b\eta} \right) \qquad (13.33)$$

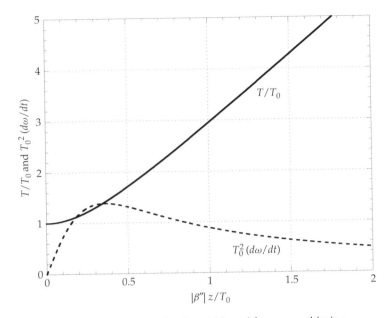

Figure 13.5   Change of pulse width and frequency chirping.

In terms of χ, (13.25) reduces to

$$A(z_n, t_n) \approx \frac{1}{\sqrt{\pi b}}\eta^{-1/3}e^P \int_0^\infty \cos(\chi^3 + Q\chi)\,d\chi \tag{13.34}$$

where P and Q are

$$P = \frac{1 + j2bz_n}{864b^3\eta^2}[(1 + j2bz_n)^2 - 72b^2t_n\eta]$$

$$Q = \frac{1}{48b^2\eta^{4/3}}[(1 + j2bz_n)^2 - 48b^2t_n\eta]$$

The integral in (13.34) can be expressed in terms of Airy function [12]

$$\text{Ai}[\pm(3a)^{-1/3}x] = \frac{(3a)^{1/3}}{\pi}\int_0^\infty \cos\,(at^3 \pm xt)\,dt \tag{13.35}$$

and the final expression is

$$A(z_n, t_n) \approx \sqrt{\frac{\pi}{b}}(3\eta)^{-1/3}\,e^P\text{Ai}(3^{-1/3}Q) \tag{13.36}$$

For most fibers of interest, η is very small and Q is very large. Therefore, we use the asymptotic expression of Airy function in lieu of Airy function itself

$$\text{Ai}(z) \approx \frac{1}{2\sqrt{\pi}} z^{-1/4} e^{-(2/3)z^{3/2}} \qquad |\arg(z)| < \pi \qquad (13.37)$$

By expending P and Q in power series of η and keeping the two leading terms of the power series, we obtain from (13.36)

$$\mathcal{A}(z_n, t_n) \approx [(1 + j2bz_n)^2 - 48b^2 t_n \eta]^{-1/4} \exp\left\{ -\frac{bt_n^2}{1 + j2bz_n} \left[1 + \frac{8b^2 t_n \eta}{(1 + j2bz_n)^2} \right] \right\} \qquad (13.38)$$

The expression is applicable for most situations. In the limit of $\eta \to 0$, that is, $\beta''' z \to 0$, (13.38) reduces to (13.26) discussed in the last subsection. On the other hand, near the zero GVD wavelength, β'' is small and β''' becomes the dominant factor affecting the pulse evolution. To focus our attention on the effects of β''', we set β'' and z_n to zero and obtain from (13.38)

$$\mathcal{A}(z_n, t_n) \approx (1 - 48b^2 t_n \eta)^{-1/4} e^{-bt_n^2(1 + 8b^2 t_n \eta)} \qquad (13.39)$$

Clearly, the pulse shape is not Gaussian. Nor is the pulse envelope symmetric with respect to its peak.

To study the effects of β'' and β''' on the pulse evolution further, we plot $|\mathcal{A}(z_n, t_n)|$ as a function of t_n for several values of z_n and η. Figure 13.6 shows

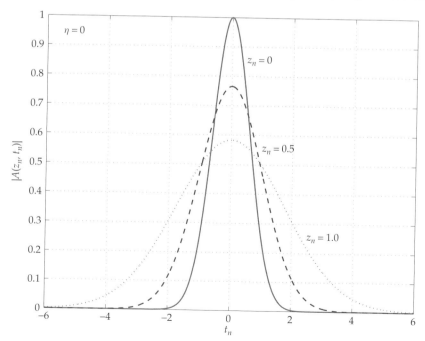

Figure 13.6   Evolution of Gaussian pulse envelope in a fiber with zero η.

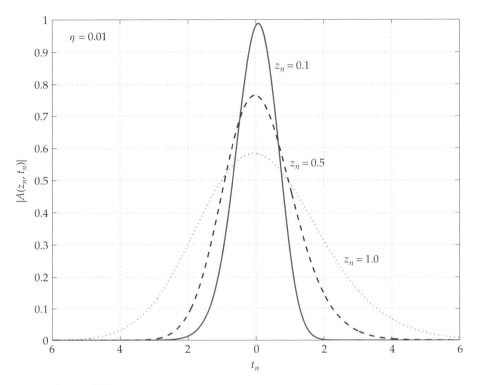

Figure 13.7 Evolution of Gaussian pulse envelope in a fiber with $\eta = 0.01$.

that the pulse spreads quickly if $\eta = 0$ and z_n is large. The envelope retains its Gaussian shape so long as η is zero. But if η is not zero, the pulse envelop is no longer Gaussian as indicated in (13.39). If η is small, the change of pulse shape is rather minor, and the main effect of a nonzero η is to shift the envelope peak very slightly to a later time. This is shown in Figure 13.7 for small η. If η is large, the distortion is quite obvious: The pulse envelop becomes asymmetric and the peak shifts to a later time. This is shown in Figure 13.8. In other words, near the zero GVD wavelength, the pulse deforms drastically if β''' and η are not very small. Once a pulse is distorted, it is meaningless to describe the pulse in terms of the pulse width. Instead, it would be meaningful to use the *moments* of the pulse power to quantify the pulse envelop [13–16].

13.5 IMPULSE RESPONSE

While the direct approach presented in Section 13.3 is easy to understand, it is cumbersome to apply. Each time a new input pulse is specified, it would be necessary to evaluate the Fourier transform anew. To circumvent the need of evaluating the

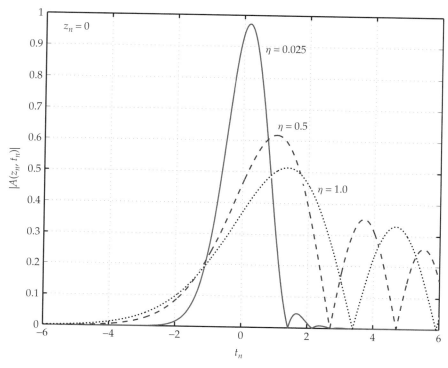

Figure 13.8 Evolution of Gaussian pulse envelope in a fiber with $\eta = 0.025$, 0.5, and 1.0.

complicated integrals, we combine (13.15) and (13.20) and obtain

$$A(z,t) = \frac{1}{2\pi} \int_{-\infty}^{\infty} \int_{-\infty}^{\infty} A(0, t - t') e^{j(\omega-\omega_0)t' - j[\beta(\omega) - \beta_0]z} \, dt' \, d\omega \qquad (13.40)$$

By interchanging the order of integration, we obtain

$$A(z,t) = \int_{-\infty}^{\infty} A(0, t - t') h(z, t') \, dt' \qquad (13.41)$$

where

$$h(z,t) = \frac{1}{2\pi} \int_{-\infty}^{\infty} e^{j(\omega-\omega_0)t - j[\beta(\omega) - \beta_0]z} \, d\omega \qquad (13.42)$$

is the *impulse response function* of the transmission medium [15]. It is clear from (13.41) that $A(z, t)$ is the convolution of the input envelop $A(0, t)$ and the impulse response function $h(z, t)$ of the transmission medium. The task of evaluating $A(0, \omega)$ is side stepped completely once $h(z, t)$ is known.

For pulses with the pulse energy confined in a narrow spectral range, $\beta(\omega)$ can be approximated by the truncated Taylor series $\beta_E(\omega)$ given in (13.7). In lieu of the exact impulse response function $h(z, t)$, we have an *approximate impulse response function*

$$h_E(z, t) = \frac{1}{2\pi} \int_{-\infty}^{\infty} e^{j(\omega - \omega_0)t - j[\beta_E(\omega) - \beta_0]z} \, d\omega \tag{13.43}$$

It is again convenient to cast expressions in terms of the four normalized variables t_n, z_n, η, and ζ defined in Section 13.3. If a pulse width T_0 has not been specified, it would be necessary to choose a characteristic time T_0. In terms of the normalized variables, the approximate impulse response function is

$$h_E(z_n, t_n) = \frac{1}{2\pi T_0} \int_{-\infty}^{\infty} e^{j[t_n \zeta - (z_n/2)\zeta^2 - \eta \zeta^3]} \, d\zeta \tag{13.44}$$

An exact analysis of (13.43) or (13.44) with all terms of $\beta_E(\omega)$ intact has been reported by Wait [18]. But the result is too complicated to be repeated here. Instead, we consider two special cases.

13.5.1   Approximate Impulse Response Function with β''' Ignored

When the last term of $\beta_E(\omega)$ is ignored, (13.44) becomes [19, 20]

$$h_E(z, t) \approx \frac{1}{2\pi T_0} \int_{-\infty}^{\infty} e^{j[t_n \zeta - (z_n/2)\zeta^2]} \, d\zeta \tag{13.45}$$

Making use of the identities,

$$\int_0^{\infty} \cos x^2 \, dx = \int_0^{\infty} \sin x^2 \, dx = \frac{1}{2}\sqrt{\frac{\pi}{2}}$$

we obtain a closed-form expression for the approximate impulse response function,

$$h_E(z_n, t_n) \approx \frac{1}{T_0}\sqrt{\frac{1}{2\pi z_n}} e^{-j\pi/4} e^{jt_n^2/(2z_n)} \tag{13.46}$$

When (13.46) is substituted in (13.41), we have the pulse shape function at an arbitrary point z:

$$A(z_n, t_n) \approx \sqrt{\frac{1}{2\pi z_n}} e^{-j\pi/4} \int_{-\infty}^{\infty} A(0, t_n - t_n') e^{jt_n'^2/(2z_n)} \, dt_n' \tag{13.47}$$

where $t_n' = (t' - \beta' z)/T_0$.

Equation (13.46) is valid for all values of z_n so long as β''' is vanishingly small. Although the magnitude of $h_E(z_n, t_n)$ is independent of t_n, the real and imaginary parts of $h_E(z_n, t_n)$ oscillate rapidly with t_n due to the presence of $e^{jt_n^2/(2z_n)}$. Because of the complex exponential function in $h_E(z_n, t_n)$, the pulse envelope $\mathcal{A}(z_n, t_n)$ at points inside the waveguide can be quite different from the initial pulse envelope.

13.5.2 Approximate Impulse Response Function with β'' Ignored

To study the effect of the second-order GVD, we set β'' and z_n to zero and obtain from (13.43)

$$h_E(z_n, t_n) \approx \frac{1}{T_0\pi} \int_0^\infty \cos(t_n\varsigma - \eta\varsigma^3)\, d\varsigma \qquad (13.48)$$

The integral again can be expressed in terms of Airy function. The result is

$$h_E(z_n, t_n) \approx \begin{cases} (3\eta)^{-1/3}\mathrm{Ai}[-(3\eta)^{-1/3}t_n] & \eta > 0 \\ (3|\eta|)^{-1/3}\mathrm{Ai}[(3|\eta|)^{-1/3}t_n] & \eta < 0 \end{cases} \qquad (13.49)$$

In Figure 13.9, we plot $h_E(z, t)$ as a function of $|(\beta''' z/2)^{-1/3}(t - \beta' z)|$. Note that the approximate impulse response function decays exponentially when β''' is positive. On the other hand, h_E is an oscillatory function and has a gradually decreasing amplitude if β''' is negative. Since $h_E(z_n, t_n)$ is highly asymmetric, the rapid evolution and degradation of $\mathcal{A}(z, t)$ is understandable.

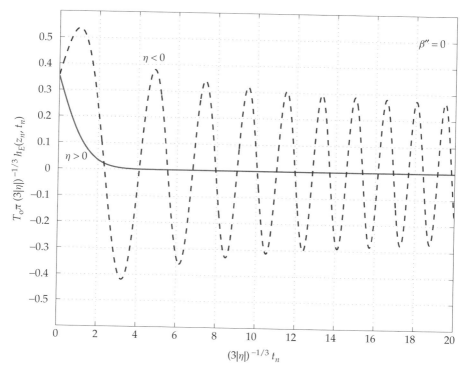

Figure 13.9 Normalized impulse response for waveguides with $\beta'' = 0$.

13.6   PROPAGATION OF RECTANGULAR PULSES IN WAVEGUIDES

As an application of the impulse response concept, we consider the transmission and evolution of rectangular pulses in waveguides. The rise and fall times of an ideal rectangular pulse are infinitesimally short. Because of the infinitesimally short rise and fall times, complication arises. But our objective here is to study the degradation of the pulse envelope in the waveguides. We will not be concerned with the causality question, forerunners, and other complications [1, 19–21].

Suppose the input is a rectangular pulse with a pulse width T_o:

$$A(0, t) = u(t) - u(t - T_o) \tag{13.50}$$

As noted earlier, the dominant influence on the pulse broadening and distortion is the first-order GVD. To consider the effect of the first-order GVD, we ignore β'''. By using the approximate impulse response function h_E given in (13.46), we obtain from (13.41) and (13.50)

$$A(z, t) \approx \sqrt{\frac{1}{2\pi \beta'' z}} e^{-j\pi/4} \int_{-\infty}^{\infty} [u(t - t') - u(t - t' - T_o)] e^{j(t' - \beta' z)^2/(2\beta'' z)} \, dt'$$

In terms of the normalized variables, the above expression becomes

$$A(z_n, t_n) \approx \sqrt{\frac{1}{2\pi z_n}} e^{-j\pi/4} \int_{t_n-1}^{t_n} e^{j t_n'^2/(2z_n)} \, dt_n'$$

The integral can be evaluated and cast in terms of well-known functions and the result is

$$A(z_n, t_n) \approx \frac{1}{\sqrt{2}} e^{-j\pi/4} \left\{ \left[C\left(\frac{t_n}{\sqrt{\pi z_n}}\right) - C\left(\frac{t_n - 1}{\sqrt{\pi z_n}}\right) \right] + j \left[S\left(\frac{t_n}{\sqrt{\pi z_n}}\right) \right. \right.$$
$$\left. \left. - S\left(\frac{t_n - 1}{\sqrt{\pi z_n}}\right) \right] \right\} \tag{13.51}$$

where
$$S(x) = \int_0^x \sin\left(\frac{\pi}{2} \varsigma^2\right) d\varsigma \qquad C(x) = \int_0^x \cos\left(\frac{\pi}{2} \varsigma^2\right) d\varsigma$$

are the sine and cosine Fresnel integrals [12]. The time-domain electric field is

$$\mathcal{E}(z_n, t_n) \approx \mathbf{e} \frac{1}{\sqrt{2}} \mathcal{F}(z_n, t_n) \cos\left[\omega_0 t - \beta_0 z - \frac{\pi}{4} + \Phi(z_n, t_n)\right] \tag{13.52}$$

The amplitude and phase functions $\mathcal{F}$ and Φ are

$$\mathcal{F}(z_n, t_n) = \left\{\left[\mathcal{C}\left(\frac{t_n}{\sqrt{\pi z_n}}\right) - \mathcal{C}\left(\frac{t_n - 1}{\sqrt{\pi z_n}}\right)\right]^2 + \left[\mathcal{S}\left(\frac{t_n}{\sqrt{\pi z_n}}\right)\right.\right.$$

$$\left.\left. - \mathcal{S}\left(\frac{t_n - 1}{\sqrt{\pi z_n}}\right)\right]^2\right\}^{1/2} \tag{13.53}$$

$$\Phi(z_n, t_n) = \tan^{-1} \frac{\mathcal{S}(t_n/\sqrt{\pi z_n}) - \mathcal{S}\left[(t_n - 1)/\sqrt{\pi z_n}\right]}{\mathcal{C}(t_n/\sqrt{\pi z_n}) - \mathcal{C}\left[(t_n - 1)/\sqrt{\pi z_n}\right]} \tag{13.54}$$

The results can also be obtained by using the Fourier transform method discussed in Section 13.3. To apply the Fourier transform method, it is necessary to evaluate the Fourier transform of $\mathcal{A}(0, t)$ given by (13.50). The Fourier transform of $\mathcal{A}(0, t)$ is

$$A(0, \omega) = \frac{T_0}{2\pi} \frac{\sin \omega_0 T_0/2}{\omega_0 T_0/2} e^{-j\omega_0 T_0/2} = \frac{T_0}{2\pi} \frac{\sin \varsigma/2}{\varsigma/2} e^{-j\varsigma/2} \tag{13.55}$$

When (13.55) is substituted into (13.11), and the integral is evaluated, (13.51)–(13.54) are reproduced [17, 19–21]. To verify the validity of (13.51), we consider the limiting case of a very short waveguide. When z is small, the arguments $t_n/\sqrt{\pi z_n}$ and $(t_n - 1)/\sqrt{\pi z_n}$ of the Fresnel integrals are very large. Thus we may approximate the cosine and sine Fresnel integrals by their asymptotic expressions for large arguments:

$$\mathcal{C}(x) \to \pm\frac{1}{2} \qquad \mathcal{S}(x) \to \pm\frac{1}{2} \qquad \text{as} \quad x \to \pm\infty \tag{13.56}$$

When these asymptotic expressions are used in (13.51), the rectangular pulse (13.50) is recovered in the limit of $z \to 0$.

Figure 13.10 shows the degradation of a 10-ps rectangular pulse in an optical fiber with $\mathcal{D}_1 = +17.5$ ps/(km·nm). At $z = 1$ m, the pulse envelope is essentially a rectangle except for the small ripples near the corners. At $z = 100$ m, the ripples become much larger. In fact, the ripples or ringing appear throughout the pulse and the pulse envelope is far from being a rectangle. The rapid degradation of rectangular pulse is due to the steep rise and fall of the input pulse.

13.7 EVOLUTION OF PULSE ENVELOPE [22, 23]

As discussed in the preceding sections, expressions for the electric field propagating in waveguides or fibers are the product of a carrier sinusoid and a pulse shape function. The carrier term is simple and well understood. The pulse envelope is more complicated except that it evolves slowly in time and position. In this section,

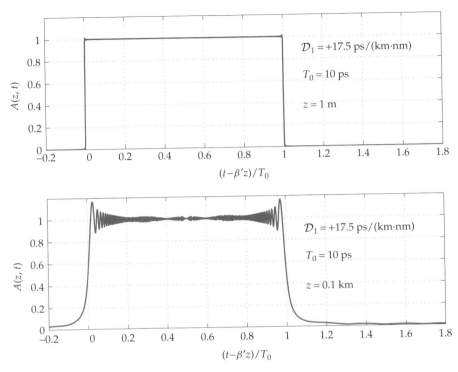

Figure 13.10 Evolution of a rectangular pulse in a single-mode fiber.

we develop a differential equation describing the slow and gradual evolution of the pulse envelope.

We begin with the time-dependent, source-free Maxwell equations (1.1)–(1.4) and the constitutive relations (1.5) and (1.6) of isotropic, nonmagnetic, dielectric materials. By combining these equations, we obtain a time-domain wave equation:

$$\nabla^2 \mathcal{E}(\mathbf{r}; t) - \mu_0 \varepsilon_0 \frac{\partial^2 \mathcal{E}(\mathbf{r}; t)}{\partial t^2} = \mu_0 \frac{\partial^2 \mathcal{P}(\mathbf{r}; t)}{\partial t^2} + \nabla[\nabla \cdot \mathcal{E}(\mathbf{r}; t)] \qquad (13.57)$$

As noted in Chapter 1, $\mathcal{P}$ depends on $\mathcal{E}$ in a complicated manner. The simplest and most obvious way to describe the relationship is to express $\mathcal{P}$ as a power series in $\mathcal{E}$:

$$\mathcal{P}(\mathbf{r}; t) = \mathcal{P}^{L}(\mathbf{r}; t) + \mathcal{P}^{NL2}(\mathbf{r}; t) + \mathcal{P}^{NL3}(\mathbf{r}; t) + \cdots \qquad (13.58)$$

where $\mathcal{P}^{L}$, $\mathcal{P}^{NL2}$, and $\mathcal{P}^{NL3}$ are linear, quadratic, and cubic in $\mathcal{E}$, and they are the *linear*, *second-order*, and *third-order polarizations*, respectively. When the electric field is weak, the leading term alone, that is, the linear polarization, would be sufficient to represent $\mathcal{P}$. The medium is a linear medium. As the electric field gets stronger, it is necessary to keep the three terms of (13.58). The medium is a nonlinear medium.

Consider a Cartesian component of the linear polarization, $\mathcal{P}_i^L$. It depends on the past history of all components of the electric field everywhere, including in particular the field near $\mathbf{r}$. Specifically, $\mathcal{P}_i^L$ can be written as a convolution integral:

$$\mathcal{P}_i^L(\mathbf{r};t) = \varepsilon_0 \sum_{j=x,y,z} \int_{-\infty}^{\infty} \int_{-\infty}^{t} \chi_{ij}(\mathbf{r} - \mathbf{r}';t - t')\,\mathcal{E}_j(\mathbf{r}';t')\,dt'\,d^3\mathbf{r}' \qquad (13.59)$$

where i and j stand for x, y, or z, and $\chi_{ij}(\mathbf{r} - \mathbf{r}';t - t')$ is the proportionality or weighting factor relating $\mathcal{P}_i^L$ to $\mathcal{E}_j$, [24, 25]. To consider the convolution integrals further, we make three assumptions. The medium in question is *uniform* and the material response to the field is *local* and *fast* or *instantaneous*. Under the uniformity and locality assumptions, $\chi_{ij}(\mathbf{r} - \mathbf{r}';t - t')$ is a delta function in position, that is, $\chi_{ij}(\mathbf{r} - \mathbf{r}';t - t') = \chi_{ij}(t - t')\delta(\mathbf{r} - \mathbf{r}')$. Thus the integration with respect to $\mathbf{r}'$ can be carried out explicitly and $\mathcal{P}_i^L(\mathbf{r};t)$ becomes

$$\mathcal{P}_i^L(\mathbf{r};t) = \varepsilon_0 \sum_{j=x,y,z} \int_{-\infty}^{t} \chi_{ij}(t - t')\,\mathcal{E}_j(\mathbf{r};t')\,dt' \qquad (13.60)$$

Combining the three Cartesian components, we have an expression for the linear polarization vector. In the tensor notation, $\boldsymbol{\mathcal{P}}^L(\mathbf{r};t)$ is

$$\boldsymbol{\mathcal{P}}^L(\mathbf{r};t) = \varepsilon_0 \int_{-\infty}^{t} \chi^{(1)}(t - t') \cdot \boldsymbol{\mathcal{E}}(\mathbf{r};t')\,dt' \qquad (13.61)$$

where $\chi^{(1)}(t)$ is a tensor function relating $\boldsymbol{\mathcal{P}}^L(\mathbf{r};t)$ to $\boldsymbol{\mathcal{E}}(\mathbf{r};t)$.

For waves in IR or visible regions, the material response is very fast and may be approximated as instantaneous. Under the assumption of instantaneous response, a further simplification becomes possible [24, 25], namely $\chi_{ij}(t - t') = \chi_{ij}\delta(t - t')$. Thus the convolution integral with respect to t' in (13.60) can be evaluated as well and we obtain

$$\mathcal{P}_i^L(\mathbf{r};t) = \varepsilon_0 \sum_j \chi_{ij}\mathcal{E}_j(\mathbf{r};t) \qquad (13.62)$$

where χ_{ij} is the tensor elements of the *linear electric susceptibility tensor*, $\chi^{(1)}$. In isotropic media, the linear susceptibility tensor reduces to a linear electric susceptibility constant and $\boldsymbol{\mathcal{P}}^L(\mathbf{r};t)$ and $\boldsymbol{\mathcal{E}}(\mathbf{r};t)$ are in parallel.

If the material response is truly instantaneous, the tensor elements are frequency independent. In real materials, the response, while fast, is not truly instantaneous. Then χ_{ij} vary slowly with frequency. Depending on the applications, the accuracy desired, and the spectral width of fields involved, we may take these tensor elements and the linear electric susceptibility as frequency dependent or independent.

Similar and more complicated expressions can be written for the second- and third-order polarizations. In this chapter, the fields are weak enough that the second- and higher-order polarizations are very small and ignored. It is only necessary to consider the linear effects only. We will come back and examine the effects of second- and third-order polarizations in Chapter 14 when we study the nonlinear interactions in fibers.

To consider the linear effects, we use $\mathcal{P}^{L}(\mathbf{r};t)$ in lieu of $\mathcal{P}(\mathbf{r};t)$. If the medium is source-free, $\nabla \cdot \varepsilon_{o}\mathcal{E}(\mathbf{r};t)$ is zero. Then the wave equation (13.57) reduces to

$$\nabla^{2}\mathcal{E}(\mathbf{r};t) - \mu_{o}\varepsilon_{o}\frac{\partial^{2}\mathcal{E}(\mathbf{r};t)}{\partial t^{2}} = \mu_{o}\frac{\partial^{2}\mathcal{P}^{L}(\mathbf{r};t)}{\partial t^{2}} \tag{13.63}$$

Since $\mathcal{P}^{L}(\mathbf{r};t)$ is linear in $\mathcal{E}(\mathbf{r};t)$, so is (13.63). Then we can apply the Fourier transform method to solve for $\mathcal{E}(\mathbf{r};t)$. To proceed, we define a Fourier transform pair:

$$\mathbf{E}(\mathbf{r};\omega) = \frac{1}{2\pi}\int_{-\infty}^{\infty}\mathcal{E}(\mathbf{r};t)e^{-j\omega t}\,dt \tag{13.64}$$

$$\mathcal{E}(\mathbf{r};t) = \int_{-\infty}^{\infty}\mathbf{E}(\mathbf{r};\omega)e^{j\omega t}\,d\omega \tag{13.65}$$

Then, $\mathcal{P}^{L}(\mathbf{r};t)$ given in (13.61) can be written in terms of the transform of $\mathbf{E}(\mathbf{r};\omega)$:

$$\mathcal{P}^{L}(\mathbf{r};t) = \varepsilon_{o}\int_{-\infty}^{\infty}\tilde{\chi}^{(1)}(\omega)\mathbf{E}(\mathbf{r};\omega)e^{j\omega t}\,d\omega \tag{13.66}$$

where

$$\tilde{\chi}^{(1)}(\omega) = \int_{0}^{\infty}\chi^{(1)}(t)e^{-j\omega t}\,dt \tag{13.67}$$

13.7.1   Monochromatic Waves

For monochromatic waves of angular frequency ω_{o}, the spectral width of $\mathbf{E}(\mathbf{r};\omega)$ is infinitesimally sharp. The integral in (13.66) can be evaluated readily, and we obtain $\mathcal{P}^{L}(\mathbf{r};t) = \varepsilon_{o}\tilde{\chi}_{0}\mathbf{E}(\mathbf{r};t)$ where $\tilde{\chi}_{0} = \tilde{\chi}^{(1)}(\omega_{o})$. Then, (13.63) is simplified to

$$\nabla^{2}\mathcal{E}(\mathbf{r};t) - \mu_{o}\varepsilon_{o}(1 + \tilde{\chi}_{0})\frac{\partial^{2}\mathcal{E}(\mathbf{r};t)}{\partial t^{2}} = 0 \tag{13.68}$$

To determine $\mathcal{E}(\mathbf{r};t)$, we write

$$\mathcal{E}(\mathbf{r};t) = \mathrm{Re}\{\mathcal{A}_{0}\mathbf{e}(x,y)e^{j(\omega_{o}t - \beta_{0}z)}\} \tag{13.69}$$

where $\mathbf{e}(x,y)$ is the electric field vector in the transverse plane. $\mathcal{A}_{0}$ is an amplitude constant and β_{0} is a yet undetermined propagation constant. Upon substituting (13.69) into (13.68), we obtain a differential equation for $\mathbf{e}(x,y)$:

$$[\nabla_{t}^{2} - \beta_{0}^{2} + \omega_{o}^{2}\mu_{o}\varepsilon_{o}(1 + \tilde{\chi}_{0})]\mathbf{e}(x,y) = 0 \tag{13.70}$$

where ∇_t^2 is the transverse Laplacian operator. The propagation constant and the electric field vector are determined when (13.70) is solved subject to the appropriate boundary conditions. In particular, β_0 is of the form of

$$\beta_0^2 = \omega_o^2 \mu_o \varepsilon_o [1 + \tilde{\chi}^{(1)}(\omega_o)] - \Lambda \tag{13.71}$$

where Λ is the eigenvalue of the boundary value problem and it depends on the waveguide geometry, materials, and the mode in question.

For future references and by a straightforward differentiation with respect to the angular frequency, we obtain

$$\beta_0 \beta' = \mu_o \varepsilon_o \omega_o \left(1 + \tilde{\chi}_0 + \tfrac{1}{2}\,\omega_o \tilde{\chi}_1\right) \tag{13.72}$$

$$\beta_0 \beta'' + \beta'^2 = \mu_o \varepsilon_o \left(1 + \tilde{\chi}_0 + 2\,\omega_o \tilde{\chi}_1 + \tfrac{1}{2}\,\omega_o^2 \tilde{\chi}_2\right) \tag{13.73}$$

$$\beta' \beta'' + \tfrac{1}{3}\beta_0\beta''' = \mu_o \varepsilon_o \left(\tilde{\chi}_1 + \omega_o \tilde{\chi}_2 + \tfrac{1}{6}\omega_o^2 \tilde{\chi}_3\right) \tag{13.74}$$

where

$$\tilde{\chi}_i = \frac{d^i}{d\omega^i} \tilde{\chi}^{(1)}(\omega)\Big|_{\omega=\omega_o}$$

and $i = 0, 1, 2,$ and 3.

13.7.2 Envelope Equation

Now we consider quasi-monochromatic waves in the visible or IR regions. As noted earlier, the material response to the waves in visible and IR regions is very fast, but not instantaneous. Then $\tilde{\chi}^{(1)}(\omega)$ varies slowly with ω and can be approximated by the four leading terms of the Taylor series expansion:

$$\tilde{\chi}^{(1)}(\omega) = \tilde{\chi}_0 + (\omega - \omega_o)\tilde{\chi}_1 + \tfrac{1}{2}(\omega - \omega_o)^2 \tilde{\chi}_2 + \tfrac{1}{6}(\omega - \omega_o)^3 \tilde{\chi}_3 + \cdots \tag{13.75}$$

By substituting (13.75) into (13.66) and making use of the properties of the Fourier transform, we obtain

$$\begin{aligned}
\mathcal{P}^{\mathrm{L}}(\mathbf{r};t) \approx \varepsilon_o \Bigg\{ &\tilde{\chi}_0 \mathcal{E}(\mathbf{r};t) - \tilde{\chi}_1 \left[j\frac{\partial \mathcal{E}(\mathbf{r};t)}{\partial t} + \omega_o \mathcal{E}(\mathbf{r};t) \right] \\
&- \tfrac{1}{2}\tilde{\chi}_2 \left[\frac{\partial^2 \mathcal{E}(\mathbf{r};t)}{\partial t^2} - j2\omega_o \frac{\partial \mathcal{E}(\mathbf{r};t)}{\partial t} - \omega_o^2 \mathcal{E}(\mathbf{r};t) \right] \\
&+ \tfrac{1}{6}\tilde{\chi}_3 \left[j\frac{\partial^3 \mathcal{E}(\mathbf{r};t)}{\partial t^3} + 3\omega_o \frac{\partial^2 \mathcal{E}(\mathbf{r};t)}{\partial t^2} \right. \\
&\left. - j3\omega_o^2 \frac{\partial \mathcal{E}(\mathbf{r};t)}{\partial t} - \omega_o^3 \mathcal{E}(\mathbf{r};t) \right] \Bigg\}
\end{aligned} \tag{13.76}$$

An exact expression for $\mathcal{E}(\mathbf{r};t)$ is either unknown or too complicated to use; we look for an approximate representation instead. We expect that $\mathcal{E}(\mathbf{r};t)$ can be expressed as the product of a carrier term $\mathbf{e}(x, y)e^{j(\omega_0 t - \beta_0 z)}$ and a slowly varying pulse shape function $\mathcal{A}(z, t)$,

$$\mathcal{E}(\mathbf{r}; t) = \text{Re}[\mathcal{A}(z, t)\mathbf{e}(x, y)e^{j(\omega_0 t - \beta_0 z)}] \tag{13.77}$$

where $\mathbf{e}(x, y)$ and β_0 are determined from (13.70). The pulse shape function $\mathcal{A}(z, t)$ is yet undetermined. In terms of $\mathbf{e}(x, y)$ and $\mathcal{A}(z, t)$, $\mathcal{P}^{\text{L}}(\mathbf{r}; t)$ is, from (13.76)

$$\mathcal{P}^{\text{L}}(\mathbf{r}; t) \approx \varepsilon_0 \left[\tilde{\chi}_0 \mathcal{A}(z; t) - j\tilde{\chi}_1 \frac{\partial \mathcal{A}(z; t)}{\partial t} - \frac{1}{2}\tilde{\chi}_2 \frac{\partial^2 \mathcal{A}(z; t)}{\partial t^2} + j\frac{1}{6}\tilde{\chi}_3 \frac{\partial^3 \mathcal{A}(z; t)}{\partial t^3} \right]$$

$$e^{j(\omega_0 t - \beta_0 z)}\mathbf{e}(x, y)$$

To determine $\mathcal{A}(z, t)$, we substitute the above expression and (13.77) into (13.63) and obtain

$$\mathcal{A}(z, t)e^{j(\omega_0 t - \beta_0 z)}\{[\nabla_t^2 - \beta_0^2 + \omega_0^2 \mu_0 \varepsilon_0(1 + \tilde{\chi}_0)]\mathbf{e}(x, y)\}$$

$$\approx \left[-\frac{\partial^2 \mathcal{A}(z, t)}{\partial z^2} + j2\beta_0 \frac{\partial \mathcal{A}(z, t)}{\partial z} + j2\omega_0 \mu_0 \varepsilon_0 \left(1 + \tilde{\chi}_0 + \frac{1}{2}\omega_0 \tilde{\chi}_1 \right) \frac{\partial \mathcal{A}(z, t)}{\partial t} \right.$$

$$+ \mu_0 \varepsilon_0 \left(1 + \tilde{\chi}_0 + 2\omega_0 \tilde{\chi}_1 + \frac{1}{2}\omega_0^2 \tilde{\chi}_2 \right) \frac{\partial^2 \mathcal{A}(z, t)}{\partial t^2}$$

$$\left. - j\mu_0 \varepsilon_0 \left(\tilde{\chi}_1 + \omega_0 \tilde{\chi}_2 + \frac{1}{6}\omega_0^2 \tilde{\chi}_3 \right) \frac{\partial^3 \mathcal{A}(z, t)}{\partial t^3} \right] e^{j(\omega_0 t - \beta_0 z)}\mathbf{e}(x, y) \tag{13.78}$$

In the manipulation, we have ignored the third- and higher-order derivatives of $\mathcal{A}(z, t)$. This is justified since the pulse envelope evolves slowly in time and position. Because of (13.70), the left-hand side of (13.78) vanishes. So does the right-hand side of the equation. Thus, we obtain

$$-\frac{\partial^2 \mathcal{A}(z, t)}{\partial z^2} + j2\beta_0 \frac{\partial \mathcal{A}(z, t)}{\partial z} + j2\omega_0 \mu_0 \varepsilon_0 \left(1 + \tilde{\chi}_0 + \frac{1}{2}\omega_0 \tilde{\chi}_1 \right) \frac{\partial \mathcal{A}(z, t)}{\partial t}$$

$$+ \mu_0 \varepsilon_0 \left(1 + \tilde{\chi}_0 + 2\omega_0 \tilde{\chi}_1 + \frac{1}{2}\omega_0^2 \tilde{\chi}_2 \right) \frac{\partial^2 \mathcal{A}(z, t)}{\partial t^2}$$

$$- j\mu_0 \varepsilon_0 \left(\tilde{\chi}_1 + \omega_0 \tilde{\chi}_2 + \frac{1}{6}\omega_0^2 \tilde{\chi}_3 \right) \frac{\partial^3 \mathcal{A}(z, t)}{\partial t^3} = 0 \tag{13.79}$$

To further simplify (13.79), we make use of the relations between β_0, β', β'', and $\tilde{\chi}_i$ established in (13.72)–(13.74). The result is

$$\left[-\frac{\partial^2 \mathcal{A}(z, t)}{\partial z^2} + \beta'^2 \frac{\partial^2 \mathcal{A}(z, t)}{\partial t^2} \right] + \beta_0 \beta'' \frac{\partial^2 \mathcal{A}(z, t)}{\partial t^2} + j2\beta_0 \frac{\partial \mathcal{A}(z, t)}{\partial z}$$

$$+ j2\beta_0 \beta' \frac{\partial \mathcal{A}(z, t)}{\partial t} - j\left(\beta'\beta'' + \frac{1}{3}\beta_0\beta''' \right) \frac{\partial^3 \mathcal{A}(z, t)}{\partial t^3} = 0 \tag{13.80}$$

From the discussion presented in previous sections, we expect the pulse envelope to travel with the group velocity $1/\beta'$. In other words,

$$\frac{\partial A(z, t)}{\partial z} - \beta' \frac{\partial A(z, t)}{\partial t}$$

is negligibly small. In addition,

$$\frac{\partial^2 A(z, t)}{\partial z^2} - \beta'^2 \frac{\partial^2 A(z, t)}{\partial t^2}$$

is also small since

$$\frac{\partial^2 A(z, t)}{\partial z^2} - \beta'^2 \frac{\partial^2 A(z, t)}{\partial t^2} = \left(\frac{\partial}{\partial z} + \beta' \frac{\partial}{\partial t} \right) \left[\frac{\partial A(z, t)}{\partial z} - \beta' \frac{\partial A(z, t)}{\partial t} \right]$$

When these terms are dropped from (13.80), the equation becomes

$$\frac{\partial A(z, t)}{\partial z} + \beta' \frac{\partial A(z, t)}{\partial t} - \frac{j}{2} \beta'' \frac{\partial^2 A(z, t)}{\partial t^2} - \frac{1}{2} \left(\frac{\beta' \beta''}{\beta_0} + \frac{1}{3} \beta''' \right) \frac{\partial^3 A(z, t)}{\partial t^3} = 0$$

(13.81)

This is the desired envelope equation describing the slow evolution of the pulse envelope. It is worthwhile to recap the basic assumptions and approximations made in the derivation. First, we assume that the spectral width $\Delta \omega$ is narrow compared to the center angular frequency ω_0. Therefore we are justified to use the leading terms of the Taylor series expansion (13.75) as an approximation for $\tilde{\chi}^{(1)}(\omega)$. We also ignore the third- and higher-order derivatives of $A(z, t)$ since the pulse envelope evolves slowly in time and position. Lastly, we drop two more groups of terms since the pulse envelope moves with the group velocity $1/\beta'$.

13.7.3 Pulse Envelope in Nondispersive Media

If the group velocity is frequency independent, then β'' and β''' are zero, and all terms of the envelope equation except the first two terms disappear. The solution to the partial differential equation (13.81) is simply

$$A(z, t) = f(t - \beta' z)$$

(13.82)

where $f(t)$ is an arbitrary function specified by the input waveform. In other words, the pulse envelope moves with the group velocity $1/\beta'$ and retains its original shape if the group velocity is frequency independent.

13.7.4 Effect of the First-Order Group Velocity Dispersion

When operating away from the zero GVD wavelength, the first-order GVD is the dominant term affecting the evolution of the pulse envelope. By ignoring the forth and fifth terms of (13.81), we obtain

$$\frac{\partial A(z, t)}{\partial z} + \beta' \frac{\partial A(z, t)}{\partial t} - \frac{j}{2} \beta'' \frac{\partial^2 A(z, t)}{\partial t^2} = 0 \tag{13.83}$$

In terms of the normalized time and distance introduced in Section 13.3, (13.83) becomes

$$\frac{\partial A(z_n, t_n)}{\partial z_n} - \frac{j}{2} \frac{\partial^2 A(z_n, t_n)}{\partial t_n^2} = 0 \tag{13.84}$$

Equations (13.83) and (13.84) are the parabolic partial differential equations describing the motion and evolution of the pulse envelope. $A(z, t)$ or $A(z_n, t_n)$ may be solved once the initial pulse envelope $A(0, t)$ is specified.

Tajima and Washio [26] showed if the input is a Gaussian pulse with a linearly chirped frequency, the solution to (13.84) can be written as

$$A(z_n, t_n) = \frac{A_1}{\sqrt{c_1 + j(z_n + c_2)}} \exp\{-t_n^2/(2[c_1 + j(z_n + c_2)])\} \tag{13.85}$$

where c_1, c_2, and A_1 are constants determined by the initial pulse shape function and frequency chirping [26]. A straightforward substitution would show that (13.85) is indeed a solution of (13.84). From the analytic expression, we can find the pulse width and frequency chirping at an arbitrary point z. For this purpose, we rewrite the exponent of (13.85) in terms of the real and imaginary parts explicitly and obtain

$$A(z_n, t_n) = \frac{A_1}{\sqrt{c_1 + j(z_n + c_2)}} \exp\left[-\frac{t_n^2}{W^2(z_n)} + jC(z_n)t_n^2\right] \tag{13.86}$$

where

$$W^2(z_n) = 2\frac{c_1^2 + (z_n + c_2)^2}{c_1} \tag{13.87}$$

$$C(z_n) = \frac{1}{2}\frac{z_n + c_2}{c_1^2 + (z_n + c_2)^2} \tag{13.88}$$

Thus the *pulse width* and *frequency chirping* are, respectively,

$$T(z_n) = T_o\sqrt{b}W(z_n) = T_o\sqrt{2\ln 2}\,W(z_n) \tag{13.89}$$

$$\omega(z_n, t_n) - \omega_o = 2t_n C(z_n)/T_o \tag{13.90}$$

Note that the pulse width depends only on $W(z_n)$, and the frequency chirping is a function of $C(z_n)$. In Figure 13.11, we plot $W(z_n)/[2c_1]^{1/2}$ and $2c_1C(z_n)$ as functions of $(z_n + c_2)/c_1$. Note that the pulse width decreases to a minimum at

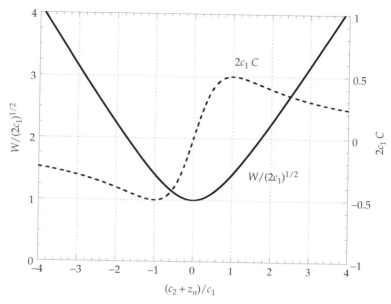

Figure 13.11 Evolution of pulse width and frequency chirping in a waveguide having a linearly chirped Gaussian pulse as the input.

$z_n + c_2 = 0$, that is, $z = -c_2 T_o^2/\beta''$ if $c_2 T_o^2/\beta''$ is negative. At this location, the frequency chirping is zero.

As a specific example, we suppose the input field is a unchirped Gaussian pulse with $c_1 = 1/(2b)$, $c_2 = 0$, and $\mathcal{A}_1 = 1/\sqrt{2b}$. Then we obtain from (13.85)

$$\mathcal{A}(0, t_n) = e^{-bt_n^2} = e^{-bt^2/T_o^2} \tag{13.91}$$

where T_o is the pulse width. For an arbitrary point inside the waveguide $z > 0$,

$$\mathcal{A}(z_n, t_n) = \frac{1}{\sqrt{1 + j2bz_n}} \exp[-bt_n^2/(1 + j2bz_n)] \tag{13.92}$$

Thus, we rederive the Gaussian pulse shape function given in (13.26). We also show that the FWHP pulse width increases as $T_o[1 + 4b^2 z_n^2]^{1/2}$. In terms of the physical parameters, the FWHP pulse width increases as $[T_o^2 + 4b^2 \beta''^2 z^2/T_o^2]^{1/2}$, as discussed in Section 13.4.1.

If the input pulse is a linearly chirped Gaussain pulse, c_2 is not zero. Depending on the signs of c_2 and β'', the pulse width may increase or decrease as the pulse moves forward. If c_2 and β'' have the same sign, $|c_2 + z_n|$ increases with z and the pulse broadens as it moves forward. The frequency chirping also changes as the pulse moves. On the other hand, if c_2 and β'' have opposite signs, the pulse width decreases initially until $c_2 + z_n = 0$. At this point, the pulse width reaches to a minimum value. As an example, consider a fiber of length L. We suppose the input is a linearly chirped Gaussian pulse with $c_2 = -\beta'' L/(2T_0^2)$. As the pulse

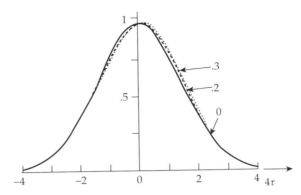

Figure 13.12   Pulse envelope shapes in a fiber with and without the third-order dispersion. The second-order dispersion is zero. The value on each curve is 192η. (From Ref. [9].)

moves forward in the fiber, the frequency chirping decreases. At the midpoint of the fiber $(z = L/2)$, $c_2 + z_n = 0$, the frequency chirping vanishes and the pulse width narrows to the minimum value. As the pulse moves beyond the midpoint, the frequency chirping reappears and the pulse width spreads as well.

13.7.5   Effect of the Second-Order Group Velocity Dispersion

For fibers operating near or at the zero GVD wavelength, β'' is vanishingly small or zero. Then the fifth term of (13.81) involving β''' is the dominant term affecting the pulse envelope. Under this condition, (13.83) becomes

$$\frac{\partial \mathcal{A}(z,t)}{\partial z} + \beta'\frac{\partial \mathcal{A}(z,t)}{\partial t} - \frac{1}{6}\beta'''\frac{\partial^3 \mathcal{A}(z,t)}{\partial t^3} = 0 \qquad (13.93)$$

It is difficult to solve a third-order partial differential equation for an arbitrary input. However, the effect of β''' on the pulse evolution has been studied by Miyagi and Nishida [9, 10]. They showed that the envelope peak shifts away from the origin of the moving time frame. For large β''', the envelope shape changes drastically as shown in Figures 13.12 and 13.13 [9]. The pulse distortion also depends on the spectral width [10].

13.8   DISPERSION COMPENSATION

It is clear from the previous sections that pulses spread in fibers for all values of β'', positive or negative. It is natural to inquire if the pulse broadening can be compensated somehow. For this purpose, we consider the concatenation of two fibers with β'' of opposite polarities (Fig. 13.14) and study the pulse evolution in the fiber sections. Let the lengths and propagation constants of the two fiber sections be L_1 and L_2, β_1' and β_2', and β_1'' and β_2'', respectively. We further suppose that the input pulse is an unchirped Gaussian pulse with a pulse width T_0. The pulse shape

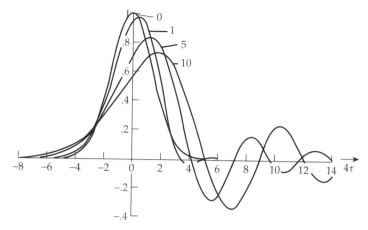

Figure 13.13 Pulse envelope shapes in a fiber with large third-order dispersion. The second-order dispersion is zero. The value on each curve is 192η. (From Ref. [9].)

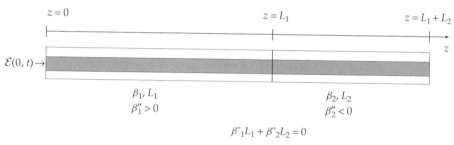

Figure 13.14 Simple dispersion compensation technique.

function in the first fiber section is, in terms of β_1' and β_1''

$$A(z,t) = \frac{1}{\sqrt{1 + j\dfrac{2b\beta_1'' z}{T_o^2}}} \exp\left[-\frac{2b(t - \beta_1' z)^2/T_o^2}{2\left(1 + j\dfrac{2b\beta_1'' z}{T_o^2}\right)} \right] \qquad 0 \le z \le L_1 \qquad (13.94)$$

As expected, the pulse width increases with z:

$$T_1(z) = \sqrt{T_o^2 + 4b^2(\beta_1'' z)^2} \qquad 0 \le z \le L_1 \qquad (13.95)$$

At the end of the first fiber section, the pulse shape function is

$$A(L_1^-, t) = \frac{1}{\sqrt{1 + j\dfrac{2b\beta_1'' L_1}{T_o^2}}} \exp\left[-\frac{2b(t - \beta_1' L_1)^2/T_o^2}{2\left(1 + j\dfrac{2b\beta_1'' L_1}{T_o^2}\right)} \right] \qquad (13.96)$$

This is the input pulse to the second fiber section. For the second fiber section, $c_1 = 1/(2b)$ and $c_2 = (\beta_1'' - \beta_2'')L_1/T_0^2$. We also shift the time origin and obtain, from (13.92), the pulse shape function in the second section, $L_1 \leq z \leq L_1 + L_2$:

$$\mathcal{A}(z, t)$$
$$= \frac{1}{\sqrt{1 + j2b\dfrac{\beta_2''(z - L_1) + \beta_1''L_1}{T_0^2}}} \exp\left\{ -2b\frac{[t - \beta_2'(z - L_2) - \beta_1'L_1]^2/T_0^2}{2\left[1 + j2b\dfrac{\beta_2''(z - L_1) + \beta_1''L_1}{T_0^2}\right]} \right\}$$

$$(13.97)$$

Note that $\mathcal{A}(L_1^+, t)$ matches with $\mathcal{A}(L_1^-, t)$. In the second fiber section the pulse width is

$$T_2(z) = \sqrt{T_0^2 + 4b^2[\beta_2''(z - L_1) + \beta_1''L_1]^2} \quad L_1 \leq z \leq L_1 + L_2 \qquad (13.98)$$

If β_1'' and β_2'' have the opposite signs, $T_2(z)$ decreases as z increases. At $z = L_1(1 + |\beta_1''/\beta_2''|)$, $T_2(z)$ reduces to T_0. In other words, if we choose the fiber lengths such that $|\beta_1''L_1| = |\beta_2''L_2|$, then the pulse broadening occurred in the first fiber section is completely compensated for in the second fiber section. The pulse regains its initial width and shape at the output of the second section. In short, by choosing the GVD and fiber lengths properly, the pulse distortion in one fiber section is offset completely by that in the second fiber section. This is a simple *dispersion compensation* technique [27].

REFERENCES

1. L. Brillouin, *Wave Propagation and Group Velocity*, Academic, New York, 1960.

2. A. Bjarklev, T. Rasmussen, O. Lumholt, K. Rottwitt, and M. Helmer, "Optimal design of single-cladded dispersion-compensating optical fibers," *Opt. Lett.*, Vol. 19, pp. 457–459 (1994).

3. A. M. Vengsarkar and W. A. Reed, "Dispersion-compensating single mode fibers: Efficient designs for first- and second- order compensation," *Opt. Lett.*, Vol. 18, pp. 924–926 (1993).

4. T. L. Wu and H. C. Chang, "An efficient numerical approach for determining the dispersion characteristics of dual mode elliptical core optical fibers," *IEEE J. Lightwave Technol.*, Vol. 13, pp. 1926–1934 (1995).

5. C. D. Poole, J. M. Wiesenfeld, D. J. DiGiovanni, and A. M. Vengsarkar, "Optical fiber-based dispersion compensation using higher order modes near cutoff," *IEEE J. Lightwave Technol.*, Vol. 12, pp. 1745–1758 (1994).

6. S. L. Dvorak, "Exact, closed-form expressions for transient fields in homogeneously filled waveguides," *IEEE Trans. Microwave Theory Techniques*, Vol. 42, pp. 2164–2170 (1994).

7. M. P. Forrer, "Analysis of millimicrosecond RF pulse transmission," *Proc. IRE*, 46, pp. 1830–1835 (1958).

8. C. G. B. Garrett and D. E. McCumber, "Propagation of a Gaussian light pulse through an anomalous dispersion medium," *Phys. Rev. A*, Vol. 1, pp. 305–313 (Feb. 1970).

9. M. Miyagi and S. Nishida, "Pulse spreading in a single-mode fiber due to third-order dispersion," *Appl. Opt.*, Vol. 18, pp. 678–682 (1979).

10. M. Miyagi and S. Nishida, "Pulse spreading in a single-mode fiber due to third-order dispersion: Effect of optical source bandwidth," *Appl. Opt.*, Vol. 18, pp. 2237–2240 (1979).

11. W. Magnus and F. Oberhettinger, *Formulas and Theorems for the Functions of Mathematical Physics*, Chelsea, New York, 1949 p. 116.

12. M. Abramowitz and I. A. Stegun, *Handbook of Mathematical Functions with Formulas, Graphs and Mathematical Tables*, Dover, New York, 1965.

13. D. Marcuse, "Pulse distortion in single-mode fibers," *Appl. Opt.*, Vol. 19, pp. 1653–1660 (May 15, 1980).

14. D. Marcuse, "Pulse distortion in single-mode fibers, Part 2," *Appl. Opt.*, Vol. 20, pp. 2969–2974 (Sept. 1, 1981).

15. D. Marcuse, "Pulse distortion in single-mode fibers, Part 3," *Appl. Opt.*, Vol. 20, pp. 3573–3579 (Oct. 15, 1981).

16. D. Marcuse, "Propagation of pulse fluctuations in single-mode fibers," *Appl. Opt.*, Vol. 19, pp. 1856–1861 (June 1, 1980).

17. C. M. Knop, "Pulsed electromagnetic wave propagation in dispersive media," *IEEE Trans. Antennas Propagation*, Vol. AP-12, pp. 494–496 (1964).

18. J. R. Wait, "Propagation of pulses in dispersive media," *Radio Sci.*, Vol. 69D, pp. 1387–1401 (Nov. 1965).

19. C. M. Knop and G. I. Cohn, "Comments on 'Pulse waveform degradation due to dispersion in waveguides'," *IRE Trans. Microwave Theory Technology*, Vol. MTT-11, pp. 445–447 (Sept. 1963).

20. C. M. Knop and G. I. Cohn, "Further comments on 'Pulse waveform degradation due to dispersion in waveguides'," *IRE Trans. Microwave Theory Technology*, Vol. MTT-18, pp. 663–665 (Sept. 1970).

21. R. E. Elliott, "Pulse waveform degradation due to dispersion in waveguides," *IRE Trans. Microwave Theory Technology*, Vol. MTT-5, pp. 254–257 (1957).

22. A. E. Siegman, *Lasers*, University Science Books, Mill Valley, CA, 1986.

23. H. A. Haus, *Waves and Fields in Optoelectronics*, Prentice-Hall, Englewood Cliffs, NJ, 1984.

24. Y. R. Shen, *The Principles of Nonlinear Optics*, Wiley, New York, 1984.

25. D. L. Mills, *Nonlinear Optics, Basic Concepts*, Springer, Berlin, 1991.

26. K. Tajima and K. Washio, "Generalized view of Gaussian pulse-transmission characteristics in single-mode optical fibers," *Opt. Lett.*, Vol. 10, pp. 460–462 (Sept. 1985).

27. C. L. Lin, H. Kogelnik, and L. G. Cohen, "Optical-pulse equalization of low dispersion transmission in single-mode fibers in the 1.3–1.7 μm spectral region," *Opt. Lett.*, Vol. 5, pp. 476–478 (Nov. 1980).

14

OPTICAL SOLITONS IN OPTICAL FIBERS

14.1   INTRODUCTION

As discussed in the last chapter, narrow pulses deform and spread as they propagate in a linear and dispersive fiber. The distortion and broadening depend on the shape and temporal width of the pulses and the group velocity dispersion (GVD) of the fiber. But the distortion and broadening are independent of the pulse amplitude. Furthermore, the pulse spectrum does not change as the pulses propagate. These features change drastically if the fiber is nonlinear. In a nonlinear fiber, both the pulse shape and spectrum may change and the change is dependent on the pulse amplitude. However, if the input pulse shape, pulse width, and amplitude satisfy a well-defined relationship, and if the nonlinear fiber has *anomalous group velocity dispersion*, the pulses either propagate indefinitely without changing their shape, width, and amplitude, or they regain the pulse shape, width, and peak amplitude periodically. These pulses are known as *solitary waves* or *solitons*. Clearly, solitons are of interest to fiber communications and they are discussed in this chapter. Specifically, we study the formation, propagation, and evolution of the optical solitons in *weakly guiding, dispersive*, and *weakly nonlinear fibers*.

   The generation, propagation, and evolution of solitons in weakly guiding, dispersive, and weakly nonlinear single-mode fibers, or dispersive and weakly nonlinear

Foundations for Guided-Wave Optics, by Chin-Lin Chen
Copyright © 2007 John Wiley & Sons, Inc.

media, are governed by a *nonlinear envelope equation*. To derive the nonlinear envelope equation, we begin with the wave equation in terms of the electric field intensity and the electric polarization. The electric field intensity and electric polarization are related through the *electric susceptibility tensor* or *intensity-dependent refractive index*. Both relationships may be used to set up the nonlinear envelope equation (Section 14.3). Finally, we introduce normalized parameters and reduce the nonlinear envelope equation to a standard or canonical form: the *normalized nonlinear envelope equation*, which is often referred to as the *nonlinear Schrödinger equation* (NLSE).

In general, the NLSE can be solved rigorously by the *inverse scattering transform* (IST) *method* or by *numerical techniques*. The IST method is complicated, indirect, and tedious. Instead of examining solitons in general, we concentrate on the fundamental and lower-order solitons. We rely on a simple and straightforward method to derive an expression for the *fundamental solitons* (Section 14.5). From the expression for fundamental solitons, we extract the key properties of, and the basic parameters describing, the fundamental solitons. Higher-order solitons are briefly discussed in Section 14.6.

As noted earlier, optical solitons have a specific pulse shape and amplitude. It is natural to inquire how are optical solitons generated. The generation of solitons is discussed in Section 14.7. When an arbitrary pulse is fed to a fiber, soliton and nonsoliton parts of the pulse may be launched. In Section 14.7, we estimate the percentage of input energy used to form the soliton part of the pulse for a given input pulse. In digital communication systems, information is transmitted as pulse streams, not a single and isolated pulse. To increase the information capacity, closely spaced pulses are launched. The interaction between adjacent and possibly overlapping pulses is an important factor limiting the time separation between pulses and thus the data rate of a communication system. This is discussed in Section 14.9.

14.2 OPTICAL KERR EFFECT IN ISOTROPIC MEDIA

As discussed in Chapters 1 and 13, the constitutive relation of a medium is

$$\mathcal{D} = \varepsilon_0 \mathcal{E} + \mathcal{P} \tag{14.1}$$

where $\mathcal{E}$, $\mathcal{D}$, and $\mathcal{P}$ are the electric field intensity, electric flux density, and electric polarization, respectively. $\mathcal{P}$ relates to $\mathcal{E}$ in a complicated manner. The simplest way to describe the relationship is to express $\mathcal{P}$ as a power series in $\mathcal{E}$:

$$\mathcal{P}(\mathbf{r};t) = \mathcal{P}^{L}(\mathbf{r};t) + \mathcal{P}^{NL2}(\mathbf{r};t) + \mathcal{P}^{NL3}(\mathbf{r};t) + \cdots \tag{14.2}$$

where $\mathcal{P}^{L}$, $\mathcal{P}^{NL2}$, and $\mathcal{P}^{NL3}$ are linear, quadratic, and cubic in $\mathcal{E}$, and they are the *linear, second-,* and *third-order polarizations*, respectively. When the electric field intensity is weak, the leading term alone would be accurate enough to approximate $\mathcal{P}$. The medium is a *linear medium*. As the electric field intensity gets stronger, it is necessary to keep the second and third terms of (14.2), in addition to the first term. The medium is a *nonlinear medium*.

14.2.1 Electric Susceptibility Tensor

As discussed in Section 13.7, a Cartesian component of the linear polarization can be expressed as a convolution integral of the three components of the electric field intensity

$$\mathcal{P}_i^{\mathrm{L}}(\mathbf{r};t) = \varepsilon_\mathrm{o} \sum_{j=x,y,z} \int_{-\infty}^{\infty} \int_{-\infty}^{t} \chi_{ij}(\mathbf{r}-\mathbf{r}';t-t')\mathcal{E}_j(\mathbf{r}';t')\,dt'\,d^3\mathbf{r}' \tag{14.3}$$

where i stand for x, y, or z, and $\chi_{ij}(\mathbf{r}-\mathbf{r}';t-t')$ are the proportionality factors connecting $\mathcal{E}_j$ to $\mathcal{P}_i^{\mathrm{L}}$ [1, 2].

The Cartesian components of $\mathcal{P}^{\mathrm{NL2}}$ and $\mathcal{P}^{\mathrm{NL3}}$ can be written as convolution integrals as well

$$\mathcal{P}_i^{\mathrm{NL2}}(\mathbf{r};t) = \varepsilon_\mathrm{o} \sum_{j,k} \int_{-\infty}^{\infty} \int_{-\infty}^{t} \chi_{ijk}(\mathbf{r}-\mathbf{r}',\mathbf{r}-\mathbf{r}'';t-t',t-t'')\mathcal{E}_j(\mathbf{r}';t')$$

$$\times\, \mathcal{E}_k(\mathbf{r}'';t'')\,dt'\,dt''\,d^3\mathbf{r}'\,d^3\mathbf{r}'' \tag{14.4}$$

$$\mathcal{P}_i^{\mathrm{NL3}}(\mathbf{r};t) = \varepsilon_\mathrm{o} \sum_{j,k,l} \int_{-\infty}^{\infty} \int_{-\infty}^{t} \chi_{ijkl}(\mathbf{r}-\mathbf{r}',\mathbf{r}-\mathbf{r}'',\mathbf{r}-\mathbf{r}''';t-t',t-t'',t-t''')$$

$$\times\, \mathcal{E}_j(\mathbf{r}';t')\mathcal{E}_k(\mathbf{r}'';t'')\mathcal{E}_l(\mathbf{r}''';t''')\,dt'\,dt''\,dt'''\,d^3\mathbf{r}'\,d^3\mathbf{r}''\,d^3\mathbf{r}''' \tag{14.5}$$

where $\chi_{ijk}(\mathbf{r}-\mathbf{r}',\mathbf{r}-\mathbf{r}'';t-t',t-t'')$ and $\chi_{ijkl}(\mathbf{r}-\mathbf{r}',\mathbf{r}-\mathbf{r}'',\mathbf{r}-\mathbf{r}''';t-t',t-t'',t-t''')$ are also the proportionality factors connecting $\mathcal{E}_j$ to $\mathcal{P}_i^{\mathrm{NL2}}$ and $\mathcal{P}_i^{\mathrm{NL3}}$. The two nonlinear polarizations describe the *Pockels* and *Kerr effects* in materials.

If the medium is *uniform* and if the material response is *local* and *instantaneous*, the integrals in (14.3)–(14.5) may be evaluated, as discussed in Section 13.7. By combining the three Cartesian components, we have algebraic expressions for $\mathcal{P}^{\mathrm{L}}$, $\mathcal{P}^{\mathrm{NL2}}$ and $\mathcal{P}^{\mathrm{NL3}}$ in terms of $\mathcal{E}$. In the tensor notation, the three polarizations are

$$\mathcal{P}_i^{\mathrm{L}}(\mathbf{r};t) = \varepsilon_\mathrm{o} \sum_j \chi_{ij}\mathcal{E}_j(\mathbf{r};t) \tag{14.6}$$

$$\mathcal{P}_i^{\mathrm{NL2}}(\mathbf{r};t) = \varepsilon_\mathrm{o} \sum_{j,k} \chi_{ijk}\mathcal{E}_j(\mathbf{r};t)\mathcal{E}_k(\mathbf{r};t) \tag{14.7}$$

$$\mathcal{P}_i^{\mathrm{NL3}}(\mathbf{r};t) = \varepsilon_\mathrm{o} \sum_{j,k,l} \chi_{ijkl}\mathcal{E}_j(\mathbf{r};t)\mathcal{E}_k(\mathbf{r};t)\mathcal{E}_1(\mathbf{r};t) \tag{14.8}$$

In the above expressions, χ_{ij}, χ_{ijk}, and χ_{ijkl} are tensor elements of the *linear, second-order*, and *third-order electric susceptibility tensors*, $\chi^{(1)}$, $\chi^{(2)}$, and $\chi^{(3)}$, respectively. $\chi^{(1)}$, $\chi^{(2)}$, and $\chi^{(3)}$ are the second-, third-, and forth-rank tensors. If the material response to fields is truly instantaneous, χ_{ij}, χ_{ijk}, and χ_{ijkl} are frequency-independent tensor elements. In many materials, the response is mainly due to the interaction of fields with electrons. Then the response time is on order of femtoseconds. For microwave and millimeter-wave fields, femtosecond responses can be treated as instantaneous and χ_{ij}, χ_{ijk}, and χ_{ijkl} are approximately independent of frequency. For visible and near-infrared fields, a response on order of femtoseconds is fast, but not truly instantaneous. Then the electric susceptibility tensor elements vary slowly with frequency. Depending on the applications and the bandwidth involved, we may take these tensor elements as frequency dependent or independent.

Most optical waveguides and fibers are made of isotropic materials. Table 14.1 lists typical values of the electric susceptibility tensor elements of isotropic media. We note in particular that all tensor elements χ_{ijk} of isotropic media are zero. In other words, Pockels effect does not exist in isotropic media. There is no need to discuss the second-order electric polarization and Pockels effect in isotropic media further. Even though the optical Kerr effect is the dominant nonlinear effect in isotropic materials, the optical Kerr effect is very weak and χ_{ijkl} is very small. In short, most isotropic materials are weakly nonlinear.

14.2.2   Intensity-Dependent Refractive Index

To derive a nonlinear envelope equation, we begin with the wave equation in $\mathcal{E}$ and $\mathcal{P}$. As noted in the last section, $\mathcal{P}$ relates to $\mathcal{E}$ through the electric susceptibility tensor. However, the final expression is simpler if the intensity-dependent refractive index is used in lieu of the electric susceptibility tensor elements. In this subsection, we establish a relationship between the electric susceptibility tensor elements and the intensity-dependent refractive index. To simplify manipulations, we assume that the electric field is linearly polarized in the x direction. So are the linear and third-order polarizations. From (14.6) and (14.8), we have

$$\mathcal{P}_x^{\mathrm{L}}(\mathbf{r};t) = \varepsilon_0 \chi_{xx} \mathcal{E}_x(\mathbf{r};t) \tag{14.9}$$

$$\mathcal{P}_x^{\mathrm{NL3}}(\mathbf{r};t) = \varepsilon_0 \chi_{xxxx}(\mathcal{E}_x(\mathbf{r};t))^3 \tag{14.10}$$

Suppose the electric field is a monochromatic wave of frequency ω,

$$\mathcal{E}_x = E_0 \cos \omega t \tag{14.11}$$

TABLE 14.1   Electric Susceptibility Tensor Elements of Isotropic Media

	Tensor Elements	Order of Magnitude of Nonzero Elements
$\chi^{(1)}$	$\chi_{ii} = \chi_{jj}$ $\chi_{ij} = 0$ if $i \neq j$	~ 1
$\chi^{(2)}$	$\chi_{ijk} = 0$ for all i, j, and k $\chi_{iiii} = \chi_{jjjj}$	
$\chi^{(3)}$	$\chi_{iijj} = \chi_{jjii}, \chi_{ijij} = \chi_{jiji}, \chi_{ijji} = \chi_{jiij}$ $\chi_{iiii} = \chi_{iijj} + \chi_{ijij} + \chi_{ijji}$	10^{-21} m^2/V^2

Then the linear and third-order polarizations are

$$\mathcal{P}_x^L = \varepsilon_0 \chi_{xx} E_0 \cos \omega t \tag{14.12}$$

$$\mathcal{P}_x^{NL3} = \varepsilon_0 \chi_{xxxx} (E_0 \cos \omega t)^3 = \varepsilon_0 \chi_{xxxx} E_0^3 \left(\tfrac{3}{4} \cos \omega t + \tfrac{1}{4} \cos 3\omega t \right) \tag{14.13}$$

and $\mathcal{P}_x^{NL3}$ contains a fundamental frequency term and a third-harmonic term. In particular, the contribution by $\mathcal{P}_x^{NL3}$ to $\mathcal{D}_x$ at the fundamental frequency is $\tfrac{3}{4} \varepsilon_0 \chi_{xxxx} E_0^3$ $\cos \omega t$. This term can be written as $\tfrac{3}{4} \varepsilon_0 \chi_{xxxx} E_0^2 (E_0 \cos \omega t)$, that is, $\tfrac{3}{4} \varepsilon_0 \chi_{xxxx} |\mathcal{E}_x(\mathbf{r}; t)|^2$ $\mathcal{E}_x(\mathbf{r}; t)$. Thus the electric flux density at the fundamental frequency is

$$\mathcal{D}_x(\mathbf{r}; t) = \varepsilon_0 \left[1 + \chi_{xx} + \tfrac{3}{4} \chi_{xxxx} |\mathcal{E}_x(\mathbf{r}; t)|^2 \right] \mathcal{E}_x(\mathbf{r}; t) \tag{14.14}$$

Let the refractive index of the medium be n. Then $\mathcal{D}_x = \varepsilon_0 n^2 \mathcal{E}_x$ and n^2 is

$$n^2 = 1 + \chi_{xx} + \tfrac{3}{4} \chi_{xxxx} |E_x(\mathbf{r}; t)|^2 \tag{14.15}$$

When the field is weak, we ignore the intensity-dependent part and obtain the refractive index of the linear medium

$$n_0^2(\omega) = 1 + \chi_{xx} \tag{14.16}$$

A subscript 0 is used to stress the fact that n_0 is independent of the electric field intensity; n_0 is the only term needed when fields are weak and the medium is a linear medium. Since n_0 is the dominant part of the refractive index, we have to describe n_0 as accurately as possible. Accordingly, we take n_0 as a frequency-dependent quantity. In (14.16), we indicate explicitly that n_0 is a function of ω.

If the field is strong, it would be necessary to keep the last term in (14.15). Then n becomes intensity dependent. As indicated in Table 14.1, χ_{xxxx} is very small. The third term of (14.15) is very small in comparison to 1 even for strong electric fields. Then the refractive index has intensity-independent and intensity-dependent parts:

$$n \approx n_0(\omega) + \frac{3}{8 n_0} \chi_{xxxx} |\mathcal{E}_x(\mathbf{r}; t)|^2 = n_0(\omega) + n_{2E} |\mathcal{E}_x(\mathbf{r}; t)|^2 \tag{14.17}$$

So far we have considered the x component of the electric field only. If $\mathcal{E}(\mathbf{r}; t)$ has x, y, and z components, the refractive index becomes

$$n \approx n_0(\omega) + n_{2E} |\mathcal{E}(\mathbf{r}; t)|^2 \tag{14.18}$$

As listed in Table 14.1, χ_{xxxx} is on the order of 10^{-21} m^2/V^2. Therefore n_{2E} is about 10^{-22} m^2/V^2. For silica glass, n_0 is approximately 1.5 and n_{2E} is approximately 6×10^{-23} m²/V² (i.e., 6×10^{-19} cm²/V²).[1] As a numerical example, we consider a typical single-mode silica fiber with a core diameter of 10 μm. Suppose an optical beam of 100 mW is fed to the fiber. Then the irradiance and the electric field intensity in the fiber core are

$$I \approx 100 \times 10^{-3}/(\pi \times 25 \times 10^{-12}) = 1.27 \times 10^8 \text{ W/m}^2$$

$$|\mathcal{E}| \approx \sqrt{2\sqrt{\frac{\mu_o}{\varepsilon_o}\frac{I}{n_0}}} = \sqrt{2 \times 120\pi \times 1.27 \times 10^8/1.5} = 2.5 \times 10^5 \text{V/m}$$

Note that $n_{2E}|\mathcal{E}|^2$ is about 3.8×10^{-12}, which is 12 orders smaller than n_0. Since $n_{2E}|\mathcal{E}|^2$ is very small, we are justified to treat n_{2E} as a frequency-independent parameter as an approximation.

14.3 NONLINEAR ENVELOPE EQUATION

We are ready to consider the pulse propagation in a nonlinear isotropic medium. In addition to the linear polarization, we also keep the third-order polarization to account for the optical Kerr effect of the medium. The wave equation (13.63) becomes

$$\nabla^2 \mathcal{E}(\mathbf{r};t) - \mu_o \varepsilon_o \frac{\partial^2 \mathcal{E}(\mathbf{r};t)}{\partial t^2} - \mu_o \frac{\partial^2 \mathcal{P}^{\mathrm{L}}(\mathbf{r};t)}{\partial t^2} = \mu_o \frac{\partial^2 \mathcal{P}^{\mathrm{NL3}}(\mathbf{r};t)}{\partial t^2} \qquad (14.19)$$

To proceed, we express $\mathcal{P}^{\mathrm{L}}$ and $\mathcal{P}^{\mathrm{NL3}}$ in terms of $\mathcal{E}$.

14.3.1 Linear and Third-Order Polarizations

Since the dominant material response is represented by the linear polarization, we have to treat it accurately. In other words, we take the linear susceptibility tensor $\chi^{(1)}$ as a frequency-dependent parameter. We use (14.7) to relate $\mathcal{P}^L$ with $\mathcal{E}$, apply the Fourier transform (13.58) and (13.59) to (13.61), and obtain

$$\mathcal{P}^{\mathrm{L}}(\mathbf{r};t) = \varepsilon_o \int_{-\infty}^{\infty} \tilde{\chi}^{(1)}(\omega)\mathbf{E}(\mathbf{r};\omega)e^{j\omega t}\, d\omega \qquad (14.20)$$

---

[1] In some literature, the refractive index is cast in term of irradiance $I(\mathbf{r};t)$

$$n \approx n_0(\omega) + n_{2I} I(\mathbf{r};t)$$

To avoid confusion, we use n_{2I} here to distinguish it from n_{2E} used in (14.18). n_{2I} and n_{2E} are related in that $n_{2E} = n_0 n_{2I}/(2\eta_o)$ and η_o is the intrinsic impedance of free space. For silica glass, $n_{2I} \approx 3.2 \times 10^{-20}$ m²/W (i.e., 3.2×10^{-16} cm²/W).

where

$$E(\mathbf{r}; \omega) = \int_{-\infty}^{\infty} \mathcal{E}(\mathbf{r}; t) e^{-j\omega t} \, dt \qquad (14.21)$$

and

$$\tilde{\chi}^{(1)}(\omega) = \int_{-\infty}^{\infty} \chi^{(1)}(t) e^{-j\omega t} \, dt \qquad (14.22)$$

For monochromatic waves, the spectral width of the field is infinitesimally sharp. Then $\mathcal{E}(\mathbf{r}; t) = \mathbf{E}(\mathbf{r}; \omega) e^{j\omega t}$ and $\mathcal{P}^{\mathrm{L}}(\mathbf{r}; t) = \varepsilon_0 \tilde{\chi}^{(1)}(\omega) \mathbf{E}(\mathbf{r}; \omega) e^{j\omega t}$. By ignoring $\mathcal{P}^{\mathrm{NL3}}$, we obtain from (14.19), the propagation constant of plane waves in the linear isotropic medium

$$\beta = k_0 n_0 = k_0 \sqrt{1 + \tilde{\chi}^{(1)}(\omega)} \qquad (14.23)$$

where $k_0 = \omega \sqrt{\mu_0 \varepsilon_0}$. For future references, we differentiate β with respect to ω and write

$$\beta \beta' = \omega \mu_0 \varepsilon_0 \left(1 + \tilde{\chi}_0 + \tfrac{1}{2} \omega \tilde{\chi}_1\right) \qquad (14.24)$$

$$\beta \beta'' + \beta'^2 = \mu_0 \varepsilon_0 \left(1 + \tilde{\chi}_0 + 2\omega \tilde{\chi}_1 + \tfrac{1}{2} \omega^2 \tilde{\chi}_2\right) \qquad (14.25)$$

where

$$\tilde{\chi}_0 = \tilde{\chi}^{(1)}(\omega) \qquad \tilde{\chi}_1 = \frac{d\tilde{\chi}^{(1)}(\omega)}{d\omega} \qquad \tilde{\chi}_2 = \frac{d^2 \tilde{\chi}^{(1)}(\omega)}{d\omega^2}$$

For quasi-monochromatic waves with a narrow bandwidth $\Delta\omega$ centering at or near ω_0, $\tilde{\chi}^{(1)}(\omega)$ can be approximated by the leading terms of the Taylor series expansion. We keep the first three terms of the Taylor series and write

$$\tilde{\chi}^{(1)}(\omega) \approx \tilde{\chi}_0 + (\omega - \omega_0)\tilde{\chi}_1 + \tfrac{1}{2}(\omega - \omega_0)^2 \tilde{\chi}_2 \qquad (14.26)$$

where $\tilde{\chi}_i$ are evaluated at ω_0. Upon substituting (14.26) into (14.20) and making use of the properties of the Fourier transform, we obtain an expression for $\mathcal{P}^{\mathrm{L}}$ in terms of $\mathcal{E}$,

$$\mathcal{P}^{\mathrm{L}}(\mathbf{r}; t) \approx \varepsilon_0 \left\{ \tilde{\chi}_0 \mathcal{E}(\mathbf{r}; t) - \tilde{\chi}_1 \left[j \frac{\partial \mathcal{E}(\mathbf{r}; t)}{\partial t} + \omega_0 \mathcal{E}(\mathbf{r}; t) \right] \right.$$
$$\left. - \frac{1}{2} \tilde{\chi}_2 \left[\frac{\partial^2 \mathcal{E}(\mathbf{r}; t)}{\partial t^2} - j 2\omega_0 \frac{\partial \mathcal{E}(\mathbf{r}; t)}{\partial t} - \omega_0^2 \mathcal{E}(\mathbf{r}; t) \right] \right\} \qquad (14.27)$$

As the optical Kerr effect is rather weak, a simple approximation for $\mathcal{P}^{\mathrm{NL3}}$ would be sufficient for the most applications. Accordingly, we take n_{2E} as a frequency-independent parameter. Then the third-order polarization at the fundamental frequency is, from (14.18),

$$\mathcal{P}^{\mathrm{NL3}}(\mathbf{r}; t) \approx 2\varepsilon_0 n_0 n_{2E} |\mathcal{E}(\mathbf{r}; t)|^2 \mathcal{E}(\mathbf{r}; t) \qquad (14.28)$$

14.3.2   Nonlinear Envelope Equation in Nonlinear Media

Upon substituting (14.27) and (14.28) in (14.19), we obtain a complicated differential equation for $\mathcal{E}$. To simplify the expression, we suppose that the field is
x polarized and has a slowly varying envelope:

$$\mathcal{E}(\mathbf{r};t) = \mathcal{A}(z,t)e^{j(\omega_o t - \beta_o z)}\hat{\mathbf{x}} \qquad (14.29)$$

where $\mathcal{A}(z, t)$ is a slowly varying *pulse shape function* and β_o is the propagation constant of plane waves of frequency ω_o. We also assume that the field is
independent of the two transverse coordinates y and z. The linear and third-order
polarizations due to the x-polarized electric field are also in the x direction and they
are, from (14.27) and (14.28),

$$\mathcal{P}_x^{\text{L}}(\mathbf{r};t) \approx \varepsilon_o \left[\tilde{\chi}_0 \mathcal{A}(z;t) - j\tilde{\chi}_1 \frac{\partial \mathcal{A}(z;t)}{\partial t} - \frac{\tilde{\chi}_2}{2} \frac{\partial^2 \mathcal{A}(z;t)}{\partial t^2} \right] e^{j(\omega_o t - \beta_o z)} \quad (14.30)$$

$$\mathcal{P}_x^{\text{NL3}}(\mathbf{r};t) \approx 2\varepsilon_o n_0 n_{2E} |\mathcal{A}(z;t)|^2 \mathcal{A}(z;t) e^{j(\omega_o t - \beta_o z)} \qquad (14.31)$$

By substituting these expressions into (14.19), we obtain

$$\frac{\partial^2 \mathcal{A}(z;t)}{\partial z^2} - j2\beta_o \frac{\partial \mathcal{A}(z;t)}{\partial z} - j2\omega_o \mu_o \varepsilon_o \left(1 + \tilde{\chi}_0 + \frac{\omega_o \tilde{\chi}_1}{2} \right) \frac{\partial \mathcal{A}(z;t)}{\partial t}$$

$$- \mu_o \varepsilon_o \left(1 + \tilde{\chi}_0 + 2 \, \omega_o \tilde{\chi}_1 + \frac{\omega_o^2 \tilde{\chi}_2}{2} \right) \frac{\partial^2 \mathcal{A}(z;t)}{\partial t^2} \approx -2k_o^2 n_0 n_{2E} |\mathcal{A}(z;t)|^2 \mathcal{A}(z;t)$$

$$\qquad (14.32)$$

In terms of β' and β'' introduced in (14.24) and (14.25), the above equation
becomes

$$\frac{\partial^2 \mathcal{A}(z;t)}{\partial z^2} - j2\beta_o \frac{\partial \mathcal{A}(z;t)}{\partial z} - j2\beta_o \beta_o' \frac{\partial \mathcal{A}(z;t)}{\partial t} - (\beta_o \beta_o'' + \beta_o'^2) \frac{\partial^2 \mathcal{A}(z;t)}{\partial t^2}$$

$$\approx -2k_o^2 n_0 n_{2E} |\mathcal{A}(z;t)|^2 \mathcal{A}(z;t) \qquad (14.33)$$

To simplify the equation further, we recall that the pulse envelope moves with
the group velocity $1/\beta_o'$. Then

$$\frac{\partial \mathcal{A}(z;t)}{\partial z} \approx \beta_o' \frac{\partial \mathcal{A}(z;t)}{\partial t}$$

Therefore, we expect

$$\frac{\partial^2 \mathcal{A}(z;t)}{\partial z^2} - \beta_o'^2 \frac{\partial^2 \mathcal{A}(z;t)}{\partial t^2} = \left(\frac{\partial}{\partial z} + \beta_o' \frac{\partial}{\partial t} \right) \left[\frac{\partial \mathcal{A}(z;t)}{\partial z} - \beta_o' \frac{\partial \mathcal{A}(z;t)}{\partial t} \right]$$

to be small. When the small terms are ignored, (14.33) becomes

$$\frac{\partial A(z;t)}{\partial z} + \beta_o' \frac{\partial A(z;t)}{\partial t} - j\frac{\beta_o''}{2}\frac{\partial^2 A(z;t)}{\partial t^2} \approx -jk_o n_{2E}|A(z;t)|^2 A(z;t) \qquad (14.34)$$

This is the *nonlinear envelope equation* describing the evolution of the slowly varying pulse envelope in a dispersive and weakly nonlinear medium. It is predicated on the assumption that the pulse envelope moves and evolves very slowly. Therefore we have ignored the derivatives of $A(z;t)$ higher than the second order. Since we have ignored the third- and higher-order derivative of $A(z;t)$, we have in fact ignored the second-order GVD effects. The effect of the first-order GVD is given by the third term on the left-hand side of (14.34). We only keep the leading term of the nonlinear term on the right-hand side. We also take n_{2E} as a frequency-independent parameter since n_{2E} is very small.

14.3.3 Self-Phase Modulation

Before considering the nonlinear envelope equation further, we examine two special cases. First we treat a linear medium as a special case of a nonlinear medium with a vanishing n_{2E}. Since the optical Kerr effect is absent, we drop the right-hand side term of (14.34). Then (14.34) is the same as (13.84) without the third derivative term. In short, we recover the differential equation for a slowly varying pulse envelope propagating in a linear and dispersive medium derived in Section 13.7. As noted earlier, we have ignored the third- and higher-order derivatives of $A(z;t)$ and the second-order GVD effects in deriving (14.34).

Next, we suppose the medium is *nonlinear* but not *dispersive*. Then β_o'' vanishes and (14.34) becomes

$$\frac{\partial A(z;t)}{\partial z} + \beta_o' \frac{\partial A(z;t)}{\partial t} \approx -jk_o n_{2E}|A(z;t)|^2 A(z;t) \qquad (14.35)$$

The solution to the first-order differential equation is very simple and can be written down immediately:

$$A(z;t) = \mathbf{A}_o(t - \beta_o'z)e^{-jk_o n_{2E}|\mathbf{A}_o(t-\beta_o'z)|^2 z} \qquad (14.36)$$

where $\mathbf{A}_o(t)$ is the input pulse shape function. However, the phase term is intensity, time, and position dependent. It arises from the optical Kerr effect of the medium. This effect is commonly referred to as the *self-phase modulation* (SPM). The time variation of the phase term is equivalent to a frequency change. To consider the frequency change further, we suppose the input electric field is

$$\mathcal{E}_x(0,t) = E_0 \operatorname{sech}\left(\frac{t}{T_o}\right) e^{j\omega_o t} \qquad (14.37)$$

At z, the electric field is, from (14.29) and (14.36),

$$\mathcal{E}_x(z,t) = E_0 \operatorname{sech}\left(\frac{t - \beta_o' z}{T_o}\right) e^{-jk_o n_{2E} z E_0^2 \operatorname{sech}^2(t-\beta_o' z/T_o)} e^{j(\omega_o t - \beta_o z)} \qquad (14.38)$$

Let the exponent of the above equation be $j\phi(z,t)$, then $\phi(z,t)$ is

$$\phi(z,t) = \omega_o t - \beta_o z - k_o n_{2E} z E_0^2 \operatorname{sech}^2\left(\frac{t - \beta_o' z}{T_o}\right)$$

The instantaneous frequency is

$$\omega(z,t) = \frac{\partial \phi(z,t)}{\partial t} = \omega_o + 2\frac{k_o n_{2E} z E_0^2}{T_o} \operatorname{sech}^2\left(\frac{t - \beta_o' z}{T_o}\right) \tanh\left(\frac{t - \beta_o' z}{T_o}\right) \quad (14.39)$$

Clearly, the instantaneous frequency changes with time and position and the frequency change arises from the amplitude-dependent phase term. The instantaneous frequency is smaller than ω_o prior to the arrival of the pulse peak and greater than ω_o following the arrival of the pulse peak. This is shown in Figure 14.1.

14.3.4 Nonlinear Envelope Equation for Nonlinear Fibers

Equation (14.34) is the nonlinear envelope equation describing the evolution of pulses in a dispersive and nonlinear medium that extends indefinitely in the transverse directions. To describe pulses propagating in nonlinear fibers, we have to

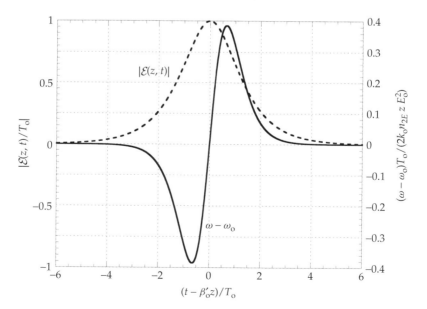

Figure 14.1 Frequency change due to self-phase modulation.

account for the nonuniform field distribution in the transverse planes in fibers. For this purpose, we introduce an *effective cross-section factor* S_{eff} on the right-hand side of (14.34). Then the nonlinear envelope equation for nonlinear and dispersive fibers is

$$\frac{\partial A(z;t)}{\partial z} + \beta_0'\frac{\partial A(z;t)}{\partial t} - j\frac{\beta_0''}{2}\frac{\partial^2 A(z;t)}{\partial t^2} \approx -j\frac{k_0 n_{2E}}{S_{eff}}|A(z;t)|^2 A(z;t) \quad (14.40)$$

Detailed derivation of S_{eff} and related discussion can be found in Refs. [3, 4].

14.3.5 Nonlinear Schrödinger Equation

Equations (14.34) and (14.40) can be reduced to a standard and normalized form by using the *normalized parameters*. For this purpose, we introduce a characteristic time T_0 that may be the temporal width of the input pulse. In terms of T_0, we define the *normalized time* $t_n = (t - \beta_0'z)/T_0$, the *normalized distance* $z_n = |\beta_0''|z/T_0^2$, and the *normalized amplitude* $u(z_n, t_n) = \sqrt{q}A(z_n;t_n)$, where

$$q = \frac{k_0 n_{2E} T_0^2}{|\beta_0''|S_{eff}}$$

The three normalized parameters are dimensionless quantities. Furthermore, t_n and z_n are the same normalized time and distance introduced in Section 13.3. In terms of the three normalized parameters, (14.40) becomes

$$j\frac{\partial u(z_n, t_n)}{\partial z_n} = -\frac{\text{sgn}(\beta_0'')}{2}\frac{\partial^2 u(z_n, t_n)}{\partial t_n^2} + |u(z_n, t_n)|^2 u(z_n, t_n) \quad (14.41)$$

where $\text{sgn}(\beta_0'')$ stands for the sign of β_0''. Equation (14.41) is the normalized nonlinear envelope equation describing the evolution of pulse envelope in weakly dispersive and weakly nonlinear fibers. It can also be used to describe the evolution of pulse envelopes in nonlinear dispersive media of infinite extent by setting S_{eff} to 1. Equations (14.40) and (14.41) have the same form as the one-dimensional time-dependent Schrödinger equation[2] provided the temporal and spatial variables are interchanged. The nonlinear term on the right-hand side corresponds to the potential function of the Schrödinger equation. Equations (14.40) and (14.41) are often referred to as the nonlinear Schrödinger equation (NLSE).

Depending on the sign of β_0'', the solutions of (14.41) are quite different. For fibers or media with an anomalous GVD, β_0'' is negative, $\mathcal{D}_1$ is positive and (14.41) becomes

$$j\frac{\partial u(z_n, t_n)}{\partial z_n} = \frac{1}{2}\frac{\partial^2 u(z_n, t_n)}{\partial t_n^2} + |u(z_n, t_n)|^2 u(z_n, t_n) \quad (14.42)$$

---

[2]The one-dimensional time-dependent Schrödinger equation with a potential function V is

$$\frac{\hbar^2}{2m}\frac{\partial^2 \psi}{\partial z^2} + i\hbar\frac{\partial \psi}{\partial t} = V\psi$$

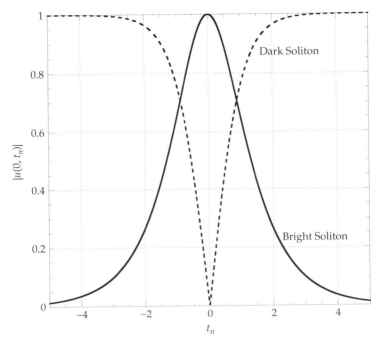

Figure 14.2 Examples of bright and dark solitons.

A well-known solution of (14.42) is a narrow and intense pulse shown as the solid curve in Figure 14.2. Such a pulse is referred to as a *bright soliton*.

On the other hand, for fibers or media with a *normal GVD*, β_0'' is positive, $\mathcal{D}_1$ is negative, and (14.41) becomes

$$j\frac{\partial u(z_n, t_n)}{\partial z_n} = -\frac{1}{2}\frac{\partial^2 u(z_n, t_n)}{\partial t_n^2} + |u(z_n, t_n)|^2 u(z_n, t_n) \qquad (14.43)$$

The solution corresponds to a sharp dip in a constant background. The narrow dip is referred to as a *dark soliton*. A typical example is shown as the dashed curve in Figure 14.2. Of current interest in fiber communications is the bright solitons, which will be discussed in the following sections.

14.4 QUALITATIVE DESCRIPTION OF SOLITONS

There are two general methods of solving the NLSE. The most basic and exact method is to apply the inverse scattering transform (IST) to the NLSE [5–8]. But the IST method is rather indirect and involved. We are not going into the details of the IST method. Instead, we describe the essential steps of the IST method. The first step of the IST method is to convert the *nonlinear differential equation* with given *initial conditions* to a set of linear differential equations. Then we find the eigenvalues and eigenfunctions of the linear differential equations. Finally we construct the solutions of the NLSE from the eigenvalues and eigenfunctions of the linear differential equations. The construction usually involves two integral equations. In a few special cases, the integral equations reduce to a set of linear algebraic equations. Alternatively, we study the nonlinear differential equation numerically [9, 10]. As expected, the numerical approach is tedious and computationally intensive.

Fortunately, for communication applications, we are interested in a particular set of solitons for which a simple analytical method exists. As a prelude to the discussion of this particular set of solitons, we present a qualitative description of the results obtained by the IST method and numerical calculations.

In Figures 14.3, 14.4, and 14.5, we plot the evolution of three pulses with increasing amplitudes. In each plot, we plot the pulse envelope as a function of t_n and z_n. The pulse shown in Figure 14.3 corresponds to a *fundamental soliton* or a *one-soliton*. There is no change in the pulse amplitude, width, and shape as the pulse moves in a dispersive and weakly nonlinear medium. There is no change in the pulse spectrum either, although this is not shown in Figure 14.3.

If the input is sufficiently strong, several fundamental solitons may be launched and they move with the same speed. However, the phase difference between the fundamental solitons varies from location to location. The interference of these fundamental solitons leads to the periodic variation of pulses. Examples are shown in Figures 14.4 and 14.5. In particular, the peak value of the input pulse shown in Figure 14.4 is *twice* as intense as the input pulse depicted in Figure 14.3. The pulse depicted in Figure 14.4 is a *second-order soliton* or a *two-soliton*. As a second-order soliton propagates, it becomes narrower and stronger. Then two minor lobes appear, one on each side of the main peak. Two dips also appear between the main peak and

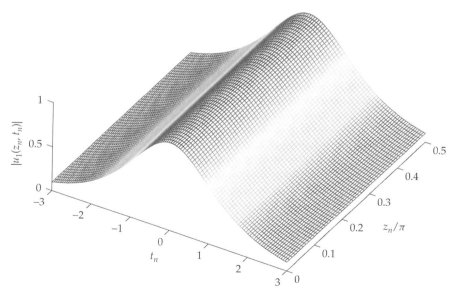

Figure 14.3 Propagation of a fundamental soliton.

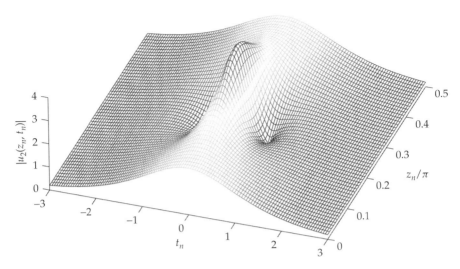

Figure 14.4 Propagation and evolution of a second-order soliton.

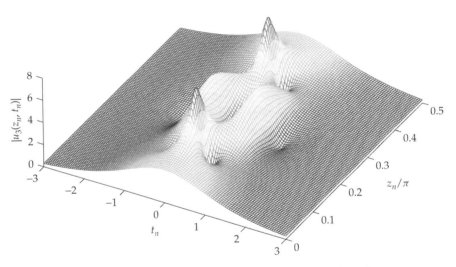

Figure 14.5 Propagation and evolution of a third-order soliton.

minor lobes. Farther into the medium or down the fiber, the minor lobes recombine with the main peak and the pulse regains its initial peak, width, and pulse shape. The evolution process repeats itself and the pulse envelope changes periodically.

A *third-order soliton* or a *three-soliton* is shown in Figure 14.5. The peak value of the input pulse required to excite a third-order soliton is *thrice* as strong as that required to launch a fundamental soliton shown in Figure 14.3. A third-order soliton becomes narrower and stronger initially as it propagates. Then it splits into two pulses of equal amplitudes. The two pulses combine and form a narrower and much stronger pulse before it returns to the original peak, width, and pulse shape.

Similar periodic evolution of pulses is also found in higher-order solitons. It is interesting to note that the spatial periodicity of the second-, third-, and higher-order solitons is the same. The spatial periodicity is commonly referred to as the *soliton period*. In the unit of the normalized distance, the soliton period is $\pi/2$.

14.5 FUNDAMENTAL SOLITONS

In this section, we derive an expression for the fundamental soliton in a straightforward manner. First, we label the *canonical expression* for the fundamental solitons as $u_{1c}(z_n, t_n)$. Then we transform it to other equivalent forms to account for possible variations. Finally, we arrive at a general expression for the fundamental solitons and refer to it as $u_1(z_n, t_n)$.

14.5.1 Canonical Expression

One of the basic techniques of solving boundary value problems is the method of separation of variables. In this section, we apply the method of separation of variables to derive an expression for the fundamental soliton. Toward this end, we write $u_{1c}(z_n, t_n)$ as a product of two functions

$$u_{1c}(z_n, t_n) = f(t_n)e^{-jz_n/2} \tag{14.44}$$

As we expect $u_{1c}(z_n, t_n)$ and its time derivative to vanish as t_n approaches $\pm\infty$, we require $f(t_n)$ and its derivative to vanish as t_n tends to $\pm\infty$. Upon substituting (14.44) into the (14.42), we obtain an ordinary differential equation for $f(t_n)$,

$$\frac{d^2 f}{dt_n^2} - f + 2f^3 = 0 \tag{14.45}$$

In general, solutions of (14.45) can be expressed in terms of Jacobi elliptic functions [11]. Here, we are only interested in a particular solution. To determine this particular solution, we multiply (14.45) by df/dt_n, rearrange the terms, and obtain

$$\frac{d}{dt_n}\left[\left(\frac{df}{dt_n}\right)^2 - f^2 + f^4\right] = 0 \tag{14.46}$$

A simple integration with respect to t_n leads to

$$\left(\frac{df}{dt_n}\right)^2 - f^2 + f^4 = C \tag{14.47}$$

The constant of integration C is zero since f and df/dt_n vanish as $t_n \to \pm\infty$. By setting C to zero, we obtain

$$\frac{df}{dt_n} = \pm f\sqrt{1 - f^2} \tag{14.48}$$

We take the minus sign and obtain[3]

$$f(t_n) = \text{sech}(t_n) \tag{14.49}$$

Thus, the *canonical expression* for a fundamental soliton is

$$u_{1c}(z_n, t_n) = \text{sech}(t_n)\, e^{-jz_n/2} \tag{14.50}$$

It describes a pulse envelope that is stationary in t_n. Recall that t_n is the normalized time referenced to the moving time frame. The reference frame moves with the group velocity in the laboratory frame. Thus $u_{1c}(z_n, t_n)$ represents a pulse envelope that moves in the laboratory frame at the speed of the group velocity $1/\beta_o'$ and the envelope reaches to a peak value of 1 at $t_n = 0$.

14.5.2 General Expression

The canonical expression can be transformed to other forms to describe fundamental solitons in different situations. Toward this end, we note if $u_{1c}(z_n, t_n)$ satisfies (14.42), so do

$$u_{1a}(z_n, t_n) = e^{-j\theta} u_{1c}(z_n, t_n) \tag{14.51}$$

$$u_{1b}(z_n, t_n) = \eta u_{1c}(\eta^2 z_n, \eta t_n) \tag{14.52}$$

and

$$u_{1d}(z_n, t_n) = e^{-j\Lambda t_n + j\Lambda^2 z_n/2} u_{1c}(z_n, t_n - \Lambda z_n) \tag{14.53}$$

for an arbitrary θ, η, and Λ. A simple substitution will show that $u_{1a}(z_n, t_n)$, $u_{1b}(z_n, t_n)$, and $u_{1d}(z_n, t_n)$ are indeed solutions of (14.42). Equation (14.51) indicates that $u_{1c}(z_n, t_n)$ can be multiplied by an arbitrary phase term $e^{-j\theta}$ so long as θ is a real quantity. However, an amplitude scaling η has to be accompanied by changes in t_n and z_n as indicated in (14.52). This is not surprising since (14.42) is a nonlinear differential equation. In (14.53), Λ specifies the motion of the pulse envelope relative to the moving reference frame. Since η and Λ are mutually independent, the pulse motion and the pulse amplitude are also mutually independent.

By combining (14.52) and (14.53), we obtain yet another expression for the fundamental soliton:

$$u_{1e}(z_n, t_n) = \eta\,\text{sech}[\eta(t_n - \Lambda z_n)]e^{-j\Lambda t_n + j(\Lambda^2 - \eta^2)z_n/2} \tag{14.54}$$

To allow for the possibility that the pulse envelope at $z_n = 0$ reaches to the peak value at a time other than $t_n = 0$, we shift the time origin to t_{n0} and obtain

$$u_{1f}(z_n, t_n) = \eta\,\text{sech}[\eta(t_n - t_{n0} - \Lambda z_n)]e^{-j\Lambda(t_n - t_{n0}) + j(\Lambda^2 - \eta^2)z_n/2} \tag{14.55}$$

$$^3 \int \frac{dx}{x\sqrt{a^2 - x^2}} = -\frac{1}{a}\,\text{sech}^{-1}\frac{x}{a}$$

Finally, we introduce an arbitrary phase term as indicated in (14.51) and obtain the most *general expression* for fundamental solitons:

$$u_1(z_n, t_n) = \eta \, \text{sech}[\eta(t_n - t_{n0} - \Lambda z_n)]e^{-j\Lambda t_n - j\theta + j(\Lambda^2 - \eta^2)z_n/2} \qquad (14.56)$$

We have combined Λt_{n0} and θ in the exponent of (14.55) into one term. Equation (14.56) represents a fundamental soliton moving with a finite velocity relative to the moving reference frame and having a peak value of η at $t_n = t_{n0} + \Lambda z_n$ and a phase term $e^{-j\theta}$ at $z_n = 0$ and $t_n = 0$.

14.5.3 Basic Soliton Parameters

A fundamental soliton is completely specified when η, Λ, t_{n0}, and θ are specified. In other words, η, Λ, t_{no}, and θ are the *four basic soliton parameters*. η specifies the peak amplitude and the pulse width. Λ quantifies the motion of the pulse envelope relative to the moving reference frame; t_{n0} and θ identify the timing and the phase of the input pulse. The four basic soliton parameters are determined or specified by the input pulse since

$$u_1(0, t_n) = \eta \, \text{sech}[\eta(t_n - t_{n0})]e^{-j\Lambda t_n - j\theta} \qquad (14.57)$$

In other word, a fundamental soliton given by (14.56) is excited when a pulse of (14.57) is fed to the fiber.

We state, without proof, that if we apply the IST method to derive (14.56), then $(\Lambda + j\eta)/2$ is the eigenvalue of the linear equations deduced from the NLSE.

14.5.4 Basic Soliton Properties

Five basic properties of the fundamental solitons can be deduced from (14.56).

1. The *normalized peak value* of $|u_1(z_n, t_n)|$ is η and the peak occurs at $t_n = t_{n0} + \Lambda z_n$.
2. At

$$t_n = t_{n0} + \Lambda z_n + \frac{\ln(\sqrt{2} \pm 1)}{\eta}$$

the envelope $|u_1(z_n, t_n)|$ decreases to $\eta/\sqrt{2}$. Thus the *normalized pulse width* [full width between half maximum (FWHM)] is

$$(\Delta t_n)_{\text{FWHM}} = 2\frac{\ln(\sqrt{2}+1)}{\eta} \approx \frac{1.7627}{\eta} \qquad (14.58)$$

and the pulse width varies inversely with η.

3. The *area* under the pulse envelope is a constant independent of η, Λ, and z_n since

$$\int_{-\infty}^{\infty} |u_1(z_n, t_n)| \, dt_n = \pi \qquad (14.59)$$

4. The *total energy* contained in a fundamental soliton is η and it varies inversely with the pulse width $(\Delta t_n)_{\text{FWHP}}$:

$$\frac{1}{2} \int_{-\infty}^{\infty} |u_1(z_n, t_n)|^2 \, dt_n = \eta = \frac{2 \ln(\sqrt{2}+1)}{(\Delta t_n)_{\text{FWHM}}} \approx \frac{1.7627}{(\Delta t_n)_{\text{FWHM}}} \qquad (14.60)$$

But the total energy is independent of Λ and z_n.

5. The *frequency spectrum* of $u_1(z_n, t_n)$ is also a hyperbolic secant function[4]

$$U_1(z_n, \Omega) = \int_{-\infty}^{\infty} u_1(z_n, t_n) e^{-j\Omega t_n} \, dt_n$$

$$= \pi \, \text{sech} \left[\frac{\pi(\Omega + \Lambda)}{2\eta} \right] e^{-j[\theta + (\Omega + \Lambda)(t_o + \Lambda z_n) - (\Lambda^2 - \eta^2) z_n / 2]} \quad (14.61)$$

Note that $|U_1(z_n, \Omega)|$ is also independent of z_n. The *spectral width* of a fundamental soliton is

$$\Delta \Omega_{\text{FWHM}} = 2\pi \, \Delta F_{\text{FWHM}} = \frac{4\eta}{\pi} \ln(\sqrt{2}+1) \qquad (14.62)$$

In summary, the peak value, pulse width, area, total energy, and spectral width of a fundamental soliton are independent of z_n. These parameters remain unchanged as the fundamental soliton moves in a fiber. The fundamental soliton amplitude, width, and the total energy are closely related as given by properties 1, 2, and 5. As soliton width decreases, the pulse amplitude increases as shown in Figure 14.6. The total energy per soliton pulse also increases as the soliton pulse width narrows. If we use a fundamental soliton to represent a bit of digital signals, the total energy per bit increases as the pulse width decreases. We also see from properties 2 and 5, that the *(pulse width)(spectral width) product* is a constant:

$$(\Delta t_n)_{\text{FWHM}} \, \Delta F_{\text{FWHM}} = \frac{4}{\pi^2} [\ln(\sqrt{2}+1)]^2 = 0.3148 \qquad (14.63)$$

---

[4] $U_1(z_n, \Omega)$ can be deduced by making use of an integral identity:

$$\int_{-\infty}^{\infty} \text{sech}(ax) e^{ixy} \, dx = \frac{\pi}{a} \, \text{sech} \left(\frac{\pi y}{2a} \right)$$

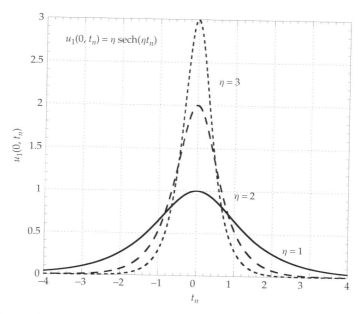

Figure 14.6   Fundamental solitons with different peak value and width.

14.6   HIGHER-ORDER SOLITONS

Equation (14.56) discussed in the last section is applicable to N fundamental solitons if the N fundamental solitons are far apart temporally. Suppose N fundamental solitons are excited. Let the amplitude, velocity relative to the moving reference frame, position of the peak, and initial phase of each fundamental soliton be η_i, Λ_i, t_{n0i}, and θ_i. Then the superposition of N fundamental solitons is

$$u(z_n, t_n) = \sum_{i=1}^{N} \eta_i \, \text{sech}[\eta_i(t_n - t_{n0i} - \Lambda_i z_n)]e^{-j\Lambda_i t_n - j\theta_i + j(\Lambda_i^2 - \eta_i^2)z_n/2} \quad (14.64)$$

As noted previously, the motion of the envelope is independent of the peak value. Therefore, it is possible that the fundamental solitons with different peak values move with the same velocity. However, due to the exponential function $e^{+j(\Lambda_i^2 - \eta_i^2)z_n/2}$, the phase difference between the fundamental solitons varies with z_n. The interference of these solitons leads to a periodic variation of the pulse shape as the N fundamental solitons move forward. The interference of two or more fundamental solitons leads to higher-order solitons. Two examples are given in the following subsections.

14.6.1   Second-Order Solitons

A well-known example of a second-order soliton has been given by Satsuma and Yajima [7]. They showed that if the input pulse is $2 \, \text{sech}(t_n)$, two fundamental solitons are launched, and they are combined to form a second-order soliton:

$$u_2(z_n, t_n) = 4e^{-jz_n/2} \frac{\cosh(3t_n) + 3e^{-j4z_n} \cosh(t_n)}{\cosh(4t_n) + 4\cosh(2t_n) + 3\cos(4z_n)} \qquad (14.65)$$

The above expression can also be written as [12]

$$u_2(z_n, t_n) = \frac{4}{Q}[\text{sech}(t_n)e^{-jz_n/2} + 3\,\text{sech}(3t_n)e^{-j9z_n/2}] \qquad (14.66)$$

where

$$Q = 5 - 3\,[\tanh(t_n)\tanh(3t_n) - \text{sech}(t_n)\,\text{sech}(3t_n)\cos(4z_n)] \qquad (14.67)$$

To demonstrate the connection between (14.64) and (14.66), we note that for large t_n, $Q \to 2$, and (14.66) approaches, asymptotically,

$$u_2(z_n, t_n) \sim e^{-jz_n/2}\,\text{sech}(t_n - \ln 2) + 3e^{-j9z_n/2}\,\text{sech}\,3\left(t_n - \frac{\ln 2}{3}\right) \qquad (14.68)$$

This is of the form of (14.64). Clearly, for large t_n, the second-order soliton given by (14.65) and (14.66) is the superposition of two fundamental solitons. The amplitude ratio of the two fundamental solitons is $1:3$ and the pulse width ratio is $3:1$. The phase difference between the two fundamental solitons is e^{-j4z_n}. Thus, the pulse shape of the second-order soliton changes periodically as shown in Figure 14.4. Due to the presence of e^{-j4z_n} and $\cos(4z_n)$ in (14.65), $u_2(z_n, t_n)$ is periodic in z_n, and it returns to its original value when $4z_n$ changes by 2π. Thus the spatial periodicity of a second-order soliton is $\pi/2$ in the normalized distance.

As a final check, we examine the pulse shape function at the input. By setting z_n to zero, we deduce from (14.65) and (14.66) that

$$u_2(0, t_n) = 2\,\text{sech}(t_n)$$

which is precisely a pulse of the form of $2\,sech(t_n)$.

14.6.2 Third-Order Solitons

When three fundamental solitons of the same speed are excited, a third-order soliton is formed. A closed-form expression for a third-order soliton excited by an input of $3\,\text{sech}\,t_n$ has been derived by Schrader [8],

$$u_3(z_n, t_n) = j3e^{-jz_n/2}\frac{Z(z_n, t_n)}{N(z_n, t_n)} \qquad (14.69)$$

where

$$Z(z_n, t_n) = 2\,\cosh(8t_n) + 32\,\cosh(2t_n) + 5e^{+j8z_n} + 45e^{-j8z_n} + 20e^{-j16z_n}$$
$$+ e^{-j4z_n}[36\,\cosh(4t_n) + 16\,\cosh(6t_n)]$$
$$+ e^{-j12z_n}[20\,\cosh(4t_n) + 80\,\cosh(2t_n)] \tag{14.70}$$
$$N(z_n, t_n) = \cosh(9t_n) + 9\,\cosh(7t_n) + 64\,\cosh(3t_n) + 36\,\cosh(t_n)$$
$$+ 36\,\cosh(5t_n)\cos(4z_n) + 20\,\cosh(3t_n)\cos(12z_n)$$
$$+ 90\,\cosh(t_n)\cos(8z_n) \tag{14.71}$$

The input pulse is

$$u_3(0, t_n) = 3\,\text{sech}(t_n)$$

The spatial periodicity of the third-order soliton in terms of normalized distance is also $\pi/2$. In fact, the spatial periodicity of all higher-order solitons is the same and it is $\pi/2$ in the normalized distance. This spatial periodicity is commonly referred to as the *soliton period*.

14.7 GENERATION OF SOLITONS

The discussion given in the last section for the second- and third-order solitons can be extended to higher-order solitons. In particular if an input pulse $N\,\text{sech}(t_n)$ is fed to a fiber and if N is an integer, an Nth-order soliton is excited. It is instructive to inquire if a soliton is generated if N is not an integer or if the input pulse shape function is not a hyperbolic secant function. In other words, would a soliton be generated if the input pulse shape and/or amplitude were not ideal? The answer is yes, provided the pulse amplitude is sufficiently intense. In fact, a soliton is generated by an input of any shape and width, so long as the input pulse amplitude is sufficiently high. In the process, a soliton part and a nonsoliton part are generated. The percentage of input energy launched into the soliton part depends on the input pulse shape and amplitude. In other words, the efficiency of soliton generation is different for different input pulse shape and amplitude. The nonsoliton part, also known as the *dispersive tail*, spreads and decays continuously as it propagates. Study shows the dispersive tail decays as $z_n^{-1/2}$. As the nonsoliton part becomes weaker, the SPM effect gets smaller. Toward the distal end of the fiber, the only observable field is the soliton part. An example is shown in Figure 14.7 [13–15]. It shows the evolution from a rectangular pulse at the input to a hyperbolic secant pulse shape at the output. Note in particular the ripples on both sides of the main peak at points near the input. These ripples are due to the interference between the soliton and the nonsoliton parts. Away from the input points, the pulse envelope has a smooth bell shape and there is no ripple at all. It clearly shows that a fundamental soliton is formed by feeding a fiber with a rectangular pulse. While a significant

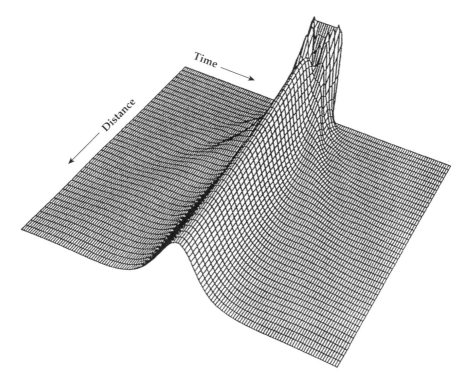

Figure 14.7   Evolution of a rectangular input pulse to a fundamental soliton and dispersive tails. (From [14].)

fraction of energy in the rectangular pulse is converted to the fundamental soliton, the nonsoliton part is not negligible either.

To examine the soliton generation problem further, we present the results reported by Satsuma and Yajima [7]. They suppose that the input is of the form of

$$u_{\text{in}}(t_n) = A \ \text{sech}(t_n) \tag{14.72}$$

where the amplitude A is not necessarily an integer. Then the input to the fiber is a pulse shape function

$$u(0, t_n) = A \ \text{sech}(t_n)$$

The total energy carried by the input pulse is proportional to A^2. Depending on the value of A, one or more fundamental solitons and possibly a dispersive tail are generated. The number of fundamental solitons, N, is given by

$$A - \tfrac{1}{2} < N \le A + \tfrac{1}{2} \tag{14.73}$$

where N is also the number of eigenvalues if the IST method is used to solve the soliton excitation problem. The eigenvalues are

$$\varsigma_i = j\left(A - i + \tfrac{1}{2}\right) \tag{14.74}$$

where $i = 1, 2, \ldots, N$. All eigenvalues are purely imaginary for an input pulse given by (14.72). All fundamental solitons excited by the input pulse (14.72) are stationary with respect to the moving reference frame, and they can be combined to form higher-order solitons.

14.7.1 Integer A

If A is exactly 1.00, there is one eigenvalue and it is $j/2$. All input energy is launched into the fundamental soliton and no dispersive tail is excited. The expression for the fundamental soliton is given by (14.56) with Λ, t_{n0}, and θ setting to zero. Figure 14.3 shows the evolution of a fundamental soliton.

If A is exactly 2.00, there exist two eigenvalues and they are $j3/2$ and $j/2$. The two fundamental solitons are also stationary with respect to the moving reference frame. When combined, they form a second-order soliton described by (14.65) or (14.66) (Fig. 14.4). Again, no dispersive tail is launched and all input energy is in the second-order soliton.

If A is 3.00, all energy is used to form a third-order soliton. In fact, this is true for all integer A's. All input energy is used to form the fundamental or higher-order soliton and no dispersive tail is generated if A is an integer.

14.7.2 Noninteger A

The situation is more complicated if A is not an integer. If A is less than $\tfrac{1}{2}$, no soliton is generated. For $\tfrac{1}{2} < A < \tfrac{3}{2}$, a part of input energy is used to form a fundamental soliton and the rest is in the dispersive tail. Let $A = 1 + \Delta$ and $|\Delta| < \tfrac{1}{2}$. Then there is one eigenvalue:

$$\varsigma_1 = j\frac{\eta_1}{2} = j\left(\Delta + \tfrac{1}{2}\right) \tag{14.75}$$

The soliton envelope is

$$u(z_n, t_n) = (1 + 2\Delta)\,\text{sech}[(1 + 2\Delta)t_n] \tag{14.76}$$

Energy contained in the fundamental soliton is proportional to $(1 + 2\Delta)$. Recall that the total input energy is proportional to $(1 + \Delta)^2$. Thus all input energy is not in the fundamental soliton if Δ is not zero. A fraction, $\Delta^2/(1 + \Delta)^2$, of the total input energy is in the dispersive tail.

These results can be extended to inputs with larger and noninteger A's. Suppose $A = N + \Delta$ and $|\Delta| < \tfrac{1}{2}$, then there are N eigenvalues. N fundamental solitons are generated, and energy contained in each fundamental soliton is proportional to η_i:

$$\eta_i = 2N + 2\Delta - 2i + 1 \tag{14.77}$$

Since η_i ranges from $\eta_1 = 2N + 2\Delta - 1$ to $\eta_N = 2\Delta + 1$, the total energy in the N fundamental solitons is proportional to

$$\sum_{i=1}^{N} \eta_i = \frac{N}{2}[(2N + 2\Delta - 1) + (2\Delta + 1)] = N(N + 2\Delta) \tag{14.78}$$

The fraction of input energy converted into the N fundamental solitons is

$$\frac{N(N + 2\Delta)}{(N + \Delta)^2} = 1 - \frac{\Delta^2}{(N + \Delta)^2} \tag{14.79}$$

The fraction of input energy in the dispersive tail is $\Delta^2/(N + \Delta)^2$. If A is an integer, $\Delta = 0$, all input energy is converted to the solitons. In Figure 14.8, we plot the fraction of input energy in each fundamental solitons (dashed curves) and the total energy in all fundamental solitons (solid curve) [16, 17].

The total energy given by (14.78) corresponds to the energy in an input pulse of the form of

$$u(t_n, 0) = (N + 2\Delta) \operatorname{sech}\left(\frac{N + 2\Delta}{N} t_n\right) \tag{14.80}$$

It excites an Nth-order soliton [16, 17]. The maximum pulse width of the Nth-order soliton is $1.763N/(N+2\Delta)$. Comparing (14.72) with (14.80), we see that the output pulse is broadened by a factor $N/(N + 2\Delta)$. A plot of the ratio of the maximum

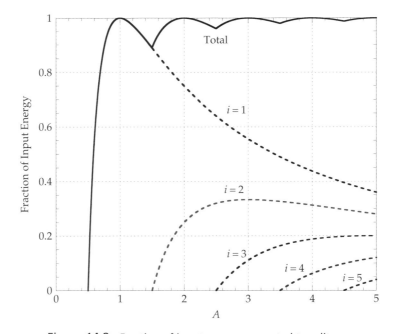

Figure 14.8   Fraction of input energy converted to solitons.

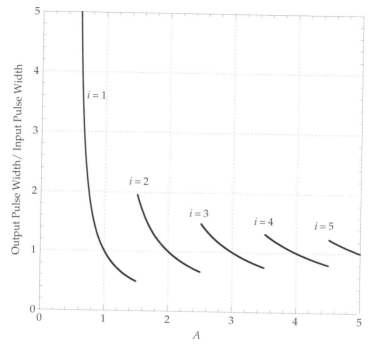

Figure 14.9 Pulse broadening.

output pulse width to the input pulse width is shown in Figure 14.9. Note that the curve jumps discontinuously at $A = N + \left(\frac{1}{2}\right)$.

14.8   SOLITON UNITS OF TIME, DISTANCE, AND POWER

Our discussion so far is couched in terms of the normalized time and distance. In this section, we revert to the physical parameters. In terms of z and t, the electric field intensity of a fundamental soliton is, from (14.56),

$$\mathcal{E}(z;t) = \sqrt{\frac{|\beta_0''|S_{\text{eff}}}{k_0 n_{2E}}} \frac{\eta}{T_0} \text{sech} \left[\eta \left(\frac{t - \beta_0' z}{T_0} - t_{n0} - \Lambda \frac{|\beta_0''|z}{T_0^2} \right) \right]$$

$$\times e^{j[\omega_0 - (\Lambda/T_0)]t} e^{-j[\beta_0 - (\Lambda\beta_0'/T_0)]z} e^{j(\Lambda^2 - \eta^2)|\beta_0|z/(2T_0^2)} \hat{\mathbf{x}} \qquad (14.81)$$

It represents a pulse moving in the laboratory frame with a velocity $1/[\beta_0' - (\Lambda|\beta_0''|/T_0)]$ that may differ slightly from the group velocity $(1/\beta_0')$. The velocity difference depends on the first-order GVD and Λ. The instantaneous frequency of the pulse is, from the time-dependent phase term,

$$\omega = \omega_0 - \frac{\Lambda}{T_0} \qquad (14.82)$$

If the pulse moves at a speed different from the group velocity, the instantaneous frequency ω also differs from the center frequency ω_0. In other words, Λ describes the frequency difference as well.

To examine the pulse further, we simplify the equation by setting Λ and t_{n0} to zero. Thus we consider a fundamental soliton moving with the group velocity and having a peak at $t = 0$ at $z = 0$. Under these conditions, we obtain from (14.81)

$$\mathcal{E}(z;t) = \sqrt{\frac{|\beta_0''|S_{\text{eff}}}{k_0 n_{2E}}} \frac{\eta}{T_0} \, \text{sech}\left[\eta\left(\frac{t - \beta_0'z}{T_0}\right)\right] e^{j(\omega_0 t - \beta_0 z)} e^{-j\eta^2|\beta_0''|z/(2T_0^2)} \hat{\mathbf{x}} \qquad (14.83)$$

Note that the pulse envelope varies with $\eta(t - \beta_0')/T_0$ and the phase term is $\eta^2|\beta_0''|z/T_0^2$. In arriving at (14.56), the variables t and z have been scaled twice, once in defining t_n and z_n, once in applying the transformation (14.52). Therefore, the independent variables are $\eta t/T_0$ and $\eta^2|\beta_0''|z/T_0^2$ instead of t and z. It is convenient to rewrite the two variables as $t = t'T_{\text{sol}}$ and $z = z'Z_{\text{sol}}$ where

$$T_{\text{sol}} = \frac{T_0}{\eta} \qquad (14.84)$$

$$Z_{\text{sol}} = \frac{T_0^2}{\eta^2|\beta_0''|} \qquad (14.85)$$

are *soliton units of time* and *distance*, respectively [18]. To relate the soliton units to physical quantities, we recall the normalized full-width between half-power points of a fundamental soliton given by (14.58). The FWHM pulse width is

$$\Delta t_{\text{FWHM}} = 2\ln(\sqrt{2} + 1)\frac{T_0}{\eta} \approx 1.7627\frac{T_0}{\eta} = 1.7627 T_{sol} \qquad (14.86)$$

Thus the soliton unit of time relates to the pulse width Δt_{FWHM},

$$T_{\text{sol}} = \frac{T_0}{\eta} \approx \frac{\Delta t_{\text{FWHM}}}{1.7627} \qquad (14.87)$$

Similarly, the soliton unit of distance is, in terms of β_0'' and $\mathcal{D}_1$,

$$Z_{\text{sol}} = \frac{T_0^2}{\eta^2|\beta_0''|} \approx \left(\frac{\Delta t_{\text{FWHM}}}{1.7627}\right)^2 \frac{2\pi c}{\lambda^2 \mathcal{D}_1} \qquad (14.88)$$

The peak electric field intensity of a fundamental soliton is, from (14.83),

$$\mathcal{E}_{\text{pk}} = \sqrt{\frac{|\beta_0''|S_{\text{eff}}}{k_0 n_{2E}}} \frac{1}{T_{\text{sol}}} \qquad (14.89)$$

The peak irradiance is

$$I_{\text{sol}} = \frac{n_0}{2}\sqrt{\frac{\varepsilon_o}{\mu_o}}|E_{\text{pk}}|^2 = \frac{n_0}{2}\sqrt{\frac{\varepsilon_o}{\mu_o}}\frac{|\beta_o''|S_{\text{eff}}}{k_o n_{2E}}\frac{1}{T_{\text{sol}}^2} \qquad (14.90)$$

In terms of Δt_{FWHM}, the peak irradiance is

$$I_{\text{sol}} = \frac{\lambda^2}{4\pi c}\sqrt{\frac{\varepsilon_o}{\mu_o}}\frac{n_0}{k_o n_{2E}}\frac{\mathcal{D}_1 S_{\text{eff}}}{T_{\text{sol}}^2} = \frac{[\ln(\sqrt{2}+1)]^2}{2\pi^2}\sqrt{\frac{\varepsilon_o}{\mu_o}}\frac{\lambda^3}{c}\frac{n_0\mathcal{D}_1}{n_{2E}}\frac{S_{\text{eff}}}{(\Delta t_{\text{FWHM}})^2} \qquad (14.91)$$

As a rough estimate, we assume that the field distributes uniformly in the core and the field is zero in the cladding. Then power contained in a fundamental soliton pulse in fiber is[5]

$$\mathcal{P}_{\text{sol}} = \pi a^2 I_{\text{sol}} = \frac{[\ln(\sqrt{2}+1)]^2}{2\pi^2}\sqrt{\frac{\varepsilon_o}{\mu_o}}\frac{\lambda^3}{c}\frac{n_0\mathcal{D}_1}{n_{2E}}\frac{S_{\text{eff}}a^2\pi}{(\Delta t_{\text{FWHM}})^2}$$

$$= 1.04 \times 10^{-4}\frac{\lambda^3}{c}\frac{n_0\mathcal{D}_1}{n_{2E}}\frac{S_{\text{eff}}a^2\pi}{(\Delta t_{\text{FWHM}})^2} \qquad (14.92)$$

where a is the core radius. $\mathcal{P}_{\text{sol}}$ is the *soliton unit of power*. Note the product of the soliton unit of power, and the pulse width square is a constant depending only on the fiber parameters and wavelength:

$$\mathcal{P}_{\text{sol}}(\Delta t_{\text{FWHM}})^2 = \frac{[\ln(\sqrt{2}+1)]^2}{240\pi^3}\frac{\lambda^3}{c}\frac{n_0\mathcal{D}_1}{n_{2E}}S_{\text{eff}}a^2\pi = 1.04 \times 10^{-4}\frac{\lambda^3}{c}\frac{n_0\mathcal{D}_1}{n_{2E}}S_{\text{eff}}a^2\pi$$
$$(14.93)$$

As a specific example, we consider a fundamental soliton propagating in a fiber with the following parameters at 1.55 μm:

$$a \sim 4.6\,\mu\text{m} \qquad S_{\text{eff}} \sim 1.5$$

$$n_0 \sim 1.45 \qquad \mathcal{D}_1 \sim 16 \text{ ps/(nm·km)}$$

$$n_{2E} \sim 0.6 \times 10^{-22}(\text{m/V})^2 \qquad n_{2I} \sim 3.2 \times 10^{-20}\text{m}^2/\text{W}$$

---

[5]In terms of n_{2I}, the power contained in a fundamental soliton pulse is

$$\mathcal{P}_{\text{sol}} = [\ln(\sqrt{2}+1)]^2\frac{\lambda^3\mathcal{D}_1}{\pi^2 c n_{2I}}\frac{S_{\text{eff}}a^2\pi}{(\Delta t_{\text{FWHM}})^2} = 0.777\frac{\lambda^3\mathcal{D}_1}{\pi^2 c n_{2I}}\frac{S_{\text{eff}}a^2\pi}{(\Delta t_{\text{FWHM}})^2}$$

and the (soliton power) (pulse width squared) product is

$$\mathcal{P}_{\text{sol}}(\Delta t_{\text{FWHM}})^2 = 0.777\frac{\lambda^3\mathcal{D}_1}{\pi^2 c n_{2I}}S_{\text{eff}}a^2\pi$$

For a fundamental soliton with $\Delta t_{\text{FWHP}} = 7$ ps, we obtain from (14.87), (14.88), and (14.92),

$$T_{\text{sol}} = 3.97 \text{ ps}$$

$$Z_{\text{sol}} = 773 \text{ m}$$

$$P_{\text{sol}} = 1.02 \text{ W}$$

The peak power required to generate a second-order soliton is $4P_{\text{sol}}$. In general, the peak power needed to launch a Nth-order soliton is $N^2 P_{\text{sol}}$.

14.9 INTERACTION OF SOLITONS

In digital communications, pulse streams are transmitted in the fibers. Suppose a time interval of 1 s is partitioned into bit slots and each bit slot has at most a pulse. The absence and presence of a pulse in a bit slot signifies a bit of 0 and 1, respectively. Let the duration of a bit slot be T_B (Fig. 14.10). Then the bit rate or pulse rate is $1/T_B$. Clearly, the temporal pulse width and the time separation between neighboring pulses must be shorter than T_B. For high-speed data transmission systems, we like to keep the bit slots as narrow as possible so that the bit rate is as high as possible. Thus, the pulse width and separation between pulses decrease with T_B as the bit rate increases. As the temporal spacing between adjacent pulses decreases, the tails of neighboring pulses overlap and the interaction between pulses increases. The increased interaction due to pulse overlapping could lead to additional pulse deformation and broadening. In preceding sections, we have discussed the evolution of isolated pulses in fibers. In this section, we consider the evolution of closely spaced and possibly overlapping pulses. In other words, we consider the effects of temporal separation between pulses on the evolution and degradation of pulses.

Qualitatively, we can understand the effects as follows. The interaction between pulses is due to the nonlinear term of the normalized nonlinear envelope equation (14.42). Consider two pulses, u_a and u_b, propagating in a fiber. If u_a alone is

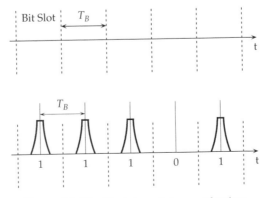

Figure 14.10   Bit slots and stream of pulses.

present, the nonlinear term is $|u_a|^2 u_a$. If u_a and u_b exist simultaneously and the two pulses overlap partially in time and space, the nonlinear term is $|u_a + u_b|^2 (u_a + u_b)$. So far as pulse u_a is concerned, the nonlinear interaction changes due to the presence of u_b. The dominant part of the change in nonlinear interaction is $|u_a|^2 u_b$. If the two pulses are in time phase, the total field is stronger when the two pulses overlap. Then the two pulses are "pushed" closer, as reported by Desem and Chu [12]. If the two pulses are out of phase by π, the total field becomes weaker when the two pulses are superimposed. Then the two pulses are "pulled" apart [12]. In other words, the interaction between two in-phase pulses is equivalent to an "attractive force" and the interaction between two out-of-phase pulses is equivalent to a "repulsive force" [12].

To study the pulse interaction quantitatively, we consider an input of two pulses separated by $2\tau_0$ in the normalized time:

$$u_{in}(t) = ae^{j\varphi} \operatorname{sech}(t_n + \tau_0) + \operatorname{sech}(t_n - \tau_0) \tag{14.94}$$

where a and φ are the amplitude ratio of and the phase difference between the two input pulses. This problem has been studied by several authors [12, 19–21]. In particular, Desem and Chu [12] showed that the evolution of the pulse envelope of the two in-phase pulses ($\varphi = 0$) is

$$u(z_n, t_n) = \frac{\mu_b^2 - \mu_a^2}{Q_{ab}} e^{j\mu_a^2 z_n/2} [\mu_a \operatorname{sech} \mu_a(t_n + \gamma_0)$$

$$+ \mu_b \operatorname{sech} \mu_b(t_n - \gamma_0)e^{j(\mu_b^2 - \mu_a^2)z_n/2}] \tag{14.95}$$

where

$$Q_{ab} = (\mu_a^2 + \mu_b^2) - 2\mu_a\mu_b \{\tanh \mu_a(t_n + \gamma_0) \tanh \mu_b(t_n - \gamma_0)$$

$$- \operatorname{sech} \mu_a(t_n + \gamma_0) \operatorname{sech} \mu_b(t_n - \gamma_0)$$

$$\cos[(\mu_b^2 - \mu_a^2)z_n/2]\} \tag{14.96}$$

and μ_a, μ_b and γ_0 are functions of a and τ_0.

If the amplitudes of the two input pulses are same, $a = 1$, then

$$\mu_{a,b} \approx 1 + \frac{2\tau_0}{\sinh 2\tau_0} \pm \operatorname{sech} \tau_0 \tag{14.97}$$

$$\gamma_0 = 0 \tag{14.98}$$

In the case of $\tau_0 = 0$, $a = 1$, and $\varphi = 0$, (14.95) and (14.96) reduces to a second-order soliton given by (14.66) and (14.67). The situation becomes complicated when τ_0 increases. A typical plot of $|u(z_n, t_n)|$ with $a = 1$, $\varphi = 0$, and $\tau_0 = 3$ is shown in Figure 14.11. As seen from the figure, the two pulses move toward each other, merge into a stronger and narrower pulse, and then split into two distinct pulses again. The process repeats itself periodically. Mathematically, the periodic spatial variation of $|u(z_n, t_n)|$ can be traced to the exponential function in

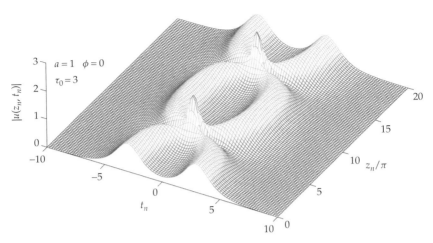

Figure 14.11 Evolution of two closely spaced and in-phase fundamental solitons of equal amplitudes ($\tau_0 = 3.0$, $a = 1.0$, and $\phi = 0$). (After [12].)

the numerator and the cosine term in denominator of (14.95). The *spatial periodicity* z_{np} is deduced from the condition

$$\frac{z_{np}}{2}(\mu_b^2 - \mu_a^2) = 2\pi \tag{14.99}$$

By substituting (14.97) into the above expression, we have

$$z_{np} \approx \frac{\pi \sinh 2\tau_0 \cosh \tau_0}{2\tau_0 + \sinh 2\tau_0} \tag{14.100}$$

For $\tau_0 \geq 3$, we ignore $2\tau_0$ in favor of $\sinh(2\tau_0)$, and approximate $\cosh(\tau_0)$ by $e^{\tau_0}/2$. Then we obtain an approximate expression for z_{np}:

$$z_{np} \approx \frac{\pi}{2}e^{\tau_0} \qquad (\tau_0 \geq 3) \tag{14.101}$$

Recall that the soliton period in the normalized distance is $\pi/2$ (see Section 14.6). Thus the spatial periodicity due to the interaction between two fundamental solitons is longer than the soliton period by a factor of e^{τ_0} if $\tau_0 \geq 3$. Alternatively, we interpret the result in the following manner. When two pulses are launched into a fiber, the two pulses coalesce into a sharp and strong pulse at $z_{np}/2$. This distance is known as the *coalesce length*. Further down the fiber, the two pulses split again. At the normalized distance $z_n = z_{np}$, the two pulses return to the original pulse separation. In other words, the coalesce distance is half of spatial periodicity z_{np}. In terms of soliton unit of distance, the coalesce length is

$$L_{coal} \approx Z_{sol}e^{0.881(T_B/\Delta t_{FWHM})} \tag{14.102}$$

Table 14.2 tabulates the soliton period and the coalesce length of a typical communication grade fiber with $\mathcal{D}_1 = 16$ ps/(nm·km) for several soliton pulse widths.

TABLE 14.2 Soliton Period and Coalesce Length

Soliton Width (ps) Δt_{FWHP}	Soliton Period (m) $\pi Z_{\text{sol}}/2$	Coalesce Length (km) L_{coal}			
		$\tau_0 = 4$	$\tau_0 = 6$	$\tau_0 = 8$	$\tau_0 = 10$
2	99.2	3.36	19.6	114.	664.
4	397.	13.4	78.3	456.	2.66×10^3
6	892.	30.3	176.	1.03×10^3	5.98×10^3
8	1.59×10^3	53.8	313.	1.82×10^3	1.06×10^4
10	2.48×10^3	84.1	489.	2.85×10^3	1.66×10^4

For $\Delta t_{\text{FWHM}} > 8$ ps and $\tau_0 > 8$, the coalesce length is longer than 1000 km. In conclusion, for system links shorter than 1000 km, a temporal separation of 8 times the pulse width would be sufficient if we wish to minimize the nonlinear interaction arising from pulse overlaps.

For input pulses with unequal amplitudes, μ_a, μ_b, and γ_0 are

$$\mu_{a,b} \approx \frac{a+1}{2} + \frac{2a^{1/2}\tau_0}{\sinh(2a^{1/2}\tau_0)} + (a^{1/2} - 1)\,\text{sech}(a\tau_0) \pm \left[\frac{a-1}{2} + \text{sech}(a\tau_0)\right]$$
(14.103)

$$\gamma_0 \approx \tau_0 - \frac{a+1}{2a}\left[1 - \frac{2\,\text{sech}(a\tau_0)}{a}\right]\ln\left(\frac{a+1}{a-1}\right)$$
(14.104)

if $a \neq 1$ and $\tau_0 > 3.5$ [12].

A plot of $|u(z_n, t_n)|$ with $a = 1.1$, $\varphi = 0$, and $\tau_0 = 3.0$ is shown in Figure 14.12. The two pulses vacillate around the initial separation. But they do not collapse into

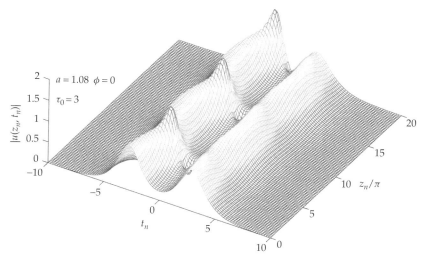

Figure 14.12 Evolution of two closely spaced and in-phase fundamental solitons of unequal amplitudes ($\tau_0 = 3.0$, $a = 1.1$, and $\phi = 0$).

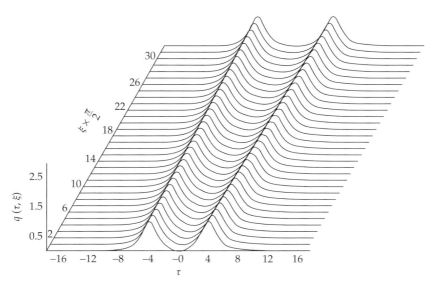

Figure 14.13   Evolution of two closely spaced fundamental solitons with 180° phase difference. The half-width and separation are, respectively, 4.5 and 13.6 ps. (From [21].)

one pulse. In other words, one way to prevent the collapse of solitons is to keep the
amplitudes of adjacent solitons different.

For out-of-phase pulses, $\varphi = \pi$, the two pulses repel and move away from each
other. An example is shown in Figure 14.13 [21].

REFERENCES

1. Y. R. Shen, *The Principles of Nonlinear Optics*, Wiley, New York, 1984.

2. D. L. Mills, *Nonlinear Optics, Basic Concepts*, Springer, Berlin, 1991.

3. Y. Kodama and A. Hasegawa, "Nonlinear pulse propagation in a monomode dielectric
   guide," *IEEE J. Quant. Electron.*, Vol. QE-23, pp. 510–524 (May, 1987).

4. Y. Kodama and A. Hasegawa, "Theoretical foundation of optical soliton concept in
   fibers," in *Progress in Optics*, E. Wolf (ed.), Vol. XXX, North-Holland, Amsterdam,
   1992, pp. 207–259.

5. P. G. Drazin and R. S. Johnson, *Solitons: an Introduction*, Cambridge University Press,
   Cambridge, 1988.

6. V. E. Zakharov and A. B. Shabat, "Theory of two-dimensional self-focusing and one-
   dimensional self-modulation of waves in nonlinear media," *Soviet Physics, JETP*, Vol. 34.
   pp. 62–69 (Jan. 1972).

7. J. Satsuma and N. Yajima, "Initial value problems of one-dimensional self-modulation
   of nonlinear waves in dispersive media," *Suppl. Progress Theoretical Physics*, No. 55,
   pp. 284–306 (1974).

8. D. Schrader, "Explicit calculation on N-soliton solutions of the nonlinear Schrödinger
   equation," *IEEE J. Quant. Electron.*, Vol. 31, pp. 2221–2225 (Dec. 1995).

9. J. A C. Weideman and B. M. Herbst, "Split-step method for the solution of the nonlinear Schrodinger equation," *SIAM J. Numer. Anal.*, Vol. 23, pp. 458–507 (June 1986).

10. H. Ghasfouri-Shiraz, P. Shum, and M. Nagata, "A novel method for analysis of soliton propagation in optical fibers," *IEEE J. Quantum Electron.*, Vol. 31, pp. 190–200 (Jan. 1995).

11. J. M. Cervero, "Unveiling the solitons mystery: The Jacobi elliptical functions," *Am. J. Phys.*, Vol. 54, No. 1, pp. 35–38 (Jan. 1986).

12. C. Desem and P. L. Chu, "Reducing soliton interaction in single-mode optical fibres," *IEE Proc., Pt. J.*, Vol. 134, pp. 145–151 (June, 1987).

13. H. A. Haus, "Optical fiber solitons, their properties and use," *Proc. IEEE*, Vol. 81, pp. 970–983 (July, 1993).

14. H. A. Haus, "Molding light into solitons," *IEEE Spectrum*, pp. 48–53 (March, 1993).

15. H. A. Haus and W. S. Wong, "Solitons in optical communications," *Rev. Modern Phys.*, Vol. 68, pp. 423–444 (April, 1996).

16. N. J. Doran and K. J. Blow, "Solitons in optical communications," *IEEE J. Quant. Electron.*, Vol. QE-19, pp. 1883–1888 (Dec. 1983).

17. K. J. Blow and N. J. Doran, "Non-linear propagation effects in optical fibers: Numerical studies," in *Optical Solitons—Theory and Experiment*, J. R. Taylor (ed.), Cambridge Studies in Modern Optics, Vol. 10, Cambridge University, Cambridge, 1992, pp. 73–106.

18. L. F. Mollenauer, J. P. Gordon, and P. V. Mamyshev, "Solitons in high bit-rate, long-distance transmission," in *Optical Fiber Telecommunications*, IIIA, I. P. Kaminow and T. L. Koch (eds.), Academic, San Diego, CA, 1997, pp. 373–460.

19. V. E. Zakharov and A. B. Shabat, "Interaction between solitons in a stable medium," *Soviet Physics, JETP*, Vol. 37. pp. 823–838 (Nov. 1973).

20. J. P. Gordon, "Interaction forces among solitons in optical fibers," *Opt. Lett.*, Vol. 8, pp. 596–598 (Nov. 1983).

21. B. Hermansson and D. Yevick, "Numerical investigation of soliton interaction," *Electron. Lett.*, Vol. 19, pp. 570–571 (July 21, 1983).

BIBLIOGRAPHY

1. R.-J. Essiambre and G. P. Agrawal, "Soliton communication systems," in *Progress in Optics*, vol. XXXVII, E. Wolf (ed.), North-Holland, Amsterdam, 1997, pp. 185–256.

2. A. Hasegawa, "Optical solitons in fibers: Theoretical review," in *Optical Solitons—Theory and Experiment*, J. R. Taylor (ed.), Cambridge Studies in Modern Optics, Vol. 10, Cambridge University, Cambridge, 1992, pp. 1–28.

3. A. Hasegawa and F. Tappert, "Transmission of stationary nonlinear optical pulses in dispersive dielectric fibers, I. Anomalous dispersion," *Appl. Phys. Lett.*, Vol. 23, pp. 142–144 (Aug. 1, 1973).

4. L. F. Mollenauer, "Solitons in optical fibers: The experimental account," in *Optical Solitons—Theory and Experiment*, J. R. Taylor (ed.), Cambridge Studies in Modern Optics, Vol. 10, Cambridge University, Cambridge, 1992, pp. 30–37.

5. M. Remoissenet, *Wave Called Solitons, Concepts and Experiments*, Springer, Berlin, 1994.

Appendix A

BROWN IDENTITY

In optical fiber communications, signals in the form of optical pulses are transmitted through fibers. The propagation and evolution of pulses in fibers depends on the *group velocity*, first-, second-, and higher-order derivatives of the group velocity. Therefore we are interested in the group velocity and its derivatives. The group velocity is $d\omega/d\beta$ where β is the propagation constant. To determine β, we solve Maxwell equations with suitable boundary conditions and obtain a dispersion relation for modes guided by the waveguide or fiber in question. This is discussed in Chapters 2, 3, 5, 9, and 10. By solving the dispersion relation, we obtain β. In principle, the group velocity can be calculated once β is known. Although the analytic expressions for dispersion relations exist for many waveguides and fibers, no analytical expression for β exists. In the absence of a closed-form expression for β, it is difficult to evaluate the derivatives analytically. In most cases, β is solved numerically. Numerical differentiation is vulnerable to error. Thus the group velocity is hard to come by. It would be useful to establish a relation between the group velocity and fields even if we have to make a few simplifying assumptions. The *Brown identity* [1, 2] is such a relation. It is an equation relating the product of the phase and group velocities to integrals of the *weighted power density* in waveguide regions, assuming the waveguide materials are nondispersive. To establish the

Foundations for Guided-Wave Optics, by Chin-Lin Chen
Copyright © 2007 John Wiley & Sons, Inc.

Brown identity, we follow the proof given originally by Kawakami [3]. An alternate derivation can be found in Haus and Kogelnik [4].

All materials are dispersive. Effects of the material dispersion must be taken into account if an accurate group velocity is desired. This is difficult to do. On the other hand, many waveguide materials of interest to optics are slightly dispersive in the spectral range of interest. For these materials, the material dispersion can be ignored as an approximation. To appreciate the order of magnitude involved, we consider uniform plane waves propagating in a homogeneous dielectric medium with index $n(\omega)$. The propagation constant of uniform plane waves in the homogeneous medium is $\beta = \omega n(\omega)/c$. The phase and group velocities are

$$v_{\text{ph}} = \frac{\omega}{\beta} = \frac{c}{n(\omega)} \tag{A.1}$$

and

$$v_{\text{gr}} = \frac{d\omega}{d\beta} = \frac{c}{n(\omega)\left[1 + \dfrac{\omega}{n(\omega)}\dfrac{dn(\omega)}{d\omega}\right]} = \frac{c}{n(\lambda)\left[1 - \dfrac{\lambda}{n(\lambda)}\dfrac{dn(\lambda)}{d\lambda}\right]} \tag{A.2}$$

Thus

$$\frac{1}{v_{\text{ph}}v_{\text{gr}}} = \frac{n^2(\lambda)}{c^2}\left[1 - \frac{\lambda}{n(\lambda)}\frac{dn(\lambda)}{d\lambda}\right] \tag{A.3}$$

In the above equations, the effect of material dispersion is represented by $[\lambda/n(\lambda)][dn(\lambda)/d\lambda]$. Take fused quartz in the visible and near IR regions as an example. $|[\lambda/n(\lambda)][dn(\lambda)/d\lambda]|$ is about 0.01. Thus effects of material dispersion are small and may be ignored unless an accurate determination of the group velocity is required. In most of our discussions, we take materials as nondispersive. In the last section, an expression is given to account for the material dispersion.

A.1   WAVE EQUATIONS FOR INHOMOGENEOUS MEDIA

We begin with the time-harmonic ($e^{j\omega t}$) Maxwell equations (1.7)–(1.10) for source-free regions. The media may be *inhomogeneous* and μ and ε [(1.12)–(1.15)] are possibly functions of position. To stress the medium inhomogeneity, we write $\mu(x, y, z)$ and $\varepsilon(x, y, z)$ in lieu of μ and ε. We assume, however, that $\mu(x, y, z)$ and $\varepsilon(x, y, z)$ are *frequency independent*. From (1.7) and (1.8), and by simple substitution, we obtain the wave equations for inhomogeneous media:

$$\nabla^2\mathbf{E} + \omega^2\mu(x, y, z)\varepsilon(x, y, z)\mathbf{E} - \nabla(\nabla \cdot \mathbf{E}) + \frac{\nabla\mu(x, y, z)}{\mu(x, y, z)} \times (\nabla \times \mathbf{E}) = 0 \quad (A.4)$$

$$\nabla^2\mathbf{H} + \omega^2\mu(x, y, z)\varepsilon(x, y, z)\mathbf{H} - \nabla(\nabla \cdot \mathbf{H}) + \frac{\nabla\varepsilon(x, y, z)}{\varepsilon(x, y, z)} \times (\nabla \times \mathbf{H}) = 0 \quad (A.5)$$

We concentrate on waveguides that are very long in the z direction. We also assume that the waveguide cross section, permittivity, and permeability of the waveguide materials do not change with z. Thus, we write $\mu(x, y)$ and $\varepsilon(x, y)$ instead of $\mu(x, y, z)$ and $\varepsilon(x, y, z)$ in the following equations. For waves guided by these waveguides and propagating in $+z$ direction, all field components behave like $e^{-j\beta z}$. Thus operator ∇ can be written as $\nabla = \nabla_t - j\beta\hat{\mathbf{z}}$. Then, we cast (A.4) as

$$\nabla_t^2 \mathbf{E} - \nabla(\nabla \cdot \mathbf{E}) + \frac{\nabla\mu(x, y)}{\mu(x, y)} \times (\nabla \times \mathbf{E}) = [\beta^2 - \omega^2\mu(x, y)\varepsilon(x, y)]\mathbf{E} \qquad (A.6)$$

Similar expression can be obtained for **H**.

A.2 BROWN IDENTITY

Next we consider waves with an angular frequency ω' that differs from ω. Let fields and the propagation constant at ω' be $\mathbf{E}'$, $\mathbf{H}'$, and β'. A wave equation for $\mathbf{H}'$ is obtained from (A.5) by changing $\mathbf{H}$, β, and ω to $\mathbf{H}'$, β', and ω':

$$\nabla_t^2 \mathbf{H}' - \nabla(\nabla \cdot \mathbf{H}') + \frac{\nabla\varepsilon(x, y)}{\varepsilon(x, y)} \times (\nabla \times \mathbf{H}') = [\beta'^2 - \omega'^2\mu(x, y)\varepsilon(x, y)]\mathbf{H}' \quad (A.7)$$

Recall that we assume that the materials are nondispersive; $\mu(x, y)$ and $\varepsilon(x, y)$ at ω' are the same as at ω. By making the cross product of $\mathbf{H}'^*$ with (A.6), $\mathbf{E}$ with the complex conjugation of (A.7), and adding the two equations, we obtain

$$\mathbf{E} \times \nabla_t^2 \mathbf{H}'^* + \mathbf{H}'^* \times \nabla_t^2 \mathbf{E} - \mathbf{H}'^* \times \nabla(\nabla \cdot \mathbf{E}) - \mathbf{E} \times \nabla(\nabla \cdot \mathbf{H}'^*)$$

$$+ \mathbf{E} \times \left[\frac{\nabla\varepsilon(x, y)}{\varepsilon(x, y)} \times (\nabla \times \mathbf{H}'^*) \right] + \mathbf{H}'^* \times \left[\frac{\nabla\mu(x, y)}{\mu(x, y)} \times (\nabla \times \mathbf{E}) \right] \qquad (A.8)$$

$$= -[(\beta^2 - \beta'^2) - (\omega^2 - \omega'^2)\mu(x, y)\varepsilon(x, y)]\mathbf{E} \times \mathbf{H}'^*$$

On the other hand, we obtain directly from (1.9) and (1.10)

$$\mathbf{E} \cdot \nabla\varepsilon(x, y) = -\varepsilon(x, y)\nabla \cdot \mathbf{E} \qquad (A.9)$$

$$\mathbf{H} \cdot \nabla\mu(x, y) = -\mu(x, y)\nabla \cdot \mathbf{H} \qquad (A.10)$$

Making use of the two equations, we deduce that

$$- \mathbf{H}'^* \times \nabla(\nabla \cdot \mathbf{E}) + \mathbf{E} \times \left[\frac{\nabla\varepsilon(x, y)}{\varepsilon(x, y)} \times (\nabla \times \mathbf{H}'^*) \right]$$

$$= -\mathbf{H}'^* \times \nabla(\nabla \cdot \mathbf{E}) + \frac{\nabla\varepsilon(x, y)}{\varepsilon(x, y)}[\mathbf{E} \cdot (\nabla \times \mathbf{H}'^*)] - \frac{\mathbf{E} \cdot \nabla\varepsilon(x, y)}{\varepsilon(x, y)}(\nabla \times \mathbf{H}'^*)$$

$$= -\mathbf{H}'^* \times \nabla(\nabla \cdot \mathbf{E}) + \frac{\nabla\varepsilon(x, y)}{\varepsilon(x, y)}[\mathbf{E} \cdot (\nabla \times \mathbf{H}'^*)] + (\nabla \cdot \mathbf{E})(\nabla \times \mathbf{H}'^*)$$

$$= \nabla \times ((\nabla \cdot \mathbf{E})\mathbf{H}'^*) + \frac{\nabla\varepsilon(x, y)}{\varepsilon(x, y)}[\mathbf{E} \cdot (\nabla \times \mathbf{H}'^*)]$$

Similarly, we make use of (A.10) and deduce

$$
-\mathbf{E} \times \nabla(\nabla \cdot \mathbf{H}'^*) + \mathbf{H}'^* \times \left[\frac{\nabla \mu(x, y)}{\mu(x, y)} \times (\nabla \times \mathbf{E}) \right]
$$

$$
= \nabla \times [(\nabla \cdot \mathbf{H}'^*)\mathbf{E}] + \frac{\nabla \mu(x, y)}{\mu(x, y)} [\mathbf{H}'^* \cdot (\nabla \times \mathbf{E})]
$$

Substituting the two equations in (A.8), we obtain

$$
(\mathbf{E} \times \nabla_t^2 \mathbf{H}'^* + \mathbf{H}'^* \times \nabla_t^2 \mathbf{E}) + \nabla \times [(\nabla \cdot \mathbf{E})\mathbf{H}'^* + (\nabla \cdot \mathbf{H}'^*)\mathbf{E}]
$$

$$
+ \left\{ \frac{\nabla \varepsilon(x, y)}{\varepsilon(x, y)} [\mathbf{E} \cdot (\nabla \times \mathbf{H}'^*)] + \frac{\nabla \mu(x, y)}{\mu(x, y)} [\mathbf{H}'^* \cdot (\nabla \times \mathbf{E})] \right\} \qquad (A.11)
$$

$$
= -[(\beta^2 - \beta'^2) - (\omega^2 - \omega'^2)\mu(x, y)\varepsilon(x, y)]\mathbf{E} \times \mathbf{H}'^*
$$

By integrating the z component of both sides of (A.11) over the entire waveguide cross section S, we obtain

$$
\int_S \{\mathbf{E} \times \nabla_t^2 \mathbf{H}'^* + \mathbf{H}'^* \times \nabla_t^2 \mathbf{E}\} \cdot \hat{\mathbf{z}} \, dA + \int_S \nabla \times \{(\nabla \cdot \mathbf{E})\mathbf{H}'^* + (\nabla \cdot \mathbf{H}'^*)\mathbf{E}\} \cdot \hat{\mathbf{z}} \, dA
$$

$$
+ \int_S \left\{ \frac{\nabla \varepsilon(x, y)}{\varepsilon(x, y)} [\mathbf{E} \cdot (\nabla \times \mathbf{H}'^*)] + \frac{\nabla \mu(x, y)}{\mu(x, y)} [\mathbf{H}'^* \cdot (\nabla \times \mathbf{E})] \right\} \cdot \hat{\mathbf{z}} \, dA \qquad (A.12)
$$

$$
= - \int_S [(\beta^2 - \beta'^2) - (\omega^2 - \omega'^2)\mu(x, y)\varepsilon(x, y)](\mathbf{E} \times \mathbf{H}'^*) \cdot \hat{\mathbf{z}} \, dA
$$

where dA is the surface element in the transverse plane.

In the following, we shall show the three integrals in the left-hand side of (A.12) are zero. Since $\hat{\mathbf{z}} \cdot \nabla \mu(x, y) = 0$ and $\hat{\mathbf{z}} \cdot \nabla \varepsilon(x, y) = 0$, the third integral of the left-hand side of (A.12) is obviously zero.

By applying the Stokes theorem, we transform the second integral to a contour integral:

$$
\int_S \nabla \times \{(\nabla \cdot \mathbf{E})\mathbf{H}'^* + (\nabla \cdot \mathbf{H}'^*)\mathbf{E}\} \cdot \hat{\mathbf{z}} \, dA = \oint_C \{(\nabla \cdot \mathbf{E})\mathbf{H}'^* + (\nabla \cdot \mathbf{H}'^*)\mathbf{E}\} \cdot d\ell
$$

$$
(A.13)
$$

For optical waveguides, the entire cross-sectional area S extends indefinitely in all directions in the transverse plane, the contour C is at infinity. For waves guided by optical waveguides, $|\mathbf{E}|$ and $|\mathbf{H}|$ decrease as $1/r$ or faster, and $|\nabla \cdot \mathbf{E}|$ and $|\nabla \cdot \mathbf{H}|$ decay as $1/r^2$ or faster, for points far away from the central waveguide region. Thus the contour integral vanishes as the contour C recedes to ∞. For microwave waveguides with perfectly conducting boundaries, the waveguide cross section is finite. The contour integral goes to zero because of the boundary conditions at the surface of perfect conductors.

To consider the first integral on the left-hand side of (A.12), we note that by straightforward differentiation,

$$[\mathbf{E} \times \nabla_t^2 \mathbf{H}'^* + \mathbf{H}'^* \times \nabla_t^2 \mathbf{E}] \cdot \hat{\mathbf{z}} = E_x \nabla_t^2 H_y'^* - E_y \nabla_t^2 H_x'^* + H_x'^* \nabla_t^2 E_y - H_y'^* \nabla_t^2 E_x$$

$$= \nabla_t \cdot (E_x \nabla_t H_y'^*) - (\nabla_t E_x) \cdot (\nabla_t H_y'^*) - \nabla_t \cdot (E_y \nabla_t H_x'^*) + (\nabla_t E_y) \cdot (\nabla_t H_x'^*)$$

$$+ \nabla_t \cdot (H_x'^* \nabla_t E_y) - (\nabla_t H_x'^*) \cdot (\nabla_t E_y) - \nabla_t \cdot (H_y'^* \nabla_t E_x) + (\nabla_t H_y'^*) \cdot (\nabla_t E_x)$$

$$= \nabla_t \cdot (E_x \nabla_t H_y'^* - E_y \nabla_t H_x'^* + H_x'^* \nabla_t E_y - H_y'^* \nabla_t E_x) \qquad (A.14)$$

By using the two-dimensional divergence theorem (Appendix B), we transform the first integral on the left-hand side of (A.12) to a contour integral:

$$\int_S \{\mathbf{E} \times \nabla_t^2 \mathbf{H}'^* + \mathbf{H}'^* \times \nabla_t^2 \mathbf{E}\} \cdot \hat{\mathbf{z}} \, dA$$

$$= \int_S \nabla_t \cdot \{E_x \nabla_t H_y'^* - E_y \nabla_t H_x'^* + H_x'^* \nabla_t E_y - H_y'^* \nabla_t E_x\} \, dA \qquad (A.15)$$

$$= \oint_C \{E_x \nabla_t H_y'^* - E_y \nabla_t H_x'^* + H_x'^* \nabla_t E_y - H_y'^* \nabla_t E_x\} \cdot \hat{\mathbf{n}} \, d\ell$$

Again, since the surface integral is over the entire waveguide cross section S, the boundary C is at infinity for optical waveguides. Thus the line integral vanishes as the contour C approaches infinity. For microwave waveguides, the contour integral goes to zero because of the boundary conditions at the surface of perfect conductors.

Since all integrals in the left-hand side of (A.12) are zero, the integral in the right-hand side of the same equation must also be zero. Thus we obtain

$$\frac{\beta^2 - \beta'^2}{\omega^2 - \omega'^2} = \frac{\int_S \mu(x, y)\varepsilon(x, y)(\mathbf{E} \times \mathbf{H}'^* \cdot \hat{\mathbf{z}}) \, dA}{\int_S (\mathbf{E} \times \mathbf{H}'^* \cdot \hat{\mathbf{z}}) \, dA} \qquad (A.16)$$

The above equation is valid for two arbitrary angular frequencies, ω and ω'. Now we suppose that the two angular frequencies are infinitesimally close. As ω' approaches ω, β', $\mathbf{E}'$, and $\mathbf{H}'$ approach β, $\mathbf{E}$, and $\mathbf{H}$, respectively. In the limit of ω' approaching ω, (A.16) becomes

$$\frac{\beta \, d\beta}{\omega \, d\omega} = \frac{1}{v_{\text{ph}} v_{\text{gr}}} = \frac{\int \mu(x, y)\varepsilon(x, y)(\mathbf{E} \times \mathbf{H}^* \cdot \hat{\mathbf{z}}) \, dA}{\int (\mathbf{E} \times \mathbf{H}^* \cdot \hat{\mathbf{z}}) \, dA} \qquad (A.17)$$

This is the *Brown identity* [1, 2]. It is instructive to review the assumptions made in the derivation. Specifically, we have assumed that the waveguide cross section extends indefinitely in all transverse directions and that the materials are not dispersive.

A.3   TWO SPECIAL CASES

As an application of the Brown identity, we consider two special cases. First, we consider uniform plane waves propagating in a single and homogeneous medium with constant ε and μ. Since ε and μ are constants independent of position, the right-hand side of (A.17) is simply $\mu\varepsilon$. Thus, for all plane waves propagating in a homogeneous medium

$$\frac{1}{v_{ph}v_{gr}} = \mu\varepsilon = \frac{n^2}{c^2}$$

Next we consider waveguides having two homogeneous media. Step-index fibers are good examples. The two homogeneous media are the core and cladding. Thus, we obtain from (A.17)

$$\frac{\beta\,d\beta}{\omega\,d\omega} = \frac{1}{v_{ph}v_{gr}} = \frac{1}{c^2}(n_{co}^2\Gamma_{co} + n_{cl}^2\Gamma_{cl}) \tag{A.18}$$

where Γ_{co} and Γ_{cl} are the fractional power transported in the core and cladding, respectively:

$$\Gamma_{co} = \frac{\int_{co}(\mathbf{E}\times\mathbf{H}^*\cdot\hat{\mathbf{z}})\,dA}{\int_{co+cl}(\mathbf{E}\times\mathbf{H}^*\cdot\hat{\mathbf{z}})\,dA} \tag{A.19}$$

$$\Gamma_{cl} = \frac{\int_{cl}(\mathbf{E}\times\mathbf{H}^*\cdot\hat{\mathbf{z}})\,dA}{\int_{co+cl}(\mathbf{E}\times\mathbf{H}^*\cdot\hat{\mathbf{z}})\,dA} \tag{A.20}$$

where co and cl stand for the core and cladding regions of the fiber. The integrals have to be, and can be, evaluated.

A.4   EFFECT OF MATERIAL DISPERSION

In the above discussion, we have ignored the material dispersion completely. If we wish to include the material dispersion into account, we could use the equation given by Adams [5]:

$$
\begin{aligned}
\frac{\beta\,d\beta}{\omega\,d\omega} &= \frac{1}{v_{ph}v_{gr}} \\
&\approx \frac{\int\{\mu(x,y,\omega)\varepsilon(x,y,\omega) + \dfrac{\omega}{2}\dfrac{\partial}{\partial\omega}[\mu(x,y,\omega)\varepsilon(x,y,\omega)]\}(\mathbf{E}\times\mathbf{H}^*\cdot\hat{\mathbf{z}})\,dA}{\int(\mathbf{E}\times\mathbf{H}^*\cdot\hat{\mathbf{z}})\,dA}
\end{aligned}
\tag{A.21}
$$

REFERENCES

1. J. Brown, "Electromagnetic momentum associated with waveguide modes," *Proc. IEE*, Vol. 113, pp. 27–34 (1966).
2. J. A. Arnaud, *Beam and Fiber Optics*, Academic, New York, San Francisco and London, 1976, Chapter 3.
3. S. Kawakami, "Relation between dissipation and power-flow distribution in a dielectric waveguide," *J. Opt. Soc. Am.*, Vol. 65, No. 1, pp. 41–45 (1975).
4. H. A. Haus and H. Kogelnik, "Electromagnetic momentum and momentum flow in dielectric waveguides," *J. Opt. Soc. Am.*, Vol. 66, pp. 320–327 (April 1976).
5. M. J. Adams, *An Introduction to Optical Waveguides*, Wiley, New York, 1981.

Appendix B

TWO-DIMENSIONAL DIVERGENCE THEOREM AND GREEN'S THEOREM

Divergence and Stokes theorems are two vector identities used extensively in establishing vector relations in electromagnetic theory. While the general theorems in the three-dimensional space are well known and can be found in many textbooks on electromagnetism, a special case particularly useful for vectors in the two-dimensional space is less well known. In this appendix, we consider the divergence theorem specialized in the two-dimensional space. We need the special divergence theorem for two dimension vectors to derive the Brown identity in Appendix A and the coupled-mode equations in Chapters 6 and 7.

We begin with the general *divergence theorem* in a three-dimensional space:

$$\int_V \nabla \cdot \mathbf{A}(x, y, z)\, dv = \oint_S \mathbf{A}(x, y, z) \cdot d\mathbf{s} \qquad (B.1)$$

where $\mathbf{A}$ is an arbitrary vector and V is an arbitrary volume bounded by a closed surface S; $d\mathbf{s}$ is a differential vector in the direction normal to the differential surface element at the point in question; dv is the volume element.

If vector $\mathbf{A}$ is either a function of two variables, or it has no component in the direction of the third coordinates, then the divergence theorem appears in a

Foundations for Guided-Wave Optics, by Chin-Lin Chen
Copyright © 2007 John Wiley & Sons, Inc.

different form. For convenience, we choose the third variable as z. Either vector $\mathbf{A}$ is independent of z,

$$\mathbf{A}(x, y, z) = \mathbf{A}(x, y) \tag{B.2}$$

or it has no component in $\hat{\mathbf{z}}$

$$\mathbf{A}(x, y, z) = \mathbf{A}_t(x, y, z) \tag{B.3}$$

where $\mathbf{A}_t$ is a vector in the xy plane. In both cases,

$$\nabla \cdot \mathbf{A}(x, y, z) = \nabla_t \cdot \mathbf{A}_t(x, y, z) \tag{B.4}$$

where

$$\nabla_t = \hat{\mathbf{x}}\frac{\partial}{\partial x} + \hat{\mathbf{y}}\frac{\partial}{\partial y}$$

is an operator in the plane transverse to the z axis. We apply the divergence theorem (B.1) to a thin volume element with a cross section S_1 and infinitesimal length Δz as shown in Figure B.1. The volume integral on the left-hand side of (B.1) is

$$\int_V \nabla \cdot \mathbf{A}(x, y, z)\,dv = \int_V \nabla_t \cdot \mathbf{A}_t(x, y, z)\,ds\,dz = \Delta z \int_S \nabla_t \cdot \mathbf{A}_t(x, y, z)\,ds \tag{B.5}$$

The surface integral on the right-hand side of (B.1) is

$$\oint_S \mathbf{A}(x, y, z) \cdot d\mathbf{s} = \int_{S_1} \mathbf{A}(x, y, z) \cdot (-\hat{\mathbf{z}}\,ds) + \int_{S_2} \mathbf{A}(x, y, z + \Delta z) \cdot (\hat{\mathbf{z}}\,ds)$$
$$+ \Delta z \oint_C A(x, y, z) \cdot \hat{\mathbf{n}}\,d\ell \tag{B.6}$$

where C is a closed contour enclosing surface S_1 and $\hat{\mathbf{n}}$ is a unit vector normal to the closed contour C. S_2 is the surface at $z + \Delta z$. As Δz tends to zero, S_2 tends to S_1.

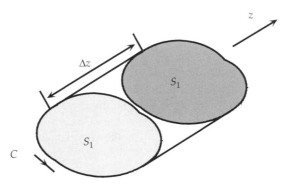

Figure B.1   A thin volume element.

In view of (B.2) or (B.3), either the first two integrals of (B.6) are zero or they cancel each other exactly. Then the third integral is the only term left in (B.6). Upon substituting (B.5) and (B.6) into (B.1), and let Δz approach 0, we obtain

$$\int_S \nabla_t \cdot \mathbf{A}(x, y, z) \, ds = \oint_C \mathbf{A}(x, y, z) \cdot \hat{\mathbf{n}} \, d\ell \tag{B.7}$$

This is the *divergence theorem* in a *two-dimensional* space.

As an application, we consider a special case where

$$\mathbf{A}(x, y, z) = \Phi(x, y) \nabla_t \Psi(x, y) - \Psi(x, y) \nabla_t \Phi(x, y) \tag{B.8}$$

where $\Phi(x, y)$ and $\Psi(x, y)$ are two arbitrary functions of x and y. The only constraint is that they have continuous second-order derivatives. Then

$$\nabla_t \cdot \mathbf{A}(x, y, z) = \Phi(x, y) \nabla_t^2 \Psi(x, y) - \Psi(x, y) \nabla_t^2 \Phi(x, y) \tag{B.9}$$

Using (B.8) and (B.9) in (B.7), we obtain the *Green's theorem*:

$$\int_S [\Phi(x, y) \nabla_t^2 \Psi(x, y) - \Psi(x, y) \nabla_t^2 \Phi(x, y)] \, ds$$
$$= \oint_C [\Phi(x, y) \nabla_t \Psi(x, y) - \Psi(x, y) \nabla_t \Phi(x, y)] \cdot \hat{\mathbf{n}} \, d\ell \tag{B.10}$$

It is also known as *Green's second identity* [1].

REFERENCE

1. J. A. Stratton, *Electromagnetic Theory*, McGraw-Hill Book Company, Inc., New York and London, 1941.

Appendix C

ORTHOGONALITY AND ORTHONORMALITY OF GUIDED MODES

C.1   LORENTZ RECIPROCITY

In this section, we study the Lorentz reciprocity theorem [1, 2] that will be used shortly to establish the orthogonality and orthonormality of modes guided by microwave and optical waveguides. Consider a source-free, lossless medium. In a lossless medium, μ and ε are real quantities. But the medium may be inhomogeneous. Suppose a particular field, $\mathbf{E}_a$ and $\mathbf{H}_a$, exists in the medium. Then, $\mathbf{E}_a$ and $\mathbf{H}_a$ satisfy the time-harmonic ($e^{j\omega t}$) Maxwell equations and the boundary conditions, if boundaries exist. For convenience, we repeat Maxwell equations (1.7) and (1.8) here:

$$\nabla \times \mathbf{E}_a(x, y, z) = -j\omega\mu\mathbf{H}_a(x, y, z) \tag{C.1}$$

$$\nabla \times \mathbf{H}_a(x, y, z) = j\omega\varepsilon\mathbf{E}_a(x, y, z) \tag{C.2}$$

Suppose $\mathbf{E}_b$ and $\mathbf{H}_b$ also exist in the same medium with the same μ and ε and at the same frequency ω. Then $\mathbf{E}_b$ and $\mathbf{H}_b$ also satisfy Maxwell equations:

$$\nabla \times \mathbf{E}_b(x, y, z) = -j\omega\mu\mathbf{H}_b(x, y, z) \tag{C.3}$$

$$\nabla \times \mathbf{H}_b(x, y, z) = j\omega\varepsilon\mathbf{E}_b(x, y, z) \tag{C.4}$$

Foundations for Guided-Wave Optics, by Chin-Lin Chen
Copyright © 2007 John Wiley & Sons, Inc.

By performing the dot products of $\mathbf{H}_b^*$ with (C.1), $\mathbf{E}_a$ with the complex conjugation of (C.4), and subtracting the resulting equations, we obtain

$$\nabla \cdot [\mathbf{E}_a(x, y, z) \times \mathbf{H}_b^*(x, y, z)] = -j\omega\mu\mathbf{H}_a(x, y, z) \cdot \mathbf{H}_b^*(x, y, z)$$
$$+ j\omega\varepsilon\mathbf{E}_a(x, y, z) \cdot \mathbf{E}_b^*(x, y, z) \qquad (C.5)$$

Similarly, from (C.2) and (C.3), we obtain

$$\nabla \cdot [\mathbf{E}_b^*(x, y, z) \times \mathbf{H}_a(x, y, z)] = +j\omega\mu\mathbf{H}_a(x, y, z) \cdot \mathbf{H}_b^*(x, y, z)$$
$$- j\omega\varepsilon\mathbf{E}_a(x, y, z) \cdot \mathbf{E}_b^*(x, y, z) \qquad (C.6)$$

Combining (C.5) and (C.6), we have

$$\nabla \cdot [\mathbf{E}_a(x, y, z) \times \mathbf{H}_b^*(x, y, z) + \mathbf{E}_b^*(x, y, z) \times \mathbf{H}_a(x, y, z)] = 0 \qquad (C.7)$$

This is the *Lorentz reciprocity relation* in the differential form. It is applicable to all media, homogeneous or inhomogeneous, so long as the media are lossless. In the integral form, the reciprocity relation is

$$\oint_S [\mathbf{E}_a(x, y, z) \times \mathbf{H}_b^*(x, y, z) + \mathbf{E}_b^*(x, y, z) \times \mathbf{H}_a(x, y, z)] \cdot d\mathbf{s} = 0 \qquad (C.8)$$

C.2   ORTHOGONALITY OF GUIDED MODES

Now we study modes guided by a waveguide. Suppose the waveguide has a uniform cross section and is infinitely long in the z direction. We consider time-harmonic waves propagate in the $+z$ direction. For many waveguides, we often need two mode numbers, represented by two subscripts, to identify a mode. To simplify writing, we use one, instead of two, subscript to identify a mode. For mode v, the field may be written as

$$\mathbf{E}_v(x, y, z) = \mathbf{e}_v(x, y)e^{-j\beta_v z} \qquad (C.9)$$
$$\mathbf{H}_v(x, y, z) = \mathbf{h}_v(x, y)e^{-j\beta_v z} \qquad (C.10)$$

where β_v is the propagation constant of mode v. The field distributions $\mathbf{e}_v$ and $\mathbf{h}_v$ are functions of x and y only. Similarly for mode μ, we have

$$\mathbf{E}_\mu(x, y, z) = \mathbf{e}_\mu(x, y)e^{-j\beta_\mu z} \qquad (C.11)$$
$$\mathbf{H}_\mu(x, y, z) = \mathbf{h}_\mu(x, y)e^{-j\beta_\mu z} \qquad (C.12)$$

We assume the two modes in question are nondegenerate in that $\beta_v \neq \beta_\mu$. From the Lorentz reciprocity theorem (C.7), we have

$$\nabla \cdot \{[\mathbf{e}_v(x, y) \times \mathbf{h}_\mu^*(x, y) + \mathbf{e}_\mu^*(x, y) \times \mathbf{h}_v(x, y)]e^{-j(\beta_v - \beta_\mu)z}\} = 0 \qquad (C.13)$$

After a straightforward manipulation, we convert the above equation to

$$\nabla_t \cdot [\mathbf{e}_\nu(x, y) \times \mathbf{h}_\mu^*(x, y) + \mathbf{e}_\mu^*(x, y) \times \mathbf{h}_\nu(x, y)]_t$$
$$= j(\beta_\nu - \beta_\mu)[\mathbf{e}_\nu(x, y) \times \mathbf{h}_\mu^*(x, y) + \mathbf{e}_\mu^*(x, y) \times \mathbf{h}_\nu(x, y)]_z \qquad \text{(C.14)}$$

where the subscripts t and z signify the transverse and longitudinal components of vectors. By integrating both sides of the equation over the entire waveguide cross section S, we obtain

$$\int_S \{\nabla_t \cdot [\mathbf{e}_\nu(x, y) \times \mathbf{h}_\mu^*(x, y) + \mathbf{e}_\mu^*(x, y) \times \mathbf{h}_\nu(x, y)]_t\} ds$$
$$= j(\beta_\nu - \beta_\mu)\int_S [\mathbf{e}_\nu(x, y) \times \mathbf{h}_\mu^*(x, y) + \mathbf{e}_\mu^*(x, y) \times \mathbf{h}_\nu(x, y)] \cdot \hat{z} ds \qquad \text{(C.15)}$$

By applying the two-dimensional divergence theorem [(B.7) of Appendix B], we can rewrite the integral on the left-hand side as

$$\int_S \{\nabla_t \cdot [\mathbf{e}_\nu(x, y) \times \mathbf{h}_\mu^*(x, y) + \mathbf{e}_\mu^*(x, y) \times \mathbf{h}_\nu(x, y)]_t\} ds$$
$$= \oint_C [\mathbf{e}_\nu(x, y) \times \mathbf{h}_\mu^*(x, y) + \mathbf{e}_\mu^*(x, y) \times \mathbf{h}_\nu(x, y)]_t \cdot \hat{n} d\ell \qquad \text{(C.16)}$$

where C is a closed contour enclosing the entire waveguide cross section. For metallic waveguides used in microwave or millimeter wave applications, the boundary conditions require the tangential components of the electric field, $(\mathbf{e}_\nu \times \hat{n})$, to vanish. Thus the integral on the right-hand side of (C.16) is zero. For optical waveguides such as thin-film waveguides, channel waveguides or optical fibers, $|\mathbf{e}_\nu|$, $|\mathbf{h}_\nu|$, $|\mathbf{e}_\mu|$, and $|\mathbf{h}_\mu|$ decay exponentially for points outside the film or core region. Thus the contour integral on the right-hand side of (C.16) vanishes as C recedes to infinite. Since the left-hand side of (C.15) vanishes, so does the right-hand side:

$$(\beta_\nu - \beta_\mu)\int_S [\mathbf{e}_\nu(x, y) \times \mathbf{h}_\mu^*(x, y) + \mathbf{e}_\mu^*(x, y) \times \mathbf{h}_\nu(x, y)] \cdot \hat{z} ds = 0 \qquad \text{(C.17)}$$

As noted earlier, the two modes are not degenerate, $\beta_\nu \neq \beta_\mu$. Then (C.17) requires

$$\int_S [\mathbf{e}_\nu(x, y) \times \mathbf{h}_\mu^*(x, y) + \mathbf{e}_\mu^*(x, y) \times \mathbf{h}_\nu(x, y)] \cdot \hat{z} ds = 0 \qquad \text{(C.18)}$$

This is the *orthogonality relation* of guided modes [3–5]. Only the transverse components of the field are involved in (C.18). More explicitly, the orthogonality relation can be written as

$$\int_S [\mathbf{e}_{\nu t}(x, y) \times \mathbf{h}_{\mu t}^*(x, y) + \mathbf{e}_{\mu t}^*(x, y) \times \mathbf{h}_{\nu t}(x, y)] \cdot \hat{z} ds = 0 \qquad \text{(C.19)}$$

C.3   ORTHONORMALITY OF GUIDED MODES

The orthogonality relation does not apply if modes ν and μ are degenerate or they are the same mode. Suppose mode ν propagates in the $+z$ direction. Then the total time-average power carried by mode ν is

$$
\begin{aligned}
P_z &= \int_S \frac{1}{2} \mathrm{Re}[\mathbf{E}_\nu(x, y, z) \times \mathbf{H}_\nu^*(x, y, z)] \cdot \hat{\mathbf{z}}\, ds \\
&= \frac{1}{4} \int_S [\mathbf{e}_{\nu t}(x, y) \times \mathbf{h}_{\nu t}^*(x, y) + \mathbf{e}_{\nu t}^*(x, y) \times \mathbf{h}_{\nu t}(x, y)] \cdot \hat{\mathbf{z}}\, ds
\end{aligned}
\tag{C.20}
$$

We can choose the amplitude constants of $\mathbf{e}_\nu$ and $\mathbf{h}_\nu$ such that the total time-average power carried by a mode is 1 W. By setting P_z to 1, the above equation becomes

$$
\frac{1}{4} \int_S [\mathbf{e}_{\nu t}(x, y) \times \mathbf{h}_{\nu t}^*(x, y) + \mathbf{e}_{\nu t}^*(x, y) \times \mathbf{h}_{\nu t}(x, y)] \cdot \hat{\mathbf{z}}\, ds = 1
\tag{C.21}
$$

On the other hand, if mode ν propagates in the $-z$ direction, we can choose the amplitude constants such that the right-hand side term is -1. Combining (C.19) and (C.21), we have the *orthonormality relation*

$$
\frac{1}{4} \int_S [\mathbf{e}_{\nu t}(x, y) \times \mathbf{h}_{\mu t}^*(x, y) + \mathbf{e}_{\mu t}^*(x, y) \times \mathbf{h}_{\nu t}(x, y)] \cdot \hat{\mathbf{z}}\, ds = \begin{cases} 0 & \text{if } \beta_\nu \neq \beta_\mu \\ 1 & \text{if } \beta_\nu = \beta_\mu > 0 \\ -1 & \text{if } \beta_\nu = \beta_\mu < 0 \end{cases}
\tag{C.22}
$$

It should be stressed that the orthogonality and orthonormality relations pertain to the transverse field components only, and the longitudinal field component is not involved in any way.

REFERENCES

1. R. E. Collins, *Field Theory of Guided Waves*, 2nd ed., IEEE Press, New York, 1991, Chapter 1.
2. S. Ramo, J. R. Whinnery, and T. Van Duzer, *Fields and Waves in Communication Electronics*, 3rd ed., Wiley, New York, 1994, Chapter 11.
3. D. Marcuse, *Light Transmission Optics*, Van Norstrand Reinhold, New York, 1972.
4. D. Marcuse, *Theory of Dielectric Optical Waveguides*, Academic, Boston, 1991, Chapter 3.
5. H. Nishihara, M. Haruna, and T. Suhara, *Optical Integrated Circuits*, McGraw-Hill, New York, 1985, Chapter 3.

Appendix D

ELASTICITY, PHOTOELASTICITY, AND ELECTROOPTIC EFFECTS

D.1 STRAIN TENSORS

In this brief discussion of elasticity, we adopt a macroscopic viewpoint and consider materials as a continuum of matter. We begin with one-dimensional objects. In considering one-dimensional objects, we ignore the two lateral dimensions and examine the longitudinal deformation and tensile stress only. Then our discussion turns to two- and three- dimensional objects.

D.1.1 Strain Tensors in One-Dimensional Objects

Consider a long bar having a uniform cross-sectional area A_z. The bar is fixed at one end and a force is applied to the other end, as shown in Figure D.1(a). Focus our attention to two arbitrary points in the bar. The coordinates of the two points are z and z' in the absence of an applied force. When a force F_z in the z direction is applied to the bar, the two points move to $z + u_z(z)$ and $z' + u_z(z')$, respectively [Fig. D.1(b)]. The distance between the two points changes from $z' - z$ to $z' - z + [u_z(z') - u_z(z)]$ in response to the applied force. The fractional length

Foundations for Guided-Wave Optics, by Chin-Lin Chen
Copyright © 2007 John Wiley & Sons, Inc.

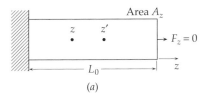

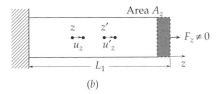

Figure D.1   One-dimensional strain and stress.

change is

$$\frac{\{[z' + u_z(z')] - [z + u_z(z)]\} - (z' - z)}{z' - z} = \frac{u_z(z') - u_z(z)}{z' - z} \tag{D.1}$$

As z' approaches z, the limiting value of the *fractional length change* in the z direction is

$$S_{zz} = \lim_{z' \to z} \frac{u_z(z') - u_z(z)}{z' - z} = \frac{\partial u_z}{\partial z} \tag{D.2}$$

This is the *tensile strain* component in a one-dimensional object and it is a dimensionless quantity [1–3]. It represents a percentage elongation if S_{zz} is positive and a fractional contraction if S_{zz} is negative.

D.1.2   Strain Tensors in Two-Dimensional Objects

In two-dimensional objects, there are *rotation* and *shearing distortion* in addition to the tensile strain. Consider a rectangle in the xy plane as shown in Figure D.2(a). As in one-dimensional objects, a tensile strain component [Fig. D.2(b)] is the fractional change of length. For an elongation or contraction in the x and y directions, the tensile strain components are

$$S_{xx} = \frac{\partial u_x}{\partial x} \tag{D.3}$$

$$S_{yy} = \frac{\partial u_y}{\partial y} \tag{D.4}$$

A rotation [Fig. D.2(d)] is the turning of the entire rectangle by an angle α' without changing the shape. In a rotation, the distance between the two arbitrary points in the rectangle remains unchanged. In a shearing distortion [Fig. D.2(e)], the rectangle becomes a rhomboid. The distance between two arbitrary points also changes. As depicted schematically in Figure D.2(e), the diagonal of the rectangle is stretched.

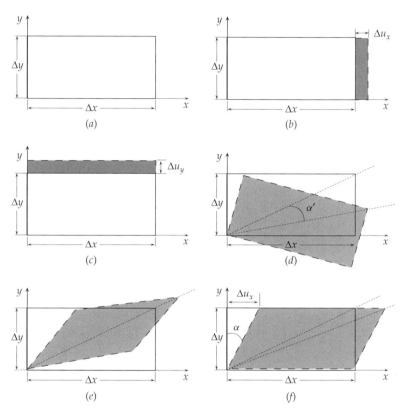

Figure D.2 Deformation of a rectangle: (a) undisturbed rectangle, (b) and (c) elongation in x and y directions, (d) rotation, (e) shearing direction, and (f) superposition of shear and rotation.

Consider a deformation shown in Figure D.2(f). We use the angle α as a measure to quantify the deformation. In the limit of $\Delta y \to 0$, α is given by

$$\tan \alpha = \lim_{\Delta y \to 0} \frac{\Delta u_x}{\Delta y} = \frac{1}{2}\left(\frac{\partial u_x}{\partial y} + \frac{\partial u_y}{\partial x}\right) + \frac{1}{2}\left(\frac{\partial u_x}{\partial y} - \frac{\partial u_y}{\partial x}\right) \qquad (D.5)$$

The terms in the first bracket represent the *shearing strain* and those in the second bracket the *rotation*. The shearing strain component corresponding to the distortion shown in Figure D.2(e) is

$$S_{xy} = S_{yx} = \frac{1}{2}\left(\frac{\partial u_x}{\partial y} + \frac{\partial u_y}{\partial x}\right) \qquad (D.6)$$

If we consider rectangles in yz and zx planes, we have three additional strain components:

$$S_{zz} = \frac{\partial u_z}{\partial z} \qquad (D.7)$$

$$S_{yz} = S_{zy} = \frac{1}{2}\left(\frac{\partial u_z}{\partial y} + \frac{\partial u_y}{\partial z}\right) \tag{D.8}$$

$$S_{zx} = S_{xz} = \frac{1}{2}\left(\frac{\partial u_z}{\partial x} + \frac{\partial u_x}{\partial z}\right) \tag{D.9}$$

D.1.3   Strain Tensors in Three-Dimensional Objects

In three-dimensional objects, all strain components discussed in the last two sub-sections may be present simultaneously. It is convenient to assemble all strain components in a matrix form, and we write

$$\boldsymbol{S} = \begin{bmatrix} S_{xx} & S_{xy} & S_{xz} \\ S_{yx} & S_{yy} & S_{yz} \\ S_{zx} & S_{zy} & S_{zz} \end{bmatrix} \tag{D.10}$$

The tensile strain components S_{xx}, S_{yy}, and S_{zz} are the diagonal matrix elements. The shearing strain components S_{xy}, S_{yz}, and S_{zx} are the off-diagonal matrix elements. Figure D.3 gives a pictorial representation of the deformation of a three-dimensional object corresponding to each strain component.

D.2   STRESS TENSORS

Stress is the force per unit cross-sectional area [1–3]. Again, we begin with a one-dimensional object [Fig. D.1(*b*)]. The *z* component of the *tensile stress* is the limiting value of the force in the *z* direction per unit area normal to the *z* axis:

$$T_{zz} = \lim_{A_z \to 0} \frac{F_z}{A_z}$$

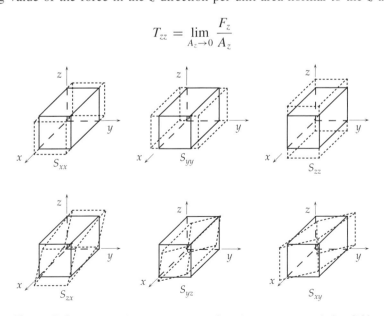

Figure D.3   Graphical representation of strain components. (After [2].)

The tensile stress component has a dimension of newtons per meters squared.

For two- and three-dimensional objects, stress components are defined in the same manner. However, it is necessary to specify the orientation of the differential area. Consider a differential area A_x normal to the x axis shown in Figure D.4(a). The force acting on A_x has three components $\mathbf{F} = \hat{\mathbf{a}}_x F_x + \hat{\mathbf{a}}_y F_y + \hat{\mathbf{a}}_z F_z$. Stress acting on the differential area A_x also has three components and they are

$$T_{xx} = \lim_{A_x \to 0} \frac{F_x}{A_x} \qquad (D.11)$$

$$T_{yx} = \lim_{A_x \to 0} \frac{F_y}{A_x} \qquad (D.12)$$

$$T_{zx} = \lim_{A_x \to 0} \frac{F_z}{A_x} \qquad (D.13)$$

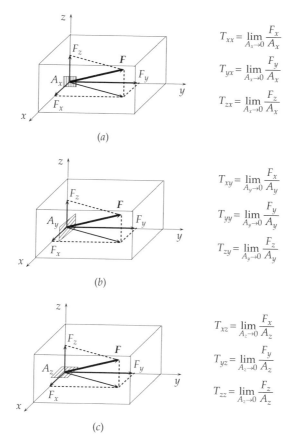

Figure D.4 Definition of stress tensor components: Force acting on elementary area (a) A_x, (b) A_y, and (c) A_z.

Similarly, we define stress components for the differential areas normal to the y axis and the z axis [Fig. D.4(b) and D.4(c)]. In general, a stress tensor component is of the form of

$$T_{ij} = \lim_{A_j \to 0} \frac{F_i}{A_j} \tag{D.14}$$

where subscripts i and j stand for x, y, or z. The first subscript i designates the force component and the second subscript j denotes the orientation of the differential area. T_{ii} is a *tensile stress* component and T_{ij} with $i \neq j$ is a *shearing stress* component. The unit of a shearing stress component is also newtons per meters squared.

It is also convenient to assemble all stress tensor components in a matrix form and write

$$\boldsymbol{T} = \begin{bmatrix} T_{xx} & T_{xy} & T_{xz} \\ T_{yx} & T_{yy} & T_{yz} \\ T_{zx} & T_{zy} & T_{zz} \end{bmatrix} \tag{D.15}$$

In the matrix form, tensile stress components are the diagonal matrix elements and the shearing stress components are the off-diagonal matrix elements. Also note that the stress matrix is symmetric with respect to the diagonal. Figure D.5 gives a pictorial representation of each stress tensor components.

D.3   HOOKE'S LAW IN ISOTROPIC MATERIALS

As the stress acting on a material increases, the strain in the material also increases. So long as the strain is within the elastic limit of the material, the strain varies linearly with the applied stress. When the applied stress is removed, the strain vanishes and the material returns to its original size and shape. The linear stress–strain relationship is known as **Hooke's law**. For most materials, Hooke's law holds if strain components are on the order of 10^{-4} to 10^{-3} or less.

For one-dimensional objects, Hooke's law is simply

$$T_{xx} = YS_{xx} \tag{D.16}$$

The proportionality constant Y is the *Young's modulus* and it has the dimension of newtons per meters squared.

When a three-dimensional object is stretched, the elongation in one direction is accompanied by contractions in the two lateral dimensions. In other words, a tensile stress T_{xx} produces S_{yy} and S_{zz} in addition to S_{xx}. Within the elastic limit, S_{yy} and S_{zz} also vary linearly with T_{xx}. For isotropic elastic materials,

$$S_{yy} = S_{zz} = -\frac{\sigma}{Y} T_{xx} \tag{D.17}$$

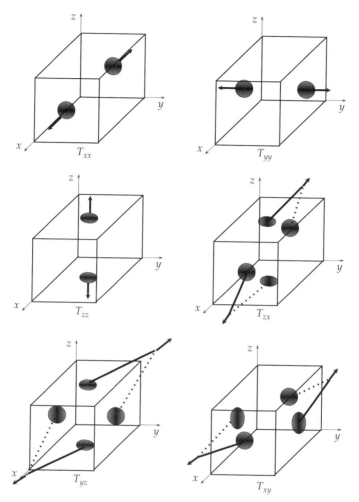

Figure D.5 Pictorial representation of stress components. (After [2].)

where σ is the *Poisson ratio*. Physically, the Poisson ratio is the ratio of the longitudinal elongation to the lateral contraction produced by the same tensile stress T_{xx},

$$\sigma = -\frac{S_{yy}}{S_{xx}} = -\frac{S_{zz}}{S_{xx}} \qquad (D.18)$$

The need for a minus sign in (D.17) and (D.18) is obvious. For most materials, σ is about $\frac{1}{3}$. For isotropic materials and within the elastic limit, the Young's modulus and Poisson's ratio are independent of stress directions and magnitude.

In the presence of three tensile strain components, the total strain is the superposition of contributions from each stress component. For isotropic materials, the relation between the tensile strain and stress can be expressed as

$$\begin{bmatrix} S_{xx} \\ S_{yy} \\ S_{zz} \end{bmatrix} = \frac{1}{Y} \begin{bmatrix} 1 & -\sigma & -\sigma \\ -\sigma & 1 & -\sigma \\ -\sigma & -\sigma & 1 \end{bmatrix} \begin{bmatrix} T_{xx} \\ T_{yy} \\ T_{zz} \end{bmatrix} \tag{D.19}$$

Conversely, it is also possible to express tensile stress components in terms of tensile strain components. By solving for T_{ii} from (D19), we obtain

$$\begin{bmatrix} T_{xx} \\ T_{yy} \\ T_{zz} \end{bmatrix} = \begin{bmatrix} \lambda + 2\mu & \lambda & \lambda \\ \lambda & \lambda + 2\mu & \lambda \\ \lambda & \lambda & \lambda + 2\mu \end{bmatrix} \begin{bmatrix} S_{xx} \\ S_{yy} \\ S_{zz} \end{bmatrix} \tag{D.20}$$

where μ and λ are *Lame's coefficients*

$$\lambda = \frac{Y\sigma}{(1+\sigma)(1-2\sigma)} \tag{D.21}$$

$$\mu = \frac{Y}{2(1+\sigma)} \tag{D.22}$$

Lame's coefficient μ is also known as the *shear modulus* or *modulus of rigidity*. Note that the shearing stress and shear strain components are absent in (D.19) and (D.20). In isotropic materials, the shearing stress components are related to the shearing strain components only,

$$T_{ij} = 2\mu S_{ij} \tag{D.23}$$

where $i \neq j$.

So far, we have introduced four elastic constants: $Y, \sigma, \lambda,$ and μ. For isotropic materials, two of the four elastic constants are independent. For example, Y is related to μ and σ,

$$Y = 2\mu(1+\sigma) \tag{D.24}$$

D.4   STRAIN AND STRESS TENSORS IN ABBREVIATED INDICES

The stress and strain matrices given in (D.10) and (D.15) are 3×3 matrices. The two matrices are symmetric with respect to the diagonal elements. In each matrix, six of the nine matrix elements are independent. By taking advantage of the matrix symmetry, it is possible to convert the 3×3 matrices to 1×6 column matrices. To this end, we introduce the *abbreviated* or *contracted indices* [4, 5]

11 or $xx \rightarrow 1$	23 and 32 or yz and $zy \rightarrow 4$
22 or $yy \rightarrow 2$	13 and 31 or xz and $zx \rightarrow 5$
33 or $zz \rightarrow 3$	12 and 21 or xy and $yx \rightarrow 6$

It is customary to use capital letters, I or J for example, as subscripts in the abbreviated notations and to keep the lowercase subscripts for the full or unabbreviated symbols. Following this prescription, we can convert the 3×3 strain and stress matrices given in (D.10) and (D.15) to 1×6 column matrices:

$$S = \begin{bmatrix} S_1 \\ S_2 \\ S_3 \\ S_4 \\ S_5 \\ S_6 \end{bmatrix} = \begin{bmatrix} S_{xx} \\ S_{yy} \\ S_{zz} \\ 2S_{yz} \\ 2S_{zx} \\ 2S_{xy} \end{bmatrix} = \begin{bmatrix} S_{xx} \\ S_{yy} \\ S_{zz} \\ 2S_{zy} \\ 2S_{xz} \\ 2S_{yx} \end{bmatrix} \tag{D.25}$$

$$T = \begin{bmatrix} T_1 \\ T_2 \\ T_3 \\ T_4 \\ T_5 \\ T_6 \end{bmatrix} = \begin{bmatrix} T_{xx} \\ T_{yy} \\ T_{zz} \\ T_{yz} \\ T_{zx} \\ T_{xy} \end{bmatrix} = \begin{bmatrix} T_{xx} \\ T_{yy} \\ T_{zz} \\ T_{zy} \\ T_{xz} \\ T_{yx} \end{bmatrix} \tag{D.26}$$

Note the presence of a factor of 2 in the definition for S_4, S_5, and S_6 in (D.25).

When expressed in terms of abbreviated indices, Hooke's law (D.20) and (D.23) can be combined into a compact form:

$$T = cS \tag{D.27}$$

where c is a 6×6 matrix

$$c = \begin{bmatrix} c_{11} & c_{12} & c_{13} & c_{14} & c_{15} & c_{16} \\ c_{21} & c_{22} & c_{23} & c_{24} & c_{25} & c_{26} \\ c_{31} & c_{32} & c_{33} & c_{34} & c_{35} & c_{36} \\ c_{41} & c_{42} & c_{43} & c_{44} & c_{45} & c_{46} \\ c_{51} & c_{52} & c_{53} & c_{54} & c_{55} & c_{56} \\ c_{61} & c_{62} & c_{63} & c_{64} & c_{65} & c_{66} \end{bmatrix} \tag{D.28}$$

Many matrix elements of c are either zero or can be expressed in terms of other matrix elements. Many nonzero matrix elements are also identical. For isotropic solids, 14 matrix elements are exactly zero. The nonzero matrix elements can be expressed in terms of the Lame's coefficients:

$$c_{11} = c_{22} = c_{33} = \lambda + 2\mu \tag{D.29}$$

$$c_{44} = c_{55} = c_{66} = \mu \tag{D.30}$$

$$c_{12} = c_{21} = c_{13} = c_{31} = c_{23} = c_{32} = \lambda \tag{D.31}$$

Conversely, we can express a strain tensor in terms of a stress tensor:

$$S = dT \tag{D.32}$$

where

$$d = \begin{bmatrix} d_{11} & d_{12} & d_{13} & d_{14} & d_{15} & d_{16} \\ d_{21} & d_{22} & d_{23} & d_{24} & d_{25} & d_{26} \\ d_{31} & d_{32} & d_{33} & d_{34} & d_{35} & d_{36} \\ d_{41} & d_{42} & d_{43} & d_{44} & d_{45} & d_{46} \\ d_{51} & d_{52} & d_{53} & d_{54} & d_{55} & d_{56} \\ d_{61} & d_{62} & d_{63} & d_{64} & d_{65} & d_{66} \end{bmatrix} \tag{D.33}$$

For isotropic solids,

$$d_{11} = d_{22} = d_{33} = \frac{1}{Y} \tag{D.34}$$

$$d_{12} = d_{21} = d_{13} = d_{31} = d_{23} = d_{32} = -\frac{\sigma}{Y} \tag{D.35}$$

$$d_{44} = d_{55} = d_{66} = \frac{2(1+\sigma)}{Y} \tag{D.36}$$

Other matrix elements of d are zero.

D.5   RELATIVE DIELECTRIC CONSTANT TENSORS AND RELATIVE DIELECTRIC IMPERMEABILITY TENSORS

The constitutive relation of a dielectric medium is

$$\mathbf{D} = \varepsilon_0 \mathbf{E} + \mathbf{P} \tag{D.37}$$

where $\mathbf{P}$ is the *electric polarization* of the medium. In vacuum or free space, $\mathbf{P} = 0$ and $\mathbf{D} = \varepsilon_0 \mathbf{E}$. In linear isotropic media, $\mathbf{P}$ is in the same direction of, and is proportional to, $\mathbf{E}$. The proportionality constant is $\varepsilon_0 \chi_e$ and χ_e is the *electric susceptibility*. Therefore, $\mathbf{D}$ can be written as

$$\mathbf{D} = \varepsilon_0(1 + \chi_e)\mathbf{E} = \varepsilon_0 \varepsilon_r \mathbf{E} \tag{D.38}$$

where ε_r is the *relative permittivity* or *relative dielectric constant*. In optics literature, it is customary to write ε_r as n^2 and n is the *refractive index*. In many applications, it is convenient to express $\mathbf{E}$ in terms of $\mathbf{D}$. Then, in lieu of (D.38), we write

$$\mathbf{E} = \frac{1}{\varepsilon_0}\frac{1}{\varepsilon_r}\mathbf{D} \tag{D.39}$$

and $1/\varepsilon_r$ is the *dielectric impermeability constant*.

In anisotropic media, $\mathbf{P}$ is not necessarily in parallel with $\mathbf{E}$. Therefore the electric susceptibility, relative dielectric constant, and dielectric impermeability are tensors. In other words, (D.38) and (D.39) become

$$\mathbf{D} = \varepsilon_0 \varepsilon_r \mathbf{E} \tag{D.40}$$

$$\mathbf{E} = \frac{1}{\varepsilon_0}(\varepsilon_r)^{-1}\mathbf{D} \tag{D.41}$$

where ε_r is the *relative dielectric constant tensor* and it is a 3×3 matrix:

$$\varepsilon_r = \begin{bmatrix} \varepsilon_{11} & \varepsilon_{12} & \varepsilon_{13} \\ \varepsilon_{21} & \varepsilon_{22} & \varepsilon_{23} \\ \varepsilon_{31} & \varepsilon_{32} & \varepsilon_{33} \end{bmatrix} \tag{D.42}$$

We follow Nye's [4] and Kaminow's [5] practice and write the *relative dielectric impermeability tensor* as

$$B = (\varepsilon_r)^{-1} = \begin{bmatrix} B_{11} & B_{12} & B_{13} \\ B_{21} & B_{22} & B_{23} \\ B_{31} & B_{32} & B_{33} \end{bmatrix} \tag{D.43}$$

For isotropic media, $\varepsilon_{ii} = n^2$, $B_{ii} = n^{-2}$ for $i = 1, 2$, or 3. In addition, $\varepsilon_{ij} = 0$ and $B_{ij} = 0$ for $i \neq j$.

By using abbreviated indices, we can express B as a 1×6 column matrix:

$$\begin{bmatrix} B_1 \\ B_2 \\ B_3 \\ B_4 \\ B_5 \\ B_6 \end{bmatrix} = \begin{bmatrix} B_{11} \\ B_{22} \\ B_{33} \\ B_{23} \\ B_{13} \\ B_{12} \end{bmatrix} \tag{D.44}$$

As noted by Nye [4], the tensor components B_J are related directly to the *index ellipsoid*. Specifically, the equation for the index ellipsoid is

$$B_1 x^2 + B_2 y^2 + B_3 z^2 + 2B_4 yz + 2B_5 zx + 2B_6 xy = 1 \tag{D.45}$$

The change in refractive index due to applied electric fields is known as the *electrooptic effect*. The change in refractive index by stress and strain is referred to as the *acoustooptic effects* or *photoelastic effects*. In this book, we are mainly interested in cases where the change of B_J is linearly proportional to the stress or strain. The refractive index and relative dielectric impermeability also change in response to the applied electric field. In isotropic materials, the change is proportional to the square of applied electric field. This is the *quadratic electrooptic effect*. Since the effect was first observed by J. Kerr in 1875, it is also known as the *Kerr effect*. In crystals without inversion symmetry, the index change is mainly proportional to the applied electric field. This is the *Pockels effect.*

D.6 PHOTOELASTIC EFFECT AND PHOTOELASTIC CONSTANT TENSORS

When a material is stressed mechanically, the refractive index changes. This is known as the acoustooptic effects or photoelastic effects. In solids, it is most convenient to express the effects in terms of changes in B_J. For small or moderate strain, B_J varies linearly with S_J. As B and S have six independent elements each, the photoelastic tensor p has 36 components. In the matrix form and in terms of abbreviated indices, the photoelastic effects can be expressed as

$$\begin{bmatrix} \Delta B_1 \\ \Delta B_2 \\ \Delta B_3 \\ \Delta B_4 \\ \Delta B_5 \\ \Delta B_6 \end{bmatrix} = \begin{bmatrix} p_{11} & p_{12} & p_{13} & p_{14} & p_{15} & p_{16} \\ p_{21} & p_{22} & p_{23} & p_{24} & p_{25} & p_{26} \\ p_{31} & p_{32} & p_{33} & p_{34} & p_{35} & p_{36} \\ p_{41} & p_{42} & p_{43} & p_{44} & p_{45} & p_{46} \\ p_{51} & p_{52} & p_{53} & p_{54} & p_{55} & p_{56} \\ p_{61} & p_{62} & p_{63} & p_{64} & p_{65} & p_{66} \end{bmatrix} \begin{bmatrix} S_1 \\ S_2 \\ S_3 \\ S_4 \\ S_5 \\ S_6 \end{bmatrix} \tag{D.46}$$

where ΔB_J is the change of B_J.

Many matrix elements of p are zero. The nonzero tensor elements are also very small. In isotropic solids, p is particularly simple

$$p = \begin{bmatrix} p_{11} & p_{12} & p_{12} & 0 & 0 & 0 \\ p_{12} & p_{11} & p_{12} & 0 & 0 & 0 \\ p_{12} & p_{12} & p_{11} & 0 & 0 & 0 \\ 0 & 0 & 0 & p_{44} & 0 & 0 \\ 0 & 0 & 0 & 0 & p_{44} & 0 \\ 0 & 0 & 0 & 0 & 0 & p_{44} \end{bmatrix} \tag{D.47}$$

and $p_{44} = (p_{11} - p_{12})/2$. For liquid, $p_{11} = p_{12}$ and p_{44} is zero.

D.7 INDEX CHANGE IN ISOTROPIC SOLIDS: AN EXAMPLE

As an example, we consider the effect of a tensile stress on the refractive index of an isotropic solid. Let n be the refractive index of the material in the absence of stress and strain. The relative dielectric impermeability tensor components are $B_1 = B_2 = B_3 = 1/n^2$ and $B_4 = B_5 = B_6 = 0$. If the isotropic solid is subjected to a tensile stress with stress components T_1, T_2, and T_3. Then the induced strain components are S_1, S_2, and S_3, which can be calculated from (D.32). No shearing strain is induced by the tensile stress, that is, $S_J = 0$ for $J \geq 4$. A glance of the photoelastic tensor (D.47) reveals immediately that $\Delta B_J = 0$ for $J \geq 4$. We also note

$$\Delta B_1 = \frac{1}{(n + \Delta n_1)^2} - \frac{1}{n^2} = p_{11}S_1 + p_{12}S_2 + p_{12}S_3 \qquad (D.48)$$

Since p_{11} and p_{12} are very small, so is Δn_1. Therefore,

$$\Delta n_1 \approx -\frac{1}{2}n^3(p_{11}S_1 + p_{12}S_2 + p_{12}S_3) \qquad (D.49)$$

Similarly, we obtain from ΔB_2

$$\Delta n_2 \approx -\frac{1}{2}n^3(p_{12}S_1 + p_{11}S_2 + p_{12}S_3) \qquad (D.50)$$

In connection with the fiber birefringence, we are interested in

$$\Delta n_1 - \Delta n_2 \approx -\frac{1}{2}n^3(p_{11} - p_{12})(S_1 - S_2) \qquad (D.51)$$

We can cast the index difference in terms of stress tensor components directly

$$\Delta n_1 - \Delta n_2 \approx -\frac{1}{2}n^3 \frac{(1 + \sigma)(p_{11} - p_{12})}{Y}(T_1 - T_2) \qquad (D.52)$$

D.8 LINEAR ELECTROOPTIC EFFECT

The change in refractive index by an applied electric field is known as the electrooptic effect. In crystals that have no center of inversion, the change of B_J is linearly proportional to the electric field intensity. This is the *linear electrooptic effect*. It is also known as the Pockels effect. In liquid, isotropic solids and crystals with inversion symmetry, there is no Pockels effect. Let $\mathbf{E}_m$ be the electric field applied to the medium, the change of a relative dielectric impermeability tensor component is given by

$$\Delta B_{ij} = \sum_k r_{ijk} E_{mk} \qquad (D.53)$$

where E_{mk} is the k component of the applied electric field and r_{ijk} is the *linear electroopitc* (EO) *coefficients*. Many linear EO coefficients are zero because of the symmetry property the material. All nonzero electrooptic coefficients are very small and they are in the range of 10^{-12} to 10^{-11} m/V range. In a matrix form and in terms of abbreviated indices, (D.53) becomes

$$
\begin{bmatrix} \Delta B_1 \\ \Delta B_2 \\ \Delta B_3 \\ \Delta B_4 \\ \Delta B_5 \\ \Delta B_6 \end{bmatrix} =
\begin{bmatrix} r_{11} & r_{12} & r_{13} \\ r_{21} & r_{22} & r_{23} \\ r_{31} & r_{32} & r_{33} \\ r_{41} & r_{42} & r_{43} \\ r_{51} & r_{52} & r_{53} \\ r_{61} & r_{62} & r_{63} \end{bmatrix}
\begin{bmatrix} E_{m1} \\ E_{m2} \\ E_{m3} \end{bmatrix} \qquad (D.54)
$$

In GaAs, InP, ZnTe, CdTe, and other cubic crystals of type 23 or $\bar{4}3m$ symmetry, all but three of the linear EO coefficients are zero. In addition, the nonzero linear EO coefficients are equal. Thus the linear EO coefficient matrix of these materials is of the form of

$$
\begin{bmatrix}
0 & 0 & 0 \\
0 & 0 & 0 \\
0 & 0 & 0 \\
r_{41} & 0 & 0 \\
0 & r_{41} & 0 \\
0 & 0 & r_{41}
\end{bmatrix}
\tag{D.55}
$$

For LiNbO$_3$, LiTaO$_3$, and other trigonal crystals with $3m$ symmetry, the linear EO coefficient matrix is of the form of

$$
\begin{bmatrix}
0 & -r_{22} & r_{13} \\
0 & r_{22} & r_{13} \\
0 & 0 & r_{33} \\
0 & r_{51} & 0 \\
r_{51} & 0 & 0 \\
-r_{22} & 0 & 0
\end{bmatrix}
\tag{D.56}
$$

D.9   QUADRATIC ELECTROOPTIC EFFECT

In liquid, isotropic solids, and crystals with inversion symmetry, the change of B_J is proportional to the square or product of electric field intensities. This is the quadratic electrooptic effect. It is also known as the Kerr effect. The change of the relative dielectric impermeability tensor elements due to an applied electric field E_m is given by

$$
\Delta B_{ij} = \sum_{kl} s_{ijkl} E_{mk} E_{ml}
\tag{D.57}
$$

where s_{ijkl} is the quadratic electrooptic coefficients or simply the *Kerr coefficients*. There are 81 quadratic electrooptic coefficients. Fortunately, many of the coefficients are zero. Several nonzero components are also equal. In isotropic solids and in terms of abbreviated indices, (D.57) can be written as

$$
\begin{bmatrix}
\Delta B_1 \\
\Delta B_2 \\
\Delta B_3 \\
\Delta B_4 \\
\Delta B_5 \\
\Delta B_6
\end{bmatrix}
=
\begin{bmatrix}
s_{11} & s_{12} & s_{12} & 0 & 0 & 0 \\
s_{12} & s_{11} & s_{12} & 0 & 0 & 0 \\
s_{12} & s_{12} & s_{11} & 0 & 0 & 0 \\
0 & 0 & 0 & s_{44} & 0 & 0 \\
0 & 0 & 0 & 0 & s_{44} & 0 \\
0 & 0 & 0 & 0 & 0 & s_{44}
\end{bmatrix}
\begin{bmatrix}
E_{m1}^2 \\
E_{m2}^2 \\
E_{m3}^2 \\
E_{m3} E_{m2} \\
E_{m3} E_{m1} \\
E_{m1} E_{m2}
\end{bmatrix}
\tag{D.58}
$$

where $s_{44} = (s_{11} - s_{12})/2$. Although the Kerr effect also exists in crystals without inversion symmetry, it is much weaker than the index change due to the Pockels effect.

As an example, suppose an isotropic solid is subjected to an electric field $\mathbf{E}_m = \hat{\mathbf{x}} E_{m1}$. Then, we have from (D.58)

$$\Delta B_1 = s_{11} E_{m1}^2 \tag{D.59}$$

$$\Delta B_2 = \Delta B_3 = s_{12} E_{m1}^2 \tag{D.60}$$

The x-directed electric field has no effect on other relative dielectric impermeability tensor components. By writing

$$\Delta B_1 = \frac{1}{(n + \Delta n_1)^2} - \frac{1}{n^2} \tag{D.61}$$

we have

$$\Delta n_1 \approx -\frac{1}{2} n^3 s_{11} E_{m1}^2 \tag{D.62}$$

Similarly,

$$\Delta n_2 \approx -\frac{1}{2} n^3 s_{12} E_{m1}^2 \tag{D.63}$$

Although Δn_3 also changes due to the presence of $\hat{\mathbf{x}} E_{m1}$, the change in n_3 is of no interest so far as the birefringence is concerned.

From the above equations, we obtain

$$\Delta n_1 - \Delta n_2 \approx \frac{1}{2} n^3 (s_{12} - s_{11}) E_{m1}^2 \tag{D.64}$$

Since $\Delta n_1 - \Delta n_2$ is proportional to the square of the electric field, $\Delta n_1 - \Delta n_2$ is independent of the polarity of the applied electric field. It is customary to rewrite the equation as

$$\Delta n_1 - \Delta n_2 = K \lambda E_{m1}^2 \tag{D.65}$$

where

$$K = \frac{n^3 (s_{12} - s_{11})}{2\lambda} \tag{D.66}$$

is the *Kerr constant* and λ is the wavelength. In all materials, the Kerr constant is very small. Typically, K is in the 10^{-16}–10^{-14} m/V^2 range. For high refractive index glasses, the Kerr constant can be as high as 1.6×10^{-14} m/V^2 for wavelength at 0.633 μm [6]. For glasses, K depends critically on the material composition, particularly the concentration of heavy-metal ions like Tl^+, Pb^{2+}, and Bi^{3+} [6].

REFERENCES

1. A. Yariv and P. Yeh, *Optical Waves in Crystals*, Wiley, New York, 1984.
2. W. R. Beam, *Electronics of Solids*, McGraw-Hill, New York, 1965.

3. B. A. Auld, *Acoustic Fields and Waves in Solids*, Wiley, New York, 1973.

4. J. F. Nye, *Physical Properties of Crystals*, Oxford University Press, Oxford, UK, 1957.

5. I. P. Kaminow, *An Introduction to Electrooptical Devices*, Academic, New York, 1974.

6. N. F. Borrelli, B. G. Aitken, M. A. Newhouse, and D. W. Hall, "Electric-field induced birefringence properties of high-refractive-index glasses exhibiting large Kerr nonlinearities," *J. Appl. Phys.*, Vol. 70, No. 5, pp. 2774–2779 (Sept. 1, 1991).

Appendix E

EFFECT OF MECHANICAL TWISTING ON FIBER BIREFRINGENCE

E.1 RELATIVE DIELECTRIC CONSTANT TENSOR OF A TWISTED MEDIUM

In this appendix, we consider the effect of a *mechanical twisting* on an isotropic cylinder. The cylinder axis is chosen as the z axis. One end ($z = 0$) of the cylinder is fixed, while the other end ($z = L$) is twisted counterclockwise by an angle Θ as shown in Figure E.1. The rate of twisting is Θ/L. We are interested in the propagation of electromagnetic waves in the twisted cylinder. In particular, we show that the transverse field components of eigenpolarization modes guided by a twisted cylinder are right- and left-hand circularly polarized fields and the two circularly polarized modes propagate with different phase velocities. In other words, a twisted medium is circularly birefringent.

Suppose a cylinder is initially isotropic and has a refractive index n. It is convenient to use a relative dielectric constant tensor to describe the constitutive relation in a twisted medium. It is therefore prudent to begin by casting the scalar relative dielectric constant of the untwisted cylinder in a matrix form. A *scalar relative dielectric constant* ε_r is simply n^2. In the matrix form, a scalar dielectric

Foundations for Guided-Wave Optics, by Chin-Lin Chen
Copyright © 2007 John Wiley & Sons, Inc.

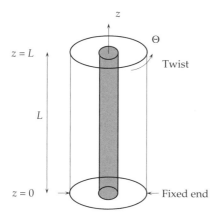

Figure E.1   Mechanical twisting.

constant is

$$\boldsymbol{\varepsilon_r} = \begin{bmatrix} n^2 & 0 & 0 \\ 0 & n^2 & 0 \\ 0 & 0 & n^2 \end{bmatrix} \tag{E.1}$$

As discussed in Appendix D, the *relative dielectric permeability tensor* is the inverse of the *relative dielectric tensor*. Thus,

$$\boldsymbol{B} = \begin{bmatrix} \dfrac{1}{n^2} & 0 & 0 \\ 0 & \dfrac{1}{n^2} & 0 \\ 0 & 0 & \dfrac{1}{n^2} \end{bmatrix} \tag{E.2}$$

In the absence of twisting, the cylinder is stress free. All strain components are zero. When the cylinder is twisted, two shearing strain components are no longer zero, and they are $S_4 = \Theta x/L$ and $S_5 = -\Theta y/L$ [1]. Other strain components S_1, S_2, S_3, and S_6 remain zero. Due to the change in S_4 and S_5, two relative dielectric impermeability tensor components are no longer zero. They are $B_4 = p_{44}\Theta x/L$ and $B_5 = -p_{44}\Theta y/L$ [2, 3]. Thus, the relative dielectric impermeability tensor of a twisted medium becomes

$$\boldsymbol{B} = \begin{bmatrix} \dfrac{1}{n^2} & 0 & -p_{44}\Theta y/L \\ 0 & \dfrac{1}{n^2} & p_{44}\Theta x/L \\ -p_{44}\Theta y/L & p_{44}\Theta x/L & \dfrac{1}{n^2} \end{bmatrix} \tag{E.3}$$

Then the relative dielectric constant tensor of a twisted medium is

$$\boldsymbol{\varepsilon}_r = \boldsymbol{B}^{-1} = \begin{bmatrix} \varepsilon_{11} & \varepsilon_{12} & \varepsilon_{13} \\ \varepsilon_{21} & \varepsilon_{22} & \varepsilon_{23} \\ \varepsilon_{31} & \varepsilon_{32} & \varepsilon_{33} \end{bmatrix} \tag{E.4}$$

where

$$\varepsilon_{11} = n^2 \frac{1 - n^4 (p_{44}\Theta x/L)^2}{1 - n^4 (p_{44}\Theta/L)^2 (x^2 + y^2)}$$

$$\varepsilon_{22} = n^2 \frac{1 - n^4 (p_{44}\Theta y/L)^2}{1 - n^4 (p_{44}\Theta/L)^2 (x^2 + y^2)}$$

$$\varepsilon_{33} = n^2 \frac{1}{1 - n^4 (p_{44}\Theta/L)^2 (x^2 + y^2)}$$

$$\varepsilon_{13} = \varepsilon_{31} = +n^4 \frac{p_{44}\Theta y/L}{1 - n^4 (p_{44}\Theta/L)^2 (x^2 + y^2)}$$

$$\varepsilon_{23} = \varepsilon_{32} = -n^4 \frac{p_{44}\Theta x/L}{1 - n^4 (p_{44}\Theta/L)^2 (x^2 + y^2)}$$

$$\varepsilon_{12} = \varepsilon_{21} = -n^6 \frac{(p_{44}\Theta/L)^2 xy}{1 - n^4 (p_{44}\Theta/L)^2 (x^2 + y^2)}$$

Since $p_{44}\Theta/L$ is very small, we neglect terms on the order of, or smaller than, $(p_{44}\Theta/L)^2$ and obtain $\varepsilon_{11} \approx \varepsilon_{22} \approx \varepsilon_{33} \approx n^2$, $\varepsilon_{12} = \varepsilon_{21} \approx 0$, $\varepsilon_{13} = \varepsilon_{31} \approx -Qy$, and $\varepsilon_{23} = \varepsilon_{32} \approx Qx$, where $Q = -n^4 p_{44}\Theta/L$. Thus, the relative dielectric tensor of a twisted medium becomes

$$\boldsymbol{\varepsilon}_r \approx \begin{bmatrix} n^2 & 0 & -Qy \\ 0 & n^2 & Qx \\ -Qy & Qx & n^2 \end{bmatrix} \tag{E.5}$$

A minus sign is incorporated in the definition for Q so that Q is positive for a counterclockwise twist.

When expressed in terms of Q explicitly, one of the Maxwell equations is

$$\nabla \times \mathbf{H} = j\omega\varepsilon_o \boldsymbol{\varepsilon}_r \cdot \mathbf{E} \approx j\omega\varepsilon_o n^2 \mathbf{E} + j\omega\varepsilon_o Q \mathbf{r} \times (\hat{\mathbf{z}} E_z - \mathbf{E}_t) \tag{E.6}$$

where $\mathbf{r} = x\hat{\mathbf{x}} + y\hat{\mathbf{y}}$ is a position vector in the transverse plane. $\mathbf{E}_t$ is the transverse electric field vector and E_z is the longitudinal electric field components.

E.2   LINEARLY POLARIZED MODES IN WEAKLY GUIDING, UNTWISTED FIBERS

Now we consider a field propagating in an untwisted medium. Let the field of a propagation mode be

$$\mathbf{E}(x, y, z) = \mathbf{e}(x, y)e^{-j\beta_0 z} \tag{E.7}$$

$$\mathbf{H}(x, y, z) = \mathbf{h}(x, y)e^{-j\beta_0 z} \tag{E.8}$$

where β_0 is the propagation constant in the untwisted medium, then $\mathbf{e}$ and $\mathbf{h}$ satisfy the following equation:

$$-j\beta_0 \hat{\mathbf{z}} \times \mathbf{h} + \nabla \times \mathbf{h} = j\omega\varepsilon_o n^2 \mathbf{e} \tag{E.9}$$

Suppose the untwisted fiber is a weakly guiding fiber. We use the LP mode designation to describe modes guided by the weakly guiding fiber. There are two mutually orthogonal and linearly independent LP_{01} modes. Explicit expressions for $\mathbf{e}$ and $\mathbf{h}$ of untwisted, weakly guiding fibers are given in Chapter 9. Pertinent equations are repeated here for convenience. For the x-polarized LP_{01} mode, $\mathbf{e}$ and $\mathbf{h}$ can be expressed as

$$\mathbf{e}^{(x)} = e_0(r)\hat{\mathbf{x}} - \frac{j}{kn}\frac{de_0(r)}{dr}\cos\phi\hat{\mathbf{z}} \tag{E.10}$$

$$\mathbf{h}^{(x)} = \frac{n}{\eta_o}e_0(r)\hat{\mathbf{y}} - \frac{j}{\omega\mu_o}\frac{de_0(r)}{dr}\sin\phi\hat{\mathbf{z}} \tag{E.11}$$

In the above equations, $e_0(r)$ is

$$e_0(r) = \begin{cases} E_o \dfrac{J_o(V\sqrt{1-b}\,r/a)}{J_o(V\sqrt{1-b})} & 0 \le r \le a \\[3mm] E_o \dfrac{K_o(V\sqrt{b}\,r/a)}{K_o(V\sqrt{b})} & a \le r < \infty \end{cases} \tag{E.12}$$

where E_o is the amplitude constant, V and b are the *generalized frequency* and *normalized guide index*. For the y-polarized LP_{01} mode, $\mathbf{e}$ and $\mathbf{h}$ can also be expressed in terms of $e_0(r)$:

$$\mathbf{e}^{(y)} = e_0(r)\hat{\mathbf{y}} - \frac{j}{kn}\frac{de_0(r)}{dr}\sin\phi\hat{\mathbf{z}} \tag{E.13}$$

$$\mathbf{h}^{(y)} = -\frac{n}{\eta_o}e_0(r)\hat{\mathbf{x}} + \frac{j}{\omega\mu_o}\frac{de_0(r)}{dr}\cos\phi\hat{\mathbf{z}} \tag{E.14}$$

E.3   EIGENPOLARIZATION MODES IN TWISTED FIBERS

In an untwisted fiber, the two LP_{01} polarization modes are not coupled. In the presence of twisting, the two LP_{01} polarization modes are coupled, and this is due to the second term in the right-hand side of (E.6). Since the coupling is weak, fields in twisted fibers can be approximated as the superposition of the two orthogonal LP_{01} polarization modes. Thus we write, as an approximation

$$\mathbf{E} = (m_1\mathbf{e}^{(x)} + m_2\mathbf{e}^{(y)})e^{-j\beta z} \tag{E.15}$$

$$\mathbf{H} = (m_1\mathbf{h}^{(x)} + m_2\mathbf{h}^{(y)})e^{-j\beta z} \tag{E.16}$$

where the propagation constant β and two constants m_1 and m_2 are to be determined. By substituting (E.15) and (E.16) into (E.6) and making use (E.9), we obtain

$$-j(\beta - \beta_0)\hat{\mathbf{z}} \times (m_1\mathbf{h}^{(x)} + m_2\mathbf{h}^{(y)}) = j\omega\varepsilon_o Q\mathbf{r} \times \hat{\mathbf{z}}(m_1 e_z^{(x)} + m_2 e_z^{(y)})$$
$$- j\omega\varepsilon_o Q\mathbf{r} \times (m_1\mathbf{e}_t^{(x)} + m_2\mathbf{e}_t^{(y)}) \tag{E.17}$$

By making the scalar product of the above equation with $\mathbf{e}_t^{(x)*}$ and integrating the resulting expression over the entire fiber cross section, we obtain

$$(\beta - \beta_0)m_1 + j\xi m_2 = 0 \tag{E.18}$$

where

$$\xi = \frac{Q}{2n^2}F_{\text{tw}}(V) = -\frac{n^2 p_{44}\Theta}{2L}F_{\text{tw}}(V) \tag{E.19}$$

and

$$F_{\text{tw}}(V) = \frac{\int r^2 e_0^*(r)\dfrac{de_0(r)}{dr}\,dr}{\int r|e_0(r)|^2\,dr} \tag{E.20}$$

Although $F_{\text{tw}}(V)$ does depend on V, it varies slowly.

Similarly, by making the scalar product with $\mathbf{e}^{(y)*}$ and integrating the resulting equation, we obtain

$$-j\xi m_1 + (\beta - \beta_0)m_2 = 0 \tag{E.21}$$

Nontrivial solution for m_1 and m_2 exists only if $(\beta - \beta_0)^2 - \xi^2 = 0$. Thus, we obtain $\beta = \beta_0 \pm \xi$ and $m_2/m_1 = \pm j$. The two eigenpolarization modes guided by a twisted fiber are

$$\mathbf{E}_L = \frac{1}{\sqrt{2}}(\mathbf{e}^{(x)} + j\mathbf{e}^{(y)})e^{-j\beta_L z} \tag{E.22}$$

and

$$\mathbf{E}_R = \frac{1}{\sqrt{2}}(\mathbf{e}^{(x)} - j\mathbf{e}^{(y)})e^{-j\beta_R z} \tag{E.23}$$

where the propagation constants are $\beta_R = \beta_0 - \xi$ and $\beta_L = \beta_0 + \xi$. Explicitly, the transverse field vectors are

$$\mathbf{E}_{Lt} = \frac{\hat{\mathbf{x}} + j\hat{\mathbf{y}}}{\sqrt{2}} e_0(r) e^{-j\beta_L z} \tag{E.24}$$

and

$$\mathbf{E}_{Rt} = \frac{\hat{\mathbf{x}} - j\hat{\mathbf{y}}}{\sqrt{2}} e_0(r) e^{-j\beta_R z} \tag{E.25}$$

Clearly, the transverse field components of the two polarization modes correspond to the left- and right-hand circularly polarized fields. Since $\beta_R \neq \beta_L$, the twisted fiber is *circularly birefringent*. The circular birefringence is

$$\beta_{\text{tw}} = \beta_R - \beta_L = -2\xi = n^2 p_{44} \frac{\Theta}{L} F_{\text{tw}}(V) \tag{E.26}$$

Suppose the field at the input ($z = 0$) is a linearly polarized field in x direction:

$$\mathbf{E}(0) = \hat{\mathbf{x}} E_{\text{in}} = \frac{E_{\text{in}}}{\sqrt{2}} \left(\frac{\hat{\mathbf{x}} - j\hat{\mathbf{y}}}{\sqrt{2}} + \frac{\hat{\mathbf{x}} + j\hat{\mathbf{y}}}{\sqrt{2}} \right) \tag{E.27}$$

Then, at the output, $z = L$, the transverse electric field is

$$\mathbf{E}(L) = \frac{E_{\text{in}}}{\sqrt{2}} \left(\frac{\hat{\mathbf{x}} - j\hat{\mathbf{y}}}{\sqrt{2}} e^{-j\beta_R L} + \frac{\hat{\mathbf{x}} + j\hat{\mathbf{y}}}{\sqrt{2}} e^{-j\beta_L L} \right)$$

$$= E_{\text{in}} e^{-j\beta_0 L} (\hat{\mathbf{x}} \cos \xi L + \hat{\mathbf{y}} \sin \xi L) \tag{E.28}$$

At the end of a fiber of length L, $\mathbf{E}(L)$ is a linearly polarized field oriented at an angle ξL relative to the x direction. In other words, when the fiber is twisted by an angle Θ, the linearly polarized field rotates by ξL. Recall that ξL is linearly proportional to Θ as given in (E.19).

REFERENCES

1. S. P. Timoshenko and J. N. Goodier, *Theory of Elasticity*, McGraw-Hill, New York, 3rd ed., 1970, p. 282.

2. I. P. Kaminow, *An Introduction to Electrooptic Devices*, Academic, New York, 1974, Chapter 1.

3. Y. Namihira, M. Kudo, and Y. Mushiake, "Polarization characteristics of twisted single-mode optical fibers," *Electron. Communi. Japan*, Vol. 64-C, No. 3, pp. 89–97 (1981).

Appendix F

DERIVATION OF (12.7), (12.8), AND (12.9)

For signals of a finite spectral width $\Delta\lambda$, the pulse broadening per unit fiber length is

$$\Delta\tau_{gr} = \frac{d}{d\lambda}\left(\frac{d\beta}{d\omega}\right)\Delta\lambda \qquad (F.1)$$

The *broadening per unit fiber length per unit spectral width* is

$$\frac{\Delta\tau_{gr}}{\Delta\lambda} = \frac{d}{d\lambda}\left(\frac{d\beta}{d\omega}\right) \qquad (F.2)$$

In the limit of infinitesimal $\Delta\lambda$, we write $d\tau_{gr}/d\lambda$ in lieu of $\Delta\tau_{gr}/\Delta\lambda$. Making use of the relation $k = \omega/c$, we can convert $\Delta\tau_{gr}/\Delta\lambda$ to various forms:

$$\frac{d\tau_{gr}}{d\lambda} = \frac{d}{d\lambda}\left(\frac{d\beta}{d\omega}\right) = -\frac{1}{2\pi c}\frac{d}{d\lambda}\left(\lambda^2\frac{d\beta}{d\lambda}\right) \qquad (F.3)$$

For weakly guiding fibers, the index difference given in (12.4) is approximately

$$\Delta \approx \frac{n_{co} - n_{cl}}{n_{cl}}$$

Foundations for Guided-Wave Optics, by Chin-Lin Chen
Copyright © 2007 John Wiley & Sons, Inc.

By a straightforward differentiation with respect to λ, we have

$$\Delta' \equiv \frac{d\Delta}{d\lambda} = \frac{n_{cl}n_{co}' - n_{co}n_{cl}'}{n_{cl}^2} \tag{F.4}$$

$$\Delta'' \equiv \frac{d^2\Delta}{d\lambda^2} \approx \frac{n_{cl}n_{co}'' - n_{cl}''n_{co}}{n_{cl}^2} - 2\frac{n_{cl}'}{n_{cl}}\Delta' \tag{F.5}$$

where a prime indicates a differentiation with respect to λ. For weakly guiding fibers, $V = ka\sqrt{n_{co}^2 - n_{cl}^2} \approx kan_{cl}\sqrt{2\Delta}$, then

$$V' \equiv \frac{dV}{d\lambda} = -\frac{V}{\lambda} + \frac{n_{cl}'}{n_{cl}}V + \frac{V\Delta'}{2\Delta} \tag{F.6}$$

From definition of the effective index of refraction N, we have

$$\beta = kN \approx \frac{2\pi}{\lambda}n_{cl}(1 + b\Delta) \tag{F.7}$$

By simple and tedious differentiation, we obtain

$$\lambda^2 \frac{d\beta}{d\lambda} = -\beta\lambda\left(1 - \lambda\frac{n_{cl}'}{n_{cl}}\right) - 2\pi n_{cl}\Delta\frac{d(Vb)}{dV}\left(1 - \lambda\frac{n_{cl}'}{n_{cl}}\right)$$
$$+ \pi\lambda n_{cl}\Delta'\left[b + \frac{d(Vb)}{dV}\right] \tag{F.8}$$

$$-\frac{\lambda^2}{2\pi}\frac{d\beta}{d\lambda} = (n_{cl} - \lambda n_{cl}') + (n_{cl} - \lambda n_{cl}')\Delta\frac{d(Vb)}{dV} - \frac{1}{2}\lambda n_{cl}\Delta'\left[b + \frac{d(Vb)}{dV}\right] \tag{F.9}$$

Differentiating the equation one more time, we obtain

$$-\frac{1}{2\pi}\frac{d}{d\lambda}\left(\lambda^2\frac{d\beta}{d\lambda}\right) = -\lambda n_{cl}''\left[1 + \Delta\frac{d(Vb)}{dV}\right] + (n_{cl} - \lambda n_{cl}')\Delta'\frac{d(Vb)}{dV}$$
$$+ (n_{cl} - \lambda n_{cl}')\Delta\frac{d^2(Vb)}{dV^2}\left(-\frac{V}{\lambda} + \frac{n_{cl}'}{n_{cl}}V + \frac{V\Delta'}{2\Delta}\right)$$
$$- \frac{1}{2}(n_{cl} - \lambda n_{cl}')\Delta'\left[b + \frac{d(Vb)}{dV}\right] \tag{F.10}$$
$$- \frac{1}{2}\lambda n_{cl}\Delta'\left[\frac{db}{dV} + \frac{d^2(Vb)}{dV^2}\right]\left(-\frac{V}{\lambda} + \frac{n_{cl}'}{n_{cl}}V + \frac{V\Delta'}{2\Delta}\right)$$
$$- \frac{1}{2}\lambda n_{cl}\left[b + \frac{d(Vb)}{dV}\right]\left(\frac{n_{cl}n_{co}'' - n_{co}n_{cl}''}{n_{cl}^2} - 2\frac{n_{cl}'}{n_{cl}}\Delta'\right)$$

Gambling et al. have arranged terms into three groups [1]. All terms containing the second derivative n''_{co} and n''_{cl} are collected as a group. Terms having $d^2(Vb)/dV^2$ and Δ' are collected as the second and third groups. Then (F.10) becomes

$$\frac{d\tau_{\text{gr}}}{d\lambda} \equiv \mathcal{D} = \mathcal{D}_{\text{mt}} + \mathcal{D}_{\text{wg}} + \mathcal{D}_{\text{pf}} \tag{F.11}$$

where

$$\mathcal{D}_{\text{mt}} = -\frac{\lambda}{c}n''_{\text{cl}}\left\{1 - \frac{1}{2}\left[b + \frac{d(bV)}{dV}\right] - \frac{\Delta}{2}\left[b - \frac{d(bV)}{dV}\right]\right\}$$
$$- \frac{\lambda}{c}n''_{\text{co}}\left\{\frac{1}{2}\left[b + \frac{d(bV)}{dV}\right]\right\} \tag{F.12}$$

$$\mathcal{D}_{\text{wg}} = -\frac{n_{\text{cl}}\Delta}{c\lambda}\left(1 - \frac{\lambda n'_{\text{cl}}}{n_{\text{cl}}}\right)^2 V\frac{d^2(bV)}{dV^2} \tag{F.13}$$

$$\mathcal{D}_{\text{pf}} = \frac{\Delta' n_{\text{cl}}}{c}\left\{\left(1 - \frac{\lambda n'_{\text{cl}}}{n_{\text{cl}}} + \frac{\lambda\Delta'}{4\Delta}\right)\left[V\frac{d^2(bV)}{dV^2} + \frac{d(bV)}{dV} - b\right]\right.$$
$$\left. + \frac{\lambda n'_{\text{cl}}}{n_{\text{cl}}}\left(b + \frac{d(bV)}{dV}\right)\right\} \tag{F.14}$$

The three equations are exactly (12.7), (12.8), and (12.9).

REFERENCE

1. W. A. Gambling, H. Matsumura, and C. M. Ragdale, "Mode dispersion, material dispersion and profile dispersion in graded-index single-mode fibers," *IEE J. Microwaves, Optics Acoustics*, Vol. 3, No. 6, pp. 239–246 (1979).

Appendix G

TWO HANKEL TRANSFORM RELATIONS

In this appendix, we state the two identities of Hankel transforms without proof. For proofs, refer to Ref. [13] of Chapter 13.

G.1 PARSEVAL'S THEOREM OF HANKEL TRANSFORMS

Let $F_\nu(q)$ and $G_\nu(q)$ be the Hankel transforms of order ν of functions $f(r)$ and $g(r)$:

$$\mathcal{H}_\nu[f(r)] = F(q) = \int_0^\infty f(r) J_\nu(qr) r\, dr \qquad (G.1)$$

$$\mathcal{H}_\nu[g(r)] = G(q) = \int_0^\infty g(r) J_\nu(qr) r\, dr \qquad (G.2)$$

then

$$\int_0^\infty f(r) g(r) r\, dr = \int_0^\infty F_\nu(q) G_\nu(q) q\, dq \qquad (G.3)$$

provided $\nu \geq -\frac{1}{2}$.

In the special case where the two functions $f(r)$ and $g(r)$ are identical, then

$$\int_0^\infty [f(r)]^2 r \, dr = \int_0^\infty [F_\nu(q)]^2 q \, dq \tag{G.4}$$

G.2   HANKEL TRANSFORMS OF DERIVATIVES OF A FUNCTION

Let $F_\nu(q)$ be the Hankel transform of order ν of $f(r)$ as defined in (G.1). We define the Hankel transform of $df(r)/dr$ in the same manner. Then

$$\mathcal{H}_\nu\left[\frac{df(r)}{dr}\right] = \int_0^\infty \frac{df(r)}{dr} J_\nu(qr) r \, dr$$

$$= (\nu - 1) \int_0^\infty f(r) J_\nu(qr) \, dr - q\mathcal{H}_{\nu-1}[f(r)] \tag{G.5}$$

In the case of $\nu = 1$, the above equation becomes

$$\mathcal{H}_1\left[\frac{df(r)}{dr}\right] = -q\mathcal{H}_0[f(r)] \tag{G.6}$$

AUTHOR INDEX

Foundations for Guided-Wave Optics, by Chin-Lin Chen
Copyright © 2007 John Wiley & Sons, Inc.

SUBJECT INDEX

Foundations for Guided-Wave Optics, by Chin-Lin Chen
Copyright © 2007 John Wiley & Sons, Inc.